SYMMETRIES IN NUCLEAR STRUCTURE

THE SCIENCE AND CULTURE SERIES — PHYSICS

1. Perspectives for New Detectors in Future Supercolliders, 1991
2. Data Structures for Particle Physics Experiments: Evolution or Revolution?, 1991
3. Image Processing for Future High-Energy Physics Detectors, 1992
4. GaAs Detectors and Electronics for High-Energy Physics, 1992
5. Supercolliders and Superdetectors, 1993
6. Properties of SUSY Particles, 1993
7. From Superstrings to Supergravity, 1994
8. Probing the Nuclear Paradigm with Heavy Ion Reactions, 1994
9. Quantum-Like Models and Coherent Effects, 1995
10. Quantum Gravity, 1996
11. Crystalline Beams and Related Issues, 1996
12. The Spin Structure of the Nucleon, 1997
13. Hadron Colliders at the Highest Energy and Luminosity, 1998
14. Universality Features in Multihadron Production and the Leading Effect, 1998
15. Exotic Nuclei, 1998
16. Spin in Gravity: Is It Possible to Give an Experimental Basis to Torsion?, 1998
17. New Detectors, 1999
18. Classical and Quantum Nonlocality, 2000
19. Silicides: Fundamentals and Applications, 2000
20. Superconducting Materials for High Energy Colliders, 2001
21. Deep Inelastic Scattering, 2001
22. Electromagnetic Probes of Fundamental Physics, 2003
23. Epioptics-7, 2004
24. Symmetries in Nuclear Structure, 2004

THE SCIENCE AND CULTURE SERIES — PHYSICS

Proceedings of the Highly Specialized Seminar on

SYMMETRIES IN NUCLEAR STRUCTURE

An Occasion to Celebrate the
60th Birthday of Francesco Iachello

Erice, Italy   23 – 30 March 2003

Editors

Andrea Vitturi and Richard F. Casten

Series Editor

A. Zichichi

World Scientific

NEW JERSEY • LONDON • SINGAPORE • SHANGHAI • HONG KONG • TAIPEI • CHENNAI

Published by

World Scientific Publishing Co. Pte. Ltd.

5 Toh Tuck Link, Singapore 596224

USA office: Suite 202, 1060 Main Street, River Edge, NJ 07661

UK office: 57 Shelton Street, Covent Garden, London WC2H 9HE

British Library Cataloguing-in-Publication Data
A catalogue record for this book is available from the British Library.

Cover Illustrations
Front: Roman mosaic *ca* 200 BC
Back: Seventeenth century map of Sicily by Dutch cartographer Jansson

SYMMETRIES IN NUCLEAR STRUCTURE: AN OCCASION TO CELEBRATE THE 60TH BIRTHDAY OF FRANCESCO IACHELLO

ISBN 981-238-812-5

Printed in Singapore by World Scientific Printers (S) Pte Ltd

Professor Francesco Iachello

CONTENTS

PREFACE

These Proceedings contain the papers presented at the Highly Specialized Seminar on "Symmetries in Nuclear Structure", which was held at the Ettore Majorana Centre in Erice, Italy, 23-30 March 2003. The meeting was intended to celebrate, on the occasion of his 60th birthday, the career and the remarkable achievements of Francesco Iachello. Since the development of the Interacting Boson Model in the early 1970s, the ideas of Francesco Iachello have provided a variety of frameworks for understanding collective behaviour in nuclear structure, founded in the concepts of dynamical symmetries and spectrum generating algebras. The original ideas, which were developed for the description of atomic nuclei, have now been successfully extended to cover spectroscopic behaviour in other fields, such as molecular or hadronic spectra. More recently, the suggestion by Iachello of Critical Point Symmetries to treat nuclei in shape/phase transitional regions has opened an exciting new front for both theoreticians and experimentalists.

The talks presented at the meeting covered many of the most active forefront areas of nuclear structure as well as other fields where ideas of symmetries are being explored. Topics in nuclear structure included extensive discussions of dynamical symmetries, critical point symmetries, phase transitions, statistical properties of nuclei, supersymmetry, mixed symmetry states, shears bands, pairing and clustering in nuclei, shape coexistence, exotic nuclei, dipole modes, and astrophysics, among others. In addition, important sessions focused on talks by European Laboratory Directors (or their representatives) outlining future prospects for nuclear structure, and the application of symmetry ideas to molecular phenomena. Finally, a special lecture by Nobel Laureate Alex Mueller, on s and d wave symmetry in superconductors, presented a unique insight into an allied field.

On behalf of the Organizing Committee and of all participants we thank the Ettore Majorana Centre and Professor Antonino Zichichi for providing such an ideal venue for the meeting. The location fitted perfectly into the spirit of the reunion, since precisely in this place so many of the early first successes of the Interacting Boson Model were announced in the late 1970s – early 1980s.

Finally, we thank Raffaella Ruggiu and Pino Aceto from the Majorana Centre for the perfect organization in Erice and Annarosa Spalla from INFN, Padova, for the highly professional management of the conference and her fundamental help in the collection and the editing of these proceedings.

Andrea Vitturi and Richard F. Casten

SYMMETRY: THE SEARCH FOR ORDER IN NATURE

FRANCESCO IACHELLO
Center for Theoretical Physics, Sloane Physics Laboratory,
Yale University, New Haven, CT 06520-8120

1 Symmetry

The subject that is being discussed at this Seminar is the role of symmetry in nuclear, molecular and hadronic physics. Symmetry is the unifying concept in many human endeavors. The word symmetry, from the Greek $\sigma\upsilon\mu\mu\varepsilon\tau\rho o\varsigma$, describes an object that is well-ordered, well-organized. It began to appear systematically in many writings during the Greek and Roman periods. The Greek sculptor Polykleitos in his book on the art of construction ($\Pi\varepsilon\rho\iota$ $\beta\varepsilon\lambda o\pi\iota\ddot{\iota}\kappa\omega\nu$) uses the word extensively. The Roman architect Vitruvius states that symmetry is "the result of proportions between different parts". Artifacts with symmetric properties have been present since the dawn of civilization. Figure 1 shows a portion of a decoration, dated circa 2000 B.C., which is symmetric with respect to translations along the horizontal axis. During the Greek and Roman periods symmetry was brought to perfection. Figure 2 shows a Roman mosaic from Corynth, Greece. This figure has a sixteen-fold rotation symmetry, $\mathcal{D}_{16h}$, in the outer part and a broken symmetry in the inner part. (Because of its complex structure this figure has been taken as symbol for this Workshop.) Symmetry became an essential part of all works of art and reached its peak in the Greek temple, as it can be seen here in Sicily at Agrigento, Selinunte and Segesta. The symbol of Sicily, Fig. 3, has a threefold symmetry, $\mathcal{D}_{3h}$, (similar to the symmetry structure of the Interacting Boson Model).

Figure 1. Decorative motif (Sumerian, circa 2000 B.C.).

Figure 2. A complex symmetry pattern in art (Roman mosaic, from Corynth, Greece, circa 200 B.C.).

Figure 3. The symbol of Sicily (Trinacria, from antiquity) is shown as an example of threefold symmetry.

Together with the development of art, there was in the Greek time, a development of Mathematics, especially geometry. The discovery of the five regular polyhedra, the tetrahedron, the cube, the octahedron, the dodecahedron and the icosahedron, with their symmetric shapes, led the ancient Greeks to think that

Figure 4. Dodecahedron, after Leonardo da Vinci and Piero della Francesca (from Luca Pacioli, *De Divina Proportione*, Venice, 1509).

the entire Universe is built out of these polyhedra. The tetrahedron, octahedron, cube, icosahedron were associated with fire, air, earth and water, respectively (the building blocks of the Universe), while the (penta)dodecahedron was the image of the Universe itself.

The study of symmetry took another step forward in the Italian Renaissance, when artists and mathematicians developed further the Greek concept. The five regular polyhedra were complemented with other types, such as the Archimedean polyhedra, Fig. 4. (Archimedes, 287-212 B.C. was also a native of Sicily.) The works of Piero della Francesca (*Libellus de Quinque Corporibus Regularis*, 1482), Luca Pacioli (*De Divina Proportione*, 1509) and others describe in full detail the symmetries of these bodies and begin to introduce what in modern mathematical language is called the theory of group transformations, or, simply, group theory. Once more, in these works, the basic idea is that the Universe is proportionate and thus follows strict mathematical (geometric) rules (simmetria). Symmetry is assumed to be the fundamental law of Nature, and is therefore above all (divina).

In 1595, the German astronomer Kepler, in his book *Mysterium Cosmographicum*, noticed that the planetary systems known at the time, Saturn, Jupiter, Mars, Earth, Venus and Mercury, could be reduced to regular bodies which are alternatively inscribed and circumscribed in spheres, Fig. 5. He was so impressed that he concluded his book with the famous sentence "Credo spatioso numen in orbe", that is "I believe in a geometric order of the Universe". Again, order and symmetry are used in an interchangeable way.

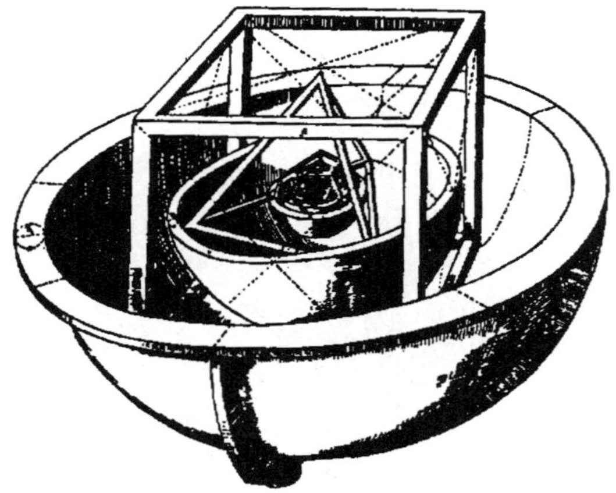

Figure 5. Construction of the Universe according to Kepler. (From *Mysterium Cosmographicum*, Tubingen, 1595).

2 Symmetry in Physics

The idea of order (and symmetry) began then its entry into Physics. By the end of the 19th Century, as Physics changed more and more from the macroscopic to the microscopic level, it became clear that symmetries play a fundamental role in Physics. Many aspects of Nature are observed to be ordered. The best example are molecules and crystals. Figure 6 shows the molecule H_3-C-C-Cl_3, whose symmetry is evident (C_3, rotations of angles multiple of 120°). The symmetries encountered in molecules and crystals are called "geometric" symmetries. They are similar to those encountered in art (rotations, reflections and translations). These symmetries were the first ones to be recognized in the microscopic world. But, as time went on, it became clear that other types of symmetry occur in Nature, not related to the geometric arrangement of atoms in a molecule or in a crystal but rather to the dynamic laws of Nature. The role of these other symmetries emerged with the development of quantum mechanics, which occurred around 1920. Contrary to what happens in classical mechanics, in quantum mechanics the bound states of a physical system are discrete. The search for order in quantum mechanics therefore becomes the search for order in the states of a physical system. The manifestation of symmetry is the observed regularity of the quantum levels of the system. As in the previous cases of "geometric" symmetries, "dynamic" symmetries are characterized by groups of transformations. The groups of transformations appropriate to describe dynamic symmetries are called Lie groups, after the Norwegian mathematician Sophus Lie who introduced them towards the end of the 19th Century. Lie groups characterize how the wave functions describing the quantum states of the system are related by symmetry transformations. If there is a dynamic symmetry, states are simply related by the action of the elements of the Lie algebra associ-

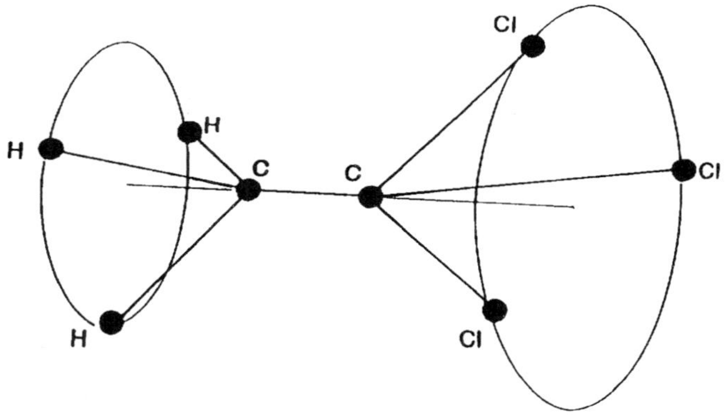

Figure 6. The molecule H_3-C-C-Cl_3 (C_3 symmetry).

ated to the Lie group, called generators. In addition to geometric and dynamic symmetries, other types of symmetry, ("kinematic" symmetries, gauge symmetries, permutation symmetries,...) have been shown to play an important role in physics. Especially gauge symmetries have played in the last 30 years a major role in Physics after the discovery that the fundamental laws of Nature appear to be governed by non-Abelian gauge symmetries. These other symmetries will not be discussed here.

In the course of the last 60 years, dynamic symmetries have been discovered at every level of quantum physics (in molecules, in atoms, in atomic nuclei and in hadrons).

In 1926, Wolgang Pauli (*Über der Wasserstoffspektrum vom Standpunkt der neuen Quantenmechanik*, Z. Physik 36, 2336 (1926)) noted that the regularity in the discrete spectrum of the hydrogen atom, Fig. 7, is due to the occurrence of a dynamic symmetry, described by the group SO(4).

In 1961, Murray Gell'Mann (*Symmetry of Baryons and Mesons*, Phys. Rev. 125, 1067 (1962)) and Yuval Ne'eman (*Derivation of Strong Interactions from a Gauge Invariance*, Nucl. Phys. 26, 222 (1961)) discovered that the spectrum of hadrons, Fig. 8, is very regular, and could be described by the group SU(3).

In 1974, Akito Arima and I (*Collective Nuclear States as a Representation of a SU(6) Group*, Phys. Rev. Lett. 35, 1069 (1975)), discovered that the discrete spectra of many atomic nuclei are regular and proposed that these spectra be described by the group of transformations U(6). An example is shown in Fig. 9.

In 1981, I (*Algebraic Methods for Molecular Rotation-vibration Spectra*, Chem. Phys. Lett. 78, 581 (1981)) observed that the discrete spectra of many molecules are also regular and suggested that these spectra be described by the group of transformations U(4). An example is shown in Fig. 10. This suggestion led to the formulation, together with Raphael D. Levine and others, of a comprehensive

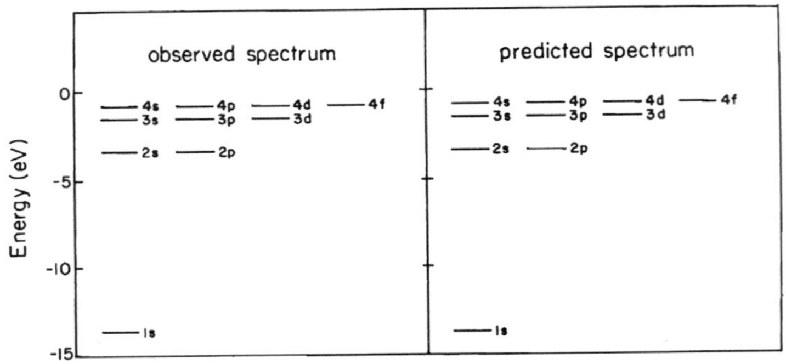

Figure 7. A portion of the discrete spectrum of the hydrogen atom. Levels are characterized by the spectroscopic notation $1s, 2s, 2p, \ldots$. The scale of energy is electron Volt (eV). On the left the experimental spectrum, on the right the calculated spectrum.

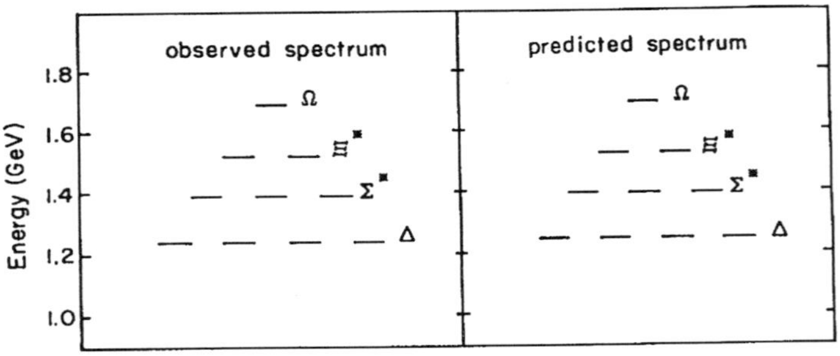

Figure 8. The energy spectrum of the baryon decuplet. States are characterized by the notation $\Delta, \ldots$The energy scale is in GeV=10^9eV. On the left the experimental spectrum, on the right the calculated spectrum.

theory of molecules.

The concept of symmetry was enlarged even further in the 1970's by Julius Wess, Bruno Zumino and others through the introduction of a more complex type of symmetry, called supersymmetry, involving simultaneously bosons and fermions. There are here again several types of supersymmetries ("kinematic" supersymmetries, gauge supersymmetries,...). In the course of the last 20 years, dynamic supersymmetries have been discovered in atomic nuclei.

In 1980, I (*Dynamical Supersymmetries in Nuclei*, Phys. Rev. Lett. 44, 772 (1980) observed that the spectra of certain even and odd nuclei could be classified in terms of the group of supersymmetry transformations U(6/4). An example is shown in Fig. 11. A mathematical theory was soon developed with Baha Balantekin and Ithzak Bars. The idea was further enlarged to include protons and neutrons. The latter type has also been found recently.

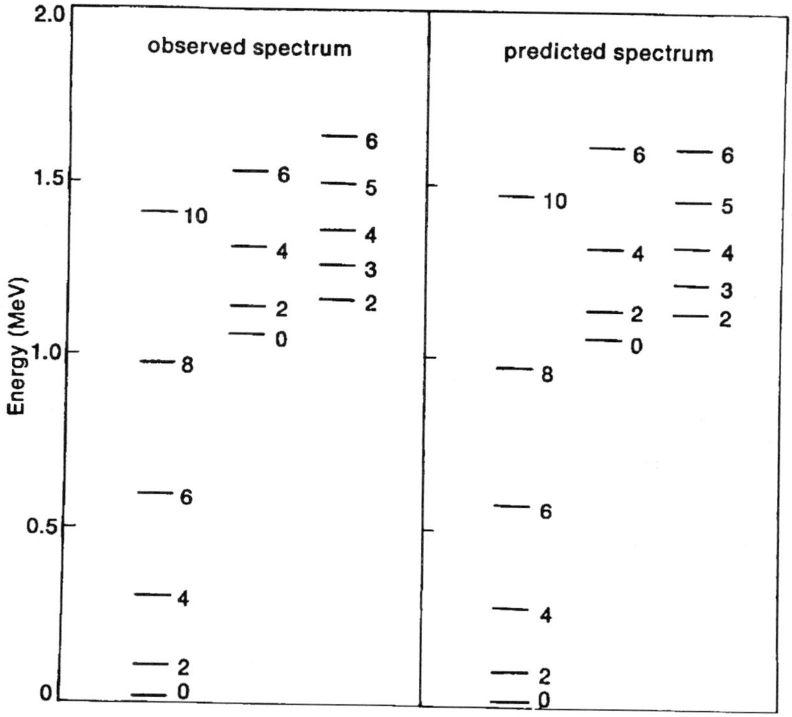

Figure 9. The energy spectrum of the nucleus ^{156}Gd is shown as an example of the dynamic symmetry $U(6) \supset SU(3)$. States are characterized by their angular momenta $0, 2, 4, \ldots$. The energy scale is in MeV=10^6eV. On the left the experimental spectrum, on the right the theoretical spectrum.

One may note in all cases shown in the figures the arrangement of the energy levels into patterns characteristic of the dynamical symmetry which governs the quantum motion of constituent particles (electrons in the case of atoms, quarks in the case of hadrons, protons and neutrons in the case of nuclei and atoms in the case of molecules). The occurrence of these regular patterns is the manifestation of dynamic symmetry.

One may wonder whether or not the study of dynamical symmetries and supersymmetries in physics has been concluded. The answer is no. The observation that these symmetries occur at all levels indicates that they may be present in other systems, including condensed matter systems.

Indeed, in 2002 Alex Muller and I (*A Model of Cuprate Superconductors Based on the Analogy with Atomic Nuclei,* Phil. Mag. Lett. 82, 279, 289 (2002)) have suggested that dynamic symmetries may be responsible for high-temperature superconductivity in cuprate materials. This idea is presently being explored.

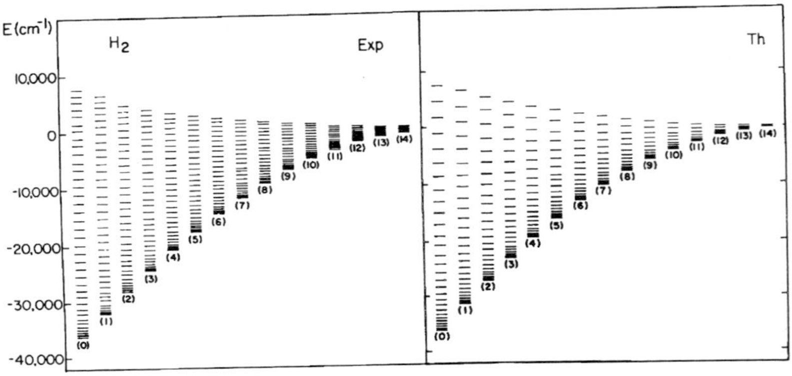

Figure 10. The energy spectrum of the H_2 molecule is shown as an example of the dynamic symmetry $U(4) \supset SO(4)$. The energy scale is in $\mathrm{cm}^{-1} = 1.24 \times 10^{-4}$eV. On the left the experimental spectrum, on the right the theoretical spectrum.

Another area where dynamic symmetries could play an important role is that of critical phenomena.

In 2000, I (*Dynamic Symmetries at the Critical Point*, Phys. Rev. Lett. 85, 3580 (2000)) have suggested that the spectra of physical systems at the critical point of phase transitions can be described by dynamic symmetries. This idea, which potentially can open the way to many applications, is presently being very actively explored (see articles by Richard Casten and others at this Workshop).

In conclusion, Nature (and Physics) appear to display at all levels an ordered structure. This is particularly evident at the quantum level in the ordered spectra of molecules, atoms, nuclei and hadrons. "Nature loves symmetry", as the German mathematician Hermann Weyl used to say. The deeper and deeper we go into the mysteries of Nature, the more and more we find symmetry. Despite the widening experience which has repeatedly tested our ideas of symmetry, we still find a mathematical harmony in the Universe.

Acknowledgments

I wish to thank all the participants and many other scientists, both theorists and experimentalists not present here, for their contributions to the study of symmetry in physics, in particular Richard Casten and Jolie Cizewski for discovering $SO(6)$, David Warner and Hans Börner for their role in establishing broken $SU(3)$ in deformed nuclei, Norbert Pietralla and Peter von Brentano for $U_\pi(5) \otimes U_\nu(5)$, Jan Jolie and Gerhard Graw for $U_\pi(6/4) \otimes U_\nu(6/12)$, Richard Casten and Victor Zamfir for $X(5)$ and $E(5)$.

I also wish to thank especially Antonino Zichichi, President of the Ettore Majorana Centre for Scientific Culture, for making it possible to have this celebration in the beautiful setting of Erice, Sicily (my homeland), and Andrea Vitturi, Director of the Course, for the excellent organization of this Workshop, and Richard

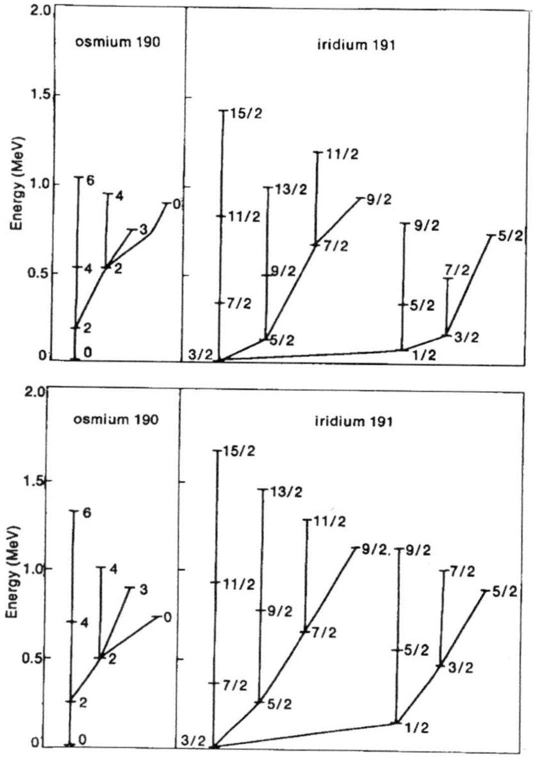

Figure 11. The energy spectra of the pair of nuclei ^{190}Os-^{191}Ir are shown as an example of $U(6/4) \supset Spin(6)$ supersymmetry. The energy scale is MeV$=10^6$eV. On the left the experimental spectrum, on the right the theoretical spectrum.

Casten for his continuing effort in searching and finding empirical manifestations of symmetry in nuclei.

I thank also Mrs. Annarosa Spalla and Mrs. Fiorella Ruggiu for their roles in making the Workshop a success.

ADDITIONAL QUANTUM NUMBERS, DYNAMICAL SYMMETRIES AND DEGENERACIES

IGAL TALMI

The Weizmann Institute of Science
Rehovot, Israel

Eigenvalues of Hamiltonians with dynamical symmetries are given by algebraic expressions of certain quantum numbers. Degeneracies occur frequently and states must be specified by additional quantum numbers. Discussion of the latter shows that such degeneracies may occur also for Hamiltonians without dynamical symmetry. These degeneracies can be removed in certain cases and some examples are presented.

Since this is a special occasion, I would like to discuss a subject related to the early work of Iachello and Arima.[1,2] It is related to the beautiful dynamical symmetries which they found for the $U(6)$ group of the interacting boson model. They expressed the most general $U(5)$, $SU(3)$ and $O(6)$ two-body Hamiltonians as linear combinations of Casimir operators of group chains I ($U(5) \supset O(5) \supset O(3)$), II ($SU(3) \supset O(3)$) and III ($O(6) \supset O(5) \supset O(3)$). Only one additional quantum number is needed in each chain for complete characterization of states. The eigenvalues are given by simple algebraic expressions of the quantum numbers which characterize the irreducible representations of the groups in the chain. They are independent of the additional quantum numbers n_Δ, K and ν_Δ. The degeneracy in these quantum numbers arises in a natural way and in papers and even in books, hardly any attention is paid to it. In the first detailed paper on $U(5)$ symmetry,[3] the expression is derived for the eigenvalues "which do not depend on ν_Δ and M". In the detailed paper on $SU(3)$ symmetry,[4] the analogous formula "shows that the eigenvalues of H do not depend on K and M but only on the angular momentum L". This situation exists also in other systems, one of which is considered in the following.

Nuclear Hamiltonians are invariant under rotations. Hence, their eigenvalues are independent of the projection M of the total spin J on the z-axis. The independence of eigenvalues on the other quantum numbers is due to a different reason. The aim of this note is to review the origin of additional quantum numbers and to discuss some of these degeneracies. The latter are clearly displayed in cases of dynamical symmetry but such degeneracies may arise also for more complicated Hamiltonians. The discussion here is limited to Hamiltonians which possess some symmetry in addition to rotational invariance. Some cases are considered in the following, along with examples where these degeneracies may be removed.

Let us look at the seniority scheme for protons and neutrons in a given shell. Seniority was introduced by Racah[5] who also showed how it can be defined by a chain of groups.[6] He first defined it for electron ℓ^n configurations but it was later adapted to jj-coupling and nucleons by Flowers[7] and by Racah himself.[8,9] The set of all allowed states of a j^n configuration with various isospins T, spins J and their projections, is the starting point. All these states are fully antisymmetric under exchange of the space coordinates and spin and isospin variables of the nucleons. Hence, all of them transform irreducibly among themselves under unitary transformations which are applied to the states of each nucleon. These are the space, spin and isospin states of each j-nucleon which span a $2(2j+1)$ dimensional space. In other words, all these states form the basis of an irreducible representation of the unitary group in $2(2j+1)$ dimensions, $U(2(2j+1))$. It is uniquely characterized by the number n of nucleons.

Unitary transformations may be applied to the 2 isospin states of a single nucleon. To obtain the total isospin of n nucleons, the same $U(2)$ transformations should be applied to all of them. The states which transform among themselves form bases of irreducible representations of $U(2)$. These are uniquely characterized by the total isospin T and n. There are $2T+1$ independent isospin states, characterized by M_T which transform among themselves under $U(2)$ operations. Hence, to introduce the isospin quantum number, separate transformations of the space and spin states of the nucleons and separate transformations of their isospin states are considered. Such transformations belong to the subgroup $U(2) \times U(2j+1)$ of $U(2(2j+1))$. The fully antisymmetric irreducible representation of $U(2(2j+1))$ splits into several irreducible representations of $U(2) \times U(2j+1)$. There is a corresponding split in bases of $U(2(2j+1))$ irreducible representations. The set of fully antisymmetric states of the j^n configuration splits into several sets of states which transform irreducibly under $U(2) \times U(2j+1)$ transformations. Each one of the latter is characterized by a definite value of the isospin T.

States with maximum isospin T, equal to $n/2$, and given M_T, are symmetric under all exchanges of isospin variables. To obtain a fully antisymmetric state, such an isospin state should be multiplied by an antisymmetric function of the space coordinates and spin variables. The antisymmetric space and spin functions transform irreducibly among themselves under operations of the $U(2j+1)$ group. Hence, they form a basis of its antisymmetric irreducible representation. Thus, a basis of the irreducible representation of $U(2) \times U(2j+1)$ with $T = n/2$, is obtained by multiplying all basis states of the antisymmetric irreducible representation of $U(2j+1)$ by all $2T+1$ isospin basis states of the symmetric irreducible representation of $U(2)$.

Structure of states with lower values of isospin is more complicated. The isospin functions have then a certain symmetry type. They are neither symmetric nor antisymmetric in the isospin variables of all nucleons. Among those, with given n, $n/2 \geq T \geq 0$ and M_T, a finite basis set of independent functions which transform irreducibly among themselves under exchanges of isospin variables, can be chosen. A state with given M_T, which is antisymmetric under exchanges of all nucleon variables, is a linear combination of basis isospin states, multiplied by corresponding basis space and spin functions. The latter belong to the *dual* symmetry type. Each of the fully antisymmetric states is characterized by n, T and M_T, as well as by

J, M and any other quantum number needed for complete characterization of the space and spin functions.

The total spins of states are introduced by going to a subgroup of $U(2j+1)$. This is the group of transformations in the $2j+1$ dimensional space induced by 3 dimensional rotations of nucleon space and spin coordinates. In the space considered here, this is the $SU(2)$ group which is a subgroup of $U(2j+1)$. The irreducible representations of the rotation group are characterized by the total spin J. The $2J+1$ states with various projections M transform irreducibly among themselves under rotations. There are usually several states with the same value of J which belong to the same irreducible representation of $U(2j+1)$, with the same value of T (and n). An irreducible representation of $U(2j+1)$ usually splits into several irreducible representations of $SU(2)$ some of which may be equivalent. Further classification of states with the same value of J, may be obtained by assigning them quantum numbers of the seniority scheme.

Seniority is introduced by considering the symplectic subgroup $Sp(2j+1)$ of the unitary group $U(2j+1)$. Operations of this subgroup leave invariant the antisymmetric wave functions of two j-nucleons in the $J=0$ state. An irreducible representation of $U(2j+1)$ may split into several irreducible ones of the $Sp(2j+1)$ subgroup. Thus, the set of wave functions with given isospin T (and n) may split into several sets which transform irreducibly under $Sp(2j+1)$ transformations. Such sets, which form bases of irreducible representations of the subgroup $Sp(2j+1)$, are characterized by the seniority v and reduced isospin t.

The meaning of these quantum numbers is simple. A state with given J and $T=t$ of the j^v configuration with no nucleon pairs coupled to $J=0$, is defined to have seniority v and reduced isospin t. To such a state, $(n-v)/2$ pairs with $J=0$, may be added to form a state of the j^n configuration with the same values of v and t. Such a state has the same value of J as in the j^v configuration but this is not the case with isospin. Since each $J=0$ pair has isospin 1, the isospin T of the j^n state need not be equal to the isospin t of the j^v state. The total spins J of states are introduced by considering the $SU(2)$ group, whose operations do not change the spin J of any wave function and hence, it is a subgroup of $Sp(2j+1)$. In this way, the introduction of the intermediate $Sp(2j+1)$ group supplies the quantum numbers v and t which help to distinguish between some states with the same values of n and T, which have the same value of J.

States which belong to bases of irreducible representations of a given group, may be eigenstates of certain Hamiltonians. The latter can be expressed as linear combinations of scalar functions of generators of the group. Dynamical symmetry is obtained by using Casimir operators. If we consider a chain of groups, like $U((2j+1)\supset Sp(2j+1)\supset SU(2)$, we may look at Hamiltonians which are linear combinations of their Casimir operators. In the present case, such a combination is[8]

$$a'C[U(2j+1)]+b'C[Sp(2j+1)]+c'C[SU(2)] \qquad (1)$$

Since the 3 Casimir operators commute, the Hamiltonian is diagonal in a basis defined by the irreducible representations of the groups in the chain. The eigenvalues of this special Hamiltonian are simply equal to the linear combination of the eigenvalues of the Casimir operators in (1)

$$a'\{(2j+1)n - n(n-1)/2 - 2[T(T+1) - 3n/4]\} + \\ b'\{v(4j+8-v)/2 - 2t(t+1)\} + c'J(J+1) \quad (2)$$

These depend only on n, T, v, t, and J. It is possible to add to (2) linear and quadratic functions of n, arising from the linear and quadratic Casimir operators of $U(2(2j{+}1))$. The eigenvalues of the two-body interaction in (2) may be then expressed in a more familiar form as

$$a\,n(n-1)/2 + b[T(T+1) - 3n/4] + \\ c\{(n-v)(4j+8-n-v)/4 - T(T+1) + t(t+1)\} + d[J(J+1) - nj(j+1)] \quad (3)$$

In the case of j=3/2, states are fully specified by the quantum numbers of the seniority scheme. For given values of n, T, v, and t, there is only one state with given value of J. Any two body Hamiltonian has dynamical symmetry for any $(3/2)^n$ configuration. Eq.(3) is an exact expression of the interaction eigenvalues of any $(3/2)^n$ state. The coefficients a, b, c, and d, are uniquely determined by the values of the interaction in the J=0,1,2,3 states of the two-nucleon configuration. This feature, however, is not shared by all states of j^n configurations with higher values of j.

The seniority scheme supplies the v and t quantum numbers, but they do not characterize uniquely all states. There may be two, or more states with the same value of J and seniority quantum numbers, as well as T, in the j^n configuration. For example, for j=11/2, there are 2 states with J=9/2, v=3 and t=3/2. For j=13/2, there are 6 states with J=15/2, v=5 and t=5/2. Additional labels for such states are necessary. There is no known subgroup of $Sp(2j{+}1)$ which includes $SU(2)$ as a subgroup, and hence, no general prescription is available for such labels. It follows from (3) that in the case of dynamical symmetry as in (1) all such states are degenerate.

The sets of $2J$+1 components of those states are bases of different but equivalent irreducible representations of $SU(2)$ which are included in one irreducible representation of $Sp(2j{+}1)$. Additional quantum numbers are required to distinguish between these equivalent representations. These states are degenerate since **J.J.** is the only scalar obtained from two generators of $SU(2)$. The degeneracy cannot be broken by adding other functions of generators of $SU(2)$. This may break the $(2J{+}1)$-fold degeneracy, but the $2J$+1 eigenvalues will be the same for every set of states with the given value of J. By choosing appropriate bases of states which belong to different equivalent irreducible representations, it is possible to obtain identical sub-matrices along the main diagonal of the generators of a group. To see this, we recall the relation between the matrices of two equivalent irreducible representations. Let $D(R)$ be the matrix which corresponds to element R of the group in one representation, and $D'(R)$ the corresponding matrix in another equivalent representation. Then there is a regular matrix which satisfies for every element R of the group the relation

$$D'(R) = AD(R)A^{-1} \tag{4}$$

A pair of matrices D and D', obeying the relation (4) have identical eigenvalues.

If the matrices D and D' are constructed from orthonormal bases, the matrix A is unitary. Transformation of one of the bases with the matrix A in (4) simplifies the matrix of the Hamiltonian constructed from generators of a group. It will have identical sub-matrices which are characterized by the two equivalent irreducible representations. Then, the matrices of the generators have identical sub-matrices along its main diagonal which belong to different equivalent irreducible representations, uniquely characterized by J. The same applies to the Hamiltonian constructed from them. Removal of the degeneracy can be obtained only by considering a group of which $SU(2)$ is a subgroup.

In general, let G be a group in a given chain and G' a sub-group of it, $G' \subset G$. An irreducible representation of G may include in its reduction several equivalent irreducible representations of G'. Some additional labels or quantum numbers, may be introduced to distinguish between sets of states which belong to bases of those equivalent irreducible representations. By proper choice of basis states, the following form of the matrices of G' generators can be obtained. The sub-matrices along the diagonal which belong to those representations are identical. Any interaction constructed from them has corresponding identical sets of eigenvalues. These are independent of any additional quantum number used. Such degeneracies may be removed only by interactions constructed from generators of the group G including some which are not generators of G'.

It is worth while to point out that the projection M of J on the z-axis is not an additional quantum number of the kind described above. Those help to distinguish between states which belong to bases of different but equivalent irreducible representations. On the other hand, M characterizes the inequivalent irreducible representations of the group of rotations around the z-axis, which is a subgroup of the group of 3-dimensional rotations, $SU(2)$ or $O(3)$. Any irreducible representation of the latter, characterized by J, is fully reducible to $2J+1$ *inequivalent* representations of $SU(1)$ or O (2)characterized by M. The degeneracy in M, as mentioned above, is due to rotational invariance of the shell model Hamiltonian. It is a scalar with respect to angular momentum, commuting with all 3 components of J. The subgroup $O(2)$ of $O(3)$ is sometimes listed as the last group in the chains I, II, II mentioned above. In contrast to the degeneracies discussed above, the degeneracy in M may be simply removed by adding the only generator of this subgroup L_z, or some function of it, to the Hamiltonian.

There is a large class of shell model Hamiltonians which are diagonal in the seniority scheme. The $Sp(2j+1)$ Casimir operator which appears in (1), is one special linear combinations of scalar products of irreducible tensor operators with odd ranks. There are, however, in j^n configurations, $(2j+1)/2$ odd rank tensor operators whose components are generators of $Sp(2j+1)$. Hence, any linear combination of their scalar products is diagonal in the seniority scheme[10]. To such linear combinations, the Casimir operator of $U(2j+1)$ may be added, as well as the scalar operator with rank $k=0$ and its square. All such Hamiltonians are diagonal in the seniority scheme.

The states with the same spin J, with given v and t, together with all other states with the same v and $t=T$ in the j^v configuration, form a basis of the same $Sp(2j+1)$ irreducible representation. Degeneracy of states with the same value of J may be removed by using a Hamiltonian constructed from scalar products of arbitrary odd tensor operators instead of just using the Casimir operator of $Sp(2j+1)$. Various choices of such Hamiltonians give rise to different sets of eigenstates with the same value of J and to corresponding different sets of eigenvalues. The spacings between energies of states with given values of v and t in the j^n configuration are equal to those in the j^v configuration[10]. Removal of the degeneracy in this case is due to the freedom in the choice of interaction which is diagonal in the seniority scheme. The price paid by removing the degeneracy is the loss of dynamical symmetry.

Similar degeneracies may arise if there are several equivalent irreducible representations of $Sp(2j+1)$ obtained by reduction of the same irreducible representation of $U(2j+1)$. In this case, all states included in such an irreducible representation of $Sp(2j+1)$, with the same values of v and t, are present with the same multiplicity. In the case of dynamical symmetry (2), the terms multiplying a' and b' have the same values, and states with the same value of J are completely degenerate. An additional quantum number is necessary in this case to distinguish between states of such different *equivalent* irreducible representations of $Sp(2j+1)$. It is not necessary to have full dynamical symmetry for this degeneracy to occur. The degeneracy of sets of states which belong to the equivalent irreducible $Sp(2j+1)$ representations obtained from the same $U(2j+1)$ irreducible representation, cannot be removed by considering more general $Sp(2j+1)$ Hamiltonians. The argument presented above shows that if the Hamiltonian is an arbitrary linear combination of $Sp(2j+1)$ generators, this degeneracy persists.

Looking at the structure of states of the seniority scheme it is possible to see some cases in which any irreducible representation of $Sp(2j+1)$ appears only once in the reduction of those of $U(2j+1)$. In other words, only one set of states with given v and t is included in the set of states with given T in the j^n configurations. This is true in the case of maximum isospin, $T=n/2$. In this case there is only one set of states with given v and t among the states with isospin $T=n/2$ in the j^n configuration. There is only one way to obtain from a state in the j^v configuration with $t=v/2$, a state of the j^n configuration with $T=n/2$. The isospin vectors of all $(n-v)/2$ pairs which should be added to the initial state must be parallel to obtain the total isospin $T=v/2+(n-v)/2=n/2$. No additional quantum number is necessary to distinguish between sets of states with the same seniority v, and $T=n/2$, in the j^n configuration. This is not always the case for lower isospin values. For example, if j=5/2, among states of n=6 nucleons with T=1, there are two sets of states with v=2 and t=1. Each set contains states with J=2 and J=4. As explained above, these two J=2 states are degenerate in the case of any Hamiltonian which is diagonal in the seniority scheme. The same is true also for the two J=4 states.

There are other general cases where no additional quantum numbers are necessary to distinguish between different sets of states with the same values of n, T, v, and t. This is the case for all states with v=0, t=0 and states with v=1, t=1/2. States with v=0, t=0 are obtained by acting with $N=n/2$ pair operators on the vacuum state (the state with no j-nucleons, v=0, t=0). Due to the symmetry properties of such states there is only one state of this kind with any isospin allowed by the Pauli principle, which is equal to $T=N$, $N-2$, $N-4$, ... To obtain states with v=1, t=1/2 (and $J=j$) in the j^n configuration, $N=(n-1)/2$ pair operators should be applied to the single j-nucleon state. The latter has isospin $T=t=1/2$ which should be coupled to T_0 of the pairs, to form the total isospin T. Due to the allowed values of T_0, there is only one state with given isospin, $T=T_0\pm1/2$ in the j^n configuration with n odd.

The fact that there is only one state with lowest seniority and given allowed T in any j^n configuration, leads to a simple formula for the corresponding eigenvalues. States with lowest seniorities are eigenstates of the scalar product of any odd tensor by itself. The eigenvalues of the v=0, t=0 states vanish for any such scalar product. Also the eigenvalue of the $Sp(2j+1)$ Casimir operator is zero for these states. For the v=1, t=1/2 state, the eigenvalue of any $Sp(2j+1)$ interaction is a linear combination of the eigenvalues of the scalar products of the various odd tensor operators. Hence, it can be expressed as the (non-vanishing) eigenvalue of the Casimir operator multiplied by a constant. The difference between this eigenvalue and the zero eigenvalue of the v=0, t=0 states yields the odd-even variation in binding energies. This variation is a major feature of seniority. Thus, the eigenvalues of any two-body Hamiltonian which is diagonal in the seniority scheme for these lowest seniority states can be expressed by (3). The term with the d coefficient is proportional to the term multiplied by c for these states and should be omitted. The a, b and c coefficients are linear combinations of two-body matrix elements of the interaction. These states, with v=0, t=0, J=0 and with v=1, t=1/2, J=j, may be taken to be ground states of even-even and even-odd nuclei respectively. The expression (3), to which single nucleon energies and Coulomb energies are added, was used in the analysis of binding energies of actual nuclei.[11] This result holds for the special case of the rather popular, T=1 pairing interaction.

There are usually several states with given higher values of the seniority quantum numbers. Their energies cannot be obtained from a simple formula like (3). Still, eigenvalues of a special set of states, those with lowest seniorities, can be obtained from a linear combination of Casimir operators of a chain of groups. This is a situation of partial dynamical symmetry.[12]

Removing the degeneracies of $Sp(2j+1)$ states with the same value of J is possible due to the large variety of two-body interactions which are diagonal in the seniority scheme. Such freedom in choice of interaction is not available in more restricted spaces. The $SU(3)$ scheme for nucleons, introduced by Elliott [13], is defined for single nucleon harmonic oscillator wave functions. The special two-body interaction in that scheme is diagonal in the LS-coupling scheme where S and L are good quantum numbers. This group has 8 generators which may form the 5 components of a special quadrupole operator $\mathbf{Q}$ and 3 components of the (orbital) angular momentum $\mathbf{L}$.

The single nucleon Hamiltonian, the kinetic energy plus the oscillator potential, commutes with the generators of the $SU(3)$ group. Its eigenvalues of all orbits in an oscillator shell are equal. The interaction term is written as a linear combination of scalar products of the $SU(3)$ generators, $\mathbf{Q}.\mathbf{Q}$ and $\mathbf{L}.\mathbf{L}$. The most general expression for a two-body interaction, which is diagonal in the $SU(3)$ scheme and includes some single nucleon terms, is

$$a\mathbf{Q}.\mathbf{Q} + b\mathbf{L}.\mathbf{L} = a[\mathbf{Q}.\mathbf{Q} + 3\mathbf{L}.\mathbf{L}] + (b-3a)\mathbf{L}.\mathbf{L} = 4aC[SU(3)] + (b-3a)C[O(3)] \quad (5)$$

The eigenvalues of the Casimir operator of $SU(3)$, which uniquely characterize its irreducible representations, are given in terms of two quantum numbers introduced by Elliott, integers λ and μ , as

$$\lambda^2 + \lambda\mu + \mu^2 + 3(\lambda + \mu) \quad (6)$$

The eigenvalues of $\mathbf{L}.\mathbf{L}$ are $L(L+1)$.

Looking at (5) we see that dynamic symmetry occurs for any two-body Hamiltonian which is constructed from $SU(3)$ generators. Within an irreducible representation of $SU(3)$, energies of eigenstates have the $L(L+1)$ dependence which is characteristic of rotational spectra. Indeed, the $SU(3)$ scheme, within the shell model, has some features of collective spectra which make it a very attractive scheme to study. The simplest representations which may occur have μ=0 and the L values of the allowed states contained in them are given by $L = \lambda, \lambda-2, \lambda-4, \ldots, 1$ or 0. In this irreducible representation describing the ground state band, no two states occur with the same value of L. This is not the case for other irreducible representations where $\mu \neq 0$. To specify states in such cases, Elliott introduced an additional quantum number K, which is the analog of the spin projection on the nuclear axis of symmetry in the collective model $(0 \leq K \leq L)$. The original definition of K did not lead to an orthogonal basis and a better choice of basis was made by Vergados[14].

The most general $SU(3)$ two-body interaction is given by (5) and it has dynamic symmetry. The Casimir operators of $SU(2)$ of isospin and $SU(2)$ of intrinsic spin may be added to it. In any case, the eigenvalues of the $SU(3)$ Hamiltonian do not depend on K. The additional quantum number K distinguishes between different equivalent irreducible $O(3)$ representations obtained from the same irreducible representation of $SU(3)$. The degeneracy of such states, with the same value of

J, cannot be removed by any two-body interaction in the $SU(3)$ scheme. More freedom is achieved if three-body terms are added to the $SU(3)$ Hamiltonian. Such terms should be constructed from the $SU(3)$ generators and a simple choice is

$$\mathbf{Q} \cdot [\mathbf{L} \times \mathbf{L}]^{(2)} \tag{7}$$

This is the scalar product of **Q** with the second rank tensor constructed from components of **L** and it contains 3-body terms in addition to two-body and single particle terms. Adding (7), multiplied by a coefficient, to the $SU(3)$ Hamiltonian removes the degeneracy in L in some cases. The dynamical symmetry, however, is lost.

The nice features of the $SU(3)$ scheme are apparent also in the interacting boson model of Arima and Iachello[1,4]. In that model there are d-bosons ($\ell = 2$) and s-bosons ($\ell = 0$). The Hamiltonian which contains two body interactions of these bosons, is constructed from $U(6)$ generators. This is the group of unitary transformations in the 6 dimensional space of 5 components of a d-boson state and 1 component of the s-boson state. Its generators can be expressed as products of creation and annihilation operators of s- and d-bosons. Interesting special cases of such Hamiltonians are obtained when they are expressed by linear combinations of scalar products of generators of sub-groups of $U(6)$. The $SU(3)$ group and its $O(3)$ subgroup are sub-groups of $U(6)$.

All $SU(3)$ Hamiltonians with two-body interactions, constructed from these generators may be expressed in the same form (5). It is possible to add to (5) the linear and quadratic Casimir operators of $U(6)$ which are equal, in the case of bosons, to linear and quadratic functions of n, the total number of s and d bosons. Hence, the eigenvalues of the most general Hamiltonian with two-boson interactions which is diagonal in the $SU(3)$ scheme are independent of K. This leads to degeneracy of states with the same value of J which belong to the same $SU(3)$ irreducible representation. This degeneracy cannot be removed in the $SU(3)$ scheme if only two-body interactions are considered. Addition of a three-body interaction like in (7) removes such degeneracies but the elegant dynamical symmetry is lost.

There are two other special cases of boson $U(6)$ Hamiltonians with analytic solutions for their eigenvalues which describe vibrational spectra[3] and gamma unstable spectra[15] of nuclei. In the first case, the Hamiltonian is constructed from $U(6)$ generators which contain only d-boson operators. These are generators of the subgroup $U(5)$ of $U(6)$ and can be expressed by the components of the irreducible tensors

$$(d^{+} \times \tilde{d})^{(k)} \qquad k = 0, 1, 2, 3, 4 \tag{8}$$

The Casimir operators of $U(6)$ may be added to such $U(5)$ Hamiltonians, contributing linear and quadratic terms in the total number of s and d bosons. The irreducible representations of $U(5)$ are uniquely characterized by n_d, the number of d-bosons. To further characterize the states, a subgroup of $U(5)$ is considered, $O(5)$, whose generators are the components of odd tensors in (8), with k=3 and

k=1. The spins of the various states are obtained when going to the group induced by 3 dimensional rotations,$O(3)$ which is a subgroup of $O(5)$. Its generators are the 3 components of the k=1 tensor which are proportional to the components of the angular momentum.

The irreducible representations of $O(5)$ are fully characterized by v, the seniority quantum number defined for d-bosons. The reduction of those to the $O(3)$ irreducible representations is not always simple. There may be several states with the same value of L which have the same seniority v. An additional label should be introduced to distinguish between such states. An integer n_Δ was defined[3] which measures in some way, in a given state, the amount of couplings of 3 d-bosons into states with L=0 . The states thus characterized are not orthogonal, nor do they have in general, well defined seniorities v . An orthogonal set of states was defined, characterized by v and by a modified quantum number $\tilde{n}_\Delta$[16,2].

The most general $U(5)$ Hamiltonian, with two-body interactions, may be expressed, apart from linear and quadratic functions of the total number of s and d bosons, by

$$\epsilon n_d + aC[U(5)] + bC[O(5)] + cC[O(3)] \tag{9}$$

Such Hamiltonians have always dynamical symmetry and their eigenvalues are given directly by the linear combination of eigenvalues of the commuting operators in (9). The eigenvalues are thus equal to

$$\epsilon n_d + an_d(n_d+4) + bv(v+3) + cL(L+1) \tag{10}$$

The eigenvalues are clearly independent of n_Δ. Since (9) is the most general $U(5)$ two-body boson Hamiltonian, the degeneracy of states with different values of n_Δ cannot be removed in the case of such Hamiltonians.

In analogy to the $SU(3)$ case, interactions of more than two nucleons, added to the $U(5)$ Hamiltonian, could possibly remove the degeneracy. Three-body terms were introduced by Van Isacker and Chen[17] in order to obtain a boson description of nuclei with triaxial deformation. These may be generally expressed as

$$[[d^+ \times d^+]^{(r)} \times d^+]^{(L)}.[[\tilde{d} \times \tilde{d}]^{(s)} \times \tilde{d}]^{(L)} \tag{11}$$

The authors pointed out that it is possible to make the choice s=r. Since there is only one independent operator for any possible value of L, it is possible to take only one value of r. For L=0, r=2 is the only possible choice. The interaction (11) may be expressed by tensor products of $U(6)$ generators. The pure three-boson interaction (11), with s=r, can be transformed by change of coupling transformations on the d^+ and $\tilde{d}$ operators.

It can be expressed as a linear combination of various two-boson terms and terms which contain the complete three-boson interaction. The latter have the form

$$\begin{Bmatrix} 2\,2\,k' \\ 2\,2\,k" \\ r\,r\,k \end{Bmatrix} [[d^+ \times \tilde{d}]^{(k')} \times [d^+ \times \tilde{d}]^{(k")}]^{(k)}.[d^+ \times \tilde{d}]^{(k)} \tag{12}$$

A three body interaction with non-vanishing eigenvalues for 3 d-boson states coupled to L=0 would have matrix elements which depend on values of n_Δ and hence, could lift the degeneracy. Such a 3-body interaction is obtained by putting L=0 in (110,

$$((d^+ \times d^+)^{(2)} \cdot d^+)([\tilde{d} \times \tilde{d}]^{(2)} \cdot \tilde{d}) \tag{13}$$

To see if such an interaction can be constructed from $O(5)$ generators, consider its expression as a linear combination of terms (12). The $9j$-symbol in the coefficient (13) has two equal columns and hence, vanishes if $(-1)^{k'+k"+k} = -1$. This implies that the 3-body part of the interaction considered, for any value of L, vanishes if all 3 tensors are generators of $O(5)$ with odd ranks. This is consistent with the results of Szpikowski and Gozdz that "the numbers n_Δ and v cannot be, in general, simultaneously defined as good quantum numbers".[16]

The reason for this vanishing is simple. If only odd tensors from (8) are used in (12), we find due to the symmetry of 3 tensors coupled to a scalar, that the only choices are $k' = k'' = k = 1$ and $k' = k'' = 3, k = 1, 3$. The coupling of two equal rank tensors into an odd tensor is antisymmetric and hence, it contains only commutation relations between the components. Since the components of odd tensors are generators of $O(5)$, their commutators are equal to linear combinations of components of odd tensors. Hence, the result is equal to a two-boson interaction. To construct real 3-body interactions of this kind it is necessary to use also even rank tensors which are generators of $U(5)$ but not of $O(5)$.

If the three-boson interaction (13) is added to the Hamiltonian (9), the latter is no longer diagonal in the $O(5)$ scheme. Still, some $O(5)$ states with n_Δ=0 may be eigenstates of (13) with vanishing eigenvalues . Hence, such states are also eigenstates of (9) with eigenvalues given by (10). The Hamiltonian (9) is a linear combination of commuting Casimir operators and hence, such states with n_Δ=0 exhibit partial dynamical symmetry [12].

A simple argument shows that there are not many such states. None exist if $n_d - v \geq 4$ ($n_d \geq 6$ if v =0). The creation operator of such a state has a factor of the form $(S^+)^2 d^+_m$ where S^+ is the creation operator of two d-bosons coupled to L =0. This factor may be transformed by changes of the order of coupling into a linear combination of terms

$$[[[d^+ \times d^+]^{(k)} \times d^+]^{(k')} \times [d^+ \times d^+]^{(k)}]^{(2)}_m \tag{14}$$

The term with k =2, k' =0 yields a component of the state for which n_Δ is at least equal to 1. Applying to this component the operator (13), the result does not vanish and the state does not have zero eigenvalue. This feature holds also for most states with $n_d = v$+2. Apart from the state with $L = 2v$, seniority v, all of them are due to creation operators which may be put in a form where $S^+[d^+ \times d^+]^{(2)}$ is a factor. This factor may be transformed into a linear combination of terms

$$[[[d^+ \times d^+]^{(2)} \times d^+]^{(k)} \times d^+]^{(2)} \tag{15}$$

The term with $k=0$ shows that the state under consideration cannot have a vanishing eigenvalue of (13) and partial dynamical symmetry.

States with seniority v with vanishing eigenvalues of (13) may be found among those with $n_d = v$ and $n_\Delta = 0$. Spins of these states are equal to [1,2] $L = v, v+1, \ldots, 2v-2, 2v$. All of them may be projected from states created by acting on the vacuum by a product of v operators $d^+_{m_i}$ with $m_i \geq 1$ for $i = 1, \ldots, v$. Hence, such states are guaranteed to contain no three boson states coupled to $L = 0$. Thus, states with partial dynamical symmetry exist for $v = n_d$ and $L = v, v+1, \ldots, 2v-2, 2v$ and also for $v = n_d - 2$, $L = 2v$.

Interactions involving more than 3 bosons may be constructed from $O(5)$ generators. Two odd rank tensors may couple into a $k=2$ tensor which could form a scalar product with another such tensor. If only $k=1$ tensors are taken, the degeneracy is not removed since these are generators of $O(3)$. Tensors with $k=3$ should also be used in the construction of such an operator. Such a four-body interaction may remove degeneracies in cases with $n \geq 6$, where states with the same seniority v and the same values of J appear.

The third special case of the $U(6)$ boson Hamiltonian is obtained[15,2] by going to the $O(6)$ subgroup of $U(6)$. The situation in the chain III is very similar to the $U(6)$ case and the discussion will not be repeated.

References

1. A. Arima and F. Iachello, Phys. Rev. Lett. **35** (1975) 1069.
2. F. Iachello and A. Arima, *The Interacting Boson Model*, Cambridge University Press (1987).
3. A. Arima and F. Iachello, Ann. Phys (NY) **99** (1976) 253.
4. A. Arima and F. Iachello, Ann. Phys (NY) **111** (1978) 201.
5. G. Racah, Phys. Rev. **63** (1943) 367.
6. G. Racah, Phys. Rev. **76** (1949) 1352.
7. B.H. Flowers, Proc. Roy. Soc. (London) **A212** (1952) 248.
8. G. Racah, in *Farkas Memorial Volume*, Res. Counc. of Israel, Jerusalem 1952, p. 294.
9. *Group Theory and Spectroscopy*, mimeographed notes, Princeton 1951, Erg. D. Exak. Wissen (Springer) **37** (1965) 27.
10. G. Racah and I. Talmi, Physica **18** (1952) 1097.
11. I. Talmi and R. Thieberger, Phys. Rev. **103** (1956)718.
12. A. Leviatan, Phys. Rev. Lett. **77**(1996) 818.
13. J.P. Elliott, Proc. Roy. Soc. (London) **A245** (1958) 128, 562.
14. J.D. Vergados, Nucl. Phys. **A111** (1968) 681.
15. A. Arima and F. Iachello, Ann. Phys. (NY) **123** (1979) 468.
16. S. Szpikowski and A. Gozdz, Nucl. Phys. **A340** (1980) 76.
17. P. Van Isacker and J-Q. Chen, Phys. rev. C **24** (1981) 684.

PARTIAL DYNAMICAL SYMMETRY AS AN INTERMEDIATE SYMMETRY STRUCTURE

A. LEVIATAN
Racah Institute of Physics, The Hebrew University, Jerusalem 91904, Israel
E-mail: ami@vms.huji.ac.il

We introduce the notion of a partial dynamical symmetry for which a prescribed symmetry is neither exact nor completely broken. We survey the different types of partial dynamical symmetries and present empirical examples in each category.

1 Introduction

Symmetries play an important role in dynamical systems. They provide quantum numbers for the classification of states, determine selection rules and facilitate the calculation of matrix elements. An exact symmetry occurs when the Hamiltonian of the system commutes with all the generators (g_i) of the symmetry-group, $[H\,,\,g_i] = 0$. In this case, all states have good symmetry and are labeled by the irreducible representations (irreps) of the group. The Hamiltonian admits a block structure so that inequivalent irreps do not mix and all eigenstates in the same irrep are degenerate. In a dynamical symmetry the block structure of the Hamiltonian is retained, the states preserve the good symmetry but in general are no longer degenerate (splitting but no mixing). When the symmetry is completely broken $[H\,,\,g_i] \neq 0$, and none of the states have good symmetry. In-between these limiting cases there may exist intermediate symmetry structures, called partial (dynamical) symmetries for which the symmetry is neither exact nor completely broken.

Models based on spectrum generating algebras, such as those developed [1,2] by F. Iachello and his colleagues, form a convenient framework for discussing these different types of symmetries. In such models the Hamiltonian is written in terms of the generators of a Lie algebra, called the spectrum generating algebra. A dynamical symmetry occurs if the Hamiltonian can be written in terms of the Casimir operators ($\hat{C}_{G_i}$) of a chain of nested algebras

$$\begin{array}{cccc} G_1 & \supset G_2 & \supset \ldots \supset & G_n \\ [\alpha_1] & [\alpha_2] & \ldots & [\alpha_n] \end{array} \tag{1}$$

in which case it has the following properties: (i) *solvability:* all states are solvable and analytic expressions are available for energies and other observables; (ii) *quantum numbers:* all states are classified by quantum numbers $\alpha_1, \alpha_2, \ldots \alpha_n$, which are the labels of the irreps of the algebras in the chain; (iii) *pre-determined structure:* the structure of wave functions is completely dictated by symmetry and is independent of the Hamiltonian's parameters a_i

$$H = a_1\,\hat{C}_{G_1} + a_2\,\hat{C}_{G_2} + \ldots + a_n\,\hat{C}_{G_n}\ . \tag{2}$$

The merits of a dynamical symmetry are self-evident. However, in most applications to realistic systems, the predictions of an exact dynamical symmetry are rarely fulfilled and one is compelled to break it. This is usually done by including in the Hamiltonian symmetry-breaking terms associated with different sub-algebra chains

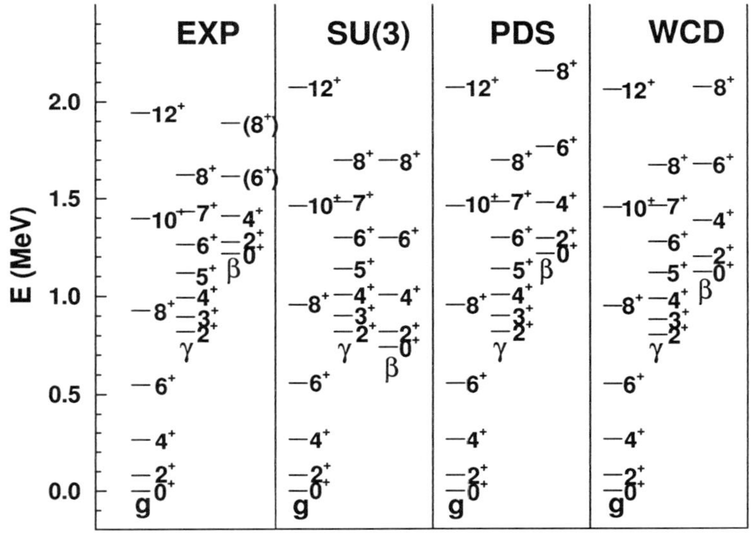

Figure 1. Spectra of ^{168}Er ($N = 16$). Experimental energies (EXP) are compared [3] with IBM calculations in an exact $SU(3)$ dynamical symmetry [$SU(3)$], in a $SU(3)$ PDS with a Hamiltonian $H + \lambda_1 L \cdot L$, Eq. (4), and parameters $t_0 = 2t_2 = 4$, $\lambda_1 = 13$ keV (PDS), and in a broken $SU(3)$ symmetry (WCD) [5] where an $O(6)$ term is added to an $SU(3)$ Hamiltonian.

of the parent spectrum generating algebra (G_1). In general, under such circumstances, solvability is lost, there are no remaining non-trivial conserved quantum numbers and all eigenstates are expected to be mixed. A partial dynamical symmetry (PDS) corresponds to a particular symmetry breaking for which some (but not all) of the above mentioned virtues of a dynamical symmetry are retained. It is then possible to identify the following types of partial dynamical symmetries:

- *type I:* **part** of the states have **all** the dynamical symmetry
- *type II:* **all** the states have **part** of the dynamical symmetry
- *type III:* **part** of the states have **part** of the dynamical symmetry

In what follows we explain each type of partial symmetry and show an empirical example of it. For that purpose we use the interacting boson model [1] (IBM) based on a $U(6)$ spectrum generating algebra. The model describes low-lying quadrupole collective states in even-even nuclei in terms of a system of N monopole (s) and quadrupole (d) bosons representing valence nucleon pairs.

2 SU(3) PDS (type I)

Partial dynamical symmetry of the first type corresponds to a situation for which *part* of the states preserve *all* the dynamical symmetry. In this case the properties of

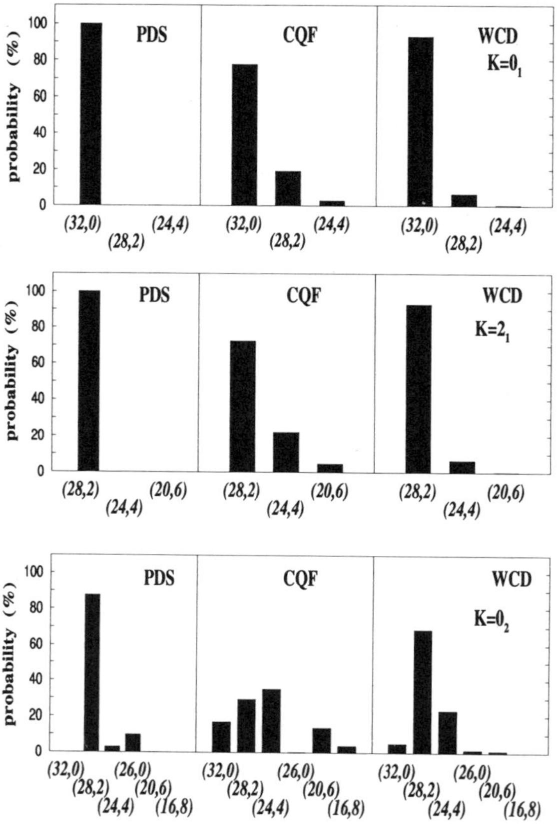

Figure 2. $SU(3)$ decomposition of wave functions of the ground ($K = 0_1$), γ ($K = 2_1$), and $K = 0_2$ bands of ^{168}Er ($N = 16$) in the reported SU(3) PDS calculation [4], and in broken-SU(3) calculations: CQF [6] with a non-$SU(3)$ quadrupole operator in the Hamiltonian, and WCD [5].

solvability, good quantum numbers, and symmetry-dictated structure are fulfilled exactly, but by only a subset of states. As an example we consider the IBM chain

$$\begin{array}{ccccc} U(6) & \supset & SU(3) & \supset & O(3) \\ [N] & & (\lambda,\mu) & K & L \end{array} \tag{3}$$

applicable to axially deformed nuclei. A rotational-invariant IBM Hamiltonian with partial $SU(3)$ symmetry has the form [3]

$$H = t_0\,\Gamma_0^\dagger\Gamma_0 + t_2\,\Gamma_2^\dagger\cdot\tilde{\Gamma}_2 \tag{4}$$

It consists of boson-pairs

$$\Gamma_0^\dagger = d^\dagger\cdot d^\dagger - 2\,(s^\dagger)^2\ ,\ \Gamma_{2,\mu}^\dagger = 2\,s^\dagger d_\mu^\dagger + \sqrt{7}(d^\dagger d^\dagger)^{(2)}_\mu \tag{5}$$

which are $SU(3)$ tensors with $(\lambda,\mu) = (0,2)$ and $L = 0,2$. For $t_0 = t_2$ the above Hamiltonian is related to the Casimir operator of $SU(3)$, hence has an exact $SU(3)$

Table 1. $B(E2)$ branching ratios from states in the γ band in ^{168}Er. The column EXP is the experimental ratios, WCD is the broken $SU(3)$ calculation [5] and PDS is the reported $SU(3)$ partial dynamical symmetry calculation [3].

L_i^π	L_f^π	EXP	PDS	WCD	L_i^π	L_f^π	EXP	PDS	WCD
2_γ^+	0_g^+	54.0	64.27	66.0	6_γ^+	4_g^+	0.44	0.89	0.97
	2_g^+	100.0	100.0	100.0		6_g^+	3.8	4.38	4.3
	4_g^+	6.8	6.26	6.0		8_g^+	1.4	0.79	0.73
3_γ^+	2_g^+	2.6	2.70	2.7		4_γ^+	100.0	100.0	100.0
	4_g^+	1.7	1.33	1.3		5_γ^+	69.0	58.61	59.0
	2_γ^+	100.0	100.0	100.0	7_γ^+	6_g^+	0.74	2.62	2.7
4_γ^+	2_g^+	1.6	2.39	2.5		5_γ^+	100.0	100.0	100.0
	4_g^+	8.1	8.52	8.3		6_γ^+	59.0	39.22	39.0
	6_g^+	1.1	1.07	1.0	8_γ^+	6_g^+	1.8	0.59	0.67
	2_γ^+	100.0	100.0	100.0		8_g^+	5.1	3.57	3.5
5_γ^+	4_g^+	2.91	4.15	4.3		6_γ^+	100.0	100.0	100.0
	6_g^+	3.6	3.31	3.1		7_γ^+	135.0	28.64	29.0
	3_γ^+	100.0	100.0	100.0					
	4_γ^+	122.0	98.22	98.5					

symmetry. For $t_0 \neq t_2$, H involves a mixture of $SU(3)$ tensors with $(\lambda,\mu) = (0,0) \oplus (2,2)$ and although it is not an $SU(3)$ scalar, it has a subset of solvable states with good $SU(3)$ symmetry. This arises from the fact that the boson pairs of Eq. (5) satisfy $\Gamma_{L,\mu}|c;\, N\rangle = 0$, where

$$|c;\, N\rangle = (N!)^{-1/2}(b_c^\dagger)^N|0\rangle \ , \ b_c^\dagger = (\sqrt{2}\, d_0^\dagger + s^\dagger)/\sqrt{3} \tag{6}$$

is the lowest weight state in the $SU(3)$ irrep $(\lambda,\mu) = (2N,0)$. In addition, $[\,\Gamma_{L,\mu}\, ,\, \Gamma_{2,2}^\dagger\,]|c;\, N\rangle \propto \delta_{L,2}\,\delta_{\mu,2}\,|c;\, N\rangle$ and $[\,[\,\Gamma_{L,\mu}\, ,\, \Gamma_{2,2}^\dagger\,]\, ,\, \Gamma_{2,2}^\dagger\,] \propto \delta_{L,2}\,\delta_{\mu,2}\,\Gamma_{2,2}^\dagger$, from which it follows that the sequence of states $|k\rangle = (\Gamma_{2,2}^\dagger)^k|c;\, N-2k\rangle$ are eigenstates of H with good $SU(3)$ symmetry $(\lambda,\mu) = (2N-4k, 2k)$. The states $|k\rangle$ are deformed and serve as intrinsic states representing the ground band ($k=0$) and γ^k bands with angular momentum projection $K = 2k$ along the symmetry axis. Since the Hamiltonian H of Eq. (4) is an $O(3)$ scalar, the rotational states projected from these intrinsic states are also solvable eigenstates of H with good $SU(3)$ symmetry. States in other bands are mixed. Adding to H $O(3)$ rotation terms produces an $L(L+1)$ splitting and lead to a $SU(3)$ PDS of type I. The corresponding spectrum is shown in Fig. 1 in comparison with ^{168}Er, and the $SU(3)$ decomposition of the lowest bands is given in Fig. 2. The ground ($K = 0_1$) and γ ($K = 2_1$) bands are solvable with good $SU(3)$ symmetry $(\lambda,\mu) = (2N,0)$ and $(2N-4,2)$ respectively. Unlike the case of an exact dynamical symmetry, the first $K = 0_2$ band is no longer degenerate with the γ-band, in agreement with the empirical situation in most deformed nuclei. Futhermore, the $K = 0_2$ band involves a mixture of $SU(3)$ irreps $(2N-4,2) \oplus (2N-8,4) \oplus (2N-6,0)$ or equivalently a mixture of a single-phonon (87.5% β) and double-phonon (12.4% $\gamma^2_{K=0}$ and 0.1% β^2) components [4].

Electromagnetic transitions provide a sensitive test for the structure of states. As shown in Table 1, the $SU(3)$ PDS E2 rates for transitions originating from the γ band are found to be in excellent agreement with experiment. The calculated values are obtained by using the general IBM E2 operator $T^{(2)} = \alpha\, Q^{(2)} + \theta\,(d^\dagger s + s^\dagger \tilde{d})$. $Q^{(2)} = d^\dagger s + s^\dagger \tilde{d} - (\sqrt{7}/2)(d^\dagger \tilde{d})^{(2)}$ is an $SU(3)$ generator, hence cannot connect the ground and γ bands which have different $SU(3)$ character. This property combined with the fact that the corresponding wave functions of these solvable bands are determined solely by symmetry, imply that the $B(E2)$ ratios for $\gamma \to g$ transitions quoted in Table 1 do not depend on parameters of the $E2$ operator nor of the Hamiltonian and therefore are parameter-free predictions of $SU(3)$ PDS. The agreement between these predictions and the data confirms the relevance of $SU(3)$ PDS to the spectroscopy of ^{168}Er.

3 O(6) PDS (type I)

It is possible to apply a similar procedure to construct a Hamiltonian with a partial symmetry for the chain

$$\begin{array}{cccc} U(6) \supset & O(6) & \supset O(5) \supset & O(3) \\ [N] & \langle 0,\sigma,0\rangle & (\tau,0) & L \end{array} \tag{7}$$

The $O(6)$ intrinsic state for the ground band

$$|c;\, N\rangle = (N!)^{-1/2}(b_c^\dagger)^N|0\rangle \ , \ b_c^\dagger = (\, d_0^\dagger + s^\dagger)/\sqrt{2} \tag{8}$$

has $\sigma = N$ and the boson pair which annihilates it, $P_0|c;\, N\rangle = 0$, has the form

$$P_0^\dagger = d^\dagger \cdot d^\dagger \,-\, (s^\dagger)^2 \ . \tag{9}$$

The resulting Hamiltonian, $H_{O(6)} = A\, P_0^\dagger P_0$ is related to the Casimir operator of $O(6)$, hence has an exact $O(6)$ symmetry. Adding to it the $O(5)$ and $O(3)$ Casimir operators induces $\tau(\tau+3)$ and $L(L+1)$ splitting and lead to an $O(6)$ dynamical symmetry. The latter has been used [7] to describe the structure of the γ-unstable deformed nucleus ^{196}Pt. The agreement is excellent for properties of the ground band ($\sigma = N$), yet the resulting fit for the observed anharmonicity of excited bands is quite poor. In the dynamical symmetry limit the lowest bands have $\sigma = N, N-2, N-4$ and the eigenvalues $A\,(N-\sigma)(N+\sigma+4)$ of $H_{O(6)}$ imply a fixed anharmonicity: $2[\,1-\frac{1}{N+1}\,]$. For ^{196}Pt with $N=6$, the predicted anharmonicity is 1.71 compared to the empirical value 1.30. One is therefore motivated to search for a Hamiltonian which will improve the fit to the intrinsic spectrum without destroying the good $O(6)$ description for the ground band. This can be accomplished [8] by the following Hamiltonian with an $O(6)$ PDS of type II

$$H = r_0\, R_0^\dagger R_0 + r_2\, R_2^\dagger \cdot \tilde{R}_2 \ . \tag{10}$$

The boson-triplets

$$R_0^\dagger = s^\dagger P_0^\dagger \quad , \quad R_{2,\mu}^\dagger = d_\mu^\dagger P_0^\dagger \tag{11}$$

are $O(6)$ tensors with $\sigma = 1$. For $r_0 = r_2$, the Hamiltonian H is proportional to $H_{O(6)}$ hence has an exact $O(6)$ symmetry. For $r_0 \neq r_2$ it involves a mixture of $O(6)$ tensors with $(\sigma = 0) \oplus (\sigma = 2)$. In general, although H is not an $O(6)$ scalar, it satisfies by construction $H|c;\, N\rangle = 0$, and therefore has an exactly solvable ground band with good $O(6)$ symmetry $\sigma = N$. Since H is an $O(5)$ scalar, states of good $O(5)$ symmetry τ and good angular momentum L projected from $|c; N\rangle$ are also eigenstates of H and form a ground band endowed with good $O(6)$ dynamical symmetry. In contrast, states in excited bands mix several σ irreps. Clearly, the Hamiltonian (10) with added $O(5)$ and $O(3)$ rotational terms exhibits $O(6)$ PDS of type I. Preliminary calculations [8] indicate that such Hamiltonian preserves the good $O(6)$ description for the ground band and is able reproduce the empirical anharmonicity of excited bands in ^{196}Pt.

It is also possible to consider a partial dynamical symmetry with respect to the third IBM chain: $U(6) \supset U(5) \supset O(5) \supset O(3)$ with quantum numbers N, n_d, τ, L respectively. A three-body Hamiltonian with a $U(5)$ PDS of type I was presented by Talmi [9]. A general algorithm how to construct Hamiltonians with partial dynamical symmetry of type I for any semi-simple group is available [10].

4 O(5) PDS (type II)

The second type of PDS corresponds to a situation for which *all* the states preserve *part* of the dynamical symmetry. In this case there are no analytic solutions, yet selected quantum numbers (of the conserved symmetries) are retained. This occurs, for example, when the Hamiltonian contains interaction terms from two different chains with a common symmetry subalgebra, as in the following IBM chains [11]

$$\left.\begin{array}{l} U(6) \supset U(5) \\ U(6) \supset O(6) \end{array}\right\} \supset O(5) \supset O(3)\ . \tag{12}$$

A realization of such an $O(5)$ PDS of type II, is given by the following Hamiltonian, typical for the $U(5)$ (spherical) to O(6) (deformed γ-unstable) transition region

$$H = \epsilon\, \hat{n}_d + A\, P_0^\dagger P_0\ . \tag{13}$$

Here $\hat{n}_d$ is the d-boson number operator which is a Casimir operator of $U(5)$ and the A-term is the $O(6)$ pairing term mentioned in Eq. (9). In this case, all eigenstates of H have good $O(5)$ symmetry but none of them have good $U(5)$ nor good $O(6)$ symmetries and hence only part of the dynamical symmetry of each chain in Eq. (12) is observed. The $E(5)$ critical point of the second order shape-phase transition, considered recently by Iachello [12], correspond to the Hamiltonian of Eq. (13) with $\epsilon = (N-1)\,A$, and falls into the present PDS category.

5 O(6) PDS (type II)

An alternative situation where PDS of type II can occur is when the Hamiltonian preserves only some of the symmetries G_i in the chain (1), and only their irreps are unmixed. Such a scenario was recently considered by Van Isacker [13] in relation to the $O(6)$ chain of Eq. (7), using the following Hamiltonian

$$H_1 \;=\; \kappa_0 P_0^\dagger P_0 + \kappa_2 \left(\Pi^{(2)} \times \Pi^{(2)}\right)^{(2)} \cdot \Pi^{(2)}\ . \tag{14}$$

The κ_0 term is the $O(6)$ pairing term mentioned in Eq. (9). The κ_2 term is constructed only from the $O(6)$ generator, $\Pi^{(2)} = d^\dagger s + s^\dagger \tilde{d}$, which is not a generator

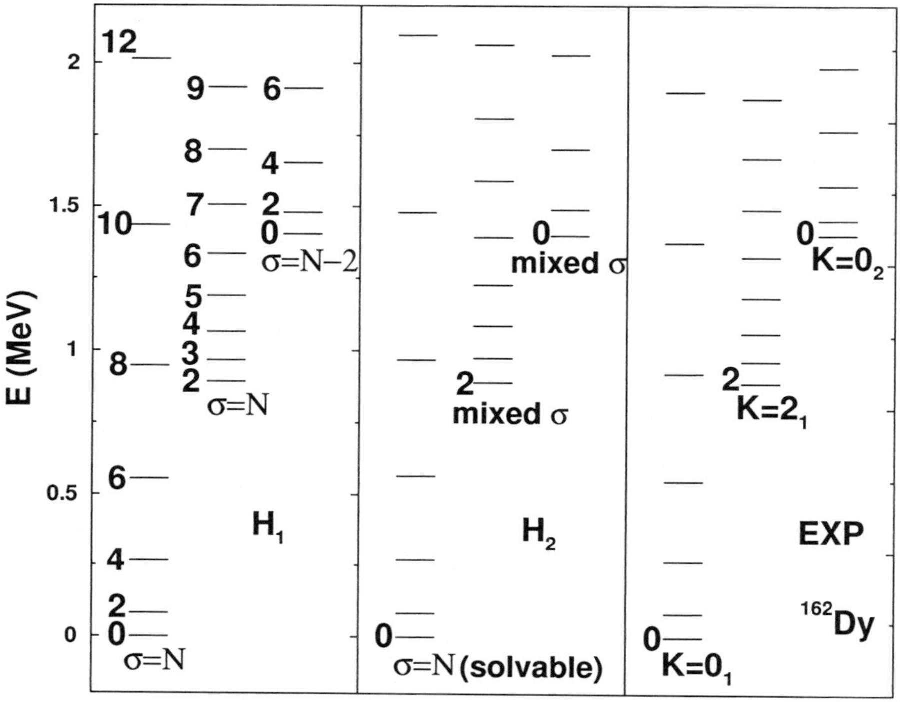

Figure 3. Experimental spectra (EXP) of ^{162}Dy compared with calculated spectra [14] of $H_1 + \lambda_1 L \cdot L$, Eq. (14), and $H_2 + \lambda_2 L \cdot L$, Eq. (15), with parameters (in keV) $\kappa_0 = 8$, $\kappa_2 = 1.364$, $\lambda_1 = 8$ and $h_0 = 28.5$, $h_2 = 6.3$, $\lambda_2 = 13.45$ and boson number $N = 15$.

of $O(5)$. Therefore, it cannot connect states in different $O(6)$ irreps but can induce $O(5)$ mixing subject to $\Delta\tau = \pm 1, \pm 3$. Consequently, H_1 preserves the $U(6)$, $O(6)$, and $O(3)$ symmetries (with good quantum numbers N, σ, L) but not the $O(5)$ symmetry (and hence leads to τ admixtures). These are the necessary ingredients of an $O(6)$ PDS of type II associated with the chain in Eq. (7).

In Fig. 3 we show the experimental spectrum of ^{162}Dy and compare with the calculated spectra of H_1 (14). The spectra display rotational bands of an axially-deformed nucleus, in particular, a ground band ($K = 0_1$) and excited $K = 2_1$ and $K = 0_2$ bands. As shown in the upper portion of Fig. 4, all bands of H_1 are pure with respect to $O(6)$. Specifically, the $K = 0_1, 2_1, 2_3$ bands have $\sigma = N$ and the $K = 0_2$ band has $\sigma = N - 2$. In this case the diagonal κ_0-term in Eq. (14) simply shifts each band as a whole in accord with its σ assignment. On the other hand, the κ_2-term in Eq. (14) is an $O(5)$ tensor with $\tau = 3$ and, therefore, all eigenstates of H_1 are mixed with respect to $O(5)$. This mixing is demonstrated in the upper portion of Fig. 5 for the $L = 0, 2$ members of the ground band.

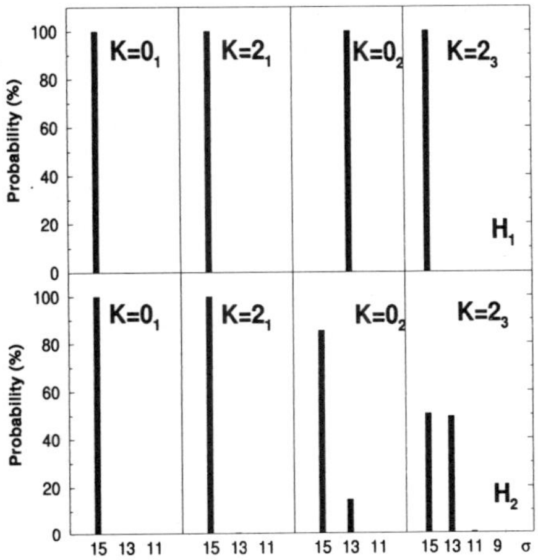

Figure 4. $O(6)$ decomposition [14] of wave functions of states in the bands $K = 0_1, 2_1, 0_2, (L = K^+)$, and $K = 2_3$, $(L = 3^+)$, for H_1 (upper portion) and H_2 (lower portion).

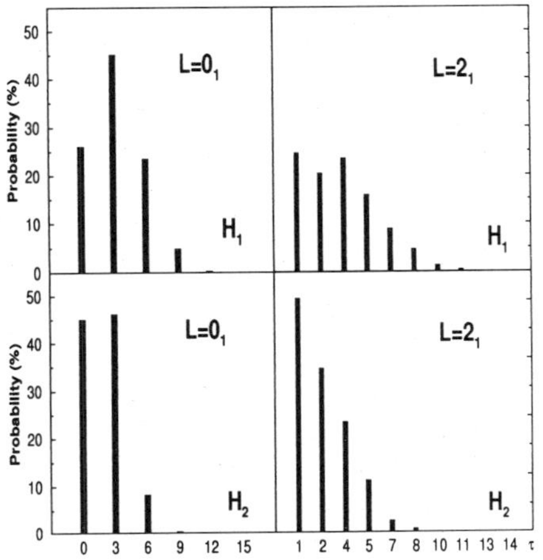

Figure 5. $O(5)$ decomposition [14] of wave functions of the $L = 0, 2$ states in the ground band $(K = 0_1)$ of H_1 (14) [upper portion] and H_2 (15) [lower portion]. Both states have $\sigma = N$.

6 O(6) PDS (type III)

The third type of partial symmetries has a hybrid character, for which *part* of the states of the system under study preserve *part* of the dynamical symmetry. Such a generalized partial symmetry associated with the $O(6)$ chain of Eq. (7), can be realized by the Hamiltonian [14]

$$H_2 \;=\; h_0 P_0^\dagger P_0 + h_2 P_2^\dagger \cdot \tilde{P}_2 \; . \tag{15}$$

Here $P_0^\dagger$ is the $\sigma = 0$ pair of Eq. (9) and the second boson-pair

$$P^\dagger_{2,\mu} = \sqrt{2}\, s^\dagger d^\dagger_\mu + \sqrt{7}(d^\dagger d^\dagger)^{(2)}_\mu \tag{16}$$

is an $O(6)$ tensor with $\sigma = 2$. For $h_0 \neq h_2$ the Hamiltonian H_2 is neither an $O(6)$-scalar nor an $O(5)$-scalar hence can induce both $O(6)$ and $O(5)$ mixing subject to $\Delta\sigma = 0, \pm 2$ and $\Delta\tau = \pm 1, \pm 3$. Although H_2 is not invariant under $O(6)$, it still has an exactly solvable ground band with good $O(6)$ symmetry. This arises from the fact that the boson pairs of Eqs. (9) and (16) annihilate the state $|c;\, N\rangle$ of Eq. (8), which is the $O(6)$ intrinsic state for the ground band with $\sigma = N$. Since H_2 is rotational invariant, states of good angular momentum L projected from $|c;\, N\rangle$ are also eigenstates of H_2 with good $O(6)$ symmetry and form its ground band. These projected states do not have good $O(5)$ symmetry and their known wave functions contain a mixture of components with different τ. It follows that H_2 has a subset of solvable states with good $O(6)$ symmetry $(\sigma = N)$, which is not preserved by other states. All eigenstates of H_2 break the $O(5)$ symmetry but preserve the $O(3)$ symmetry. These are precisely the required features of type III $O(6)$ PDS.

Table 2. Calculated [14] and observed $B(E2)$ values (in $10^{-2}e^2b^2$) for ^{162}Dy. The $E2$ operator is $T^{(2)} = e_B[\, d^\dagger s + s^\dagger \tilde{d} + \chi\,(d^\dagger \tilde{d})^{(2)}\,]$ with parameters $e_B = 0.138$ [0.127] eb and $\chi = -0.235$ [-0.557] for H_1 (14) [H_2 (15)].

Transition	H_1	H_2	Expt.	Transition	H_1	H_2	Expt.
$2^+_{K=0_1} \to 0^+_{K=0_1}$	107	107	107(2)	$2^+_{K=2_1} \to 0^+_{K=0_1}$	2.4	2.4	2.4(1)
$4^+_{K=0_1} \to 2^+_{K=0_1}$	151	152	151(6)	$2^+_{K=2_1} \to 2^+_{K=0_1}$	3.8	4.0	4.2(2)
$6^+_{K=0_1} \to 4^+_{K=0_1}$	163	165	157(9)	$2^+_{K=2_1} \to 4^+_{K=0_1}$	0.24	0.26	0.30(2)
$8^+_{K=0_1} \to 6^+_{K=0_1}$	166	168	182(9)	$3^+_{K=2_1} \to 2^+_{K=0_1}$	4.2	4.3	
$10^+_{K=0_1} \to 8^+_{K=0_1}$	164	167	183(12)	$3^+_{K=2_1} \to 4^+_{K=0_1}$	2.2	2.3	
$12^+_{K=0_1} \to 10^+_{K=0_1}$	159	163	168(21)	$4^+_{K=2_1} \to 2^+_{K=0_1}$	1.21	1.14	0.91(5)
				$4^+_{K=2_1} \to 4^+_{K=0_1}$	4.5	4.7	4.4(3)
$0^+_{K=0_2} \to 2^+_{K=0_1}$	0.16	0.23		$4^+_{K=2_1} \to 6^+_{K=0_1}$	0.59	0.61	0.63(4)
$0^+_{K=0_2} \to 2^+_{K=2_1}$	0.14	17.23		$5^+_{K=2_1} \to 4^+_{K=0_1}$	3.4	3.3	3.3(2)
$2^+_{K=0_2} \to 0^+_{K=0_1}$	0.02	0.04		$5^+_{K=2_1} \to 6^+_{K=0_1}$	2.9	3.1	4.0(2)
$2^+_{K=0_2} \to 2^+_{K=0_1}$	0.04	0.05		$6^+_{K=2_1} \to 4^+_{K=0_1}$	0.84	0.72	0.63(4)
$2^+_{K=0_2} \to 2^+_{K=2_1}$	0.03	3.69		$6^+_{K=2_1} \to 6^+_{K=0_1}$	4.5	4.7	5.0(4)

The spectra of H_2 is shown in Fig. 3, while the $O(6)$ and $O(5)$ decomposition of selected states are shown in the lower portion of Fig. 4 and Fig. 5 respectively. For H_2, the solvable $K = 0_1$ ground band has $\sigma = N$ and all eigenstates are mixed with respect to $O(5)$. However, in contrast to H_1 of Eq. (14), excited bands of H_2 can have components with different $O(6)$ character. For example, the $K = 0_2$ band of H_2 has components with $\sigma = N$ (85.50%), $\sigma = N - 2$ (14.45%), and $\sigma = N - 4$ (0.05%). These σ-admixtures can in turn be interpreted in terms of multi-phonon excitations. Specifically, we find that the $K = 0_2$ band is composed of 36.29% β, 63.68% $\gamma^2_{K=0}$, and 0.03% β^2 modes, *i.e.*, it is dominantly a double-gamma phonon excitation with significant single-β phonon admixture. The $K = 2_1$ band has only a small $O(6)$ impurity and is an almost pure single-gamma phonon band. The combined results of Figs. 4 and 5 constitute a direct proof that H_2 (15) possesses a type III $O(6)$ PDS which is distinct from the type II $O(6)$ PDS of H_1 (14).

In Table 2 we compare the presently known experimental $B(E2)$ values for transitions in ^{162}Dy with PDS calculations. The $B(E2)$ values predicted by H_1 and H_2 for $K = 0_1 \to K = 0_1$ and $K = 2_1 \to K = 0_1$ transitions are very similar and agree well with the measured values. On the other hand, their predictions for interband transitions from the $K = 0_2$ band are very different. Future measurements of these transitions will enable one to distinguish which type of partial $O(6)$ symmetry is more suitable for ^{162}Dy.

7 Summary and Conclusions

In this contribution we have considered departures from complete dynamical symmetry by introducing the notion of a partial dynamical symmetry (PDS). The latter refers to an intermediate symmetry structure for which some (but not all) of the virtues of a dynamical symmetry (*e.g.* solvability, quantum numbers) are retained.

We have presented empirical examples of nuclei in each category of PDS. Although we have focused the discussion to partial symmetries in systems of one type of bosons (IBM-1) relevant to nuclei, there are also examples of PDS in systems of several types of bosons [15,16] (*e.g.* proton-neutron bosons in the IBM-2), in bose-fermi systems [17] (IBFM) and in purely fermionic systems [18,19]. Thus, PDS seem to be a generic feature in dynamical systems with concrete applications to nuclear and molecular [20] spectroscopy. In addition, PDS have been shown to be relevant to the study of mixed systems [21,22] with coexisting regularity and chaos.

Acknowledgments

It is a pleasure and honor for me to dedicate this contribution to F. Iachello on the occasion of his 60th birthday. Franco's innovative approach to physics, emphasizing the unity of science and uncovering of underlying symmetries, has influenced and inspired the ideas discussed here. Segments of the reported results were obtained in collaboration with I. Sinai (HU), P. Van Isacker (GANIL) and J.E. Garcia-Ramos (Huelva). This work was supported by the Israel Science Foundation.

References

1. F. Iachello and A. Arima, "The Interacting Boson Model", (Cambridge Univ. Press, Cambridge, 1987).
2. F. Iachello and R.D. Levine, "Algebraic Theory of Molecules", (Oxford Univ. Press, Oxford, 1994).
3. A. Leviatan, Phys. Rev. Lett. **77**, 818 (1996).
4. A. Leviatan and I. Sinai, Phys. Rev. C **60**, 061301 (1999).
5. D.D. Warner, R.F. Casten and W.F. Davidson, Phys. Rev. C **24**, 1713 (1981).
6. D.D. Warner and R.F. Casten, Phys. Rev. C **28**, 1798 (1983).
7. J.A. Cizewski *et. al.*, Nucl. Phys. A **323**, 349 (1979).
8. A. Leviatan, J.E. Garcia-Ramos and P. Van Isacker, work in progress.
9. I. Talmi, these proceedings.
10. Y. Alhassid and A. Leviatan, J. Phys. A. **25**, L1265 (1992).
11. A. Leviatan, A. Novoselsky, and I. Talmi, Phys. Lett. B **172**, 144 (1986).
12. F. Iachello, Phys. Rev. Lett. **85**, 3580 (2000).
13. P. Van Isacker, Phys. Rev. Lett. **83**, 4269 (1999).
14. A. Leviatan and P. Van Isacker, Phys. Rev. Lett. **89**, 222501 (2002).
15. I. Talmi, Phys. Lett. B **405**, 1 (1997).
16. A. Leviatan and J.N. Ginocchio, Phys. Rev. C **61**, 024305 (2000).
17. R.V. Jolos and P. von Brentano, Phys. Rev. C **62**, 034310 (2000).
18. J. Escher and A. Leviatan, Phys. Rev. Lett. **84**, 1866 (2000); J. Escher and A. Leviatan, Phys. Rev. C **65**, 054309 (2002).
19. D.J. Rowe and G. Rosensteel, Phys. Rev. Lett. **87**, 172501 (2001); G. Rosensteel and D.J. Rowe, Phys. Rev. C **67**, 014303 (2003).
20. J.L. Ping and J.Q. Chen, Ann. Phys. (N.Y.) **255**, 75 (1997).
21. N. Whelan, Y. Alhassid and A. Leviatan, Phys. Rev. Lett. **71**, 2208 (1993).
22. A. Leviatan and N.D. Whelan, Phys. Rev. Lett. **77**, 5202 (1996).

VECTOR COHERENT STATE THEORY: A POWERFUL TOOL FOR SOLVING ALGEBRAIC PROBLEMS IN PHYSICS

D.J. ROWE

Department of Physics, University of Toronto
Toronto, Ontario M5S 1A7, Canada

VCS theory is perhaps the simplest and most effective way known for computing the matrix elements of a Lie algebra. It is a mathematical tool that noone who is serious about using algebraic methods in physics should be without. It encorporates the mathematical theories of induced representations and geometric quantization in a physically intuitive manner that makes it easy to construct the explicit representations of a desired Lie algebra in a chosen basis in a systematic manner. Its practical utility has been confirmed in numerous applications.

1 Introduction

VCS theory was developed in nuclear physics to provide practical and efficient ways to do calculations with non-trivial algebraic models. It was designed for use with the symplectic model [1] on which the microscopic theory of nuclear collective motion is founded [2]. However, it subsequently proved capable of solving numerous other problems in physics and the mathematics it employs [3].

VCS theory is a synthesis[4,5]of the powerful mathematical theories of induced representations[6] and geometric quantization[7,8]. It is accessible to physicists and provides the explicit results they need. In addition to the standard (reducible) representations of induced-representation theory , it gives the explicit matrices of irreducible representations required for applications of symmetry in physics.

It is shown here how VCS theory is used to construct the representations of su(3) in an su(2) basis [4], the representations of su(3) in an so(3) basis [9,10], and the generic representations of so(5) in an so(3) basis [11,12]. The theory has been applied to many other Lie algebras and superalgebras (cf. ref. [4] for a list of early references) and to the computation of SU(3) Clebsch-Gordan coefficients [13]. As a theory of quantization [5], VCS theory relates the classical and quantal representations of an algebraic model and provides the maps between them. It resolves the problem with Dirac's canonical theory of quantization. It provides a physical perspective on the methods of *geometric quantization* and simple ways to implement the prescriptions of that theory. Moreover, the vector generalizations of coherent state theory provide quantizations of systems with intrinsic gauge degrees of freedom [5]. Unfortunately, there is no space to discuss these many applications here.

2 Scalar coherent state representations

Definition (COHERENT STATES): If T is a representation of a Lie group G on a Hilbert space H *and $|0\rangle$ is a fixed state in* H, *then the states*

$$\{|g\rangle = T^{\dagger}(g)|0\rangle\,,\ g \in G\} \tag{1}$$

are called coherent states [14].

If the representation T is irreducible, then the coherent states span the Hilbert space H for this irrep. Thus, any state $|\psi\rangle \in$ H is defined by its overlaps with a set of coherent states in H.

Definition (COHERENT STATE WAVE FUNCTIONS): If $|0\rangle$ is a fixed state in the Hilbert space H *for a representation T of a group G, then any $|\psi\rangle \in$ H can be represented by a coherent state wave function Ψ, i.e., a function over G with values*

$$\Psi(g) = \langle g|\psi\rangle = \langle 0|T(g)|\psi\rangle\,, \quad g \in G\,. \tag{2}$$

Suppose, for example, that $\hat{R}$ is an irrep of SO(3)[a] of angular momentum L and $|0\rangle \equiv |LK\rangle$. Then a state $|LM\rangle$ has coherent state wave function Ψ_{LM} defined as a function of Euler angles by

$$\Psi_{LM}(\Omega) = \langle LK|\hat{R}(\Omega)|LM\rangle = \mathcal{D}^L_{KM}(\Omega)\,. \tag{3}$$

Depending on the choice of the fixed state $|0\rangle$ it is generally sufficient to specify a state $|\psi\rangle$ by giving the values of its coherent state wave function at a subset of elements of G. In the above example, if $|0\rangle \equiv |L0\rangle$ then the state $|LM\rangle$ has coherent state wave function Ψ_{LM} defined over an SO(2)\SO(3) coset (the sphere) by

$$\Psi_{LM}(\theta,\varphi) = \langle L0|e^{i\hat{L}_y\theta}e^{i\varphi\hat{L}_z}|LM\rangle = \sqrt{\frac{2L+1}{4\pi}}\,Y_{LM}(\theta,\varphi)\,. \tag{4}$$

Definition (COHERENT STATE REPRESENTATION): With coherent state wave functions defined by eq. (2), the coherent state representation $\Gamma(X)$ of an infinitesimal generator X of the group G is defined by

$$[\Gamma(X)\Psi](g) = \langle g|T(X)|\psi\rangle = \langle 0|T(g)T(X)|\psi\rangle\,, \quad g \in G\,. \tag{5}$$

For example, the group SU(2) has coherent state wave functions

$$\Psi_{jm}(z) = \langle j, m{=}{-j}|e^{z\hat{J}_-}|jm\rangle\,, \quad m = -j, \ldots, +j\,, \tag{6}$$

and the definition (5) gives

$$\Gamma(J_-) = \frac{d}{dz}\,, \quad \Gamma(J_0) = z\frac{d}{dz} - j\,, \quad \Gamma(J_+) = 2jz - z^2\frac{d}{dz}\,. \tag{7}$$

This is the well-known Dyson representation of su(2).

3 Vector coherent state representations

The above shows that a good choice of the fixed state $|0\rangle$ can result in a simple coherent state representation. We now show that by choosing a vector space of intrinsic states rather than a single state, much more simplification is achieved and the theory becomes much more powerful and versatile.

Definition (VCS WAVE FUNCTIONS): If $B = \{\xi_\nu \equiv |\nu\rangle\}$ is an orthonormal basis for a fixed subspace $U \subset$ H of the Hilbert space for an irrep T of a group G and

[a] We use upper case symols for the group and lower case for its Lie algebra.

$N \subset G^c$ is a subset of elements of the complex extension of G such that the coherent states

$$\{|\nu(z)\rangle = T^\dagger(z)|\nu\rangle\,,\ z \in N\,,\ |\nu\rangle \in B\} \tag{8}$$

span the Hilbert space H*, then any state $|\psi\rangle \in$ H can be represented by a VCS wave function Ψ, with vector values in U given by*

$$\Psi(z) = \sum_\nu \xi_\nu\, \langle \nu(z)|\psi\rangle = \sum_\nu \xi_\nu\, \langle \nu|T(z)|\psi\rangle\,, \quad z \in N\,. \tag{9}$$

Definition (VCS REPRESENTATION): With VCS wave functions defined by eq. (9), the VCS representation $\Gamma(X)$ of an infinitesimal generator X of the group G is defined by

$$[\Gamma(X)\Psi](z) = \sum_\nu \xi_\nu\, \langle \nu(z)|T(X)|\psi\rangle = \sum_\nu \xi_\nu\, \langle \nu|T(z)T(X)|\psi\rangle\,, \quad z \in N\,. \tag{10}$$

4 VCS representations of su(3) in an su(2) basis

The su(3) algebra is a subalgebra of traceless Hermitian complex linear combinations of a set of matrices $\{C_{ij}\}$ with entries

$$(C_{ij})_{kl} = \delta_{ik}\delta_{jl} \tag{11}$$

and commutation relations

$$[C_{ij}, C_{kl}] = \delta_{jk}C_{il} - \delta_{il}C_{kj}\,. \tag{12}$$

The complex extension of su(3) is spanned by the matrices

$$e_2 = C_{13}\,, \quad e_3 = C_{12}\,, \tag{13}$$

$$H_1 = C_{11} - \tfrac{1}{2}(C_{22} + C_{33})\,, \quad H_2 = C_{22} - C_{33}\,, \quad e_1 = C_{23}, \quad f_1 = C_{32}\,, \tag{14}$$

$$f_2 = C_{31}\,, \quad f_3 = C_{21}\,. \tag{15}$$

These matrices are associated with the root vectors of the root diagram for su(3) shown in fig. 1. The horizontal root vectors define a u(2)⊂su(3) subalgebra.

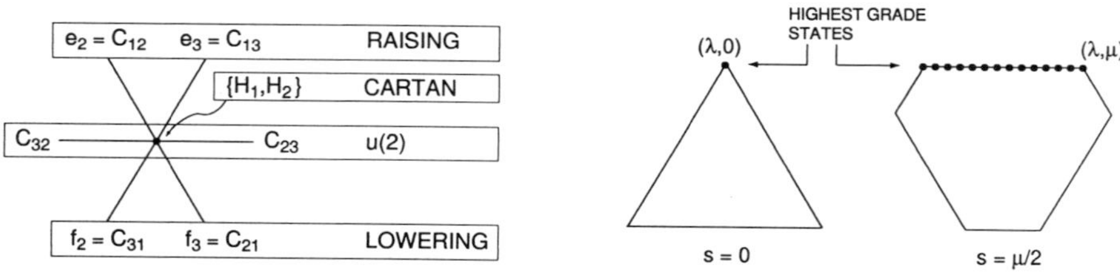

Figure 1. The root diagram for su(3) showing a u(2) subalgebra and complementary sets of raising and lowering operators. Also shown are outlines of the weight diagrams for irreps of highest weight $(\lambda, 0)$ and (λ, μ) and their highest grade states.

For an irrep T of highest weight $(\lambda, 0)$, it is best to choose the highest weight state as the fixed state for a scalar coherent state representation. This choice results in considerable simplification because the highest weight state is annihilated by the raising operators $\hat{e}_2 = T(C_{12})$, $\hat{e}_3 = T(C_{13})$, and the su(2) operators $\hat{H}_2$, $\hat{C}_{23}$, $\hat{C}_{32}$; it satisfies the equations

$$\hat{H}_1|0\rangle = \lambda|0\rangle\,, \quad \hat{H}_2|0\rangle = \hat{C}_{23}|0\rangle = \hat{C}_{32}|0\rangle = \hat{e_2}|0\rangle = \hat{e_3}|0\rangle = 0\,. \tag{16}$$

The Hilbert space for the representation T of highest weight $(\lambda, 0)$ is then spanned by the states $\{|z\rangle = e^{z_2^* \hat{f}_2 + z_3^* \hat{f}_3}|0\rangle\}$ for a suitable range of a pair of complex variables z_2 and z_3.

For a general irrep T of highest weight $(\lambda\mu)$, the states that are annihilated by the $\hat{e}_2$ and $\hat{e}_3$ raising operators

$$U = \{|\phi\rangle \in \mathrm{H} \mid \hat{e}_2|\phi\rangle = \hat{e}_3|\phi\rangle = 0\} \tag{17}$$

are not also annihilated by elements of the u(2) $\subset$ su(3) subalgebra. However, they span a u(2)–invariant subspace $U \subset \mathrm{H}$ of *highest grade* states. Moreover, if $\{\xi_\nu \equiv |s\nu\rangle\}$ is an orthornormal basis for U indexed by ν, then the Hilbert space for the su(3) representation T is spanned by the states $\{e^{z_2^* \hat{f}_2 + z_3^* \hat{f}_3}|s\nu\rangle\}$ for a suitable range of the complex variables z_2 and z_3. Thus, any state $|\psi\rangle$ in the Hilbert space is represented by the VCS wave function

$$\Psi(z) = \sum_\nu \xi_\nu \langle s\nu|e^{\hat{z}}|\psi\rangle\,, \quad \hat{z} = z_2\hat{e}_2 + z_3\hat{e}_3\,. \tag{18}$$

An element X of the su(3) algebra is then represented as a linear operator $\Gamma(X)$ on the VCS wave functions that is defined by

$$[\Gamma(X)\Psi](z) = \sum_\nu \xi_\nu \langle s\nu|e^{\hat{z}}\hat{X}|\psi\rangle = \sum_\nu \xi_\nu \langle s\nu|\hat{X}(z)e^{\hat{z}}|\psi\rangle\,, \tag{19}$$

where

$$\hat{X}(z) = e^{\hat{z}}\hat{X}e^{-\hat{z}} = \hat{X} + [\hat{z}, X] + \tfrac{1}{2}[\hat{z}, [\hat{z}, \hat{X}]]\,. \tag{20}$$

Explicit expressions for the $\Gamma(X)$ operators are obtained by first observing that $\hat{X}(z)$ is an element of $su(3)$ and that

$$\textstyle\sum_\nu \xi_\nu \langle s\nu|\hat{f}_i e^{\hat{z}}|\psi\rangle = 0\,, \quad \sum_\nu \xi^\nu \langle s\nu|\hat{e}_i e^{\hat{z}}|\psi\rangle = \partial_i\Psi(z)\,, \tag{21}$$

$$\textstyle\sum_\nu \xi_\nu \langle s\nu|\hat{H}_1 e^{\hat{z}}|\psi\rangle = (\lambda + s)\Psi(z)\,, \quad \sum_\nu \xi_\nu \langle s\nu|\hat{H}_2 e^{\hat{z}}|\psi\rangle = 2\hat{s}_0\Psi(z)\,, \tag{22}$$

$$\textstyle\sum_\nu \xi_\nu \langle s\nu|\hat{C}_{23} e^{\hat{z}}|\psi\rangle = \hat{s}_+\Psi(z)\,, \quad \sum_\nu \xi_\nu \langle s\nu|\hat{C}_{32} e^{\hat{z}}|\psi\rangle = \hat{s}_-\Psi(z)\,, \tag{23}$$

where $\hat{s}_0$ and $\hat{s}_\pm$ are intrinsic spin operators defined such that

$$\hat{s}_0\xi_\nu = \nu\,\xi_\nu\,, \quad \hat{s}_\pm\,\xi_\nu = \sqrt{(s \mp \nu)(s \pm \nu + 1)}\;\xi_{\nu\pm 1}\,, \tag{24}$$

with $s = \mu/2$. It follows that

$$\Gamma(H_1) = \lambda + s - \tfrac{3}{2}\hat{n}\,, \quad \Gamma(H_2) = 2(\hat{s}_0 + \hat{j}_0)\,, \tag{25}$$

$$\Gamma(C_{23}) = \hat{s}_+ + \hat{j}_+\,, \quad \Gamma(C_{32}) = \hat{s}_- + \hat{j}_-\,, \quad \Gamma(e_i) = \partial_i\,, \tag{26}$$

$$\textstyle\Gamma(f_2) = [\lambda - \hat{s}_0]z_2 - \hat{s}_+ z_3 - z_2\sum_i z_i\partial_i\,, \tag{27}$$

$$\textstyle\Gamma(f_3) = [\lambda + \hat{s}_0]z_3 - \hat{s}_- z_2 - z_3\sum_i z_i\partial_i\,, \tag{28}$$

where

$$\hat{n} = \sum_i z_i \partial_i \,, \quad \hat{j}_0 = \tfrac{1}{2}(z_2\partial_2 - z_3\partial_3) \,, \quad \hat{j}_+ = z_2\partial_3 \,, \quad \hat{j}_- = z_3\partial_2 \,. \tag{29}$$

It is seen that all the operators are simply expressed in terms of the elements $\hat{s}_i$ and $\hat{j}_i$ of two su(2) algebras, one of which of which is regarded as an intrinsic spin. The most complicated operators in the set are $\Gamma(f_1)$ and $\Gamma(f_2)$. However, their matrix elements are easily determined by expressing them in the form

$$\Gamma(f_i) = [\hat{\Lambda}, z_i] \,, \tag{30}$$

where

$$\hat{\Lambda} = (\lambda + s)\hat{n} - \tfrac{1}{2}\hat{n}(\hat{n} - 1) - 2\hat{\mathbf{s}} \cdot \hat{\mathbf{j}} \,. \tag{31}$$

The expressions suggest defining orthonormal basis states for the su(3) irrep in the su(2)–coupled form

$$\psi_{jJM}(z) = K_{jJ} \, [\xi \otimes \varphi_j(z)]_{JM} \,, \tag{32}$$

where

$$\varphi_{jm}(z) = \frac{(z_2)^{j+m}(z_3)^{j-m}}{\sqrt{(j+m)!(j-m)!}} \,, \quad m = -j, \ldots, +j \,, \tag{33}$$

and the norm factors $\{K_{jJ}\}$ remain to be determined. It is seen that $\Gamma(H_1)$, $\Gamma(H_2)$ and $\hat{\Lambda}$ are diagonal in this basis with eigenvalues given by

$$\Gamma(H_1)\,\psi_{jJM} = (\lambda + s - 3j)\,\psi_{jJM} \,, \quad \Gamma(H_2)\,\psi_{jJM} = 2M\,\psi_{jJM} \,, \tag{34}$$

$$\hat{\Lambda}\,\psi_{jJM} = \Omega(sjJ)\,\psi_{jJM} \,, \tag{35}$$

and

$$\Omega(sjJ) = 2(\lambda + s)j + s(s+1) - j(j-2) - J(J+1) \,. \tag{36}$$

The operators $\Gamma(C_{23})$ and $\Gamma(C_{32})$ are simply the su(2) raising and lowering operators

$$\Gamma(C_{23}) = \hat{J}_+ = \hat{s}_+ + \hat{j}_+ \,, \quad \Gamma(C_{32}) = \hat{J}_- = \hat{s}_- + \hat{j}_- \,, \tag{37}$$

with the usual su(2) actions

$$\hat{J}_\pm \psi_{jJM} = \sqrt{J \mp M)(J \pm M + 1)}\; \psi_{jJ,M\pm 1} \,. \tag{38}$$

The matrix elements of ∂_i and z_i can be evaluated explicity for the given basis wave functions. Since the z_i are components of an su(2) spin–1/2 tensor, the result is conveniently expressed in terms of reduced matrix elements. With some Racah recoupling, we obtain

$$\langle sjJ \| \hat{e} \| s, j + \tfrac{1}{2}, J' \rangle = -\sqrt{(2j+1)(2j+2)(2J+1)(2J'+1)}\; W(\tfrac{1}{2} j J' s : j + \tfrac{1}{2} J) \times \frac{K_{j+1/2,J'}}{K_{jJ}} \,, \tag{39}$$

$$\langle s, j + \tfrac{1}{2}, J' \| \hat{f} \| sjJ \rangle = (-1)^{J'-J+\frac{1}{2}} \langle sjJ \| \hat{e} \| s, j + \tfrac{1}{2}, J' \rangle \times \left(\frac{K_{jJ}}{K_{j+1/2,J'}} \right)^2 [\tfrac{1}{2}(2\lambda + \mu) + J(J+1) - J'(J'+1) - j + \tfrac{3}{4}] \,. \tag{40}$$

Thus, by setting

$$\left(\frac{K_{j+1/2,J'}}{K_{jJ}}\right)^2 = \tfrac{1}{2}(2\lambda+\mu) + J(J+1) - J'(J'+1) - j + \tfrac{3}{4}\,, \qquad (41)$$

we obtain the reduced matrix elements of a unitary representation with j and J running over all integer and half-odd integer values for which the K_{jJ} coefficients are non-zero.

5 Representation of $su(3)$ in an $so(3)$ basis

For applications in nuclear physics, one needs the su(3) representations in an angular momentum basis. They are easily constructed in coherent state theory as a result of the well-known observation [?]:

If $|0\rangle$ is a highest weight state for an su(3) irrep, then the rotated states

$$\{\hat{R}(\Omega)|\lambda\mu\rangle; \Omega \in SO(3)\} \qquad (42)$$

span the Hilbert space H *of this irrep.*

Suppose the Hilbert space H has an orthonormal basis of angular-momentum coupled states $\{|\alpha LM\rangle\}$. Then, these states are represented by coherent state wave functions of the form

$$\Psi_{\alpha LM}(\Omega) = \langle\lambda\mu|\hat{R}(\Omega)|\alpha LM\rangle = \sum_K \langle\lambda\mu|\alpha LK\rangle\, \mathcal{D}^L_{KM}(\Omega)\,. \qquad (43)$$

An element X of the $su(3)$ Lie algebra then has coherent state representation as a linear operator $\Gamma(X)$ on the coherent state wavefunctions, defined by

$$[\Gamma(X)\Psi](\Omega) = \langle\lambda\mu|\hat{R}(\Omega)\hat{X}|\psi\rangle = \langle\lambda\mu|\hat{X}(\Omega)\hat{R}(\Omega)|\psi\rangle\,, \qquad (44)$$

where (with $\tilde{\Omega}$ denoting the transpose of Ω)

$$\hat{X}(\Omega) = \hat{R}(\Omega)\hat{X}\hat{R}(\tilde{\Omega})\,. \qquad (45)$$

In an angular-momentum basis, the su(3) algebra is spanned by the angular momentum and quadrupole operators with components given in terms of the root vectors shown in fig. 1 by

$$L_0 = -\mathrm{i}(C_{23} - C_{32})\,, \quad L_\pm = \mathrm{i}(e_3 - f_3) \pm (e_2 - f_2)\,, \qquad (46)$$

$$Q_0 = 2H_1\,, \quad Q_{\pm 1} = \mp\sqrt{\tfrac{3}{2}}\,[e_2 + f_2 \pm \mathrm{i}(e_3 + f_3)]\,,$$

$$Q_{\pm 2} = \sqrt{\tfrac{3}{2}}\,[H_2 \pm \mathrm{i}(C_{23} + C_{32})]\,. \qquad (47)$$

From the definition (44), the coherent state representation of a quadrupole operator is given by

$$\begin{aligned}[\Gamma(Q_m)\Psi_{\kappa LM}](\Omega) &= \langle\lambda\mu|\hat{R}(\Omega)Q_m|\kappa LM\rangle \\ &= \sum_\nu \langle\lambda\mu|\hat{Q}_\nu|\kappa LK\rangle\, \mathcal{D}^2_{\nu m}(\Omega)\mathcal{D}^L_{KM}(\Omega)\,. \end{aligned} \qquad (48)$$

The matrix elements $\langle\lambda\mu|\hat{Q}_\nu|\kappa LK\rangle$ are inferred from the expansions (46) and (47) and the identities

$$\langle\lambda\mu|\hat{L}_0|\alpha LK\rangle = K\,\langle\lambda\mu|\hat{L}_0|\alpha LK\rangle\,, \tag{49}$$

$$\langle\lambda\mu|\hat{L}_\pm|\alpha LK\rangle = \sqrt{(L\mp K)(L\pm K+1)}\,\langle\lambda\mu|\alpha L, K\pm 1\rangle\,, \tag{50}$$

$$\langle\lambda\mu|\hat{H}_1|\alpha LK\rangle = \tfrac{1}{2}(2\lambda+\mu)\langle\lambda\mu|\alpha LK\rangle\,, \quad \langle\lambda\mu|\hat{H}_2|\alpha LK\rangle = \mu\langle\lambda\mu|\alpha LK\rangle\,, \tag{51}$$

$$\langle\lambda\mu|\hat{C}_{32}|\alpha LK\rangle = \langle\lambda\mu|\hat{f}_i|\alpha LK\rangle = 0\,. \tag{52}$$

One finds [10] that, if

$$\Psi_{\alpha LM} = \sum_K a_K(\alpha L)\,\mathcal{D}^L_{KM}\,, \tag{53}$$

then

$$[\Gamma(Q)\otimes\Psi_{\alpha L}]_{L'M} = \sum_{\kappa\kappa'} M^{(L'L)}_{\kappa'\kappa}\,a_\kappa(\alpha L)\,\mathcal{D}^{L'}_{\kappa' M} \tag{54}$$

with

$$\begin{aligned} M^{(L'L)}_{\kappa'\kappa} &= \delta_{\kappa',\kappa}\left[(2\lambda+\mu+3)+\delta_{K1}\sigma_{L'L}-\tfrac{1}{2}L'(L'+1)+\tfrac{1}{2}L(L+1)\right](L\kappa,20|L'\kappa)\\ &\quad+\delta_{\kappa',\kappa+2}\sqrt{\tfrac{3}{2}}\,(\mu-\kappa)(L\kappa,22|L'\kappa+2)\\ &\quad+\delta_{\kappa',\kappa-2}\sqrt{\tfrac{3}{2}}\,(\mu+\kappa)(L\kappa,2-2|L'\kappa-2)\,, \end{aligned} \tag{55}$$

and

$$\sigma_{L'L} = \tfrac{1}{2}(\mu+1)(-1)^{\lambda+L}\times\begin{cases} -\frac{3L(L+1)}{3-L(L+1)} & \text{for } L'=L\\ L+1 & \text{for } L'=L+1\\ -L & \text{for } L'=L-1\\ -1 & \text{for } L'=L\pm 2\,. \end{cases} \tag{56}$$

Thus, one needs only a table of Clebsch-Gordan coefficients to obtain explicit matrix elements of the su(3) quadrupole operators in any given basis of $a(\alpha L)$ vectors. However, to obtain the matrix elements of a unitary representation, one needs an orthonormal basis. An orthonormal basis is constructed as follows.

A set of vectors $\{a(\alpha L)\}$ whose components are the expansion coefficients $\{a_K(\alpha L)\}$ of an orthonormal basis $\{|\alpha LM\rangle\}$ is now constructred by use of the following three theorems:

Theorem 1 (Elliott): *An SU(3) irrep of highest weight $(\lambda\mu)$ contains a sequence of SO(3) states of angular momenta*

$$L = \begin{cases} \lambda+K, \lambda+K-1, \ldots, K & \textit{for } K\neq 0\\ \lambda, \lambda-2, \ldots, 0 \text{ or } 1 & \textit{for } K=0 \end{cases} \tag{57}$$

with K running over the range

$$K = \mu, \mu-2, \ldots 0 \text{ or } 1\,. \tag{58}$$

Theorem 2: *If Ψ_1 is a wave function from an irrep $(\lambda_1\mu_1)$ and Ψ_2 is a wave function from an irrep $(\lambda_2\mu_2)$, then Ψ defined by $\Psi(\Omega) = \Psi_1(\Omega)\Psi_2(\Omega)$ is a wave function belonging to the irrep $(\lambda_1 + \lambda_2, \mu_1 + \mu_2)$.*

Thus, we can easily build up a non-orthonormal basis starting from the wave functions for the $(1,0)$ and $(0,1)$ irreps:

$$\Psi^{(1,0)}_{01M}(\Omega) \propto \mathcal{D}^1_{0M}(\Omega)\,, \quad \Psi^{(0,1)}_{11M}(\Omega) \propto \mathcal{D}^1_{1M}(\Omega) + \mathcal{D}^1_{-1,M}(\Omega)\,. \tag{59}$$

Theorem 3: *A state $|\alpha LM\rangle$ which is an eigenfunction of the scalar operator $[L\otimes Q\otimes L]_0$ is characterized by an eigenvector of the matrix $M^{(LL)}$, i.e., a vector $a_K(\alpha L)$ satisfying the equation*

$$\sum_{K'} M^{(LL)}_{KK'} a_{K'}(\alpha L) = a_K(\alpha L)\,. \tag{60}$$

Moreover, a set of vectors $\{a(\alpha L)\}$ which are all eigenvectors of the corresponding $\{M^{(LL)}\}$ matrices define a set of orthogonal states with coherent state wave functions given by eqn. (53).

Proof: The theorem follows from the observation that

$$[L\otimes Q\otimes L]_0|\alpha LM\rangle \propto [\Gamma(Q)\times\Psi_{\alpha L}]_{LM} \tag{61}$$

and the observation that, for an SU(3) irrep, there are no multiplicities of $L=0$ states for which $[\Gamma(Q)\times\Psi_{\alpha L}]_{LM}$ vanishes. Orthogonality of states of different L and/or different M follows automatically from the transformation properties of the states under $SO(3)$ rotations. Λ

Thus, for each of the L values in eq. (57), one can diagonalize the corresponding $\{M^{(LL)}\}$ matrix to obtain the $\{a(\alpha L)\}$ eigenvectors for a set of orthogonal states. It then remains only to normalize these vectors such that the reduced matrix elements, defined by the Wigner-Eckart theorem in an orthonormal basis by

$$[\Gamma(Q)\times\psi_{\alpha L}]_{L'M} = \frac{1}{\sqrt{2L'+1}}\sum_\beta \psi_{\beta L'M}\,\langle\beta L'\|Q\|\alpha L\rangle\,, \tag{62}$$

satisfy the hermiticity relationship of a unitary representation

$$\langle\beta L'\|Q\|\alpha L\rangle = (-1)^{L-L'}\langle\alpha L\|Q\|\beta L'\rangle^*\,. \tag{63}$$

For a multiplicity-free representation, i.e., one for which the α label is redundant, everything can be done analytically as shown explicitly in ref. [10]. For example, for the multiplicity free $(\lambda,0)$ irreps the reduced matrix elements are given by

$$\langle L\|Q\|L\rangle = \sqrt{2L+1}\,(L0,20|L0)\,(2\lambda+3)\,, \tag{64}$$

$$\langle L+2\|Q\|L\rangle = \sqrt{2L+1}\,(L0,20|L+2,0)\,\left[4(\lambda-L)(\lambda+L+3)\right]^{\frac{1}{2}}\,. \tag{65}$$

It is seen that the sequence of angular momentum states with $L=0,2,4,\ldots$ or $L=1,3,5,\ldots$ terminates with $L=\lambda$ in accordance with the branching rule (57). More details of the procedure are given in refs. [9,10,3].

VCS theory gives analytical asymptotic expressions for matrix elements of su(3) in an so(3) basis that become accurate as the dimension of the representation becomes large. This is because the su(3) algebra has the rotor model algebra as a contraction limit.

6 Representations of so(5) in an so(3) basis

I conclude by indicating that construction of the generic representations of so(5) in an so(3) basis, is also straightforward by VCS methods. A computer code for implimenting the construction, written by Peter Turner [12], and will soon be generally available.

Constructions have been given for the so-called one-rowed representations by several authors (reviewed in ref. [11]). These are the representations that appear in the space of a single particle in a five-dimensional harmonic oscillator and in the IBM1 version of the Interacting Boson Model [15]. The generic two-rowed representations occur for two or more particles in a five-dimensional oscillator and in the neutron-proton IBM2 version of the Interacting Boson Model.

It was shown a while ago by Hecht and myself [11] that the one-rowed so(5) representations are constructed simply and systematically by the methods outlined in section V for su(3). For the solution of this problem, we did not need the full power of VCS theory; the coherent state wave functions were simple scalar functions. However, for the generic representations, vector-valued coherent state wave functions are needed.

Parallels with the su(3) representation theory can be seen by comparison of the root diagrams for the two algebras, shown respectively in figs. 1 and 2. Both Lie algebras are of rank two and their irreps can be labelled by highest weights $(\lambda\mu)$. In both cases the corresponding highest weight state $|\lambda\mu\rangle$ for an irrep is an eigenstate of two mutually orthogonal Cartan operators $\hat{H}_1$ and $\hat{H}_2$ with eigenvalues

$$\hat{H}_1|\lambda\mu\rangle = \tfrac{1}{2}(2\lambda+\mu)\,|\lambda\mu\rangle\,, \quad \hat{H}_2|\lambda\mu\rangle = \mu\,|\lambda\mu\rangle\,. \tag{66}$$

Moreover, for both Lie algebras, there are subspaces of highest grade states that are annihilated by a subset of raising operators and carry irreps of the 'horizontal' u(2) algebras shown in figs. 1 and 2. Thus, for so(5), as for su(3) we can define an orthonormal basis $\{\xi_\nu \equiv |s\nu\rangle; \nu = -s, \ldots, +s\}$ of highest grade states for a u(2) irrep of spin $s = \mu/2$. Constructing an so(5) irrep in an so(3) basis by inducing from this u(2) irrep using VCS theory is now made possible by the following observation.

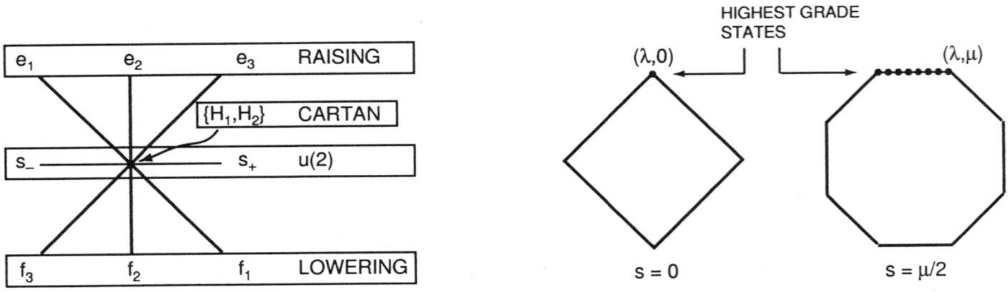

Figure 2. The root diagram for so5 showing a u(2) subalgebra and complementary sets of raising and lowering operators. Also shown are outlines of the weight diagrams for irreps of highest weight $(\lambda, 0)$ and (λ, μ) and their highest grade states.

Observation: *Provided the Cartan subalgebra for so(5) is chosen such that it contains no component of the angular momentum, then the set of states* $\{\hat{R}(\Omega)|s\nu\rangle; \Omega \in \mathrm{SO}(3)\}$ *spans the Hilbert space for the so(5) irrep of highest weight* $(\lambda\mu)$.

It follows that, if the highest grade states $\{|s\nu\rangle\}$ are assigned wave functions $\{\xi_\nu\}$ then a state $|\psi\rangle$ of the so(5) irrep of highest weight $(\lambda\mu)$ is defined by the set of overlaps

$$\Psi(\Omega) = \sum_\nu \xi_\nu \langle s\nu|\hat{R}(\Omega)|\psi\rangle \,. \tag{67}$$

Thus, the vector- valued function Ψ over the group $SO(3)$ is a VCS a wave function for the state $|\psi\rangle$. The rest of the construction follows the general prescription for such VCS representations.

7 Discussion

The above fairly detailed outline of the su(3) irreps in both su(2) and so(3) basis give the essential principles underlying VCS theory. The application to so(5) is an indication of how the theory can be applied systematically to the construction of representations by straighforward systematic methods that have traditionally proved challenging. The limitations of the theory have yet to be discovered.

References

1. G. Rosensteel and D.J. Rowe, *Phys. Rev. Letts.* **38**, 10 (1977); *Annals of Phys.* **126**, 343 (1980).
2. D.J. Rowe, *Rep. Prog. Phys.* **48**, 1419 (1985).
3. D.J. Rowe, *Prog. in Part. and Nucl. Phys.* **37**, 265 (1996).
4. D.J. Rowe and J. Repka, *J. Math. Phys.* **32**, 2614 (1991).
5. S.D. Bartlett, D.J. Rowe, and J. Repka, *J. of Phys. A: Math.* & Gen. **35**, 5599, 5625 (2002).
6. Mackey G W 1952 *Ann. of Math.* **55** 101; 1968 *Induced Representation of Groups and Quantum Mechanics* (New York: Benjamin).
7. B. Kostant, 1970 *On Certain Unitary Representations which Arise from a Quantization Theory*, in *Group Representations in Mathematics and Physics, Lecture Notes in Physics, Vol. 6* (Berlin: Springer).
8. Souriau J–M 1966 *Comm. Math. Phys.* **1** 374; 1970 *Structure des systèmes dynamiques* (Paris: Dunod)
9. D.J. Rowe, R. Le Blanc and J. Repka, *J. Phys. A: Math. Gen.* **22**, L309 (1989).
10. D.J. Rowe, M.G. Vassanji and J. Carvalho, *Nucl. Phys.* **A504**, 76-102 (1989).
11. D.J. Rowe and T. Hecht, *J. Math. Phys.* **36**, 4711 (1995).
12. P.S. Turner, D.J. Rowe, and J. Repka, (in preparation).
13. D.J. Rowe and C. Bahri *J. Math. Phys.* **41**, 6544 (2000).
14. Perelomov A 1986 *Generalized Coherent States and Their Applications* (Berlin: Springer–Verlag)
15. A. Arima and F. Iachello, *Ann. Rev. Nucl. Part.* **31**, 75 (1981).

SYMMETRIES IN STRONGLY DEFORMED NUCLEI

J.P. DRAAYER
Department of Physics and Astronomy, Louisiana State University,Baton Rouge, LA, 70808-4001 USA
E-mail: draayer@lsu.edu

G. POPA
Department of Physics, Rochester Institute of Technology, Rochester, NY 14623-5612, USA
E-mail: gxpsps@rit.edu

J.G. HIRSCH, C.E. VARGAS*
Instituto de Ciencias Nucleares, Universidad Nacional Autónoma de México, Apartado Postal 70-543 México 04510 DF, México
E-mail: hirsch@nuclecu.unam.mx, vargas@nuclecu.unam.mx
**Present address: Facultad de Física e Inteligencia Artificial, Universidad Veracruzana, Sebastián Camacho 5; Xalapa, Ver., 91000, México*

Symmetries, articulated mathematically through group theory, play a central role in nuclear physics. The Interacting Boson Model (IBM), introduced to our field by Franco Iachello, who's 60th birthday we honor through this symposium, is a clear and clever example of the use of group theory to model symmetries of atomic nuclei. While all present may not be IBM disciples, it is clear we are all at least IBM apostles and here to honor Franco and acknowledge his significant contributions to the field of nuclear physics and beyond. I will take the opportunity of this very special occasion to give a brief update on a complementary theory, the pseudo-SU(3) model, that builds on pseudo-spin symmetry in heavy nuclei, a theory that is applicable to deformed rare earth and actinide species.

1 Introduction

Symmetries, whether exact – like rotational invariance that derives from the isotropy of space, or approximate – like SU(3) that enters when an oscillator approximation is applicable, have always played a key role in nuclear physics. Starting with the pioneering work of Wigner[1] and Racah[2], group theory has became a 'hallmark' of nuclear structure with wide-ranging applications that include in addition to the compact classical groups, non-compact structures that are required for 'open' systems. The Interacting Boson Model (IBM), introduced to the nuclear physics community by Franco Iachello at the *International Conference on Nuclear Structure and Spectroscopy* that was held in Amsterdam in 1974, built on our understanding of the role of the pairing mode in nuclei with its symplectic group underpinnings[3] and the Elliott SU(3) model[4] that brought mathematical definition to our understanding of rotational motion in nuclei. While our work on pseudo-spin symmetry and the pseudo-SU(3) model may not qualify us for IBM 'discipleship' per se, our appreciation of the model and what it has meant and continues to mean to our community should deem us 'apostleship' status among the ranks of Franco's collaborators and cohorts — Happy Birthday Franco!

Experimental nuclear physicists, with improved equipment and techniques that yield higher sensitivity and greater resolving power, continue to challenge nuclear structure theorists with interesting new data. In particular, the identification of additional low-lying levels and more precise measurements of E2 and M1 transition strengths in deformed nuclei test our understanding of collective nuclear phenomena. While nuclei with atomic numbers greater than about 150 are good candidates for probing drivers of deformation, microscopic calculations for these systems remain illusive. The pseudo-SU(3) model referred to above – which capitalizes on good pseudo-spin symmetry in heavy deformed nuclei[5,6] – is applicable in this domain and has enjoyed considerable success. Here we consider some recent applications of the pseudo-SU(3) model. A question of special interest is the role the spin degree-of-freedom plays in these nuclei as the observed fragmentation of the ground state $M1$ strength in odd-mass nuclei seems to require a strong dependence upon spin while this appears not to be the case for neighboring even-even species.

2 Model space and interaction

Recall that the pseudo-SU(3) model refers to the fact that single-particle orbitals with $j = l - \frac{1}{2}$ and $j = (l-2) + \frac{1}{2}$ in the shell η lie close in energy and can therefore be labeled as pseudo-spin doublets with quantum numbers $\tilde{j} = j$, $\tilde{\eta} = \eta$ - 1, and $\tilde{l}$ = l - 1. Its origin has been traced back to the relativistic Dirac equation.[7] In the current version of the pseudo-SU(3) model, the intruder level of opposite (unique) parity in each major shell is removed from active consideration and pseudo-orbital and pseudo-spin quantum numbers are assigned to the remaining (normal parity) single-particle states.[8] There is a tacit assumption that the role of the intruder levels can be taken into account by renormalizing the normal parity results, an assumption that is easy to justify for E2 transitions but less so for M1 strengths since unlike the E2 operator, which is spin independent, the M1 operator has a strong dependence on spin.

The rare earth nuclei are considered to have a closed proton core at $N_\pi = 50$ and a closed neutron core at $N_\nu = 82$. Basis states are built by placing valence nucleons in the open $\eta_\pi = 4$ shell for protons, less the $g_{9/2}$ level which forms part of the proton core plus the unique-parity $h_{11/2}$ intruder level from the shell above, and the $\eta_\nu = 5$ shell for neutrons, less the $h_{11/2}$ level which forms part of the neutron core plus the unique-parity $i_{13/2}$ intruder level from the shell above it. As noted above, nucleons assigned to the unique-parity intruder levels are relegated to only a renomalization role. These shells have a complementary pseudo-harmonic oscillator shell structure that is given by $\tilde{\eta}_\sigma$ $(\sigma = \pi, \nu) = \eta_\sigma - 1$ plus each shell's respective intruder level. Typical applications include about 5 proton and 5 neutron pseudo-SU(3) irreducible representations (irreps) with largest values of their respective second order Casimir operators, $\tilde{C}_2^\sigma = (\tilde{Q}^\sigma \cdot \tilde{Q}^\sigma - 3\tilde{L}_\sigma^2)$, and up to about 20 proton-neutron coupled irreps, again with largest combined $\tilde{C}_2$ values where in this case $\tilde{C}_2 = (\tilde{Q} \cdot \tilde{Q} - 3\tilde{L}^2)$ where $\tilde{Q} = \tilde{Q}^\pi + \tilde{Q}^\nu$ and $\tilde{L} = \tilde{L}^\pi + \tilde{L}^\nu$.

A realistic pseudo-SU(3) Hamiltonian is used in the calculations:

$$H = \tilde{H}_{sp}^\pi + \tilde{H}_{sp}^\nu - \frac{1}{2}\,\chi\,\tilde{Q} \cdot \tilde{Q} - G_\pi\,\tilde{H}_{pair}^\pi - G_\nu\,\tilde{H}_{pair}^\nu$$

$$+ \, a \, J^2 \, + \, b \, K_J^2 \, + \, a_3 \, \tilde{C}_3 + \, a_{sym} \, \tilde{C}_2. \tag{1}$$

Strengths of the quadrupole-quadrupole ($\tilde{Q} \cdot \tilde{Q}$) and pairing interactions ($\tilde{H}^{\sigma}_{pair}$) are taken to be fixed, respectively, at values typical of those used by other authors; namely, $\chi = 35 \; A^{5/3}$ MeV, $G_\pi = 21/A$ MeV and $G_\nu = 19/A$ MeV. The spherical single-particle terms in this expression have the form

$$\tilde{H}^{\sigma}_{sp} = \sum_{i_\sigma} (C_\sigma \, \tilde{l}_{i_\sigma} \cdot \tilde{s}_{i_\sigma} \, + \, D_\sigma \, \tilde{l}^2_{i_\sigma}). \tag{2}$$

Calculations are normally carried out with the single-particle spin-orbit ($\tilde{l}_\sigma \cdot \tilde{s}_\sigma$) and orbit-orbit ($\tilde{l}^2_\sigma$) interaction strengths fixed by systematics.[9] The four 'free' parameters a, b, a_3, a_{sym} are typically fixed by requiring a best fit to the low-energy spectra. (For the odd-mass systems considered, the a_3 parameter was set to be zero because it was found to have very little effect on the overall results.) No other parameters, except for effective charges in the definition of the E2 operator, enter into the theory; E2 and M1 transitions are not part of the fitting procedure.

3 Representative results

Two sets of results have been reported in the literature for even-even systems. The earliest were for the Gd isotopes.[10] The most recent, and hence most advanced in terms of the number of SU(3) irreps included in the basis states and the nature of the Hamiltonian, are for the ^{160,162,164}Dy and ^{168}Er nuclei.[11] There are also two sets of pseudo-SU(3) shell-model calculations reported in the literature for odd-mass nuclei. The first of these was for ^{163}Dy.[12] The configuration space was restricted to the most spatially symmetric configurations with pseudo-spin 0 and $\frac{1}{2}$ and because of this limitation only the first three low-energy bands could be described. More recent results for ^{157}Gd, ^{163}Dy and ^{169}Tm take into account pseudo-spin 1 and $\frac{3}{2}$ admixing with the predominantly pseudo-spin 0 and $\frac{1}{2}$ configurations.[13] Below we present a few results for the ^{162,163}Dy cases.

3.1 Excitation spectra

The even-even Gd, Dy, and Er nuclei all exhibit well-developed ground-state rotational bands as well as bands that are built on low-lying $K^\pi = 0^+$, $K^\pi = 2^+$ and even $K^\pi = 4^+$ states. Relative excitation energies for states with angular momentum 0^+ are determined mainly by the quadrupole-quadrupole interaction. The single-particle terms and the pairing interactions mix these states. The 0^+_2 states lie close to their experimental counterparts while the 0^+_3 states (not fit to the data) usually lie slightly above their experimental counterparts. Of the four 'free' parameters in the Hamiltonian, a is adjusted to reproduce the moment of inertia of the ground state band, a_3 is varied to yield a best fit to the energy of the second 0^+ state, a_{sym} is adjusted to give a best fit to the first 1^+ state, and b is determined by the band-head of the $K^\pi = 2^+$ band.

Figure 1(a) shows the calculated and experimental[14] $K^\pi = 0^+_1$ (ground state), $K^\pi = 2^+_\gamma$ (gamma) and the first ($K^\pi = 0^+_2$) and second ($K^\pi = 0^+_3$) excited bands for ^{162}Dy. For the first three bands, the calculated numbers are within 7% of the

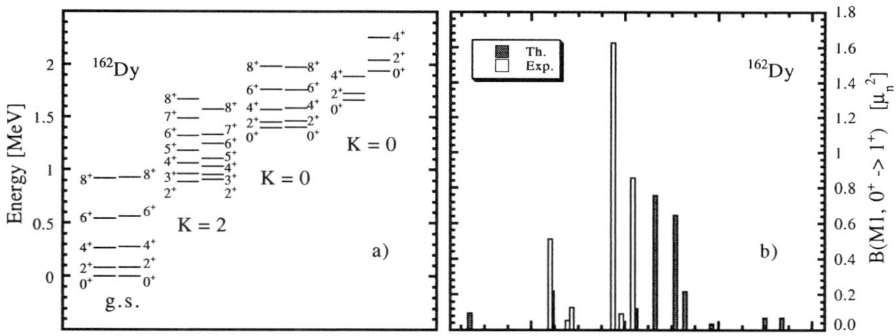

Figure 1. Insert a) gives the energy spectra of ^{162}Dy obtained using Hamiltonian (1). 'Exp.' represents the experimental results and 'Th.' the calculated ones. Insert b) gives the corresponding M1 transition strengths from the $J^\pi = 0^+$ ground state to the various $J^\pi = 1^+$ states.

measured energies. The model predicts a continuation of the various bands to higher values of the angular momentum. The calculated second excited $K^\pi = 0_3^+$ band (not included in the fitting procedure) lies about 0.5 MeV higher than experiment.

Table 1. SU(3) content of calculated eigenvectors for the ground-state band in ^{162}Dy. Only basis states that contribute more than 2% are identified.

(λ_π, μ_π)	(λ_ν, μ_ν)	(λ, μ)	0	2	4	6	8
(10, 4)	(18, 4)	(28, 8)	59.3	59.3	59.4	59.6	61.9
(10, 4)	(20, 0)	(30, 4)	20.1	19.5	18.2	16.3	13.9
(10, 4)	(18, 4)	(30, 4)	6.5	6.2	5.7	4.9	4.0
(12, 0)	(18, 4)	(30, 4)	7.1	6.9	6.7	6.3	5.7
(10, 4)	(18, 4)	(32, 0)	2.7	2.6	2.3	-	-
(12, 0)	(20, 0)	(32, 0)	3.0	2.9	2.8	2.5	2.1
(10, 4)	(16, 5)	(26, 9)	-	-	-	3.2	3.5

Details regarding the structure of the calculated eigenstates for members of the ground state band of ^{162}Dy are given in Table 1. These results, which are typical of others, show that although many more irreps were included in the analysis only few irreps (three for proton and three neutrons for ^{162}Dy) make significant contributions to the calculated eigenstates. In addition, of the many possible coupled irreps, only a relative few (four for ^{162}Dy) – all with large $\tilde{C}_2$ values – contribute to the calculated eigenstates. Note also that the SU(3) content of the eigenvectors is fairly constant across states within a band. Typically the percentages vary slowly and smoothly as one moves up a band to states of higher angular momentum. The SU(3) content of calculated eigenvectors for the four lowest band-head states of ^{162}Dy is shown in Table 2. Note that the $K^\pi = 0_1^+$ ground state is dominated by a prolate proton configuration while the first-excited $K^\pi = 0_2^+$ configuraton has a predominately oblate proton structure. The ($K^\pi = 2^+$) gamma band is purer than the ground state and the second-excited $K^\pi = 0^+$ is the most mixed.

Table 2. SU(3) content of calculated eigenvectors for the four lowest band-heads of ^{162}Dy. The percentage distribution across (λ, μ) values is given in the second column. Only states that contribute more than 2% are identified.

J^π_α	Th.	(λ_π, μ_π)	(λ_ν, μ_ν)	(λ, μ)
0^+_1	59.3	(10, 4)	(18, 4)	(28, 8)
	6.5	(10, 4)	(18, 4)	(30, 4)
	20.1	(10, 4)	(20, 0)	(30, 4)
	7.1	(12, 0)	(18, 4)	(30, 4)
	2.7	(10, 4)	(18, 4)	(32, 0)
	3.0	(12, 0)	(20, 0)	(32, 0)
0^+_2	91.8	(4, 10)	(18, 4)	(22, 14)
	4.2	(10, 4)	(20, 0)	(30, 4)
	-	(10, 4)	(18, 4)	(28, 8)
	-	(10, 4)	(20, 0)	(30, 4)
	-	(12, 0)	(18, 4)	(30, 4)
0^+_3	33.3	(10, 4)	(18, 4)	(28, 8)
	47.0	(10, 4)	(20, 0)	(30, 4)
	11.0	(12, 0)	(20, 0)	(32, 0)
	5.9	(4, 10)	(18, 4)	(22, 14)
	-	(12, 0)	(18, 4)	(30, 4)
2^+_γ	81.4	(10, 4)	(18, 4)	(28, 8)
	5.2	(10, 4)	(20, 0)	(30, 4)
	4.6	(12, 0)	(18, 4)	(30, 4)
	4.5	(4, 10)	(18, 4)	(22, 14)

For odd-mass nuclei, the calculated spectra by and large result from the interplay between the single-particle and quadrupole-quadrupole terms in the Hamiltonian. The use of realistic single-particle energies plays an important role in the appropriate ordering of the different band-heads. Figure 2 shows the normal parity bands in ^{163}Dy. The agreement between theory and experiment is reasonable for the seven (A-G) rotational bands shown. These seven bands represents nearly all of the measured levels.

An interesting feature for odd-mass nuclei is the pseudo-spin content of the rotational bands. A comparison of ^{157}Gd, ^{163}Dy and ^{169}Tm is made in Table 3. While the content is nearly constant along all the members of each band, the percentage $\tilde{S} = \frac{1}{2}$ versus $\tilde{S} = \frac{3}{2}$ is quite different for these three nuclei with ^{163}Dy showing very little pseudo-spin mixing. In each case some of the bands are dominated by $\tilde{S} = \frac{3}{2}$ configurations.

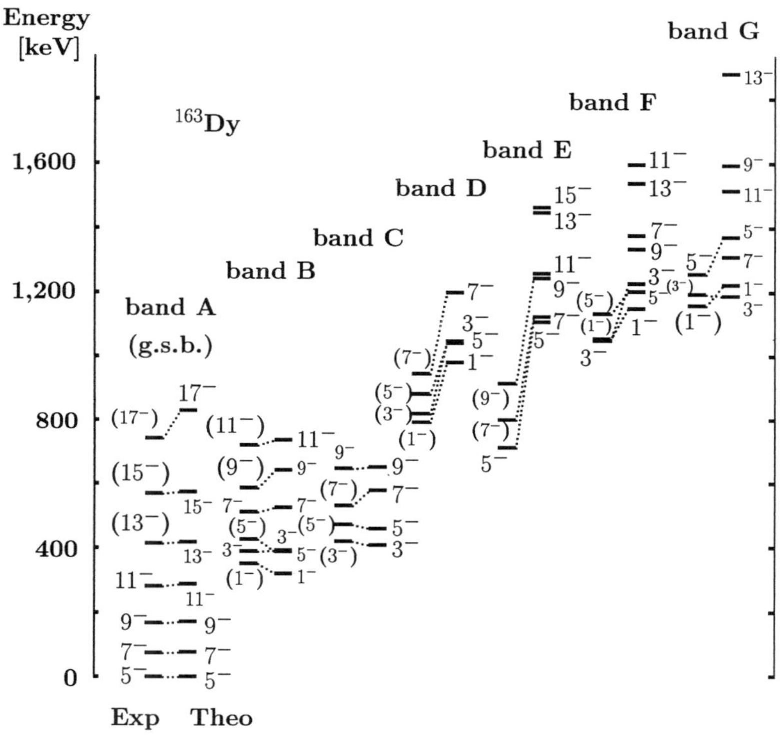

Figure 2. Normal (negative) parity bands in ^{163}Dy.

3.2 Electromagnetic transitions

Theoretical and experimental[14] $B(E2)$ transitions strengths between the states in the ground state band in ^{162}Dy are shown in Table 4. The overall agreement between the calculated and experimental numbers is also reasonable. The $B(E2; 2_1 \rightarrow 4_1)$ is within 1% of the experimental value, and the last two calculated $B(E2)$ values differ from the experimental values by less than 0.1 e^2b^2 which is well within the experimental error. As with the excitation spectra, these results are typical of what one finds for other nuclei in this region. Contributions to the quadrupole moments from the nucleons in the unique parity orbitals are parameterized through an effective charge, e_f, with $e_\nu = e_f$, and $e_\pi = 1 + e_f$, so the E2 operator is given by: $E2 \rightarrow \tilde{Q}_\mu = e_\pi \tilde{Q}_\pi + e_\nu \tilde{Q}_\nu$.[15] Predictions were also made for the inter-band E2 transitions between members of the excited bands as well as for various intra-band E2 transitions.

Table 3. Pseudo-spin content of bands in ^{157}Gd, ^{163}Dy and ^{169}Tm.

Nuclei	Band	$\tilde{S} = 1/2$ (%)	$\tilde{S} = 3/2$ (%)
^{157}Gd	A	89	11
	B	92	8
	C	92	8
	D	3	97
	E	77	23
	F	16	84
^{163}Dy	A	100	0
	B	100	0
	C	100	0
	D	100	0
	E	0	100
	F	100	0
	G	0	100
^{169}Tm	A	93	7
	B	100	0
	C	9	91
	D	100	0
	E	35	65
	F	37	63

Table 4. Experimental and theoretical $B(E2)$ strengths between members of ground-state band of ^{162}Dy.

$J_i \to J_f$	$B(E2; J_i \to J_f)$ (e^2b^2)	
	Exp.	Theory
$0_1 \to 2_1$	5.134 ± 0.155	5.134
$2_1 \to 4_1$	2.675 ± 0.102	2.635
$4_1 \to 6_1$	2.236 ± 0.127	2.325
$6_1 \to 8_1$	2.341 ± 0.115	2.201

Another test of the theory is M1 transition strengths, mediated by the operator $M^1_\mu = \sqrt{\frac{3}{4\pi}}\mu_N\{g^o_\pi L^\pi_\mu + g^s_\pi S^\pi_\mu + g^o_\nu L^\nu_\mu + g^s_\nu S^\nu_\mu\}$ where the $g^{o,s}_\alpha$ factors are the respective orbit (o) and spin (s) gyromagnetic ratios for protons ($\alpha = \pi$) and neutrons ($\alpha = \nu$). This is called the scissors mode because it can be pictured in lowest order as the rotation of the proton and neutron distributions relative to one another, like the opening and closing of the blades of a scissors.[16,17] A description of this mode within the framework of the interacting boson model (IBM) led to its detection in ^{156}Gd using high-resolution inelastic electron scattering techniques.[18,19] Systematic studies employing nuclear resonance fluorescence scattering measurements followed.[20] The non-observation of these low-energy M1 excitations in inelastic proton scattering has served to confirmed its orbital character.[21,22] Over the past two decades an impressive wealth of information about the scissors mode in even-even nuclei has been obtained and analyzed.[23]

The pseudo-SU(3) model offers a very similar, but even richer interpretation of the scissors mode.[24] According to the Littlewood rules[25] for coupling Young diagrams, the allowed product pseudo-SU(3) configuration can be expressed in mathematical terms by using three integers (m, l, k): $(\lambda_\pi, \mu_\pi) \otimes (\lambda_\nu, \mu_\nu) = \oplus_{m,l,k}(\lambda_\pi + \lambda_\nu - 2m + l, \mu_\pi + \mu_\nu - 2l + m)^k$, where the parameters l and m are defined in a fixed range given by the values of the initial SU(3) representations. In this formulation, k serves to distinguish between multiple occurrences of equivalent (λ, μ) irreps in the tensor product. The number of k values is equal to the outer multiplicity, ρ_{max} $(k = 1, 2, \ldots, \rho_{max})$. The l and m labels can be identified with excitation quanta of a two-dimensional oscillator involving relative rotations (θ, the angle between the principal axes of the proton and neutron system, and ϕ, the angle between semi-axes of the proton and neutron system) of the proton-neutron system, $m = n_\theta, l = n_\phi$.[26] These correspond to two distinct types of 1^+ motion, the scissors and twist modes, and their realization in terms of the pseudo-SU(3) model. The pseudo-SU(3) irreps obtained from the tensor product that contain a $J^\pi = 1^+$ state are those corresponding to $(m, l, k) = (1, 0, 1)$, $(0, 1, 1)$, $(1, 1, 1)$, and $(1, 1, 2)$. A pure pseudo-SU(3) picture gives rise of a maximum of four 1^+ states that are associated with the scissors, twist, and doubly degenerate scissors-plus-twist modes [(1,1,1) and (1,1,2)].[26,27] Results for the Dy isotopes, assuming a pure pseudo-SU(3) scheme, are given in Table 5.

Table 5. B(M1) transition strengths $[\mu_N^2]$ in the pure symmetry limit of the pseudo-SU(3) model. The strong coupled pseudo-SU(3) irrep $(\lambda, \mu)_{g.s.}$ for the ground state is given with its proton and neutron sub-irreps and the irreps associated with the 1^+ states, $(\lambda', \mu')_{1^+}$. In addition, each transition is labeled as a scissors (s) or twist (t) or combination mode.

Nucleus	$[(\lambda_\pi, \mu_\pi)$	(λ_π, μ_π)	$(\lambda, \mu)]_{gs}$	$(\lambda, \mu)_{1^+}$	B(M1)	mode
^{162}Dy	(10,4)	(18,4)	(28,8)	(29, 6)	0.56	t
				(26, 9)	1.77	s
				$(27, 7)^1$	1.82	s+t
				$(27, 7)^2$	0.083	t+s

The experimental results[14] shown in Figure 1(b) suggest a much larger number of 1^+ states with non-zero $B(M1)$ transition strengths from the 0^+ ground state. The SU(3) breaking residual interactions lead to a fragmentation in the M1 strength distribution, since the ground state 0^+ is then a combination of several SU(3) irreps, each with allowed M1 transitions to other SU(3) irreps. For ^{162}Dy the summed M1 strength, 4.24 μ_N^2 and 2.29 μ_N^2, respectively, for the pure SU(3) limit of theory and as determined using Hamiltonian (1), is in reasonable agreement with the experimental number (3.29 μ_N^2). It is important to note that unlike the $E2 \to \tilde{Q}$ case, real, not effective, gyromagnetic ratios are used to define the M1 operator.

Low-energy E2 and M1 transition strengths for the normal parity bands in ^{163}Dy are also available.[13] The number of calculated results in this case are far more than for even-even nuclei because they include the non-stretched $B(E2; J \to J + 1)$ as well as the stretched $B(E2; J \to J+2)$ cases. Low-energy M1 transitions in odd-mass nuclei were first reported for ^{163}Dy in 1993.[28] Unambiguous spin and parity assignments of excited states in these nuclei are difficult to make due to the half-integer spin character of the states.[29] Furthermore, M1 strengths in odd-mass nuclei are highly fragmented with intensities far smaller than in even-even systems so that their identification against the background represents a major challenge.[23,30,31] Results for E2 and M1 transitions are given elsewhere.[13] The most important conclusion coming out of these studies is that states with energies between 2 and 4 MeV realize important M1 contributions from states with proton and neutron pseudo-spin 1 and $\frac{3}{2}$.

4 Conclusions

The results presented here for ^{162}Dy and ^{163}Dy serve to illustrate the current level of applicability of the pseudo-SU(3) shell model. First and foremost, one can say that it can be used to reproduce excitation spectra: in addition to the ground-state ($K^\pi = 0_1^+$) and gamma ($K^\pi = 2_\gamma^+ = 2_1^+$) bands in even-even systems, it offers a reasonable description of the first two exited 0^+ bands ($K^\pi = 0_{2,3}^+$) as well as the $K^\pi = 4_1^+$ band; it can also be used to reproduce several (seven for ^{163}Dy) bands in odd-mass systems.

The theory, which now accommodates realistic one-body terms (single-particle energies) and two-body interactions (for example, pairing), also allows one to calculate E2 and M1 transition strengths. Agreement with existing experimental numbers is reasonable for both inter- and intra-band transitions. In the case of the E2 operator, effective charges are required because the theory as currently employed still does not include its logical symplectic [SU(3) $\to$ Sp(3,R)] extension and the so-called opposite parity intruder states are relegated to a renormalization role. Work to build both of these aspects into the theory is underway.

Acknowledgments

This work was supported in part by the U.S. National Science Foundation, Grants Numbers 9970769 and 0140300, and CONACyT (México).

References

1. E. P. Wigner, *Phys. Rev.* **51**, 106 (1937); *Group Theory and Its Application to the Quantum Mechanics of Atomic Spectra* (Academic, New York, 1959).
2. G. Racah, *Phys. Rev.* **63**, 367 (1943); in *Quantum Mechanics of Angular Momentum*, edited by L. C. Biedenharn and H. Van Dam (Academic, New York, 1965).
3. A. Bohr, B. R. Mottelson, and D. Pines, *Phys. Rev.* **110**, 936 (1958).
4. J. P. Elliott, *Proc. Roy. Soc. London* **A245**, 128, 562 (1958).
5. K. T. Hecht and A. Adler, *Nucl. Phys.* **A 137**, 129 (1969).
6. A. Arima, M. Harvey, and K. Shimizu, *Phys. Lett.* **B 30**, 517 (1969).

7. A. L. Blokhin *et. al.*, *Phys. Rev. Lett.* **74**, 4149 (1995); J. N. Ginocchio, *Phys. Rev. Lett.* **78**, 436 (1997).
8. C. Vargas, J. G. Hirsch, P. O. Hess, and J. P. Draayer, *Phys. Rev.* **C 58**, 1488 (1998).
9. P. Ring and P. Schuck, *The Nuclear Many-Body Problem*, Springer, Berlin (1979).
10. G. Popa, J. G. Hirsch, and J. P. Draayer, *Phys. Rev.* C **62**, 064313 (2000).
11. J. P. Draayer, G. Popa, J. G. Hirsch, *Acta Physica Polonica* **B 32** 2697 (2001).
12. C. E. Vargas, J. G. Hirsch, J. P. Draayer, *Phys. Rev.* C **64**, 034306 (2001).
13. C. E. Vargas, J. G. Hirsch, J. P. Draayer, *Phys. Rev.* C. **66** (2002) 064309; C. E. Vargas, J. G. Hirsch, and J. P. Draayer, *Phys. Lett.* **B 551** (2003) 98.
14. National Nuclear Data Center, (http:\\bnlnd2.dne.bnl.gov).
15. R. F. Casten, J. P. Draayer, K. Heyde, P. Lipas, T. Otsuka, and D. Warner, *Algebraic Approaches to Nuclear Structure: Interacting Boson and Fermion Models* (Harcourt, Brace, and Jovanovich, New York, 1993).
16. H. Ui, *Prog. Theor. Phys.* **44**, 153 (1970).
17. N. Lo Iudice and F. Palumbo,*Phys. Rev. Lett.* **41**, 1532 (1978).
18. A. E. L. Dieperink, *Prog. Part. Nucl. Phys.* **9**, 121 (1983); F. Iachello, *Phys. Rev. Lett.* **53**, 1427 (1984).
19. D. Bohle *et. al.*,*Phys. Lett.* **137 B**, 27 (1984).
20. U. E. P. Berg *et. al.*, *Phys. Lett.* **149 B**, 59 (1984); B. Kasten *et. al. Phys. Rev. Lett.* **63**, 609 (1989); H. H. Pitz *et. al. Nucl. Phys.* **A 492**, 411 (1989).
21. C. Djalali *et. al. Phys. Lett.* **164 B**, 269 (1985); D. Frekers *et. al. Phys. Lett.* **218 B**, 429 (1989).
22. C. Wesselborg *et. al.*, *Z. Phys.* **323 A**, 485 (1986).
23. U. Kneissl *et. al.*, *Prog. Part. Nucl. Phys.* **37**, 349 (1996); A. Richter, *Prog. Part. Nucl. Phys.* **34**, 261 (1995); D. Zawischa, *J. Phys. G: Nucl. Part. Phys.* **24**, 683 (1998).
24. O. Castaños, J. P. Draayer and Y. Leschber, *Z. Phys.* A **329**, 33 (1988).
25. J. F. Cornwell, *Techniques in Physics 7: Group Theory in Physics*, Vol. 2, (Academic Press, Orlando, 1985) 33 (1988).
26. D. Rompf, T. Beuschel, J. P. Draayer, W. Scheid, and J. G. Hirsch,*Phys. Rev.* C **57**, 1703 (1998).
27. T. Beuschel, J. P. Draayer, D. Rompf, and J. G. Hirsch, *Phys. Rev.* C **57**, 1233 (1998).
28. I. Bauske *et. al.*, *Phys. Rev. Lett.* **71**, 975 (1993).
29. J. Margraf *et. al.*, *Phys. Rev.* C **52**, 2429 (1995).
30. H. Friedrich *et. al.*, *Nucl. Phys* **A 567**, 266 (1994).
31. J. Enders, N. Huxel, P. von Neumann-Cosel and A. Richter, *Phys. Rev. Lett.* **79**, 2010 (1997).

SHAPE COEXISTENCE AND ITS SYMMETRIES

K. HEYDE AND R. FOSSION
Department of Subatomic and Radiation Physics,
Proeftuinstraat,86 B-9000 Gent, Belgium

We discuss the early evidence that led to introducing particle-hole excitations across closed shells as a means to understand low-lying shape coexisting configurations. We point out the essential features of proton-neutron forces in order to understand the salient features of these configurations and the very specific mass dependence of the excitation energy. In the central part, we discuss how to reconcile particle-hole excitations within the framework of the Interacting Boson Model (IBM) and specifically discuss the underlying symmetry structures. Applications in order to describe recently observed shape coexisting excitations in the neutron-deficient Pb nuclei are presented.

1 Introduction: the early work

It was at the International Conference on Nuclear Structure, in 1974, in Amsterdam I first met Franco Iachello [1]. At that particular meeting, a pretty young person was presenting new expressions in order to describe vibrational and transitional nuclei making use of what we would now call just d bosons. These results were really striking and started well animated discussions in the audience, that I remember very well. Soon, Akito Arima was joining in and they both studied the algebraic structures with the extensions to s bosons and studying the properties of the sd boson U(6) algebra. Soon after, two specialized seminars were set up in Erice [2,3] where some 40 people joined in to discuss the early developments in this very rapidly exploding field, attracting both theorists and experimentalist. I was present at the second of these meetings in 1980. Because in Gent, we had been studying a way to describe a class of low-lying excited states near closed shells, states that were not expected to appear at such low energy (therefore we called them intruder states), it was clear that particle-hole excitations (p-h) across the closed shells had to be at the origin, however dressed by quadrupole proton-neutron correlations. In seeing the truncation of the IBM to essentially a valence shell of nucleons, keeping closed shells inert, I was pointing out that in order to have a full description of low-lying nuclear structure phenomena, one had somehow to incorporate those new modes too. In Gent, with Piet Van Isacker we set out to study this issue much closer.

It had been shown form various spectroscopic selective experiments e.g. transfer reactions in particular, that very near to closed shells (the In and Sb nuclei at Z=50 and the Tl and Bi nuclei at Z=82) very low intruder states were observed with a conspicous energy dependence on the number of free valence neutrons hinting for 1p-1h excitations as the starting point (see [4] and references in there). If these excitations were appearing on both sides of the shell, it was a natural step to suggest that low-lying 0^+ excitations had to show up in the even-even nuclei in between eg. in the Sn and Pb nuclei. Because the Sn nuclei with a large number of valence neutrons were lying near to stability, they could be studied more easily and in 1979 the group at the Free University of Amsterdam first observed those 0^+ excitations with a full band developing on top [5]. It took a pretty long time in experimental development in order to study similar excitations in the Pb nuclei because here one had to go far away from stability in order to find the conditions favouring to observe the states at low excitation energy. The group of the IKS Leuven was successful with first results in 1984 in the nuclei $^{192-198}Pb$ [6]. An extensive discussion has been reported by J. Wood et al. [7] covering the most imporant regions of even-even nuclei. In order to illustrate the dramatic increase of quality over the early Sn experiments, I like to present results on ^{116}Sn as obtained by Savelius et al. [8] now showing the particular lowest intruder band in great detail.

2 Quadrupole force dominance

Since making a 1p-1h excitation at the closed shell at Z=50 takes about 4.5 MeV (the proton shell gap), the unperturbed energy for 2p-2h excitations comes up to about 9 MeV. Even though pairing amongst the particles and holes will lower the energy in an important way to 4-5 MeV, this is still far from the observed excitation energy of 1.7 MeV. Some essential element is missing when starting from the spherical symmetry of the shell model.

One way to come around is breaking the spherical symmetry and allowing the mean field to acquire quadrupole deformation thereby giving rise to the possibility that .spherical orbits split and the large spherical shell gap at Z=50 and also at Z=82 rapidly vanishes. Calculations have been carried out over the years using deformed mean-field studies (Nilsson model, deformed Woods-Saxon, Hartree-Fock-(Bogoliubov) studies) and we cite some studies, in particular, that addressed the Pb region [9,10,11]. I cannot refrain from showing the beautiful study of the energy surface for the nucleus ^{186}Pb showing the fact that besides a spherical mininum, also an oblate and problate shape, coexisting at low energy show up [12]. The deformed field essentially points out the need for the quadrupole component in the mean field as the agent for the increased binding energy. Knowing this, and having experimental knowledge of the fact that 1p-1h (in odd-mass nuclei) and 2p-2h (in even-even nuclei) components are present in these states, it is tempting to incorporate this in a spherical shell-model description. It is possible to evaluate the excitation energy of a 2p-2h configuration by invoking a schematic model that was discussed in detail in [13] as

$$E_{intr.}(0^+) = \langle 0_I^+ \mid H \mid 0_I^+ \rangle - \langle 0_{GS}^+ \mid H \mid 0_{GS}^+ \rangle \quad , \tag{1}$$

in which the index I denotes the nucleon distribution in the intruder state and GS the distribution in the ground state. Using a pair distribution for the neutrons, combined with a 2p-2h excitation and 0p-0h excitation for the intruder and regular state, respectively, one can derive the expression

$$E_{intr.}(2p-2h) \simeq 2(\epsilon_p - \epsilon_h) + \Delta E_M(2p-2h) - \Delta E_{pair} + \Delta E_Q(2p-2h) \quad , \tag{2}$$

where the various terms describe the unperturbed energy to create the 2p-2h configuration, a monopole correction due to a change in proton single-particle energy while changing the neutron number, the pairing-energy correction because essentially 0^+- coupled pairs are formed, and the quadrupole binding energy originating in the proton-neutron force, respectively.

In calculating the neutron number dependence of the 2p-2h intruder 0^+ configurations we have to determine the quadrupole energy contribution and this we do by using the SU(3) expression given in [13], i.e.,

$$\Delta E_Q \simeq 2\kappa \Delta N_\pi N_\nu \quad , \tag{3}$$

in which ΔN_π denotes the number of pairs excited out of the closed shell configuration at Z=50, i.e., ΔN_π=2 for a 2p-2h excitation.

This approach points out that the essential elements are the strong pairing interactions amongst the particles and the holes that make up the excited configuration and the strong quadrupole proton-neutron forces. It is precisely here that early contacts between the disconnected "spaces" of interacting boson within a valence space only and the p-h excitations of the core itself showed up. In a lowest order approximation, one can think of the 2p and the 2h parts to bring in two extra bosons increasing the active model space from N to N+2 bosons. One can then carry out separate calculations for both spaces introducing a coupling which precisely causes the 2p-2h excitation to show up and results in the simple form

$$H_{mix} = \alpha(s^\dagger s^\dagger)^0 + \beta(d^\dagger d^\dagger)^0 + h.c. \quad . \tag{4}$$

This approach was introduced by Duval and Barrett [14] and has been used in many mass regions with success (the Cd,Pd,Ru,Mo nuclei in the Sn region, the Pt,Hg and Po nuclei in the Z=82 mass region,..). The drawback to those calculations is that one gets involved with a lot of parameters and unless one has some physics guidance the detailed agreement needs some caution. An attempt to lower the number of free parameters has been suggested [15] and used in the Cd nuclei with quite some success and it was clear that something extra was needed besides the numerical configuration mixing approach.

Here, we also like to mention that the two approaches, one starting with a description in which the effects of the strong residual quadrupole forces are incorporated in a deformed mean field, and the other explicitely treating the strong quadrupole correlations within a spherical shell-model description, are essentially equivalent [16].

3 Shape coexistence and its symmetries

The starting point is that one can enlarge the group structure of U(6) by separately treating the particle and hole-pairs as described by particle and hole bosons, respectively (see fig.1)[17]. If we denote those, using shorthand notation, as $p_i^\dagger \equiv s_p^\dagger, d_p^\dagger$ for the s and d particle bosons and $h_j^\dagger \equiv s_h^\dagger, d_h^\dagger$ for the hole bosons,then both a $U_p(6)$ and $U_h(6)$ algebra results. Looking for some particular symmetries, one is led to extending the set of 72 generators of the particle and hole U(6) algebras with the 72 shift operators $\{p_i^\dagger h_j, h_j^\dagger p_i\}$, leading to a $U(12)$ group-structure. It turns out one can reduce this not just as a product of the separate $U(6)$ groups of particle and hole bosons but construct the chain

$$U(12) \supset U_{p+h}(6) \otimes SU_I(2) \otimes U_+(1) \quad , \tag{5}$$

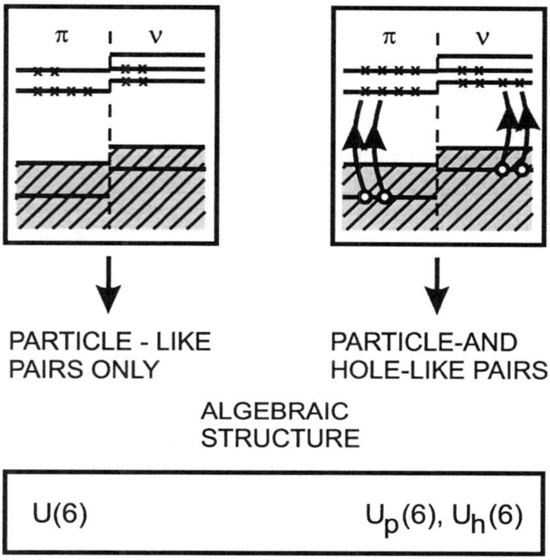

Figure 1. Extension of the s, d IBM by explicitly treating particle and hole bosons as separate $U_p(6)$ and $U_h(6)$ group structures.

in which the $U(6)$ group describes a regular s,d boson model space but now includes an $SU(2)$ group, much like one obtains by adding the proton and neutron label to s and d bosons, thereby introducing the symmetry properties of F-spin. Here, one connects a particle-type boson with some new spin-like quantum number,we called I-spin, and has projection $+1/2$ whereas the hole-type boson has the projection $-1/2$. Using the appropriate step operators one ends up with a set of I-spin multiplets characterized by the total number of particle plus hole bosons $I = N/2$ and its z-component $I_z = (N_p - N_h)/2$. This generates a "horizontal" classification and interesting multiplet structures arise in which $N_p + N_h$ is a constant with changing N_p and N_h boson number when moving through the $I = N/2$ multiplet. First hints showed up in the Sn region, where the 2p-2h intruder states in Sn and the band structure on top, was fitting rather well into a $I = 1$ multiplet completing the multiplet with the ground-state bands of Pd and Xe. Likewise, the Cd intruder states, forming the $I_z = -1/2$ member of a $I = 3/2$ multiplet, could be put into a multiplet by adding the Ru and Ba ground-state bands as well as $I_z = +1/2$ intruder members in Te nuclei. Recently, it has been shown [18] that this holds rather well for the E2 reduced transition probabilities too. This idea of I-spin has been worked out in detail by De Coster et al.[19] and Lehman et al.[20].

It is clear that the combination of particle and hole bosons could lead to more symmetries in the interacting particle-hole system. If one extends the 72 generators resulting from the separate U(6) groups by another 72 operators with the structure $\{p_i^\dagger h_j^\dagger, p_i h_j\}$, a non-compact group U(6,6) results. This structure contains mp-nh

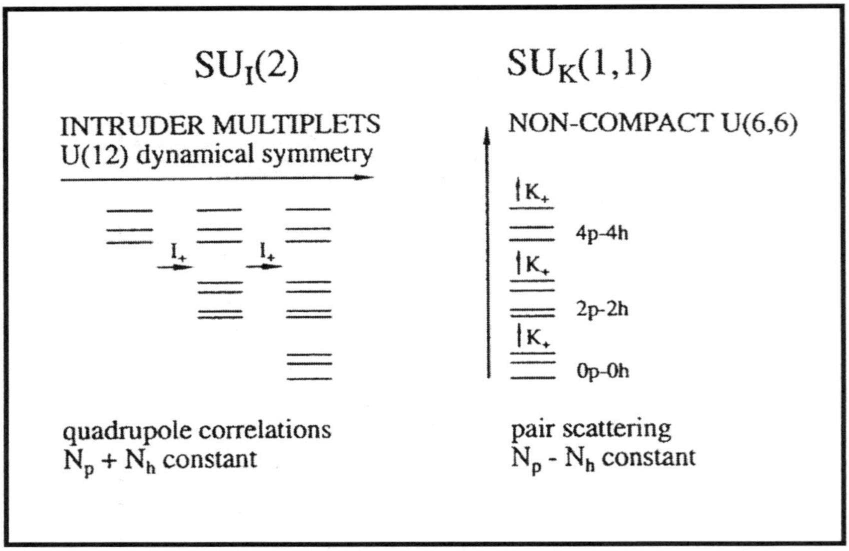

Figure 2. Classification of particle and hole bosons within intruder multiplets (U(12) dynamical symmetry) and a non-compact U(6,6) structure

excitations starting from the lowest regular state in a given nucleus and thus forms the basis of a "vertical" classification [19,20]. The group reduction results into the chain

$$U(6,6) \supset U_{p-h}(6) \otimes SU_K(1,1) \otimes U_-(1) \quad , \tag{6}$$

with now $K_z = (N_p + N_h)/2 + 3$ and with $N_p - N_h$ a constant.

These two spin-like structures are illustrated in fig.2 where on one side one notices the horizontal multiplet classification and where the quadrupole force is the underlying physics component, whereas the right-hand side shows the vertical classification which results essentially from the pair scattering force component that generates multiple p-h excitations. It is interesting to find out that one can even put the two substructures into an even larger group structure (see [19] for more details). Studies in the Pb region, concentrating mainly on symmetry aspects have been discussed in [21,22].

At this stage, we can inject information of the underlying symmetry structure of the interacting boson system where both particle and hole-type bosons are present and go back to the numerical configuration mixing idea. It is possible to reduce the number of parameters by invoking the idea of I-spin symmetry which seems to hold rather well in the Pb,Hg,Pt,.. nuclei [23] and also in the Sn,Cd,Pd,.. region.

As an application, I like to present some very recent results concerning calculations in the Pb even-even neutron-deficient nuclei (see [24] for a recent overview of experimental data in this particular region of the nuclear mass table) encompasing both 0p-0h, 2p-2h and 4p-4h excitations across the Z=82 core [25]. The data indeed

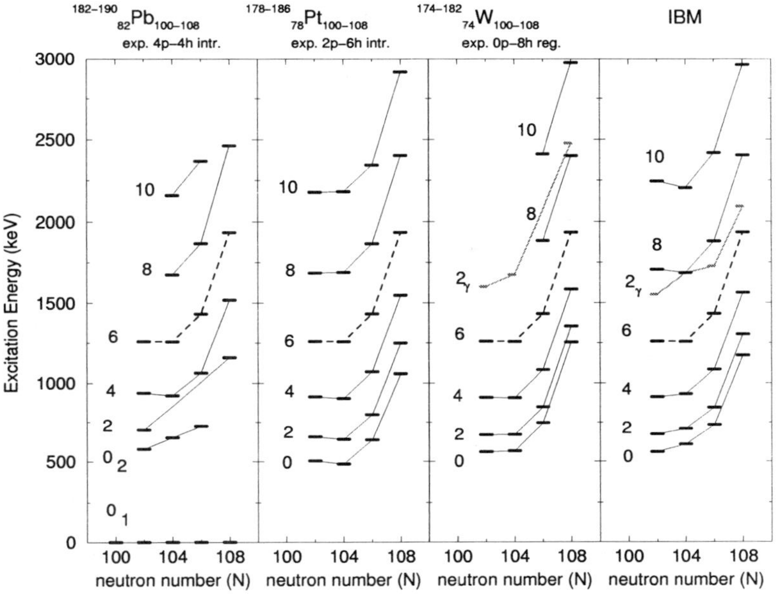

Figure 3. Illustration of the I-spin symmetry in the N=4 multiplet, connecting 4p-4h excitations in Pb nuclei, 6h-2p excitations in Pt and regular 8h configurations in W nuclei. Comparison with IBM calculations is shown

show (i) the very specific energy dependence of the intruding first 0^+ state, as described in sect.2, (ii) the presence, near the N=104 mid-shell region, of two nearby low-lying 0^+ excited states hinting for yet another structure. Most of this experimental information has been obtained using recoil and recoil-decay (in particular α-decay) tagging (RDT) techniques and of heavy-ion induced fusion-evaporation reactions [24].

In general, the configuration mixing of 3 different families leads to a big number of parameters fixing the Hamiltonians in the different model spaces as well as in the coupling Hamiltonians. Here, the idea of I-spin symmetry in relating 2p-2h excitations in Pb with regular states in Pt and of 4p-4h excitations in Pb with 6p-2h intruder configurations in Pt nuclei and with the 8h regular ground-state band in W is working rather well and allows a drastic reduction in the number of parameters [25]. After the calculations were carried out, very recent new data have became available in ^{188}Pb where now members of two distinct collective bands have been observed [26,27]. The advances in studying nuclei really far from stability do show up here and even lifetime measuremts have become possible [27]. These data, moreover, can shed light on the precise characteristics of these bands, on their mixing and the underlying microscopic structure. It is the aim to extend the present calculation such as to cover the whole set of nuclei in the Pb region encompassing both the regular and the intruding states in as consistent a way as possible, making use of the underlying symmetries in the interacting particle and hole boson system.

In the paragraphs before, we have not yet taken the charge character of the bosons - which can both be proton-particle, proton-hole or neutron-particle and neutron-hole - into account. Therefore, extensions inevitably have to come in when aiming at more realistic descriptions of the nuclei at hand. In constructing these large symmetry groups, one has to start from bosons which contain, besides the angular moment s and d information, the charge and I-spin information through extra labels. It leads eventually to a system of 576 generators of the form

$$b^{\dagger}_{\alpha,\beta;lm} b_{\alpha',\beta';l'm'} \quad , \tag{7}$$

with $\alpha, \alpha' \equiv p, h$, $\beta, \beta' \equiv \pi, \nu$ and $l, l' \equiv 0, 2$. The resulting group structure then becomes U(24) which is expected because we have added both the charge label (F-spin) and the particle-hole label (I-spin) on top of the U(6) basic s,d IBM structure. Carrying out the appropriate contractions in the U(24) generators, one can end up with (i) the IBM-2 contracting on p-h labels, (ii) the E-IBM (E means extended in the sense that the core can be excited) contracting on the π, ν labels and (iii) a $U_{I,F}(4)$ group by contracting on the l, m labels. Much of the general structure has been discussed by Decroix et al. [28]. This very rich structure has not yet been studied in its details, clearly not in order to try to generate energy spectra,transition rates, transfer reaction amplitudes. A couple of simple situations though, other then the horizontal and vertical classifications discussed before, can be treated.

Exploring the U(4) structure, one can define and SU(4) multiplet encompassing all the states in those nuclei with a constant total number of bosons such that $N = N_p^{\pi} + N_h^{\pi} + N_p^{\nu} + N_h^{\nu}$ and that are connected through 8 ladder operators, associated with I- and F-spin. The light nuclei, around ^{16}O, form an interesting region for testing this idea since shape coexistence is long known to show up in this region. In particular, in ^{16}O, there is the experimental evidence for a $K^{\pi} = 0^+$ and a $K^{\pi} = 2^+$ band associated with 4p-4h excitations. The microscopic structure then is related to a proton 2p-2h neutron 2p-2h excitation which means that these levels can belong to the SU(4) multiplet with N=4. If this symmetry is a real symmetry for this multiplet, the same structure should also be found as the ground-state band structure of ^{24}Mg. Moreover, levels associated with a proton 4p-4h and neutron 4p-4h excitations in ^{16}O would all belong to the same multiplet. We present results for both $K^{\pi} = 0^+$ and $K^{\pi} = 2^+$ bands. Even though a number of band members have been tentatively associated with the proton and neutron 4p-4h excitations in ^{16}O, the comparison hints for the possible presence of this SU(4) symmetry (see also [29]),

Another intruiging situation results from analyzing the U(24) group structure. The reduction leading to an expression of the full wave funtion as a product of an "orbital s,d" part, the charge or F-spin part times the particle-hole or I-spin part i.e. $\Psi = \phi_L . \alpha_I . \beta_F$ conform with the group chain

$$U(24) \supset (U_L(6) \supset ... \supset O_L(3)) \otimes (SU(4) \supset SU_I(2) \otimes SU_F(2)) \quad , \tag{8}$$

gives the possibility, in the appropriate limit of the U(6) limit, to produce 2 different states with the same orbital [N-1,1] symmetry. There is the well-known case of mixed-symmetry (in the charge or F-spin label) scissors states that have been extensively studied over the last 15-20 years [30] described as $\mid I = N/2, F = N/2-1\rangle$ but also the exciting possibility for a state with a mixed-symmetry character in the space of particle and hole boson excitations and described as $\mid I = N/2-1, F = N/2\rangle$ results. An analysis for 1^+ states has been carried out [31] and the ratio of M1 decay of the two different types of mixed-symmetry states to the 0^+ band-head of the particle-hole boson configuration was derived. Nothing much is known at this stage about such a new class of states and more intensive study is needed to pin down the excitation energy and decay transition probabilities in a reliable way.

4 Outlook

In the present discussion, we have shown that it has been possible to combine both low-lying collective excitations, appearing within the the valence space of nucleons moving outside closed shells and the particle-hole excitations across closed shells that have amply been shown to give rise to low-lying, shape coexisting states. Starting from the essential role played by the residual proton-neutron quadrupole forces, the underlying mechanisms within a deformed mean field and shell-model approach have been pointed out. We have discussed the rich symmetry structures that results by combining the degrees of freedom in the particle and hole subspaces of interacting s and d bosons (see fig.4 for an overview of symmetries as appearing in nuclear physics). The covering large group of U(24) has barely been studied and further exploration of the various reduction schemes and the related physical excitations needs to be carried out.

The authors like to thank the FWO-Flandersm the IWT and the DWTC for financial support. They are most grateful to C. De Coster, B.Decroix, J.E. Garcia-Ramos, P.Van Isacker and J.Jolie for constant input over the years and to M. Huyse, W. Nazarewicz,P. Van Duppen and J.L.Wood for stimulating discussions and for help in building up the various parts of the present contribution. Finally, K.H. has enjoyed the many interactions with F.Iachello which laid the seeds for the present picture in which shape coexistence near closed shells is well covered within the algebraic structure of the extended Interacting Boson model.

References

1. F. Iachello in *Proc.Int. Conf. on Nucl. Struct. and Spectroscopy*,eds. H. P. Blok and A. E. L. Dieperink (Scholar's Press, Amsterdam,1974) 163
2. *Interacting Bosons in Nuclear Physics*, ed. F. Iachello (Plenum Press,1979)
3. *Interacting Bose-Fermi Systems in Nuclei*, ed. F. Iachello (Plenum Press,1980)
4. K. Heyde, P. Van Isacker, M. Waroquier, J.L. Wood and R.A.Meyer, Phys. Repts. **102** (1983) 291
5. J. Bron et al. , Nucl. Phys. **A318** (1979) 335
6. P. Van Duppen, E. Coenen, K. Deneffe, M. Huyse, K.Heyde and P. Van Isacker, Phys. Rev. Lett. **52** (1984) 1974

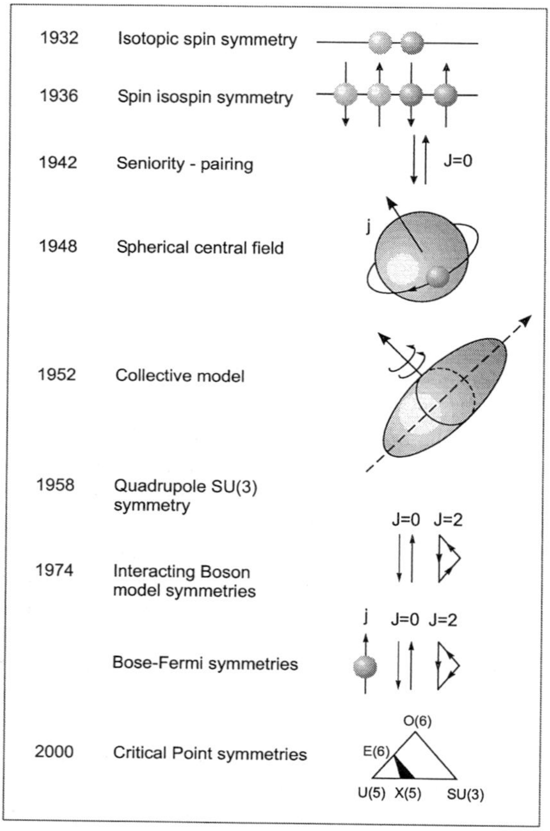

Figure 4. Symmetries in nuclear physics

7. J.L. Wood, K. Heyde, W. Nazarewicz, M. Huyse and P. Van Duppen, Phys. Repts. **215** (1992) 101
8. A. Savelius et al., Nucl. Phys. **A637** (1998) 491
9. R. Bengtsson, T. Bengtsson, J. Dudek, G. Leander, W. Nazarewicz and Jing-Ye Zhang, Phys. Lett. **B183** (1987) 1
10. R. Bengtsson and W. Nazarewicz, Z. Phys. **A334** (1989) 269
11. W. Nazarewicz, Phys. Lett. **B305** (1993) 195
12. A. N. Andreyev et al., Nature. **Vol.405** (2000) 430
13. K. Heyde, J. Jolie, J. Moreau, J. Ryckebusch, M. Waroquier, P. Van Duppen, M. Huyse and J.L. Wood, Nucl. Phys. **A446** (1987) 189
14. P. D. Duval and B.R. Barrett, Nucl. Phys. **A376** (1982) 213
15. J. Jolie and K. Heyde, Phys. Rev. **C42** (1990) 2034
16. K. Heyde, D. De Coster, J. Ryckebusch, M. Waroquier, Phys. Lett. **B218** (1989) 287

17. K. Heyde, C. De Coster, J. Jolie and J.L. Wood, Phys. Rev. **C46** (1992) 541
18. S. Yates, private communicaton
19. C. De Coster, K. Heyde, B. Decroix, P. Van Isacker, J. Jolie, H. Lehmann and J.L. Wood, Nucl. Phys. **A600** (1996) 251
20. H. Lehmann, J. Jolie, C. De Coster, K. Heyde, B. Decroix and J.L. Wood, Nucl. Phys. **A621** (1997) 767
21. C. De Coster, K. Heyde, B. Decroix, J.L. Wood, J. Jolie and H. Lehmann, Nucl. Phys. **A621** (1997) 802
22. C. De Coster, K. Heyde, B. Decroix, K. Heyde, J. Jolie, H. Lehmann and J.L. Wood, Nucl. Phys. **A651** (1999) 31
23. K.Heyde, P. Van Isacker and J.L.Wood, Phys. Rev. **C49** (1994) 559
24. R. Julin, K. Helariutta and M. Muikku, J. Phys. **G27** (2001) R109
25. R. Fossion, K. Heyde, G. Thiamova and P. Van Isacker, Phys. Rev. **C67** (2003) 024306
26. G. Dracoulis et al., *Proc.Int.Conf.Frontiers of Nuclear Structure*, ed. P. Fallon (AIP Conf. Proc., 2003), in print
27. A.Dewald, proposal at JYFL and private communication
28. B. Decroix, J. De Beule, C. De Coster, K. Heyde, A. M. Oros and P. Van Isacker, Phys. Rev. **C57** (1998) 2329
29. B. Decroix, J. De Beule, C. De Coster and K. Heyde, Phys. Lett. **B439** (1998) 237
30. A. Richter, these proceedings
31. B. Decroix, C. De Coster, K. Heyde, A. M. Oros and J. De Beule, Phys. Rev **C58** (1998) 237

SHAPE-INVARIANCE AND MANY-BODY PHYSICS

A. B. BALANTEKIN
University of Wisconsin, Department of Physics
Madison, WI 53706, USA
E-mail: baha@nucth.physics.wisc.edu

Recent developments in the study of shape-invariant Hamiltonians are briefly summarized. Relations between certain exactly solvable problems in many-body physics and shape-invariance are explored. Connection between Gaudin algebras and supersymmetric quantum mechanics is pointed out.

1 Introduction

Supersymmetric Quantum Mechanics (SSM) is the name given to the study of particular pairs of Hamiltonians [1,2]. SSM can be motivated by considering the ground ground state wavefunction, $\psi_0(x)$, for a one-dimensional bound system. Since $\psi_0(x)$ has no nodes it can be written as

$$\psi_0(x) = \exp\left(-\frac{\sqrt{2m}}{\hbar} \int W(x)dx \right), \tag{1}$$

where the function $W(x)$ is related to the potential energy of the system. Introducing the operators

$$\begin{aligned} \hat{A} &= W(\hat{x}) + \frac{i}{\sqrt{2m}}\hat{p}, \\ \hat{A}^\dagger &= W(\hat{x}) - \frac{i}{\sqrt{2m}}\hat{p}, \end{aligned} \tag{2}$$

one can write the Hamiltonian of the system as

$$\hat{H} - E_0 = \hat{A}^\dagger \hat{A}, \tag{3}$$

where E_0 is the ground state energy. The ground state wavefunction satisfies the condition

$$\hat{A}|\psi_0\rangle = 0. \tag{4}$$

It is straightforward to show that the supersymmetric partner potentials

$$\begin{aligned} \hat{H}_1 &= \hat{A}^\dagger \hat{A} \\ \hat{H}_2 &= \hat{A}\hat{A}^\dagger \end{aligned} \tag{5}$$

have the same energy spectra except the ground state of $\hat{H}_1$, the energy of which is zero. Potentials corresponding to these Hamiltonians are

$$\begin{aligned} V_1(x) &= [W(x)]^2 - \frac{\hbar}{\sqrt{2m}}\frac{dW}{dx} \\ V_2(x) &= [W(x)]^2 + \frac{\hbar}{\sqrt{2m}}\frac{dW}{dx}. \end{aligned} \tag{6}$$

The partner potentials in Eq. (6) are called shape-invariant[3] if they can be obtained from one another by changing their parameters:

$$V_2(x;a_1) = V_1(x;a_2) + R(a_1), \tag{7}$$

where a_2 is a function of a_1, and the remainder $R(a_1)$ is independent of x. Eq. (7) is equivalent to the operator relation

$$\hat{A}(a_1)\hat{A}^\dagger(a_1) = \hat{A}^\dagger(a_2)\hat{A}(a_2) + R(a_1). \tag{8}$$

1.1 Algebraic Approach

Shape-invariance problem was formulated in algebraic terms in Ref. [4]. In this formulation one introduces an operator which transforms the parameters of the potential:

$$\hat{T}(a_1)O(a_1)\hat{T}^{-1}(a_1) = O(a_2). \tag{9}$$

Defining the operators

$$\begin{aligned} \hat{B}_+ &= \hat{A}^\dagger(a_1)\hat{T}(a_1) \\ \hat{B}_- = \hat{B}_+^\dagger &= \hat{T}^\dagger(a_1)\hat{A}(a_1) \end{aligned} \tag{10}$$

one can show that the Hamiltonian can be written as

$$\hat{H} - E_0 = \hat{A}^\dagger\hat{A} = \hat{B}_+\hat{B}_-. \tag{11}$$

Using the definitions given in Eq. (10), the shape-invariance condition of Eq. (8) takes the form

$$[\hat{B}_-, \hat{B}_+] = R(a_0), \tag{12}$$

where $R(a_0)$ is defined via

$$R(a_n) = \hat{T}(a_1)R(a_{n-1})\hat{T}^\dagger(a_1). \tag{13}$$

In terms of these new operators Eq. (4) takes the form

$$\hat{B}_-|\psi_0\rangle = 0, \tag{14}$$

i.e. the ground state is annihilated by the lowering operator $\hat{B}_-$.

One can easily establish the commutation relations[4]

$$[\hat{H}, \hat{B}_+^n] = (R(a_1) + R(a_2) + \cdots + R(a_n))\hat{B}_+^n \tag{15}$$

$$[\hat{H}, \hat{B}_-^n] = -\hat{B}_-^n(R(a_1) + R(a_2) + \cdots + R(a_n)). \tag{16}$$

i.e., $\hat{B}_+^n|\psi_0\rangle$ is an eigenstate of the Hamiltonian with the eigenvalue $R(a_1) + R(a_2) + \cdots + R(a_n)$. The normalized eigenstate is

$$|\psi_n\rangle = \frac{1}{\sqrt{R(a_1) + \cdots + R(a_n)}}\hat{B}_+ \cdots \frac{1}{\sqrt{R(a_1) + R(a_2)}}\hat{B}_+ \frac{1}{\sqrt{R(a_1)}}\hat{B}_+|\psi_0\rangle. \tag{17}$$

To identify the algebra we consider the commutation relations

$$[\hat{B}_-, \hat{B}_+] = R(a_0) \tag{18}$$

$$[\hat{B}_+, R(a_0)] = (R(a_1) - R(a_0))\hat{B}_+, \tag{19}$$

$$[\hat{B}_+, (R(a_1) - R(a_0))\hat{B}_+] = \{(R(a_2) - R(a_1)) - (R(a_1) - R(a_0))\}\hat{B}_+, \tag{20}$$

and so on. In general there are an infinite number of such commutation relations. If the quantities $R(a_n)$ satisfy certain relations one of the commutators in this series may vanish. For such a situation the commutation relations obtained up to that point plus their complex conjugates form a Lie algebra with a finite number of elements. For example if the condition

$$(R(a_2) - R(a_1)) - (R(a_1) - R(a_0)) = 0 \tag{21}$$

is satisfied then the algebra is [4] either $SU(2)$ or $SU(1,1)$. Most of the exactly solvable one-dimensional problems in quantum mechanics can be described by this algebra [5]. It can be shown that this algebra also describes for example both the bound and scattering states of the Pöschl-Teller potential [6] as well as associated transfer matrices.

1.2 Outlook on future applications

Almost all exactly solvable one-dimensional potential problems encountered in quantum mechanics textbooks are shape invariant where the parameters are related by a translation [2]

$$a_2 = a_1 + \eta. \tag{22}$$

It should be emphasized that shape-invariance is not the most general integrability condition one can write for such potentials as there are exactly solvable problems which are not shape invariant [7]. There is a second class of shape invariant potentials where the parameters of the partner potentials are related by a scaling [8,9]

$$a_2 = qa_1. \tag{23}$$

In this latter class, corresponding one-dimensional potentials are defined implicitly, but explicit expressions are not given.

In searching for integrable models in two-dimensional statistical mechanics a relationship was uncovered between those models, three-dimensional Chern-Simons gauge theory and quantum groups [10]. These models, being completely integrable, can be written in a shape-invariant way [11], corresponding to a shift in the parameters

$$a_2 = qa_1 + \eta. \tag{24}$$

The associated algebras are called up-down algebras [12]. These developments suggest that there may be shape-invariant potentials where the parameters are related by linear-fractional transformations:

$$a_2 = (qa_1 + \eta)/(a_1 + \eta') \tag{25}$$

This is a completely unexplored direction of research as nothing is known about such integrable systems. Recall that the notation a_1, a_2, etc. may represent not only single parameters, but also a set of them. In general one may suggest to simply relate these parameters by the transformation

$$\hat{T}(a_1)\mathcal{O}(a_1)\hat{T}^{-1}(a_1) = \mathcal{O}(a_2). \tag{26}$$

where $\hat{T}$ is an element of any group, not just of SL(2,R) as suggested by the linear-fractional transformation and its limits that were so far employed. What kind of exactly solvable problems do we obtain? At the moment this is an open question.

The basic philosophy of this approach is to consider the parameters of the Hamiltonians as auxiliary dynamical variables. This is reminiscent of the path leading to the Interacting Boson Model [13]. To describe the quadrupole collectivity in nuclei one needs to consider a five-dimensional space. It is possible to formulate this problem in terms of boson variables [14], however the problem is nonlinear written in terms of quadrupole bosons. By considering a parameter of the problem (boson number) as an additional degree of freedom, Interacting Boson Model introduced a scalar boson as a dynamical variable. This has led to the subsequent realization [15] of s and d bosons as pairs of nucleons coupled to the angular momentum $L = 0$ and $L = 2$

So far we talked about considering parameters of the shape-invariant problem as auxiliary dynamical variables. One can imagine an alternative approach of classifying some of the dynamical variables as "parameters". An example of this is provided by the supersymmetric approach to the spherical Nilsson model of single particle states [16]. The Nilsson Hamiltonian is given by

$$H = \sum_i a_i^\dagger a_i - 2k\mathbf{L}.\mathbf{S} + k\nu\mathbf{L}^2. \tag{27}$$

Introducing the variable

$$F^\dagger = \sum_i \sigma_i a_i^\dagger \tag{28}$$

one can show that the "Hamiltonians"

$$H_1 = F^\dagger F = \sum_i a_i^\dagger a_i - \sigma.\mathbf{L} \tag{29}$$

and

$$H_2 = FF^\dagger = \sum_i a_i a_i^\dagger + \sigma.\mathbf{L} \tag{30}$$

can be considered as supersymmetric partners of each other [16]. The shape-invariance condition of Eq. (8) can be written as

$$FF^\dagger = F^\dagger F + R, \tag{31}$$

where the remainder is

$$R = \sigma.\mathbf{L} - 3/4, \tag{32}$$

i.e. in this example the radial variables are considered as the main dynamical variables and the angular variables are considered as the "parameters".

A number of applications of shape-invariance are available in the literature. These include i) Quantum tunneling through supersymmetric shape-invariant potentials [17]; ii) Study of neutrino propagation through shape-invariant electron densities [18]; iii) Investigation of coherent states for shape-invariant potentials [19,20]; and iv) As attempts to devise exactly solvable coupled-channel problems, generalization of Jaynes-Cummings type Hamiltonians to shape-invariant systems [21,22]. In this article we focus on the applications to many-body systems.

2 Many-Body Hamiltonians

One can ask if these methods can be used to search for exactly-solvable many-body systems. It has been shown that the concept of supersymmetric shape-invariance can be utilized to derive the energy spectrum of Calogero-Sutherland model [23]. Here we discuss an alternative approach and first write down multiple commutators for a shape-invariant Hamiltonian

$$[\hat{H},\hat{B}_+] = R(a_1)\hat{B}_+ \tag{33}$$

$$[[\hat{H},\hat{B}_+],\hat{B}_+] = (R(a_1) - R(a_2))\hat{B}_+^2 \tag{34}$$

$$[[[\hat{H},\hat{B}_+],\hat{B}_+],\hat{B}_+] = (R(a_1) - 2R(a_2) + R(a_3))\hat{B}_+^3 \tag{35}$$

$$[[[[\hat{H},\hat{B}_+],\hat{B}_+],\hat{B}_+],\hat{B}_+] = (R(a_1) - 3R(a_2) + 3R(a_3) - R(a_4))\hat{B}_+^4 \tag{36}$$

and so on. We wish to address the possibility of defining an exactly solvable problem through these commutation relations. We will consider $\hat{B}_+$ as a raising operator. We assume that the Hamiltonian $\hat{H}$ may or may not be in the form given by Eq. (11). We consider a generalized pairing problem with

$$\hat{B}_+ = \sum_j c_j S_j^+. \tag{37}$$

In Eq. (37) the pair creation operator in a single-j shell is defined as

$$S_j^+ = \sum_m \frac{1}{2}(-)^{j-m} a_{j,m}^\dagger a_{j,-m}^\dagger, \tag{38}$$

where $a_{j,m}^\dagger$ is the particle creation operator. If we assume that the shape-invariant Hamiltonian has only one- and two-body terms the commutator $[[\hat{H},\hat{B}_+],\hat{B}_+]$ will only involve products of four creation operators. Consequently the next nested commutator will vanish:

$$[[[\hat{H},\hat{B}_+],\hat{B}_+],\hat{B}_+] = 0 \tag{39}$$

Higher nested commutators will also vanish. This will place strong constraints on $R(a_n)$, i.e.

$$R(a_3) = -R(a_1) + 2R(a_2), \tag{40}$$

$$R(a_4) = R(a_1) - 3R(a_2) + 3R(a_3) \tag{41}$$

and so on. Consequently we can immediately write the energy eigenvalues and eigenstates of the Hamiltonian

$$\hat{H}\hat{B}_+^n|\psi_0\rangle = \left(nR(a_1) + \frac{1}{2}Wn(n-1)\right)\hat{B}_+^n|\psi_0\rangle, \tag{42}$$

where

$$W = R(a_2) - R(a_1). \tag{43}$$

A similar approach was first given by Talmi [24].

3 Connection to Gaudin Algebras

The pairing model with a constant two-body interaction was solved exactly by Richardson [25]. In a parallel development Gaudin developed an algebraic approach to solve many-body spin Hamiltonians [26,27]. Here we will explore the relationship between Gaudin's methods, algebraic methods developed to search for quasi-exactly solvable models [28] and supersymmetric quantum mechanics.

Following the notation of Ref. [29] we consider the function defined as

$$\Psi(\lambda) = \prod_i^N (\lambda - \xi_i)\, e^{-\int W d\lambda}, \tag{44}$$

where $W(\lambda)$ is an arbitrary function of λ and ξ_i are numbers to be determined. Introducing the operators

$$A = W + ip, \quad A^\dagger = W - ip, \tag{45}$$

it can be shown that the function defined in Eq. (44) satisfies the equation

$$A^\dagger A\ \Psi = \left[2\sum_{i\neq j}\frac{1}{(\lambda-\xi_i)(\lambda-\xi_j)} - 2\sum_i \frac{W(\lambda)}{(\lambda-\xi_i)}\right]\Psi. \tag{46}$$

Requiring the residue at ξ_i to vanish yields the Bethe-ansatz conditions:

$$W(\xi_i) = \sum_{i\neq j}\frac{1}{\xi_i - \xi_j}. \tag{47}$$

Inserting Eq. (47) into Eq. (46) we obtain

$$A^\dagger A\ \Psi = 2\sum_i \left(\frac{W(\lambda) - W(\xi_i)}{\lambda - \xi_i}\right)\Psi. \tag{48}$$

Provided that their superpotentials satisfy the condition given in Eq. (47), factorized supersymmetric Hamiltonians satisfy Eq. (48). Note that the right side of Eq. (48) in general depends on λ, hence we cannot interpret the term that multiplies the function Ψ as an energy eigenvalue. However, for a number of limited cases (certain functions $W(\lambda)$ such as those that correspond to a harmonic oscillator) this λ dependence drops out and one can recover the standard expressions for the energy eigenvalues [30].

The three generators of Gaudin's algebra $(J_0(\lambda), J_\pm(\lambda))$ can be defined through the commutation relations

$$[J_0(\lambda), J_+(\mu)] = -\frac{J_+(\lambda) - J_+(\mu)}{\lambda - \mu}, \tag{49}$$

$$[J_-(\lambda), J_+(\mu)] = -2\frac{J_0(\lambda) - J_0(\mu)}{\lambda - \mu}, \tag{50}$$

and

$$[J_{\pm,0}(\lambda), J_{\pm,0}(\mu)] = 0, \tag{51}$$

where λ is, in general, a continuous parameter. Gaudin studied the eigenstates of the "Hamiltonian" [27]

$$H(\lambda) = J_0(\lambda)J_0(\lambda) - \frac{1}{2}J_-(\lambda)J_+(\lambda) - \frac{1}{2}J_+(\lambda)J_-(\lambda) \tag{52}$$

If a state $\mid 0\rangle$ which is annihilated by all $J_-(\lambda)$ can be identified

$$J_-(\lambda) \mid 0\rangle = 0, \tag{53}$$

then $W(\lambda)$ is introduced as the eigenvalue of $J_0(\lambda)$ on that state:

$$J_0(\lambda) \mid 0\rangle = W(\lambda) \mid 0\rangle. \tag{54}$$

One can then find the eigenvalues and eigenstates of the "Hamiltonian" of Eq. (52):

$$H(\lambda) \mid \Phi\rangle = E(\lambda) \mid \Phi\rangle, \tag{55}$$

where the eigenstates are

$$\mid \Phi\rangle = J_+(\xi_N)J_+(\xi_{N-1})\cdots J_+(\xi_1) \mid 0\rangle, \tag{56}$$

and the eigenvalues are

$$E(\lambda) = W^2(\lambda) + W'(\lambda) + 2\sum_i \left(\frac{W(\lambda) - W(\xi_i)}{\lambda - \xi_i}\right). \tag{57}$$

In deriving the above equations the conditions

$$W(\xi_i) = -\sum_{i\neq j} \frac{1}{\xi_i - \xi_j}, \quad i, j = 1, \cdots, N \tag{58}$$

were assumed to be fulfilled.

The strategy of using Richardson-Gaudin methods to deal with many-body problems were employed by a number of authors [31,32,33,34]. Clearly there is a mapping between the solutions of the Gaudin problem (Eq. (55)) and those of the factorized supersymmetric Hamiltonians (Eq. (48)). One may ask if this correspondence can be exploited to study pairing and related problems.

The pair creation operator[16] in a single-j shell is defined in Eq. (38),

$$S_j^+ = \sum_m \frac{1}{2}(-)^{j-m} a_{j,m}^\dagger a_{j,-m}^\dagger, \tag{59}$$

its Hermitian conjugate, and the number operator span an $SU(2)$ algebra (the so-called quasi-spin algebra). One can obtain a Gaudin algebra from the quasi-spin algebra by defining

$$S_+(\lambda) = \sum_j \frac{S_j^+}{\lambda - \epsilon_i}, \tag{60}$$

(and similar formulas for the other elements). This realization of the Gaudin algebra can be very useful in many-body systems. As a simple example we consider a system with s and p bosons and define three operators that satisfy Gaudin's commutation relations

$$B_+(\lambda) = \frac{1}{2}\left[\frac{s^\dagger s^\dagger}{\lambda - \alpha_s} + \frac{(p^\dagger \cdot p^\dagger)}{\lambda - \alpha_p}\right] \tag{61}$$

$$B_-(\lambda) = [B_+(\lambda)]^\dagger, \tag{62}$$

and

$$B_0(\lambda) = \frac{1}{2}\left[\frac{\hat{n}_s}{\lambda - \alpha_s} + \frac{\hat{n}_p + 3/2}{\lambda - \alpha_p}\right]. \tag{63}$$

It is easy to show that as $\alpha_s \to \alpha_p$ the quantity $B_+(\lambda)B_-(\lambda)$ reduces to

$$\frac{1}{\lambda - \alpha_p}\hat{P}_4 \tag{64}$$

where $\hat{P}_4$ is the O(4) pairing operator. One can then study a Gaudin-type Hamiltonian which generalizes this operator

$$H(\lambda) = B_0(\lambda)B_0(\lambda) - \frac{1}{2}B_-(\lambda)B_+(\lambda)B_+(\lambda)B_-(\lambda). \tag{65}$$

Following steps above one can show that this Hamiltonian is associated with the one-dimensional potential

$$V(x) = \frac{1}{2}\left(\frac{1}{x - \alpha_s} + \frac{1}{x - \alpha_p}\right)^2. \tag{66}$$

Similar ideas could conceivably be useful in dealing with other many-body systems.

Acknowledgments

I thank F. Iachello for his support, encouragement, and friendship over many years. This work was supported in part by the U.S. National Science Foundation Grants No. PHY-0136261 and PHY-0070161.

References

1. E. Witten, *Nucl. Phys.* B **188**, 513 (1981).
2. F. Cooper, A. Khare and U. Sukhatme, *Phys. Rept.* **251**, 267 (1995). [arXiv:hep-th/9405029].
3. L. E. Gendenshtein, *JETP Lett.* **38**, 356 (1983).
4. A. B. Balantekin, *Phys. Rev.* A **57**, 4188 (1998) [arXiv:quant-ph/9712018].
5. F. Iachello and R.D. Levine, *Algebraic theory of Molecules* (Oxford University Press, New York, 1995).
6. Y. Alhassid, F. Gursey and F. Iachello, *Phys. Rev. Lett.* **50**, 873 (1983).
7. F. Cooper, J. N. Ginocchio and A. Khare, *Phys. Rev.* D **36**, 2458 (1987).
8. A. Khare and U. P. Sukhatme, *J. Phys.*A **26**, L901 (1993) [arXiv:hep-th/9212147].
9. D. T. Barclay, R. Dutt, A. Gangopadhyaya, A. Khare, A. Pagnamenta and U. Sukhatme, *Phys. Rev.* A **48**, 2786 (1993) [arXiv:hep-ph/9304313].
10. E. Witten, *Nucl. Phys.* B **330**, 285 (1990).
11. A.B. Balantekin, in preparation.
12. G. Benkart, *Contemp. Math.* **224**, 29 (1999).
13. A. Arima and F. Iachello, *Annals Phys.* **99**, 253 (1976).
14. D. Janssen, R.V. Jolos, and F. Donau, *Nucl. Phys.* A **224**, 93 (1974).
15. A. Arima, T. Otsuka, F. Iachello, and I. Talmi, *Phys. Lett.* B **66**, 205 (1977).
16. A. B. Balantekin, O. Castanos and M. Moshinsky, *Phys. Lett. B* **284**, 1 (1992).
17. A. N. Aleixo, A. B. Balantekin and M. A. Candido Ribeiro, *J. Phys. A* **33**, 1503 (2000) [arXiv:quant-ph/9910051].
18. A. B. Balantekin, *Phys. Rev. D* **58**, 013001 (1998) [arXiv:hep-ph/9712304].
19. A. B. Balantekin, M. A. Candido Ribeiro and A. N. Aleixo, *J. Phys. A* **32**, 2785 (1999) [arXiv:quant-ph/9811061].
20. A. N. Aleixo, A. B. Balantekin and M. A. Candido Ribeiro, *J. Phys. A* **35**, 9063 (2002) [arXiv:math-ph/0209033].
21. A. N. Aleixo, A. B. Balantekin and M. A. Candido Ribeiro, *J. Phys. A* **33**, 3173 (2000) [arXiv:quant-ph/0001049].
22. A. N. Aleixo, A. B. Balantekin and M. A. Candido Ribeiro, *J. Phys. A* **34**, 1109 (2001) [arXiv:quant-ph/0101024].
23. P. K. Ghosh, A. Khare and M. Sivakumar, *Phys. Rev.* A **58**, 821 (1998). [arXiv:cond-mat/9710206].
24. I. Talmi, *Simple Models of Complex Nuclei: the Shell Model and Interacting Boson Model* (Harwood Academic Publishers, Chur, 1993).
25. R.W. Richardson, *Phys. Lett.* **3**, 277 (1963).
26. M. Gaudin, *J. Phys. (Paris)* **37**, 1087 (1976).
27. M. Gaudin, *La fonction d'onde de Bethe* (Masson, Paris, 1983).
28. A.G. Ushveridze, *Quasi-Exactly Solvable Problems in Qunatum Mechanics* (IOP Publishing, Bristol, 1994).
29. A. G. Ushveridze, *Annals Phys.* **266**, 81 (1998) [arXiv:hep-th/9707151].
30. G. Akemann and A.B. Balantekin, in preparation.
31. F. Pan, J. P. Draayer and W. E. Ormand, *Phys. Lett.* B **422**, 1 (1998). [arXiv:nucl-th/9709036].

32. J. Dukelsky and S. Pittel, *Phys. Rev. Lett.* **86**, 4791 (2001). [arXiv:nucl-th/0102034].
33. J. Dukelsky, C. Esebbag and P. Schuck, *Phys. Rev. Lett.* **87**, 066403 (2001). [arXiv:cond-mat/0107477].
34. F. Pan and J. P. Draayer, *Phys. Rev.* C **66**, 044314 (2002).

SOME NEW PERSPECTIVES ON PAIRING IN NUCLEI

S. PITTEL

Bartol Research Institute, University of Delaware, Newark, Delaware 19716, USA

J. DUKELSKY

Instituto de Estructura de la Materia, Consejo Superior de Investigaciones Cientificas, Serrano 123, 28006 Madrid, Spain

Following a brief reminder of how the pairing model can be solved exactly, we describe how this can be used to address two interesting issues in nuclear structure physics. One concerns the mechanism for realizing superconductivity in finite nuclei and the other concerns the role of the nucleon Pauli principle in producing *sd* dominance in interacting boson models of nuclei.

1 Introduction

Ever since the work of Richardson in the mid-60s[1], it has been recognized that the Pairing Model (PM) is exactly solvable, even in the presence of non-degenerate single-particle levels. In recent years, there has been a resurgence of interest in the PM, with several applications reported that build on this exact solvability [2].

This talk reviews two recent applications of the PM in nuclear physics. Both build on the fact that there exists a classical electrostatic analogy for every PM. One makes use of this analogy to obtain a pictorial representation of how superconductivity arises in finite nuclear systems [3]. The other has led us to propose a new mechanism for *sd* dominance in interacting boson models of nuclei[4].

2 Richardson's solution of the Pairing Model

The PM hamiltonian for both fermions and bosons can be written as

$$H_P \;=\; \sum_l \epsilon_l \hat{N}_l \;+\; \frac{g}{2}\sum_{ll'} A_l^\dagger A_l \;, \qquad (1)$$

where

$$\hat{N}_l \;=\; \sum_m a_{lm}^\dagger a_{lm} \quad , \quad A_l^\dagger \;=\; \sum_m a_{lm}^\dagger a_{l\bar{m}}^\dagger \;. \qquad (2)$$

Here $a_{lm}^\dagger$ creates either a boson or a fermion in single-particle state lm and $l\bar{m}$ denotes the time reverse of lm.

Richardson considered the following ansatz for the ground state of a system of $2N$ particles subject to this hamiltonian:

$$|\Psi\rangle \;=\; \prod_{i=1}^{N} B_\alpha^\dagger \,|0\rangle \quad , \quad B_\alpha^\dagger \;=\; \sum_l \frac{1}{2\epsilon_l - e_\alpha}\, A_l^\dagger \;. \qquad (3)$$

He showed that it is an exact eigenstate of the pairing hamiltonian if the *pair energies* e_α satisfy the set of equations ($\Omega_l = l + \frac{1}{2}$)

$$1 \,+\, 2g \sum_l \frac{\Omega_l}{2\epsilon_l - e_\alpha} \,\mp\, 4g \sum_{\beta(\neq\alpha)} \frac{1}{e_\beta - e_\alpha} \,=\, 0 \,. \tag{4}$$

In eq. (4) and throughout the presentation, the upper sign refers to boson systems and the lower sign to fermion systems.

The coupled equations (4), one for each of the N collective pairs, are called the Richardson equations.

Once the set of Richardson equations has been solved, the total ground state energy of the system can be obtained by summing the resulting pair energies,

$$E \,=\, \sum_\alpha e_\alpha \,. \tag{5}$$

While the above discussion focused on the ground state solution, it is possible to use the same general procedure to generate *all* excited states as well.

3 An electrostatic analogy for Pairing Models

As we have seen, the eigenvalues and eigenfunctions of the hamiltonian can be obtained using the Richardson approach, both for fermion and boson systems. From this, it is straightforward to establish an exact electrostatic analogy for the quantum pairing problem. To do so, consider the energy functional

$$\begin{aligned} U = \mp \frac{1}{4g}\left[\sum_\alpha e_\alpha - \sum_j \Omega_j \epsilon_j\right] + \frac{1}{2}\sum_{j\alpha}(\pm\Omega_j)\ln|2\epsilon_j - e_\alpha| \\ - \frac{1}{2}\sum_{\alpha\neq\beta}\ln|e_\alpha - e_\beta| - \frac{1}{8}\sum_{i\neq j}\Omega_i\Omega_j\ln|2\epsilon_i - 2\epsilon_j| \,. \end{aligned} \tag{6}$$

It can be readily shown that when we differentiate U with respect to the pair energies e_α and equate to zero we recover precisely the Richardson equations (4).

To appreciate the physical meaning of U, we should remember that the Coulomb interaction between two point charges in two dimensions is

$$v\left(\mathbf{r}_1, \mathbf{r}_2\right) = -q_1 q_2 \ln\left|\mathbf{r}_1 - \mathbf{r}_2\right| \;, \tag{7}$$

where q_i is the charge and r_i the position of particle i.

Thus, U is the energy functional for a classical two-dimensional (2D) electrostatic system with the following ingredients:

- There are a set of *fixed charges*, one for each single-particle level, which are located at the positions $2\epsilon_i$ and have charges $\pm\frac{\Omega_i}{2}$. We will call them *orbitons.*
- There are N *free charges*, one for each collective pair, which are located at the positions e_α and have positive unit charge. We will call them *pairons.*
- There is a Coulomb interaction between all charges.

Table 1. Position and charges of the orbitons appropriate to a pairing treatment of $^{114-116}Sn$.

Orbiton	Position	Charge
$d_{5/2}$	0.0	−1.5
$g_{7/2}$	0.44	−2.0
$s_{1/2}$	3.80	−0.5
$d_{3/2}$	4.40	−1.0
$h_{11/2}$	5.60	−3.0

- There is a uniform electric field in the vertical direction with intensity $\pm\frac{1}{4g}$.

The existence of this exact analogy suggests that we might be able to use the positions that emerge for the pairons in the classical problem to gain insight into the quantum problem, hopefully insight that was not otherwise evident.

Some other properties of the electrostatic problem that we will be using are:

- Since the orbiton positions are given by the single-particle energies, which are real, they must lie on the vertical or real axis.
- For fermion problems, the pair energies that emerge from the Richardson equations are not necessarily real. They can either be real or they can come in complex conjugate pairs. Thus, a pairon must either lie on the vertical axis (real pair energies) or be part of a mirror pair (complex pair energies).
- For boson problems, the pair energies are of necessity real and, thus, like the orbitons lie on the real axis.

4 A new pictorial representation of nuclear superconductivity

We now apply the electrostatic analogy to the problem of identical nucleon pairing and in particular to the question of how superconductivity arises in such systems. Because of the limited number of active nucleons in a nucleus, it is extremely difficult to see evidence for the transition to superconductivity in such systems.

We will discuss what happens when we apply the electrostatic analogy to the semi-magic nuclei $^{114-116}Sn$. The calculations are done as a function of pairing strength g, using single-particle energies extracted from experiment. Table 1 shows the corresponding information on the positions and charges of the orbitons.

Fig. 1 focuses on the nucleus ^{114}Sn, showing the positions of the pairons in the 2D plane as a function of g. Since ^{114}Sn has 14 valence neutrons, there are seven pairons in the classical picture. In the limit of very weak coupling, six neutrons fill the $d_{5/2}$ orbit and eight fill the $g_{7/2}$. The corresponding electrostatic picture (Fig. 1a) has three pairons close to the $d_{5/2}$ orbiton and four close to the $g_{7/2}$. In the figure, we draw lines connecting each pairon to the one that is closest to it. These lines make clear that at very weak coupling the pairons organize themselves as artificial *atoms* around their corresponding orbitons.

What happens as we increase the magnitude of g (Figs. 1b-c)? [The physical value is roughly $-0.092\ MeV$.] As g increases, the pairons repel, causing the atoms

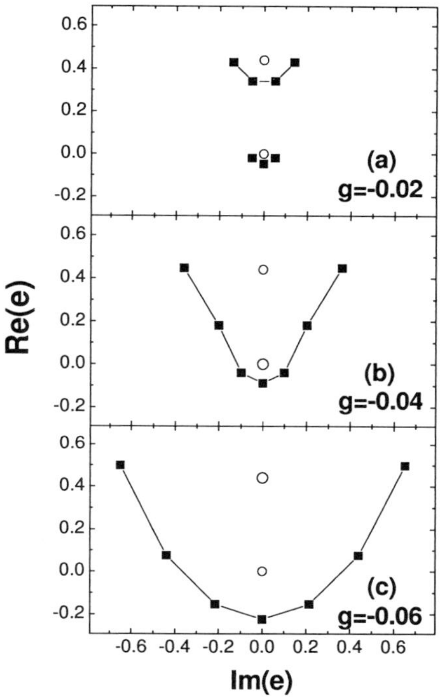

Figure 1. Two-dimensional representation of the pairon positions in ^{114}Sn for three selected values of g. The orbitons are represented by open circles; only the lowest two, the $d_{5/2}$ and $g_{7/2}$, are shown at the positions dictated by Table 1.

to expand. For $g \approx -0.04$, a transition takes place from two isolated atoms to a *cluster*, with all pairons connected to one another. We claim that this geometrical transition from atoms to clusters in the classical problem is a reflection of the superconducting transition in the quantum problem.

We have also treated the nucleus ^{116}Sn, with the same set of single-particle energies as in ^{114}Sn. What we find is that in Sn^{116} the transition to complete superconductivity occurs in two stages. For small g, the pairons distribute themselves into three atoms, surrounding the $d_{5/2}$, $g_{7/2}$ and $s_{1/2}$ orbitons. When g reaches roughly -0.06, the two lowest atoms - containing 7 pairons - merge into a cluster, as in ^{114}Sn, with the eighth still separate. When g grows to roughly -0.095 a second transition takes place, with the eighth pairon merging into a larger cluster

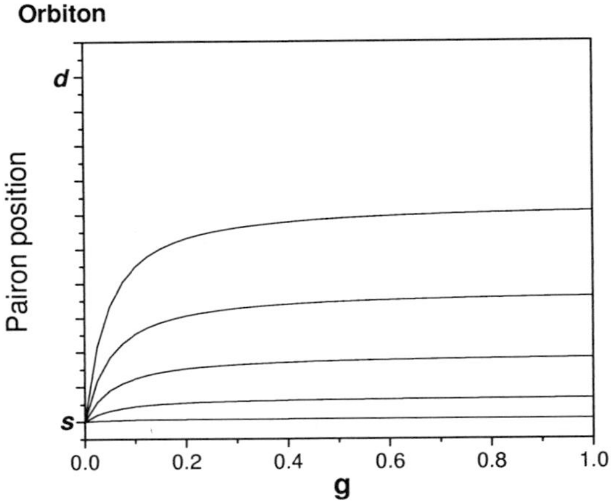

Figure 2. Evolution of pairons for a model involving 10 bosons in all even-L states up to $L = 12$ subject to a hamiltonian with linear single-boson energies and a repulsive boson pairing interaction

with the other seven. From this point on, superconductivity is complete.

5 A new mechanism for *sd* dominance in the IBM

The electrostatic analogy can also be applied to boson pairing models, with the important caveat that the pairons are now confined to the real axis. Fig. 2 shows the pairon positions for a model involving 10 bosons moving in all even-L boson states up to $L = 12$ and interacting via repulsive boson pairing with strength g. The single-boson energies are assumed to increase linearly with l.

Several points are immediately apparent. At low pairing strength, the pairons sit very near the s orbiton, reflecting the fact that the bosons are almost completely in the s state. As the pairing strength increases, a phase transition takes place to a scenario in which the pairons are no longer sitting near the s orbiton. However, even after the phase transition all pairons are confined to the region between the lowest two orbitons, the s and d. What this suggests is that after the phase transition the boson pairs that define the corresponding quantum ground state are most likely primarily of s and d character.

What is the relevance of this to the IBM? As a reminder, in the IBM the s and d bosons model the lowest two pair degrees of freedom for identical nucleons, those with $J^\pi = 2^+$ and 4^+. The key assumption of the model is that all other bosons, reflecting higher pair states, can be ignored, except for their renormalization effects. A second point to remember is that in any effort to model composite objects by structureless particles, there invariably arises a repulsive interaction between these particles, to reflect the Pauli exchange between their constituents.

The results in Fig. 2 are suggestive that in the presence of such a repulsive interaction between bosons only the two lowest boson degrees of freedom can correlate, namely the s and d. This suggests that repulsive pairing between bosons

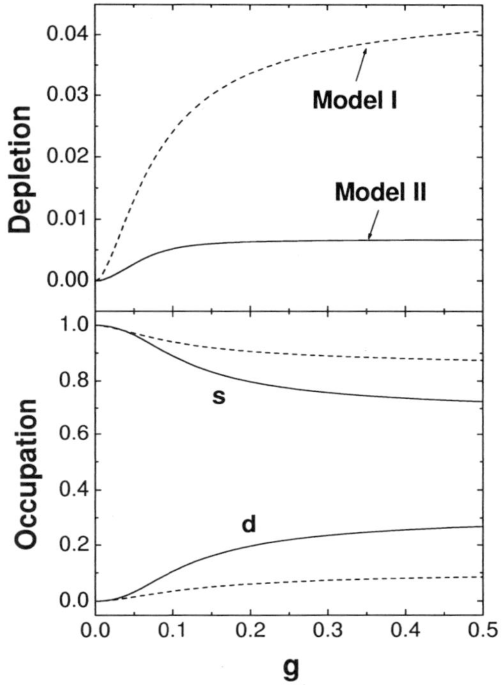

Figure 3. Occupation probabilities for the ground state of a system of 5 boson pairs and maximum angular momentum $L = 12$ as a function of the g. The upper graph shows the sum of occupation probabilities (depletion) for high-spin bosons ($l > 2$) while the lower graph gives the occupation probabilities for s and d bosons. The dashed lines refer to Model I and the solid line4s to Model II, as described in the text.

provides a new mechanism for sd dominance in interacting boson models of nuclei.

These points can be made more quantitative by looking directly at the quantum results. In Fig. 3, we show results for the same interacting boson model as above, but now with two possible choices for the single-boson spectrum. In addition to the choice $\epsilon_l = l$ used before (Model I), we also consider $\epsilon_l = l^2$ (Model II). In this way, we can assess whether sd dominance is a general feature of boson models involving repulsive pairing or is limited to the model earlier shown.

As we can see from the figure, both models show the same general features. For weak g, most of the bosons are in the s state. As g increases, there is a phase transition to a mixed or fragmented state. However, even in the fragmented state there are essentially no bosons other than those with $L = 0$ and 2.

Indeed, when we carry out the calculation as a function of boson number, we find that as N grows the number of non-sd bosons decreases, and in the thermodynamic limit there are only s and d bosons.

6 Summary

There are two key points we have tried to get across in this presentation. The first is that pairing models, even with non-degenerate levels, can be solved exactly using a method introduced by Richardson almost 40 years ago. The second is that these exactly solvable models can be used to provide interesting insight into several issues of importance in nuclear physics. The two examples we discussed concerned the mechanism for realizing superconductivity in finite nuclear systems and the role of the nucleon Pauli principle in producing sd dominance in interacting boson models of nuclei.

Acknowledgments

This work was supported in part by the US National Science Foundation under grant #s PHY-9970749 and PHY-0140036 and by the Spanish DGI under grant BFM2000-1320-C02-02.

References

1. R.W. Richardson, Phys. Lett. **3**, 277 (1963); R.W. Richardson and N. Sherman, Nucl. Phys. **52**, 221 (1964); R.W. Richardson, J. Math. Phys. **9**, 1327 (1968).
2. G. Sierra, J. Dukelsky, G.G. Dussel, J. von Delft and F. Braun, Phys. Rev. B **61**, R11890 (2000).
3. J. Dukelsky, C. Esebbag and S. Pittel, Phys. Rev. Lett. **88** (2002) 062501.
4. J. Dukelsky and S. Pittel, Phys. Rev. Lett. **86** (2001) 4791.

PAIRING AND QUARTETTING IN THE INTERACTING BOSON MODEL

P. VAN ISACKER

GANIL, B.P. 55027, F-14076 Caen Cedex 5, France

E-mail: isacker@ganil.fr

The nuclear shell model allows several analytical solutions which broadly can be divided in two classes: pairing models and rotational models. In this contribution a review is given of nuclear pairing models with an emphasis on their symmetry character. The most general SO(8) model which accommodates neutrons and protons as well as $T = 0$ and $T = 1$ pairing, is solvable in three limits: only $T = 0$ pairing, only $T = 1$ pairing and equal strengths in the two channels. In these limits, the superconducting ground-state solution of even-even $N = Z$ nuclei exhibits a quartet structure. This fermionic model is used as a starting point for a mapping onto an interacting boson model which includes $T = 0$ as well as $T = 1$ bosons.

1 A symmetry triangle for the shell model

Symmetry considerations have played an important role in the development of nuclear physics. Important landmarks in this development include Wigner's SU(4) supermultiplet model [1] which assumes invariance in spin and isospin, Racah's SU(2) pairing model [2] leading to the concept of seniority, Elliott's SU(3) model [3] which provides an understanding of rotation in the context of the spherical shell model and the U(6) interacting boson model of Arima and Iachello [4] which gives a unified description of collective structures observed in nuclei.

The shell model of the atomic nucleus is characterised by a shell structure generated by a nuclear mean field with harmonic oscillator features perturbed by a spin–orbit term and a competition between the pairing and the quadrupole residual interactions. These features can be represented algebraically as is illustrated schematically in Fig. 1 where each vertex corresponds to an analytic solution of the shell model. A hamiltonian corresponding to the top vertex yields uncorrelated Hartree–Fock type wave functions. This limit is reached if the single-particle energy

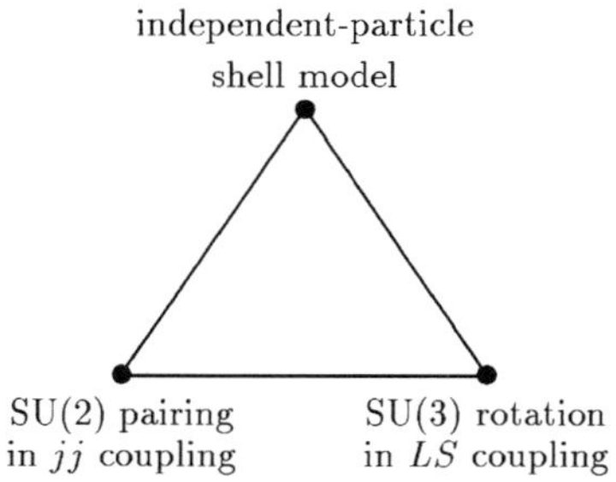

Figure 1. Schematic representation of the shell-model parameter space with three analytically solvable vertices.

spacings are large in comparison with a typical matrix element of the residual interaction. The transition between jj and LS coupling is controlled by the strength of the spin–orbit coupling (as compared to that of the residual interaction). And, finally, the transition from SU(2) pairing to SU(3) rotation requires a change of the character of the residual interaction from pairing to quadrupole. The triangle in Fig. 1 summarises schematically the basic symmetries of the shell model; many extensions are possible as discussed comprehensively in the review by Jerry Draayer [5].

In a previous meeting [6] I have focussed on the rotational vertex of Fig. 1. At the occasion of this meeting in Erice in honour of Franco Iachello, I concentrate on the pairing vertex of the symmetry triangle. First I turn to the question what superconducting solutions are possible in the presence of two kinds of particles, neutrons and protons.

2 Symmetry structure of superconducting solutions

2.1 SU(2) superconductivity

The most basic situation with regard to pairing in nuclei is encountered when n identical nucleons occupy a set of degenerate single-particle states. For convenience in the subsequent discussion, ls coupling is used (with $s=\frac{1}{2}$) to label a single-particle state. (Similar results are obtained in jj coupling.) A nucleon creation operator is then denoted as $a^\dagger_{lm_l sm_s}$. The nucleons are assumed to interact through a pairing force of the form

$$V_{\mathrm{SU}(2)} = -g_0 S^0_+ S^0_-, \tag{1}$$

with

$$S^0_+ = \sqrt{\frac{1}{2}} \sum_l \sqrt{2l+1} (a^\dagger_{ls} \times a^\dagger_{ls})^{(00)}_{00}, \qquad S^0_- = (S^0_+)^\dagger . \tag{2}$$

The notation S refers to the fact that these are nucleon pairs coupled to $L=0$ while the superscript 0 in $S^0_\pm$ refers to spin $S=0$. Since the nucleons are identical, only one pair state is allowed by the Pauli principle, that is, the spin-antiparallel state for either neutrons or protons. Therefore, this pairing mode is called *spin singlet*. The hamiltonian (1) has an SU(2) dynamical symmetry and by virtue of this symmetry can be solved analytically [7]. The ground state has a superconducting structure of the form (for even and odd n, respectively)

$$(S^0_+)^{n/2} |\mathrm{o}\rangle, \qquad a^\dagger_{lm_l sm_s} (S^0_+)^{(n-1)/2} |\mathrm{o}\rangle, \tag{3}$$

where $|\mathrm{o}\rangle$ represents the vacuum (i.e., the doubly-magic core nucleus). The conserved quantum number that emerges from these considerations is *seniority* [2], the number of nucleons not in pairs coupled to $L=0$. The superconducting solution of the pairing hamiltonian (1) leads naturally to three characteristic features of semi-magic nuclei: the constant excitation energy (independent of n) of the first-excited 2^+ state in even-even isotopes, the linear variation of pair removal or two-nucleon separation energies as a function of n and the odd-even staggering in the nuclear binding energy.

2.2 SO(5) superconductivity

This type of superconductivity arises for a system of neutrons and protons. It is assumed that the pairing interaction is isospin invariant which implies that it is the same in the three possible $T = 1$ channels, neutron-neutron, neutron-proton and proton-proton, and that (1) can be rewritten as

$$V_{\mathrm{SO}(5)} = -g_0 S^{01}_{+} \cdot S^{01}_{-}, \tag{4}$$

where the dot indicates a scalar product in isospin. In terms of the nucleon operators $a^\dagger_{lm_l sm_s tm_t}$, which now carry also isospin indices (with $t = \frac{1}{2}$), the pair operators are

$$S^{01}_{+,\mu} = \sqrt{\frac{1}{2}} \sum_l \sqrt{2l+1} (a^\dagger_{lst} \times a^\dagger_{lst})^{(001)}_{00\mu}, \qquad S^{01}_{-,\mu} = (S^{01}_{+,\mu})^\dagger, \tag{5}$$

where the superscripts 0 and 1 in $S^{01}_{\pm,\mu}$ refer to spin $S = 0$ and isospin $T = 1$. The index μ (isospin projection) distinguishes neutron-neutron ($\mu = +1$), neutron-proton ($\mu = 0$) and proton-proton ($\mu = -1$) pairs. There are thus three different pairs with $S = 0$ and $T = 1$ but they are trivially related through isospin symmetry. The dynamical symmetry of the hamiltonian (4) is SO(5) which makes the problem analytically solvable although in a much more laborious way [8,9] than in SU(2). The quantum number, besides seniority, that emerges from this analysis is *reduced isospin* [10], which is the isospin of the nucleons not in pairs coupled to $L = 0$. As a consequence of the neutron-proton quadrupole interaction which breaks seniority and is responsible for deformation, SO(5) superconductivity has not found widespread application in nuclei.

2.3 SO(8) superconductivity

For a neutron and a proton there exists a different paired state with *parallel* spins. The most general pairing interaction for a system of neutrons and protons thus involves a *spin-singlet* and a *spin-triplet* term,

$$V_{\mathrm{SO}(8)} = -g_0 S^{01}_{+} \cdot S^{01}_{-} - g_1 S^{10}_{+} \cdot S^{10}_{-}, \tag{6}$$

where the $S = 1, T = 0$ pair operators are defined as

$$S^{10}_{+,\mu} = \sqrt{\frac{1}{2}} \sum_l \sqrt{2l+1} (a^\dagger_{lst} \times a^\dagger_{lst})^{(010)}_{0\mu 0}, \qquad S^{10}_{-,\mu} = (S^{10}_{+,\mu})^\dagger. \tag{7}$$

The index μ is the spin projection in this case and distinguishes different spatial orientations of the $S = 1$ pair. The pairing hamiltonian (6) now involves two parameters g_0 and g_1, the strengths of the spin-singlet and spin-triplet interaction, respectively. While in the previous cases the single strength parameter g_0 just defines an overall scale, this is no longer so for SO(8). Superconducting solutions with an intrinsically different structure are obtained for different ratios g_0/g_1.

In general, the eigenproblem associated with the interaction (7) can only be solved numerically which, given the typical size of a nuclear shell-model space, can be a formidable task. However, for specific choices of g_0 and g_1, namely, $g_0 = 0$,

$g_1 = 0$ and $g_0 = g_1$, the solution of $V_{SO(8)}$ can be obtained analytically [11]. Since a realistic shell-model hamiltonian has values $g_0 \approx g_1$ [12], the solution obtained for $g_0 = g_1$ should be the correct starting point for nuclei. Its quartet properties are discussed in the next section.

3 Quartetting

A special kind of SO(8) solution occurs for the ground state of $N = Z$ nuclei. For example, in the limit SO(8) which corresponds to equal $T = 0$ and $T = 1$ pairing strengths, the exact ground-state solution can be written as [13]

$$\left(S_+^{10} \cdot S_+^{10} - S_+^{01} \cdot S_+^{01}\right)^{n/4} |o\rangle. \tag{8}$$

This shows that the superconducting solution acquires a quartet structure in the sense that it reduces to a condensate of bosons which each correspond to four nucleons. Since the boson in (8) is a scalar in spin and isospin, it can be thought of as an α particle; its orbital character, however, might be different from that of an actual α particle. A quartet structure is also present in the two other limits of the SO(8) model [with SO(5) symmetry], which yields a ground-state wave function of the type (8) with either the first or the second term suppressed. Thus, a reasonable *ansatz* for the $N = Z$ ground-state wave function of the SO(8) pairing interaction (6) with arbitrary strengths g_0 and g_1 is

$$\left(\cos\theta\, S_+^{10} \cdot S_+^{10} - \sin\theta\, S_+^{01} \cdot S_+^{01}\right)^{n/4} |o\rangle, \tag{9}$$

where θ is a parameter which depends on the ratio g_0/g_1.

The justification for the use of (9) as a trial state is that, for the specific values $\theta = 0, \pi/4, \pi/2$, it gives the *exact* ground-state wave function in the three limits of the SO(8) model. Outside these limits the parameter θ must be determined either by minimising the expectation value of $V_{SO(8)}$ in the trial state [13] or by maximising the overlap with the exact wave function [14]. Both procedures lead to essentially identical values for θ and to a quartet trial state which is very close to the exact wave function (deviations of only a fraction of a percent [14]). Similar results are obtained for excited states.

In summary, the α-like condensate (9) provides an excellent approximation to the $N = Z$ ground state of the pairing hamiltonian (6) for any combination of g_0 and g_1. The important (and as yet unanswered) question is now: To what extent does this quartet structure survive other terms that are present in a realistic shell-model hamiltonian, in particular, possible single-particle splittings? With that question in mind, we now turn to the discussion of superconductivity in the presence of a single-particle mean field.

4 A boson mapping of the SO(8) model

The model hamiltonian (6) has SO(8) as a dynamical algebra in the sense that for any choice of g_0 and g_1 the pairing interaction can be expressed in terms of generators of an SO(8) algebra [11]. Furthermore, it can be shown [15] that the fermion hamiltonian (6) can be mapped onto a boson hamiltonian with a U(6) algebraic

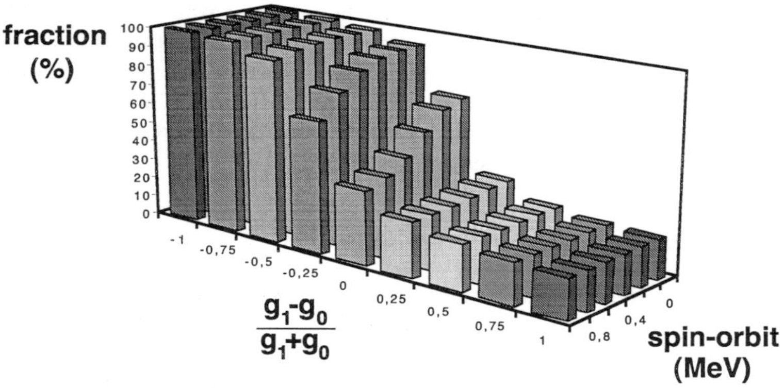

Figure 2. Fraction of p-boson states in the lowest $J = 1$, $T = 0$ state of an odd–odd system (5 neutrons and 5 protons in the pf shell) as a function of the spin–orbit strength and the relative intensity of the two pairing modes.

structure which arises from bosons with an orbital angular momentum $l = 0$ and an intrinsic spin s and an isospin t which can be either $(s,t) = (1,0)$ or $(s,t) = (0,1)$. Because the hamiltonian (6) can be expressed in terms of collective pairs that are algebraically closed, an exact reproduction of the eigenspectrum can be obtained in the boson space by use of a Dyson expansion.

Because the eigenvalue problem associated with the hamiltonian (6) is directly solvable in the fermion space, the boson mapping must be viewed as just an alternative procedure to obtain the same solution. However, if the fermion hamiltonian is less schematic, the boson mapping becomes more useful. A nice illustration was given by Juillet and Josse [16] who considered the extension

$$H = \sum_{nlj} \epsilon_{nlj} \hat{n}_{nlj} + V_{\mathrm{SO}(8)}, \tag{10}$$

thus adding single-particle energies to the original pairing hamiltonian. In particular, the effect of the spin–orbit term $-V_{\mathrm{so}}\bar{l} \cdot \bar{s}$ can be studied by choosing the single-particle energies $\epsilon_{nlj} = -\frac{1}{2}V_{\mathrm{so}}[j(j+1) - l(l+1) - \frac{3}{4}]$. The solution of the eigenvalue problem associated with the hamiltonian (10) now invokes the entire valence shell-model space which may require considerable numerical effort.

To see the influence of the spin–orbit splitting on the pair structure of $N = Z$ nuclei, the boson realisation of the hamiltonian (10) was studied for the pf shell in Ref. [16] where all the details of the calculation can be found. The results obtained are illustrated in Fig. 2 which shows the pair structure of the lowest $J = 1$, $T = 0$ state of an odd–odd system (5 neutrons and 5 protons in the pf shell) as a function of the spin–orbit strength and the relative intensity of the two pairing modes. If the pairing interaction is of isoscalar character ($g_1 = 0$), the fraction of isoscalar ('p') bosons is saturated (100%), independent of the strength of the spin–orbit splitting. Likewise, if the pairing interaction has an isovector character ($g_0 = 0$), this fraction is minimal (20%) and again independent of V_{so}. The spin–orbit splitting plays, however, a crucial rôle when the two strengths are comparable in size ($g_0 \approx g_1$) and it is seen from Fig. 2 that it consistently disfavours the formation of isoscalar $T = 0$ bosons. Since realistic nuclear interactions have isoscalar and isovector components of roughly equal size, the conclusion of this schematic model is that the spin–orbit term in the nuclear mean field hinders the formation of an isoscalar component of a superconducting nuclear condensate.

5 A mapping from the shell model to IBM-4

The IBM-4 [17] is the most elaborate version of the interacting boson model (IBM) of Arima and Iachello [18]. The bosons of IBM-4 are assigned an orbital angular momentum l which can be either $l = 0$ (s) or $l = 2$ (d). In addition, they are labelled by an intrinsic spin s and an isospin t, for which the combinations $(s,t) = (0,1)$ and $(1,0)$ are retained. Just as was the case in the schematic model of Sect. 4, the bosons represent correlated fermion pairs. The choice of bosons is dictated by the requirement that they should describe nuclear excitations at low energy.

As regards applications, the set of bosons of IBM-4 is particularly well adapted to deal with even–even and odd–odd $N \sim Z$ nuclei. This was demonstrated for nuclei in the *sd* shell [19,20]. The mass region of primary interest here is the first half of the 28–50 shell. Shell-model calculations in the full $pf_{5/2}g_{9/2}$ space are feasible for nuclei just beyond $^{56}_{28}\mathrm{Ni}_{28}$ but they become increasingly difficult for the heavier isotopes. One enters a mass region where the full shell-model calculations are, if not impossible, at least arduous and where the IBM-4 offers a viable alternative.

A crucial point in establishing a mapping between the shell model and the boson model is that the boson quantum numbers l and s do *not necessarily* correspond to the total angular momentum L and the total spin S of the corresponding fermion pairs. In fact, except in the very light nuclei, L and S are badly broken by the strong spin–orbit term in the nuclear mean field. Nuclei beyond $^{56}_{28}\mathrm{Ni}_{28}$, however, can be classified in terms of a pseudo-SU(4) symmetry [21].

The existence of this approximate shell-model symmetry and the possibility to define an SU(4) boson classification then lead to a mapping procedure which is described in detail in Ref. [22]. The aspect of this mapping to be emphasised is the important rôle played by symmetries: they provide quantum numbers through which both spaces are related. This correspondence is established for the two-particle system $A = 58$ ($N = 1$ boson) and for the four-particle system $A = 60$ ($N = 2$ bosons) after which the IBM-4 hamiltonian is fixed and can be used to make predictions for systems with boson number $N > 2$. Satisfactory results are obtained in this way but the density of levels generally is somewhat low as compared with that in the shell model. This is a truncation effect and can, in principle, be remedied by a renormalisation of the hamiltonian due to non-bosonised pair degrees of freedom. Instead of a fully microscopic renormalisation, which in the case of IBM-4 is exceedingly difficult, a simple scaling of the entire boson hamiltonian is adopted here which reproduces the 0^+–2^+ splitting (as obtained in the shell model) in the two-boson nucleus ^{60}Zn.

The first test of the IBM-4 hamiltonian thus derived is the three-boson nucleus ^{62}Ga. Figure 3 shows the experimental levels [23] together with the shell-model [23]

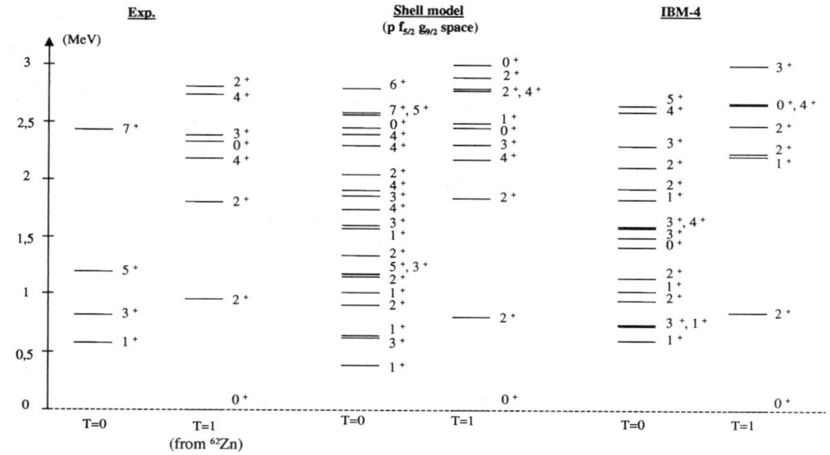

Figure 3. Experimental, shell-model and IBM-4 spectra of ^{62}Ga.

and the IBM-4 results. Both shell model and IBM-4 predict a 0^+ $(T=1)$ ground state and a 1^+ $(T=0)$ first-excited state. Note that this represents an inversion with respect to the order in ^{58}Cu which agrees with the data. Given that no free parameter is introduced in the IBM-4 calculation, the agreement for the $T=0$ levels between shell model and IBM-4 can be called remarkable and a near one-to-one correspondence between levels can be established, the exceptions being higher-spin (5^+ to 7^+) shell-model states which are absent from the IBM-4 because it does not include high-spin $T=0$ bosons. Note also a low-lying 0^+ state in the IBM-4 calculation which, since the shell-model counterpart is much higher in energy, might have an important spurious component.

6 Summary and outlook

The results presented in this contribution illustrate the usefulness and power of the algebraic approach. One typically starts from a simplified many-body hamiltonian which is analytically solvable (at least in a few limits) and which contains the essential degrees of freedom of the problem under consideration. In the present example the starting point is the SO(8) model hamiltonian which is solvable in three limits, contains the correct degrees of freedom in spin and isospin but is oversimplified as regards the orbital degrees of freedom. This model hamiltonian is sufficiently simple to illustrate some important features of pairing in $N \sim Z$ nuclei such as the hindrance of the formation of an isoscalar component of a nuclear superconducting nuclear condensate as a result of the spin–orbit coupling. However, the SO(8) hamiltonian is too simple for an adequate description of the difficult spectroscopy of nuclei at the $N \sim Z$ line for which an extension to the full IBM-4 space (in terms of orbital s and d bosons) is required. The emphasis of this application of the IBM-4 has been its microscopic justification in terms of the nuclear shell model. Through the use of an approximate pseudo-SU(4) symmetry, a mapping procedure from the fermion to the boson space can be established giving rise to a parameter-free IBM-4 hamiltonian. This hamiltonian, applied to nuclei in the 56 to 70 mass region, gives results that compare well with those of the shell model. An element that is still lacking in this study is the development of some hamiltonian phenomenology which did prove crucial in earlier versions of IBM. Now that a microscopic IBM-4 hamiltonian is available, its structure must be investigated to see what are its important components.

Acknowledgements

It is a pleasure to present this contribution at the occasion of the sixtieth birthday of Franco Iachello who has profoundly influenced my scientific thinking as should be clear from the work reported here. This work has been done in collaboration with Olivier Juillet, Stuart Pittel and Dave Warner.

References

1. E.P. Wigner, *Phys. Rev.* **51**, 106 (1937).
2. G. Racah, *Phys. Rev.* **63**, 367 (1943).
3. J.P. Elliott, *Proc. Roy. Soc. (London) A* **245**, 128 & 562 (1958).
4. A. Arima and F. Iachello, *Phys. Rev. Lett.* **35**, 1069 (1975).
5. J.P. Draayer in *Algebraic Approaches to Nuclear Structure. Interacting Boson and Fermion Models* (Harwood, New York, 1993) p 423.
6. P. Van Isacker in *Computational and Group Theoretical Methods in Nuclear Physics*, to be published.
7. A.K. Kerman, *Ann. Phys.* (N.Y.) **12**, 300 (1961).
8. K.T. Hecht, *Nucl. Phys. A* **102**, 11 (1967).
9. K.T. Hecht, *Nucl. Phys. A* **493**, 29 (1989).
10. B.H. Flowers, *Proc. R. Soc. A* **212**, 248 (1952).
11. B.H. Flowers and S. Szpikowski, *Proc. Phys. Soc.* **84**, 673 (1964).
12. M. Dufour and A.P. Zuker, *Phys. Rev. C* **54**, 1641 (1996).
13. J. Dobes and S. Pittel, *Phys. Rev. C* **57**, 688 (1998).
14. Yu.V. Palchikov, J. Dobes and R.V. Jolos, *Phys. Rev. C* **63**, 034320 (2001).
15. P. Van Isacker, J. Dukelsky, S. Pittel and O. Juillet, *J. Phys. G* **24**, 1261 (1998).
16. O. Juillet and S. Josse, *Eur. Phys. J. A* **8**, 291 (2001).
17. J.P. Elliott and J.A. Evans, *Phys. Lett. B* **101**, 216 (1981).
18. F. Iachello and A. Arima, *The Interacting Boson Model* (Cambridge University Press, Cambridge, 1987).
19. P. Halse, J.P. Elliott and J.A. Evans, *Nucl. Phys. A* **417**, 301 (1984).
20. P. Halse, *Nucl. Phys. A* **445**, 93 (1985).
21. P. Van Isacker, O. Juillet and F. Nowacki, *Phys. Rev. Lett.* **82**, 2060 (1999).
22. O. Juillet, P. Van Isacker and D.D. Warner, *Phys. Rev. C* **63**, 054312 (2001).
23. S.M. Vincent *et al.*, *Phys. Lett. B* **437**, 264 (1998).

CHALLENGES FROM SYMMETRY ON THE DRIP LINES

D. D. WARNER
CCLRC Daresbury Laboratory, Daresbury, Warrington, WA44AD, UK
E-mail: d.warner@dl.ac.uk

The crucial role played by symmetries in suggesting and defining phenomena that can be expected near the neutron and proton drip lines is demonstrated. Possible experimental signatures are discussed and new results for Coulomb energy differences in the A=53 mirror nuclei are presented.

1 Introduction

With the rapid development of radioactive beam facilities, the frontiers of nuclear structure research are pushing out towards the natural limits of the nuclear chart. Two of these limits are defined by the neutron and proton drip lines and the aim of this contribution is to demonstrate how symmetries in nuclei play a crucial role in suggesting and defining some of the phenomena which can be expected as these extremes of the nuclear chart are approached. On the proton-rich side, it is the neutron-proton exchange symmetry represented by the isospin quantum number which is relevant and which is manifest through the phenomena of neutron-proton pairing and mirror nuclei. On the neutron-rich side, the neutron drip line is unknown for all but the lightest nuclei. Nevertheless, signatures of its approach in heavier systems may be given by the build up of a neutron skin. Again, the symmetry of interest involves neutron-proton exchange but in this case between the collective degrees of freedom represented by the bosons and the associated F-spin quantum number. Dynamical symmetries can then give a unique insight into the new collective modes available.

2 The Neutron Drip Line

In recent years, experiments with radioactive beams have revealed [1] the presence of a neutron halo in several of the lightest nuclei on the neutron drip line. This is now understood to arise when the last one or two neutrons are in low angular momentum orbits very near the top of the well so that their wave functions have a very extended distribution which is manifest empirically in an anomalously large matter radius. There is, however, a distinctly different phenomenon which is predicted in Hartree–Fock calculations [2,3] to occur in heavier nuclei, in which an excess of several neutrons builds up so that the neutron density actually extends out significantly further than that of the protons, resulting in a mantle of dominantly neutron matter. This situation is illustrated in fig. 1. The presence of a neutron "skin" may then give rise to new collective modes of nuclear excitation, such as a "soft" dipole mode [4] in which the core nucleons move against the more weakly bound skin neutrons.

In the normal application of the IBM to nuclei nearer to stability, the nucleon pairs can be described microscopically in terms of the known shell structure. In the problem addressed here, the shell structure at the extremes of stability is not known and no such link can be established quantitatively at this stage. Nevertheless, the

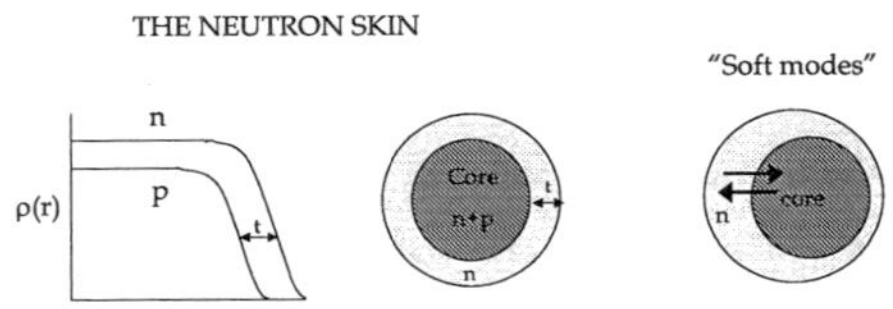

Figure 1. Schematic illustration of the neutron skin and soft dipole mode.

algebra of U(6) can still be used [5] to provide an elegant and simple way of looking at the quadrupole degrees of freedom available to the three component system which results from the development of a neutron skin sufficient in extent to be at least partially decoupled from the core nucleons.

The starting point for the quadrupole modes of a nucleus with a neutron skin is taken as a triple product involving an additional algebra $\mathrm{U}_{\nu_s}(6)$ with the core neutrons being described by $\mathrm{U}_{\nu_c}(6)$. The dynamical algebra is then

$$\begin{array}{ccccc} \mathrm{U}_\pi(6) & \otimes & \mathrm{U}_{\nu_c}(6) & \otimes & \mathrm{U}_{\nu_s}(6) \\ \downarrow & & \downarrow & & \downarrow \\ [N_\pi] & & [N_{\nu_c}] & & [N_{\nu_s}]. \end{array} , \tag{1}$$

Each U(6) algebra is characterised by a number of bosons coupled symmetrically to $[N_i]$.

The fact that the skin neutrons are assumed to interact weakly with the core neutrons and protons, which interact strongly with each other, is represented in the reduction of (1) by coupling the corresponding U(6) algebra of the neutron skin *after* those describing the core nucleons. The reduction thus proceeds as

$$\mathrm{U}_\pi(6) \otimes \mathrm{U}_{\nu_c}(6) \otimes \mathrm{U}_{\nu_s}(6) \supset \mathrm{U}_{\pi\nu_c}(6) \otimes \mathrm{U}_{\nu_s}(6) \supset \mathrm{U}_{\pi\nu_c\nu_s}(6). \tag{2}$$

The triple-sum algebra $\mathrm{U}_{\pi\nu_c\nu_s}(6)$ has a subalgebra structure familiar from IBM-1, and, specifically, the three usual limits [6], U(5), SU(3) and O(6), can be obtained as subchains of (2). However, the results derived below are not only valid in the dynamical symmetries but also for intermediate situations.

It is the coupled nature of the algebra $\mathrm{U}_{\pi\nu}(6)$ in IBM-2 that permits states with mixed symmetry [7]. In the reduction (2), $\mathrm{U}_{\pi\nu_c}(6)$ is characterised by irreducible representations $[N_c - f, f]$ where N_c is the number of nucleon pairs in the core, $N_c = N_\pi + N_{\nu_c}$. The lowest states are then contained in the representation $[N_c, 0]$, which denotes the totally symmetric coupling. The lowest states of mixed symmetry are in the next representation, $[N_c - 1, 1]$. The triple-sum algebra $\mathrm{U}_{\pi\nu_c\nu_s}(6)$ is characterised by up to three rows, with the lowest couplings arising from $[N_c, 0] \times [N_{\nu_s}]$ being $[N, 0, 0]$ and $[N-1, 1, 0]$, N denoting the total number of bosons. Hence the first non-symmetric representation resulting from the triple-sum algebra describes *symmetric* coupling of the core nucleons and *non-symmetric* coupling of the skin neutrons. However, the non-symmetric representation $[N-1, 1, 0]$ of $\mathrm{U}_{\pi\nu_c\nu_s}(6)$ may also arise from the product $[N_c - 1, 1] \times [N_{\nu_s}]$. In this case, it is the core nucleons which are coupled non-symmetrically. The result is that there are now *two* non-symmetric modes, one representing out-of-phase motion between the neutrons and protons in the core and the other denoting an oscillation between the core and the skin where, in this case, as in the soft dipole mode, the core protons carry the core neutrons with them.

The relative energies of the soft and normal mixed symmetry modes can now be estimated by considering the Hamiltonian

$$\hat{H} = A\,\hat{M}_{\pi\nu_c} + B\,\hat{M}_{\pi\nu_c\nu_s}, \tag{3}$$

where $\hat{M}_{ij}$ is the Majorana operator associated with the Casimir operator of $\mathrm{U}_{ij}(6)$

$$\hat{C}_2[\mathrm{U}_{ij}(6)] = (N_i + N_j)(N_i + N_j + 5) - 2\hat{M}_{ij}. \tag{4}$$

For $\hat{M}_{\pi\nu_c\nu_s}$ the same result holds with $N_i + N_j = N_\pi + N_{\nu_c} + N_{\nu_s}$. The energy of the normal mode is then $A(N_\pi + N_{\nu_c}) + B\,N$ while for the soft mode it is BN.

2.1 Experimental Signatures

The characteristic excitation of mixed-symmetry modes is via magnetic dipole transitions. In the case of SU(3), the mixed symmetry mode corresponds to the scissors excitation, where the data in rare earth nuclei indicate an excitation energy for the normal mode of $E_\mathrm{x} \sim 3$ MeV. The signs of A and B in eq. (3) are determined to be positive by the requirement that the fully symmetric representation of U(6) should lie lowest in energy. Assuming, in addition, that these two constants have roughly equal magnitude, then if the number of neutrons in the skin is small compared to the total, the soft mode should appear at approximately half the energy of the normal one; $E_\mathrm{x} \sim 1.5$ MeV.

In even–even nuclei the IBM-2 prediction for the M1 strength is

$$B(\mathrm{M1}; 0^+_\mathrm{G} \to 1^+_\mathrm{S}) = \frac{3}{4\pi}(g_\pi - g_\nu)^2 f(N) N_\pi N_\nu, \tag{5}$$

where g_π and g_ν are the boson g factors. The function $f(N)$ is known analytically in the three limits of the IBM, and eq. (5) is valid for the scissors state in which all the protons oscillate against all neutrons. We stress that, with an appropriate choice of interpolating function $f(N)$ (e.g., the one of [8]), (5) is also valid for a mixture of the three limits and corresponds to the sum rule M1 strength, which is concentrated in one state in the three symmetry limits but is fragmented over several states in general. Likewise, all results reported below for the soft scissors mode are valid for intermediate situations.

A corresponding expression can be derived for the dipole strength to the soft scissors state by considering the separate contributions to the M1 operator from the core and the skin neutrons,

$$\hat{T}(\mathrm{M1}) = g_\pi \hat{L}_\pi + g_\nu \hat{L}_\nu = g_\pi \hat{L}_\pi + g_\nu \hat{L}_{\nu_c} + g_\nu \hat{L}_{\nu_s}, \tag{6}$$

and this yields

$$B(\mathrm{M1}; 0^+_\mathrm{G} \to 1^+_\mathrm{SS}) = \frac{3}{4\pi}(g_\pi - g_\nu)^2 f(N) \frac{N_\pi^2 N_{\nu_s}}{N_\pi + N_{\nu_c}}. \tag{7}$$

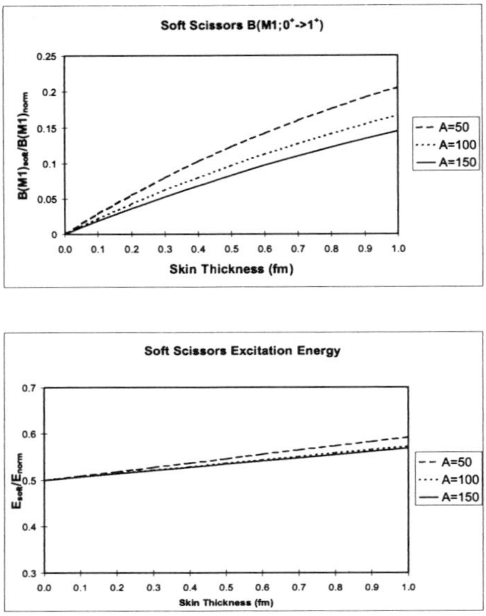

Figure 2. M1 strength (top)and excitation energy (bottom)of soft scissors relative to normal mode versus neutron skin thickness.

From (5) and (7) one finds the following simple result for the ratio of $B(\text{M1})$'s in the soft and normal scissors modes:

$$\frac{B(\text{M1}; 0^+_{\text{G}} \to 1^+_{\text{SS}})}{B(\text{M1}; 0^+_{\text{G}} \to 1^+_{\text{S}})} = \frac{N_\pi N_{\nu_s}}{(N_\pi + N_{\nu_c})N_\nu}. \tag{8}$$

This value is identical to that found previously for the E1 sum rule ratio of the soft and giant dipole resonances [9]; the result depends only on the relative number of constituents in the subsystems, rather than on the specific algebra chosen.

The expressions derived above can now be used to make an estimate of excitation energy and B(M1) strength by assuming simple Fermi distributions and calculating the number of neutrons in skin thickness t. Note that the dependence on the diffuseness is small. The results are shown in fig. 2 where it can be seen that the ratio of the soft to normal M1 strength is in the range of 10-20% as t grows to 1 fm. This would imply an (unfragmented) strength of $\simeq 0.5\mu_N^2$. As anticipated, the corresponding energy ratio stays rather constant around 0.5.

Thus, evidence for enhanced M1 strength at low excitation energies could represent a signature for the onset of a neutron skin and hence the approach of the neutron drip line. The experimental technique most likely to reveal such features would be relativistic Coulomb excitation using radioactive beams from a projectile fragmentation facility. In coulex experiments at barrier energies, the M1 excitation probability is severely hindered relative to the E2 by a factor of $(v/c)^2$. At facilities such as GSI or MSU, however, $v/c \simeq 0.5$ and the M1 cross section becomes

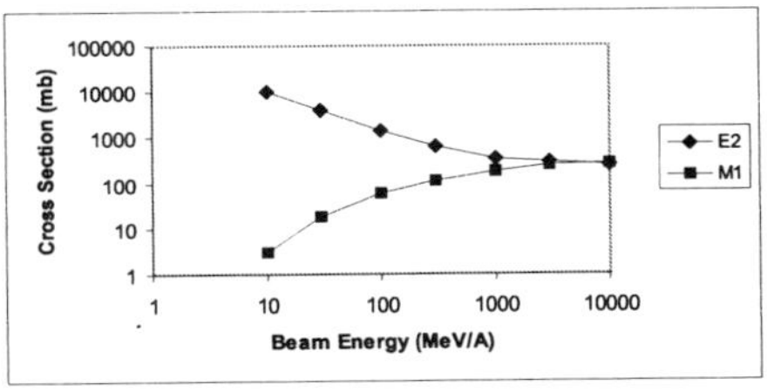

Figure 3. E2 versus M1 Coulomb excitation cross sections at relativistic energies (see text).

comparable to that for E2 excitations, as illustrated in fig. 3. The figure assumes a B(E2) value of 1 e^2b^2 for a 2^+ state at 200 keV and a B(M1) value of 1 μ_N^2 for a state at 3 MeV.

3 The Proton Drip Line

Above ^{56}Ni, the locus of N=Z nuclei rapidly approaches the proton drip line, eventually crossing it just above ^{100}Sn. The medium heavy, N=Z nuclei in this $pfg_{9/2}$ shell are of crucial importance in nuclear structure for two principal reasons: they have neutrons and protons in the same valence shell and the 28-50 shell is large enough for the nuclei to exhibit all aspects of nuclear collective behaviour. These characteristics result in a number of features unique to this class of nuclei, resulting from the role played by the neutron-proton exchange symmetry in the collective structure in this region. Perhaps the most intriguing example is the role played by neutron-proton pairing, in contrast to the "normal" nuclear pairing between like nucleons witnessed throughout the rest of the nuclear chart. The new ingredient that appears on the N=Z line is, of course, the symmetry associated with the isospin quantum number, T, and isospin-invariant versions of the IBM have been introduced which incorporate the isopin triplet by the introduction of a boson with $T = 1$ and $M_{\rm T} = 0$ (IBM-3 [10])and an additional $T = 0$ neutron–proton pair (IBM-4 [11]). In IBM-4 the $T = 1$ bosons are assigned an intrinsic spin $S = 0$ and complemented with a set of $T = 0$, $S = 1$ bosons. The rationale behind this choice is that the two-particle isospin-spin combinations $(TS) = (10)$ and (01) are the ones favoured by Wigner's SU(4) classification which is known to have physical significance in light nuclei. The recognition of dynamical symmetries appropriate to this problem has been shown [12] to reveal some crucial features linking the two descriptions and to provide an ideal tool to study the competition between $T = 0$ and $T = 1$ pairing. Many of these aspects are discussed in the contribution of Van Isacker to these Proceedings.

3.1 *Experimental Signatures*

The most obvious signature of the presence of n-p pairing will be the development of pairing gaps in the $T = 0$ and 1 states in the odd-odd system. Indeed, for the $T = 1$ states, the existence of such a gap follows naturally if the even-even isobaric analogue nucleus is sufficiently collective to exhibit one. For the $T = 0$ states, it is difficult to test this feature using reactions which favour only the yrast states. Nevertheless, the effect of the last neutron and proton forming part of a pair condensate should be quite striking. Recall that in an odd-odd nucleus with no n-p pairing, the rotational bands are characterised by $K = |\Omega_n \pm \Omega_p|$ where Ω_n and Ω_p represent the projection of the single-particle angular momenta of the "unpaired" neutron and proton on the deformation axis. The bands are spaced simply according to the summed single-quasiparticle energies of the possible combinations of orbits with Ω_n, Ω_p. For example, in ^{74}Rb, the Nilsson orbits in the vicinity of the Fermi surface would give rise to a total of 36 rotational band heads of each isospin below 2 MeV.

An additional signature could arise in β-decay. The simple isospin invariant IBM treatment cited earlier can be used to look at Gamow-Teller decay from $T = 0(1^+)$ to $T = 1(0^+)$ states. The results imply a strong enhancement of the GT matrix element with boson number, consistent with the idea that the presence of n-p correlations can lead to a collective coherence in the β-decay. However, these results assume the SU(4) generator for the GT operator; while it is clear that this will no longer be applicable in the $N = Z = 28 - 50$ shell, recent use of the pseudo-spin concept [13] has raised the possibility of a pseudo-spin symmetry in this shell which can still give rise to a degree of enhancement, albeit reduced. In practice, testing of this concept involves study of the decay of the $T_Z = -1$ even-even nucleus to the $T = 0$ odd-odd, since the ground state of the latter is invariably $T = 1$ in the region of interest.

A third identification could come from the study of band crossing phenomena where , if isospin is ignored, the coincidence of neutron and proton Fermi surfaces would give rise to simultaneous neutron and proton alignments. It has been pointed out [14] that inclusion of $T = 0$ pairing can give rise to a delay in the first alignment, and existing data on N=Z nuclei seems to confirm this feature. However, while necessary, this is not a sufficient condition, since such a delay could also be induced by shape changes in any specific case.

4 Mirror Nuclei at High Spin

The charge independence of the nuclear force is one of the fundamental tenets of nuclear structure. Until the last decade studies of the Coulomb energy differences (CED) in mirror partners were focused almost exclusively on the ground states of nuclei. However, in recent years the massive increases in sensitivity and resolving power which have resulted from the advent of large arrays of gamma-ray detectors have provided a possibility to study nuclei with N<Z to ever increasing excitation energy. In particular, in the $f_{7/2}$ shell it has now become possible to extract information on CED between excited states to a relatively high spin and to investigate how the Coulomb energy changes as the spin increases. Such changes are simply caused by changes in the spatial rearrangement of the protons.

A simple example of what is meant by Coulomb energy differences is illustrated in fig. 4. Consider two neutrons and two protons in the $f_{7/2}$ shell and the CED between the T=1 two-particle states which result. These differences are defined

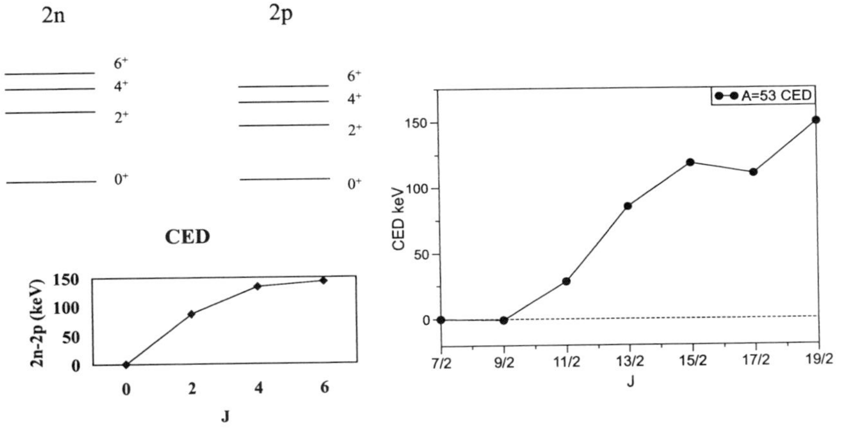

Figure 4. Left: Schematic illustration of CED for 2 nucleons in the $f_{7/2}$ orbital. Right: Measured CED for the A=53 mirror pair.

simply as a difference in excitation energies so that all features associated with the bulk Coulomb energy and with the differences in Coulomb energy of the ground states are eliminated and we are simply left with effects of the order of 10 to 100 keV. In the case illustrated below, the breaking of the nucleon pair to form states with J>0 in each case causes a decrease in the spatial overlap of the two nucleons. In the case of the neutrons, this has no effect on the Coulomb energy but the breaking of the proton pair decreases the Coulomb energy and hence gives rise to a compressed spectrum for the 2p system. A quantitative estimate of the CED for each excited state can be obtained [15] from the harmonic oscillator wave functions for the $f_{7/2}$ orbital and is included in the figure, reflecting the gradual alignment of the two nucleons and the fact that only the protons affect the CED.

4.1 Experimental methods and results

The most recent results have just been obtained for the A=53 mirror system [16]. The reaction used involved a ^{32}S beam impinging on a ^{24}Mg target to produce a ^{56}Ni compound nucleus. The reaction channels of interest were the 2pn and p2n channels leading to the A=53 pair. Since ^{53}Fe was well studied prior to the measurement, the real goal of the experiment was to extract the structure of the T_Z= -1/2 nucleus ^{53}Co . The experiment was performed using the Gammasphere array at the ATLAS facility at Argonne National Laboratory.

The measured CED are plotted on the right in Fig. 4. The effect anticipated on the left of the figure is evident, both nuclei having been studied up to the band terminating spin of $19/2^-$ which corresponds to full alignment of the three $f_{7/2}$ holes. There is thus only one pair of nucleons to break in each mirror partner.

Figure 5 also shows a comparison with a pure $f_{7/2}$ shell model calculation of Kutschera et. al. [17] and a more recent full fp shell calculation using the modified

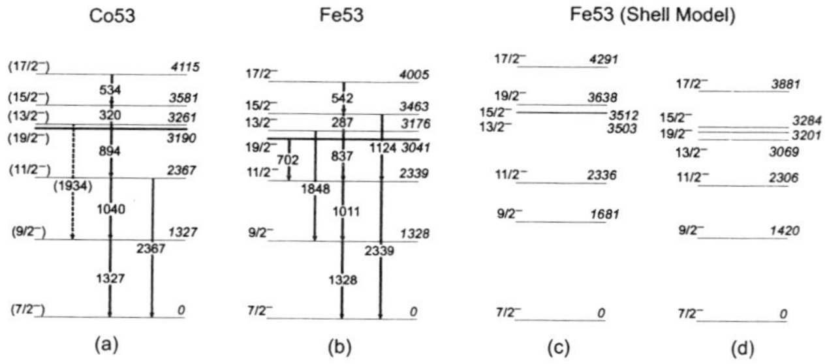

Figure 5. (a) and (b)The level schemes of ^{53}Co and ^{53}Fe from ref [16]. (c) the pure $f_{7/2}$ shell model scheme of ref [17]. (d) shell model scheme from full fp calculations (see text).

KB3 interaction and the model of Caurier et. al. [18] Not surprisingly, the smaller basis calculation does reasonably well given the small number of valence nucleons in this case. Note that neither calculation uses a Coulomb term and therefore neither calculation distinguishes between the mirror partners at this stage.

4.2 Calculating the Coulomb energy differences

The aim now is to use the shell model calculations to reproduce the observed Coulomb energy differences. To do this it is necessary to define a set of two proton Coulomb matrix elements to be used in conjunction with the shell model wave functions to estimate the Coulomb energy contribution for each level. It is valuable to consider alternative methods to extract these matrix elements:

1. Fit the Coulomb matrix elements for J = 0, 2, 4, 6 to the A=53 data using the pure $f^{-3}_{7/2}$ wave functions of Kutschera, Brown and Ogawa (KBO) [17] .

2. Obtain Coulomb matrix elements from A=42 nuclei assuming charge symmetry. i.e. using the A=42 mirror pair [19]. The difference in Coulomb energy between two states can be represented as

$$E_C(J) = BE_J(^{42}Ti) - BE_J(^{42}Ca) + constant$$

   where the constant accounts for the neutron-proton mass difference and the Coulomb interaction with the core.

3. Obtain Coulomb matrix elements from A=42 nuclei assuming charge independence. i.e. using the A=42 isobaric triplet [15]. In this case, the difference in Coulomb energies is taken as

$$E_C(J) = 2BE_J(^{42}Sc) - BE_J(^{42}Ca) - BE_J(^{42}Ti)$$

4. Use values of the Coulomb matrix elements calculated from pure harmonic oscillator wave functions [15].

Method 1 produces excellent agreement with the measured A=53 CEDs [16]. In 2, use of the excitation energies eliminates the constant as long as the interaction between core and valence protons remains constant as a function of J, while in 3 the difference in nucleon masses and the interaction of the valence protons with the core are eliminated on a state by state basis by incorporating the excitation energies of the odd-odd Sc system so that one is simply left with the pure Coulomb interaction between the last two valence protons.

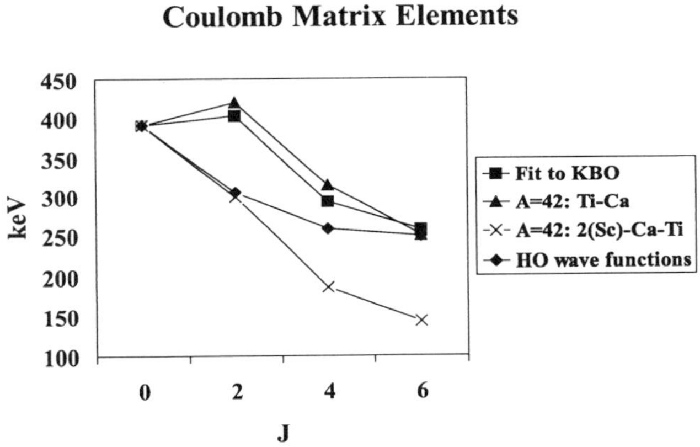

Figure 6. Values of 2-proton Coulomb matrix elements extracted using the four methods discussed in the text.

The four sets of Coulomb matrix elements extracted in this way are shown in fig. 6 and it is immediately evident that the first two produce an anomaly in that the values increase in going from the J=0 to the J=2 state in each case. This is clearly unphysical on a simple intuitive basis, since breaking a proton pair must decrease the Coulomb energy. In contrast, matrix elements extracted from the A=42 isobaric triplet behave as expected, as do the pure harmonic oscillator values. Nevertheless it is the first two methods which have produced results which are in reasonable agreement with experiment and hence a rise in the value of the matrix element for J=2 is clearly needed. In fact, the comparison of the two sets of values from the A=42 nuclei in fig. 6 suggests a possible origin for the anomaly. As mentioned earlier , if the interaction of the valence protons with the core is changing with J, part of it will still be contained in the charge symmetric values of the A=42 Coulomb matrix elements but will have been totally eliminated in the extraction of the charge independent values. In fact, similar fits to the A=47,49 and 51 systems [16,20] reveal a consistent picture arising across the entire shell of the need for an anomalously large J=2 matrix element to describe the mirror pairs.

The source of the anomaly remains unclear. It is probably not surprising that the matrix elements need to be renormalised; it is perhaps more surprising that a single set of values can give such good agreement across the entire $f_{7/2}$ shell.

In a recent study [20], the origin of this anomaly is attributed to an isospin non-conserving contribution to the nuclear interaction for J=2 for mirror pairs (another contribution is added to the J=0 matrix element to describe the triplet systems). Further details of this parameterisation are given in the contribution of Lenzi to these Proceedings.

5 Conclusions

Dynamical and mirror symmetries have been shown to highlight a number of new features which arise from the interplay of new constituents near the nucleon drip lines. There are many other intriguing possibilities , which may stem in particular from the changes in shell structure expected near the neutron drip line. Clearly, more experimental information is needed, much of which must come from the exploitation of radioactive beams. Most importantly, information on *non-yrast* states is required.

Acknowledgements

I wish to thank many colleagues who have contributed to these studies, in particular Mike Bentley, Charles Barton, Piet Van Isacker and Scott Williams.

References

1. I. Tanihata *et al.*, *Phys. Rev. Lett.* **55**, 2676 (1985) ; *Phys. Lett.* B **206**, 592 (1988).
2. N. Fukunishi, T. Otsuka and I. Tanihata, *Phys. Rev.* C **48**, 1648 (1993).
3. J. Dobazcewski, W. Nazarewicz and T.R. Werner, *Z. Phys.* A **354**, 27 (1996).
4. P.G. Hansen and B. Jonson, *Europhys. Lett.* **4**, 409 (1987).
5. D.D. Warner and P. Van Isacker, *Phys. Lett.* B **395**, 145 (1997).
6. F. Iachello and A. Arima, *The Interacting Boson Model* (Cambridge University Press, Cambridge, 1987).
7. P. Van Isacker *et al.*, *Ann. Phys.* **171**, 253 (1986).
8. P. von Neumann-Cosel *et al.*, *Phys. Rev. Lett.* **75**, 4178 (1995).
9. P. Van Isacker, M. A. Nagarajan and D. D. Warner, *Phys. Rev.* C **45**, R13 (1992).
10. J. P. Elliott and A. P. White, *Phys. Lett.* B **97**, 169 (1980).
11. J. P. Elliott and J. A. Evans, *Phys. Lett.* B **101**, 216 (1981).
12. P. Van Isacker and D. D. Warner , *Phys. Rev. Lett.* **78**, 3266 (1997).
13. P. Van Isacker, O. Juillet and F. Nowacki, *Phys. Rev. Lett.* **82**, 2060 (1999).
14. N. S. Kelsall *et al.*, *Phys. Rev.* C **64**, 24309 (2001).
15. J. A. Sheikh , D. D. Warner and P. Van Isacker , *Phys. Lett.* B **443**, 16 (1998)
16. S. J. Williams *et al.*, *in press*
17. W. Kutschera, B. A. Brown and K. Ogawa, *Riv. Nuovo Cimento* **31**, 1 (1978).
18. E. Caurier *et al.*, *Phys. Rev.* C **50**, 225 (1994).
19. A. Poves *et al.*, *Nucl. Phys.* A **694**, 157 (2001).
20. A. P. Zuker *et al.*, *Phys. Rev. Lett.* **89**, 142502 (2002).

HIGH ACCURACY ATOMIC MASS MEASUREMENTS. APPLICATION TO NEUTRON-RICH ZIRCONIUM ISOTOPES

JUHA ÄYSTÖ, V. KOLHINEN, S. RINTA-ANTILA, S. KOPECKY, J. HAKALA, J. HUIKARI, A. JOKINEN , A. NIEMINEN AND J. SZERYPO*[1]

Department of Physics, PB 35, FIN-40014 University of Jyväskylä, Finland
E-mail: juha.aysto@phys.jyu.fi

The relevance of precision measurements on nuclear ground states is discussed in the frame of current experimental developments. Cooling and trapping techniques of low-energy radioactive ion beams are shortly presented with emphasis on high-precision measurements on the ground state properties of exotic nuclei. The impact of the new generation Penning traps on mass measurements of short-lived nuclei is discussed. Examples of recent precision measurements of masses of very neutron-rich isotopes of zirconium and rhodium are given. Possible impacts of nuclear structure calculations such as the Interacting Boson Model on understanding the fine structure of the mass surface are discussed.

1 Physics of accurate mass measurements

The mass of the ground state of a nucleus results from a high order of symmetry of a complex quantum system. Accurate measurements of the nuclear mass surface can reveal additional knowledge, in addition to that obtained from excited states, on the underlying symmetries and microscopic features of nucleon systems that are derived from charge symmetry of nuclear interaction, shell effects, coexisting structures, pairing effects, spin-orbit interaction and so forth [1,2]. For this to be successful, the measurements and the theory have to be able to probe the deviations or fluctuations that are in the order of 1 to 100 keV. Global correlations are typically variations due to closed shells and broad areas of deformation where the required accuracies in mass measurements are typically of the order of 100 keV. However, detection of local correlations such as those due to the presence of closed shell discontinuities, the local zones of deformation or those due to configuration mixing or shape mixing require typically mass accuracies of the order of 10 keV.

A specific class of masses are those linked to each other by the charge symmetry of nuclear interaction. Isospin multiplets and Coulomb energy differences between mirror nuclei have been studied among light nuclei and to some extent in medium heavy nuclei with great accuracy. Typical accuracies required in such measurements are in the range of 1 keV. A

---

[1] Present address: Ludwig-Maximilian University, Munich, Germany

beautiful example of such a measurement is a recent study of the masses of the A=33 isospin quartet at ISOLTRAP [3].

The super-allowed $0^+ \rightarrow 0^+$ β-decay between isobaric analog states can be used as a test for the Conserved Vector Current hypothesis and for obtaining precise values for the up-down quark mixing matrix element of the Cabibbo-Kobayashi-Maskawa matrix. In these studies super-allowed beta decay strengths have to be measured with ultimate precision of close to 10^{-4}. This involves the measurement of the beta-decay Q-value with an accuracy better than 1 keV [4].

In nuclear astrophysics the masses of exotic, often short-lived, nuclei are crucially important in network calculations. In the rp-process calculations the binding energies are needed to an accuracy of 10 keV or better, similarly for r-process calculations a slightly less stringent requirement is presented.

In this article we describe briefly the status of nuclear mass measurements of radioactive ions employing ion trap technologies. A special emphasis is on the JYFLTRAP setup which is located at the IGISOL facility of the cyclotron laboratory of the University of Jyväskylä in Finland.

2 Ion traps in mass measurements of radioactive isotopes

The basic principle of mass determination by a Penning trap is based on the determination of the cyclotron frequency of an ion in a strong magnetic field. In the presence of an axially symmetric quadrupole field ions perform characteristic motion consisting of three independent components which are an axial motion with a frequency f_z and two radial motions, a slow magnetron motion (f_-) and a fast reduced cyclotron motion (f_+). The overall radial motion is a superposition of the two radial motions and is connected to the magnetic field via the relation,

$$f_c = f_- + f_+ = q/m \cdot B/2\pi$$

where f_c is the cyclotron frequency of an ion with a charge-to-mass ratio of q/m oscillating in the external magnetic field B. The field is determined by the aid of ions with well-known masses, preferably with the ^{12}C ion or its cluster constituents [5]. The resolving power of a Penning trap as a mass spectrometer obviously depends on how accurately the magnetic field and the cyclotron frequency can be determined. In general, the mass resolving power $R = m/\Delta m = \Delta f_c/f_c \approx f_c \cdot T_{obs}$, where T_{obs} is the time of observation. In typical conditions the mass resolving power of about 10^6 is obtained for ions with A/q=100 stored over 1 second in a few Tesla magnetic field. This allows the measurement of unknown masses with precision better than 10^{-7}, i.e. 10 keV for A=100 ions.

Ion trap technology was introduced to atomic mass measurements of radioactive isotopes at ISOLDE, CERN [6]. In the early days of these experiments, the fast and efficient injection into the trap was a serious limitation for using traps for exotic short-lived isotopes. This obstacle was recently solved by the introduction of a fast injection scheme based on the gas-filled linear Paul trap concept, see ref. [7] for further reading.

3 JYFLTRAP – Universal approach to mass measurements

The JYFLTRAP setup, shown in Fig. 1, is located at one of the beam lines of the IGISOL facility. At IGISOL - Ion Guide Isotope Separator Online - ions are produced in a thin target and then recoil out of this target. They are guided as single charged ions by helium flow and electric fields through stages of differential pumping and finally accelerated to 40 keV. After

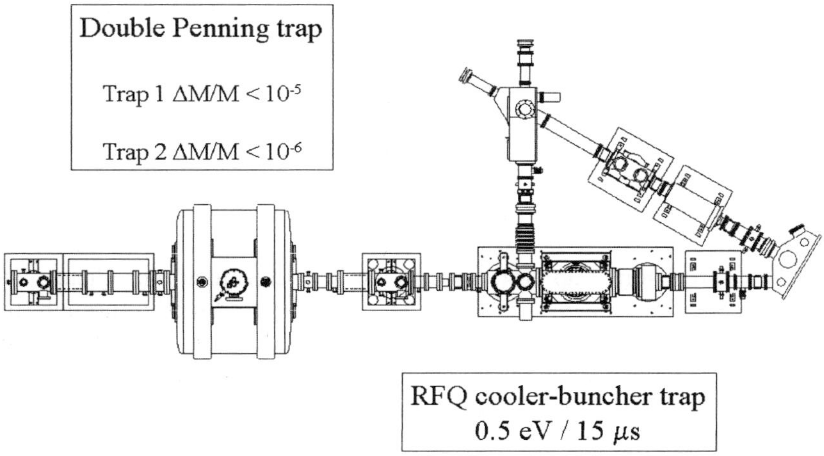

Fig. 1. JYFLTRAP setup.

acceleration the beam is mass selected by a dipole magnet, allowing for separation of nuclei with different mass numbers. The IGISOL technique allows access to exotic short-lived isotopes of all chemical elements [8] and is therefore an ideal supplier of radioactive ions for a trap system. As the energy spread of the IGISOL beam is rather large, up to 50-100 eV, the beam has to be cooled before it can be manipulated further. The cooling is accomplished in a buffer-gas filled RFQ [9]. This allows not only cooling but also bunching of ions leading to extraction of ion pulses with a duration of 15 μs and an energy spread of the order of 1 eV. Then the ions are injected into a double Penning trap system for isobaric beam purification and precise mass determination [10]. Basic components of the Jyväskylä Penning trap system are: the cooling and cleaning Penning trap (CT), with a mass resolving power $R=10^5$ having an ability to perform an isobaric purification, and the precision Penning trap (PT), with a mass resolving power $R>10^6$, serving as an instrument for the very precise (10^{-7} precision) mass measurements of radioactive nuclides

A superconducting magnet is used to create the magnetic field for the Penning trap. This 7.0 T actively screened magnet system is made by Magnex Scientific Ltd. and it has a 160 mm diameter warm bore. The magnet has two 1 cm^3 homogenic magnetic field regions along the beam axis 10 cm towards both direction from the magnet centre. In these areas the homogeneity of the magnetic field is below 10^{-6} and 10^{-7}, respectively.

Ion bunches from the RFQ Cooler are sent at low energy into the purification trap where they are captured by time-varying electric potential, thermalized and cooled in a buffer gas in the centre of the potential. Then they are excited by applying successive dipole and quadrupole RF fields which leads to mass-selective cooling and centring according to $\omega_c = qB/m$. Now the ejection through a small diaphragm allows only those ions close to the trap axis to be extracted. These ions are then transferred to the adjacent precision trap for precise mass determination. Clean and monoisotopic ion beam can also be transported directly through the precision trap to an MCP detector for ion detection. In fact, the results reported here have been obtained employing the purification trap only. The precision trap at the JYFLTRAP system will be operational only in the fall of 2003.

Two parameters are crucial when judging the performance of the purification trap: the mass resolving power and the overall transmission. Currently, for the A= 100 ions the overall transmission of the order of 20 % and mass resolving power close to 10^5 have been routinely obtained. The first radioactive ions studied were produceded in $^{58}Ni(p,n)^{58}Cu$

reactions. This reaction provided us with favourable conditions for the first test. The beam consisted mostly of sputtered ^{58}Ni ions and radioactive ^{58}Cu ions – one of them stable the other one with a half-life of 3.2s. The mass difference of these two isobars is approximately 8.5 MeV. In the first experiment a mass resolving power of approximately 45.000 was reached as shown in Fig. 2. Later on the resolving power could be increased up to 135.000 allowing to determine the (already well known) mass of ^{58}Cu with an uncertainty of about 30 keV.

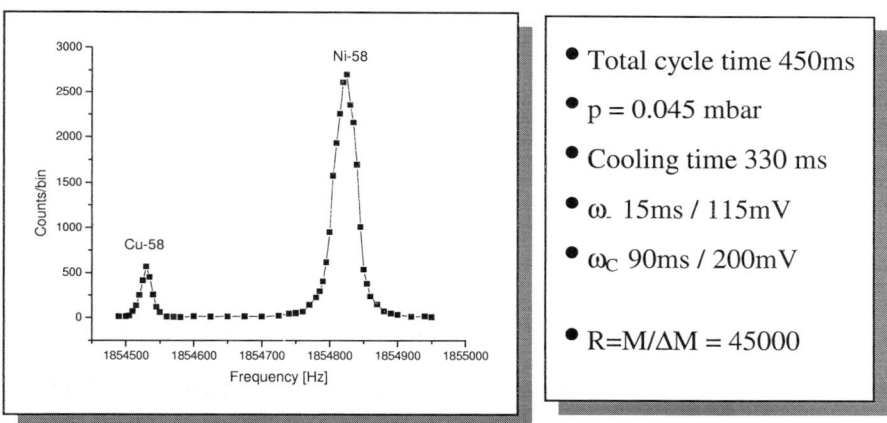

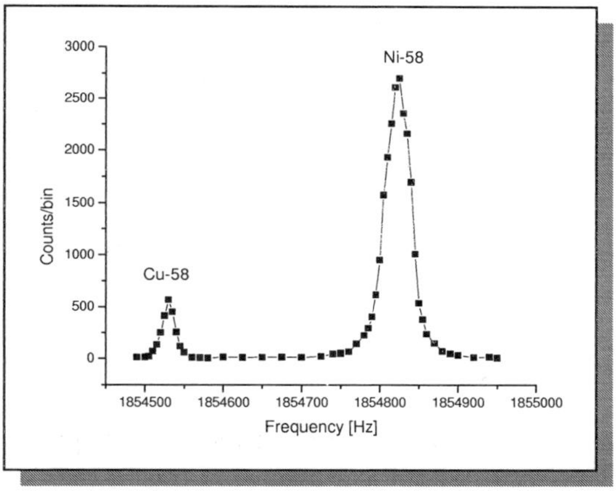

- Total cycle time 450ms
- p = 0.045 mbar
- Cooling time 330 ms
- ω_- 15ms / 115mV
- ω_C 90ms / 200mV
- R=M/ΔM = 45000

Fig. 2. Cyclotron frequency spectrum of A=58 ions produced by IGISOL.

The next milestone was approached when neutron-rich nuclides produced in 30 MeV proton induced fission of ^{238}U were successfully injected into JYFLTRAP. In the first experiments the main goal was to separate ^{112}Rh and ^{112}Ru, which have a mass difference of about 4 MeV. The mass resolving power measured as FWHM was approximately 60.000. This corresponded to the width of the mass peak of about 1.5 MeV. Therefore, it was quite not enough to separate the isomeric state known to exist in ^{112}Rh from its ground state. On the other hand, in cases where only one long-living state, i.e. the ground state, is known to exist the current precision provides a possibility for accurate mass measurement employing the purification trap only. In such cases a typical accuracy of the order of 50 – 100 keV could be obtained for the centroid of the cyclotron resonance curve [11].

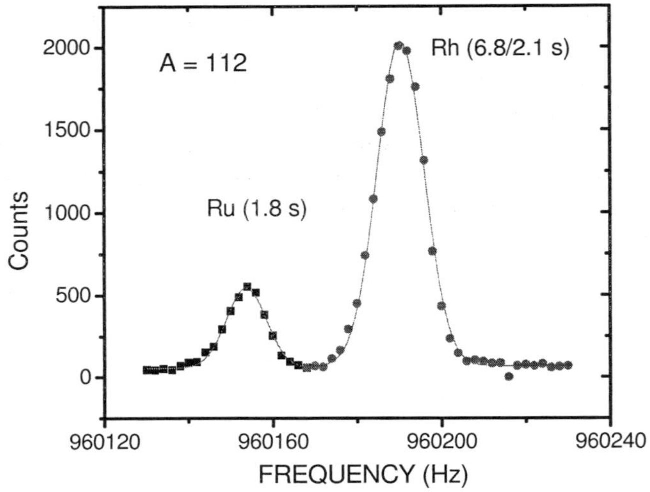

Fig. 3. Separation of adjacent isobars for A=112 fission products.

4 First results on neutron-rich Zr isotopes

Zirconium isotopes are known to possess interesting and rapidly changing nuclear structure features when the neutron number changes from 56 to 60. Extensive spectroscopic studies have been employed to understand this shape transition and its relation to structural dynamic symmetries as well as to underlying microscopic structures, proton-neutron interaction and to the role of neutron-pairing. The experiments so far have investigated these nuclei via beta decays of yttrium parent nuclei [12], via prompt fission fragment gamma-ray coincidence studies [13] as well as via collinear laser spectroscopy experiments [14]. These studies have changed the picture considerably so that the original description of the shape change [15] as a sharp phase transition from a spherical shape at N=59 to a strongly deformed ^{100}Zr, is now rather described as a gradual transition setting in already at N=56. Despite thorough spectroscopic studies of these nuclei their masses are known rather poorly. The binding energies of these neutron-rich Zr isotopes have only been determined if at all by measuring the beta end-point energies of yttrium and Zr isotopes. Often such measurements are in serious systematic error due to the fact that the mass determinations are relying on the long decay chains to the valley of stability and inadequate knowledge of the decay schemes.

Zirconium isotopes are available as primary beams at IGISOL with intensities of the order of 4000 – 200 ions/s when moving from the most copiously produced ^{98}Zr out towards ^{96}Zr or ^{104}Zr. These isotopes are attractive for direct mass measurement already with the purification trap because from decay spectroscopic studies they are known to possess only one long-living nuclear state surviving through the ion manipulation process at JYFLTRAP. This ion manipulation process consisted of a total cycle time of 450 ms within which the atomic Zr ions from IGISOL were first injected for cooling and bunching into the RFQ cooler. After this they were ejected as a 15 μs long bunch into the purification trap where they were axially cooled for 330 ms and subsequently excited into large orbit by applying a magnetron frequency in dipole mode for 15 ms. Then the trap was shined by an RF quadrupole field for 90 ms leading to centring of the orbits of those ions that corresponded to the resonant cyclotron frequency ω_c=qB/m. In fact, during the course of these experiments it was found that Zr ions were transformed into monoxide ions in the cooler and in the trap itself, both filled with helium buffer gas. This allowed a clean production of Zr isotopes and no contaminants with masses similar to ZrO were observed.

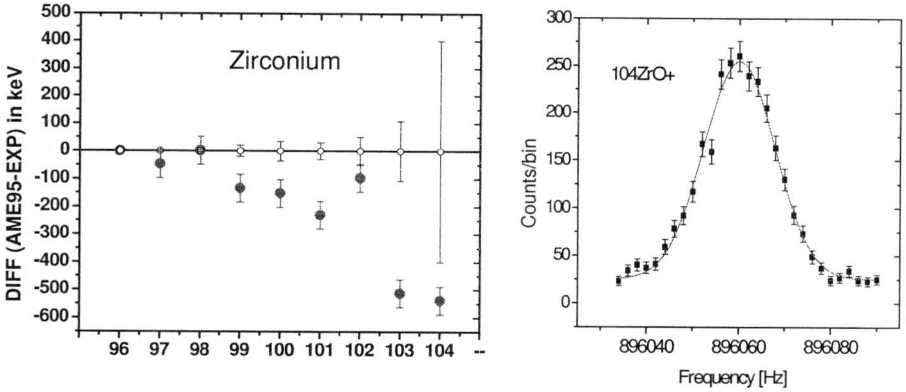

Fig. 4. Preliminary results from the first mass measurement at JYFLTRAP.

Fig. 4 shows a cyclotron frequency spectrum measured for $^{104}ZrO^+$ ions using the measuring sequency described above. Similar measurements were performed for all Zr isotopes between A=96 and 104. The mass calibration was performed by using the well known mass of ^{96}Zr and cross-checked with the mass of ^{58}Ni that was created in the IGISOL gas cell by sputtering of the Ni-foil separating the uranium targets and the stopping gas cell of the ion guide. Preliminary results of all these measurements are shown in Fig. 4. The mass difference in the left figure is taken between the recent Atomic Mass Evaluation and this experiment. A typical uncertainty of these measurements was 50 keV consisting of the fitting error of the resonance curve and the systematic error estimated from the fluctuations studied separately for $^{112}Rh^+$ ions. These preliminary values show that our measurements agree with the values from literature for the well know masses of ^{97}Zr and ^{98}Zr, but show significant differences for the more neutron-rich isotopes – all of which have been previously determined either by beta endpoint measurements or have not been measured at all (^{104}Zr). Therefore already at the present stage we can provide valuable information for some of the neutron rich isotopes.

In order to see possible local structure effects in the binding energy in a way similar to excited states or charge radii we have plotted two neutron separation energy S_{2n} as a function of the neutron number in Fig. 5. Indeed, we see a kink developing at ^{98}Zr and ending at ^{101}Zr. This coincides with the neutron numbers where a strong change in quadrupole deformation is known to occur based on the measurements of charge radii [14]. This is inherently also coinciding with the neutron numbers where different shapes coexist at low energy and consequently can mix substantially. It is clear that the shape transition in this case is very strongly exhibited in the binding of the ground states. Therefore, these nuclei would provide an interesting testing ground for similar theoretical calculation applied for neutron-rich rare earth isotopes above the closed N=82 neutron shell that employed the Interacting Boson Model, see ref. [2].

Since it was also of significant interest to know whether this effect is observed in the nuclear mean field we show in Fig. 5 a recent preliminary calculation of Stoitsov et al. for the same Zr isotopes [16]. This calculation is based on the Hartree-Fock-Bogolyubov approximation using the transformed harmonic oscillator basis. Skyrme force is Sly4 and pairing is mixed volume-surface type. Exact particle-number-projection is performed after finding the Lipkin-Nogami solution. The S_{2n} values obtained correspond to the minimum potential energy solution. In this calculation Zr isotopes have a spherical ground state up to N=56, an oblate ground state for N=58 and prolate shape with $\beta \approx 0.35$ from N=60 on. It is observed that this calculation does observe the kink in the S_{2n} systematics, even without special efforts to fit forces in this specific calculation.

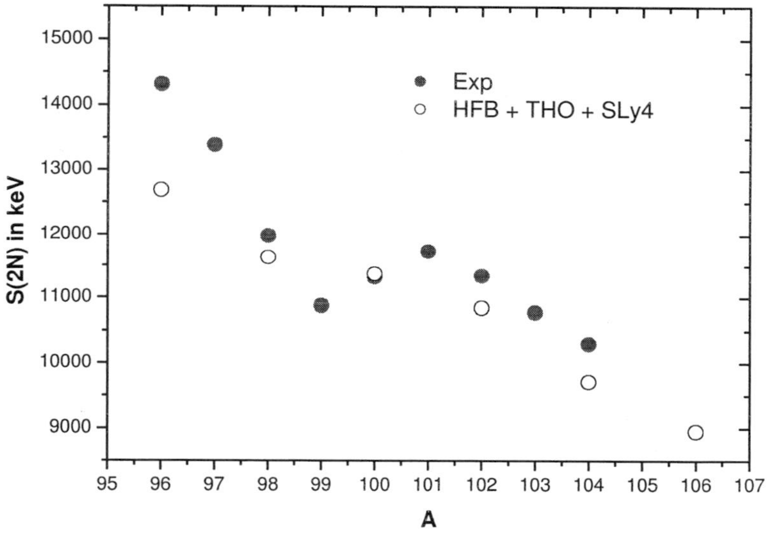

Fig. 5. Two-neutron separation energies of neutron-rich Zr isotopes.

5 Future directions

In summary, it is clear that todays's nuclear physics requests the knowledge of the masses to a precision preferably better than few 10 keV. Highest accuracies are requested by the studies of fundamental symmetries and interactions as well as by nuclear astrophysics. In nuclear structure physics the fine structure of the mass surface is a rich source of information for local structure effects, as discussed in this paper.

Successfully initiated mass measurement program at JYFLTRAP has shown that accurate mass measurements are possible also for short-lived exotic isotopes of refractory elements. At IGISOL these measurements will be continued to cover neighboring isotopes from Sr up to Pd in the near future. Utilization of the precision trap will soon become possible as well and the mass measurements will be done with 10 keV accuracy or even better for nuclei far from stability. These experiments will provide a novel way of studying the connections between the mass surface and nuclear structure far from the valley of beta stability.

6 Acknowledgements

The authors thank K. Heyde and J. Dobaczewski for valuable discussions and contributions. This work has been supported by the European Union within the NIPNET RTD project under the contract no HPRI-CT-2001-50034 and by the Academy of Finland under the Finnish Center of Excellence Programme 2000-2005 (project number 44875).

References

1. Iachello F. and Arima A., *The Interacting Boson Model*, Cambridge University Press, Cambridge, 1987.
2. Fossion R., De Coster C., Garcia-Ramos J.E., Werner T. and Heyde K., *Nucl. Phys. A* **697** (2002) 703
3. Herfurth F. et al., *Phys. Rev. Lett.* **87** (2001) 142501
4. Hardy J. and Towner I., *Phys. Rev. C* **66** (2002) 035501
5. Blaum K. et al., *Eur. Phys. J. A* **15** (2002) 245
6. Bollen G. et al., *Nucl. Instr. Meth. A* **368** (1996) 675
7. Äystö J. and Jokinen A., *J. Phys. B: At. Mol. Opt. Phys.* **36** (2003) 573
8. Äystö J., *Nucl. Phys. A* **693** (2001) 477
9. Nieminen A. et al., *Phys. Rev. Lett.* **88** (2002) 094801
10. Kolhinen V., Proc. EMIS-14 Conf., *Nucl. Instr. Meth. B*, in press.
11. Kolhinen V., *PhD thesis*, to be published.
12. Lhersonneau G. et al., *Phys. Rev. C* **56** (1997) 2445
13. Urban W. et al., *Nucl. Phys. A* **689** (2001) 605
14. CampbellP. et al., *Phys. Rev. Lett.* **89** (2002) 082501
15. Lhersonneau G. et al., *Phys. Rev. C* **49** (1994) 1379
16. Stoitsov M.V.,Dobaczewski J., and Nazarewicz W., to be published; see also Dobaczewski J. et al., *Eur. Phys. J. A* **15** (2002)21

NUCLEAR SUPERSYMMETRY: NEW TESTS AND EXTENSIONS

A. FRANK[1,2], J. BAREA[1] AND R. BIJKER[1]
[1] *ICN-UNAM, AP 70-543, 04510 México, DF, México*
[2] *CCF-UNAM, AP 139-B, 62251 Cuernavaca, Morelos, México*

Extensions of nuclear supersymmetry are discussed, together with a proposal for new, more stringent and precise tests that probe the susy classification and specific two-particle correlations among supersymmetric partners. The combination of these theoretical and experimental studies may play a unifying role in nuclear phenomena.

1 Introduction

Nuclear supersymmetry (n-susy), first proposed by Franco Iachello more than two decades ago [1], is a composite-particle phenomenon linking the properties of bosonic and fermionic systems, framed in the context of the Interacting Boson Model of nuclear structure [2]. Composite particles, such as the α-particle are known to behave as approximate bosons. As He atoms they become superfluid at low temperatures, an under certain conditions can also form Bose-Einstein condensates. At higher densities (or temperatures) the constituent fermions begin to be felt and the Pauli principle sets in. Odd-particle composite systems, on the other hand, behave as approximate fermions, which in the case of the IBFM are treated as a combination of bosons and an (ideal) fermion. In contrast to the theoretical construct of supersymmetric particle physics, where susy is postulated as a generalization of the Lorentz-Poincaré invariance at a fundamental level, n-susy has been subject to experimental verification [3]. Nuclear supersymmetry should not be confused with fundamental susy, which predicts the existence of supersymmetric particles, such as the photino and the selectron, for which no evidence has yet been found. If such particles exist, however, susy must be strongly broken, since large mass differences must exist among superpartners, or otherwise they would have been already detected. Competing susy models give rise to diverse mass predictions and are the basis for current superstring and brane theories [4].

Nuclear supersymmetry, on the other hand, is a theory that establishes precise links among the spectroscopic properties of certain neighboring nuclei. Even-even and odd-odd nuclei are composite bosonic systems, while odd-A nuclei are fermionic. It is in this context that n-susy provides a theoretical framework where bosons and fermions are treated as members of the same supermultiplet [5]. the mass difference between superpartners is thus of the order of $1/A$, but the theory goes much further and treats the excitation spectra and transition intensities of the different systems as arising from a single Hamiltonian and a single set of transition operators. Originally framed as a symmetry among pairs of nuclei [1, 2, 5], it was subsequently extended to nuclear quartets or "magic squares", where odd-odd nuclei could be incorporated in a natural way [6]. Evidence for the existence of n-susy (albeit possibly significantly broken) grew over the years, specially for the quartet of Fig. 1, but only recently more systematic evidence was found. This was achieved by means of one-nucleon transfer reaction experiments leading to the odd-odd nu-

$$
\begin{array}{ccccc}
{}^{195}_{79}\mathrm{Au}_{116} & \underset{b^\dagger_\nu a_\nu}{\overset{a^\dagger_\nu b_\nu}{\rightleftarrows}} & {}^{196}_{79}\mathrm{Au}_{117} & & \\
a^\dagger_\pi b_\pi \uparrow\downarrow b^\dagger_\pi a_\pi & & a^\dagger_\pi b_\pi \uparrow\downarrow b^\dagger_\pi a_\pi & & N = N_b + N_f = 7 \\
{}^{194}_{78}\mathrm{Pt}_{116} & \underset{b^\dagger_\nu a_\nu}{\overset{a^\dagger_\nu b_\nu}{\rightleftarrows}} & {}^{195}_{78}\mathrm{Pt}_{117} & &
\end{array}
$$

Figure 1. The magic quartet of nuclei. The one-nucleon transfer reactions among them are indicated schematically, with $a^\dagger$ corresponding to a fermion and $b^\dagger$ to a boson of the indicated particles.

cleus ^{196}Au, which, together with the other members of the susy quartet (^{194}Pt, ^{195}Au and ^{195}Pt) is considered to be the best example of n-susy in nature [6–9]. We should point out, however, that while these experiments provided the first complete energy classification for ^{196}Au (which was found to be consistent with the theoretical predictions [6–9]), the reactions involved ($^{197}\mathrm{Au}(\vec{d},t)$, $^{197}\mathrm{Au}(p,d)$ and $^{198}\mathrm{Hg}(\vec{d},\alpha)$) did not actually test directly the supersymmetric wave functions, as we shall discuss below. Furthermore, whereas these new measurements are very exciting, the dynamical susy framework is so restrictive that there was little hope that other quartets could be found and used to verify the theory [6–9]. The purpose of this paper is two-fold. On the one hand we report on an ongoing investigation of one- and two-nucleon transfer reactions [10] in the Pt-Au region that will more directly analyze the supersymmetric wave functions and measure new correlations which have not been tested up to now. On the other hand we discuss some ideas put forward several years ago, which question the need for dynamical symmetries in order for n-susy to exist [11]. We thus propose a more general theoretical framework for nuclear supersymmetry. The combination of such a generalized form of supersymmetry and the transfer experiments now being carried out [12], could provide remarkable new correlations and a unifying theme in nuclear structure physics.

2 New experiments

The quartet of nuclei of Fig. 1 was classified by means of the $U_\nu(6/12) \times U_\pi(6/4)$ dynamical supersymmetry, obtained by combining the $SO^{BF}(6)$ symmetry limit for the odd neutron (^{195}Pt) and the $Spin(6)$ symmetry limit for the odd proton (^{195}Au) [6]. The excitation spectra of the nuclei ^{194}Pt, ^{195}Au and ^{195}Pt was used to determine the Hamiltonian and subsequently the spectra of the odd-odd partner ^{196}Au was predicted, for which at the time little or no experimental data was available. One should note, however, that the great majority of tests carried out since then for the supersymmetric framework have involved one-nucleon transfer experiments leading to the nuclei in figure 1 through reactions coming from outside the quartet, such as $^{197}\mathrm{Au}(\vec{d},t)^{196}\mathrm{Au}$ and $^{196}\mathrm{Pt}(\vec{d},t)^{195}\mathrm{Pt}$ that, in first approximation, are formulated using a transfer operator of the form $a^\dagger_\nu$. The latter reactions are useful to measure energies, angular momenta and parity of the residual nucleus and in principle provide information about the systems wave functions. However, they cannot test correlations present in the quartet's wave functions and thus in the susy classification scheme, as is the case for one-nucleon transfer reactions inside the supermultiplet. The inner reactions do provide a direct test of the fermionic sector of the graded Lie Algebras $U_\nu(6/12)$ and $U_\pi(6/4)$, since they involve the nondiagonal generators (see also figure 1)

$$U_\nu(6/12) \otimes U_\pi(6/4) : \begin{pmatrix} b^\dagger_\nu b_\nu & b^\dagger_\nu a_\nu \\ & \\ a^\dagger_\nu b_\nu & a^\dagger_\nu a_\nu \end{pmatrix} \oplus \begin{pmatrix} b^\dagger_\pi b_\pi & b^\dagger_\pi a_\pi \\ & \\ a^\dagger_\pi b_\pi & a^\dagger_\pi a_\pi \end{pmatrix} . \qquad (1)$$

New experimental facilities and detection techniques [7–9] offer a unique opportunity for analyzing the supersymmetry classification in greater detail [12]. In reference [13] we pointed out a symmetry route for the theoretical analysis of such reactions, via the use of tensor operators of the algebras and superalgebras. An alternative route is the use of a semi-microscopic approach where projection techniques starting from the original nucleon pairs lead to specific forms for the operators [14, 15] which, however, are only strictly valid in the generalized seniority regime [16]. The former and latter routes may be related by a consistent operator approach, where the Hamiltonian exchange operators are made to be consistent with the one-nucleon transfer operator implying that the exchange term in the boson-fermion Hamiltonian can be viewed as an internal exchange reaction among the nucleon and the nucleon pairs.

In addition to these experiments, we are also exploring the possibility of testing susy through new transfer reactions. The two-nucleon transfer $(\alpha, \vec{d})$ reaction probes $n-p$ correlations in the nuclear wave function and constitutes a very stringent test of the supersymmetry classification. Note also that the $^{194}\mathrm{Pt}(\alpha,\vec{d})^{196}\mathrm{Au}$ reaction corresponds to a combination of the single-nucleon transfers going either through $^{195}\mathrm{Pt}$ or through $^{195}\mathrm{Au}$, and that the corresponding operator is thus a product of the fermionic components in equation (1), as schematically indicated in Fig. (2). Likewise, the reaction $^{195}\mathrm{Pt}(^3He,t)^{195}\mathrm{Au}$, is again expressible in terms of the superalgebra fermionic operators in (1) and in this case is associated to the beta-decay operator [17]. These reactions and their relation to single-nucleon transfer experiments raise the exciting possibility of testing direct correlations among transfer reaction spectroscopic factors in different nuclei, predicted by the supersymmetric classification of the magic quartet. A preliminary report on these analyses is presented in Ref. [10].

3 Susy without dynamical symmetry

The concept of dynamical algebra (not to be confused with that of dynamical symmetry) implies a generalization of the concept of symmetry algebra. The latter

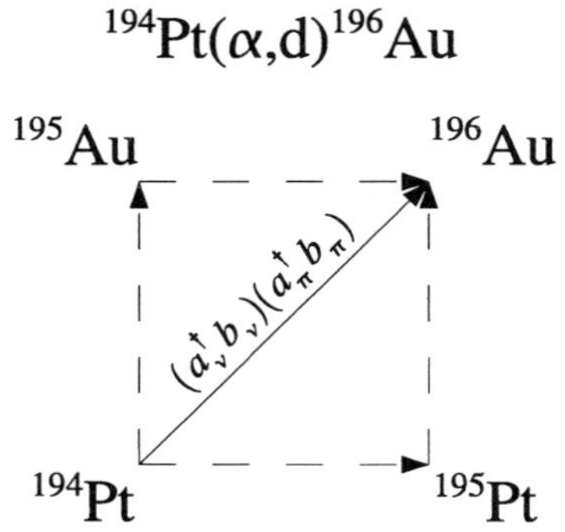

Figure 2. The "diagonal" (α, d) reaction within the susy quartet.

is defined as follows: G is the dynamical algebra of a system if all physical states considered belong to a single irreducible representation (IR) of G. (In a symmetry algebra, in contrast, each set of degenerate states of the system is associated to an IR). The best known examples of a dynamical algebra are perhaps $SO(4,2)$ for the hydrogen atom and the $U(6)$ IBM algebra for even-even nuclei. A consequence of having a dynamical algebra associated to a system is that all states can be reached using the algebra's generators or, equivalently, all physical operators can be expressed in terms of these operators [9]. Naturally, the same Hamiltonian and the same transition operators are employed for all states in the system. To further clarify this point, it is certainly true that a single H and a single set of operators are associated to a given even-even nucleus in the IBM framework, expressed in terms of the $U(6)$ (dynamical algebra) generators. It doesn't matter whether this Hamiltonian can be expressed or not in terms of the generators of a single chain of groups (a dynamical symmetry).

In the same fashion, if we now consider $U(6/12)$ to be the dynamical algebra for the pair of nuclei ^{194}Pt-^{195}Pt, it follows that the same H and operators (including in this case the transfer operators that connect states in the different nuclei) should apply to all states. No restriction is imposed on the form of H, except that it is a function of the generators of $U(6/12)$ (the enveloping space associated to it), and that it has the ususal symmetries like hermiticity, rotational and time reversal invariance. The concept of supersymmetry does not require the existence of a particular dynamical symmetry. Extending these ideas to the $\nu - \pi$ space of IBM-2 we can say that susy is equivalent to requiring that a product of the form

$$U_\nu(6/\Omega) \otimes U_\pi(6/\Omega') \tag{2}$$

plays the role of dynamical (super) algebra for a quartet of even-even, even-odd,

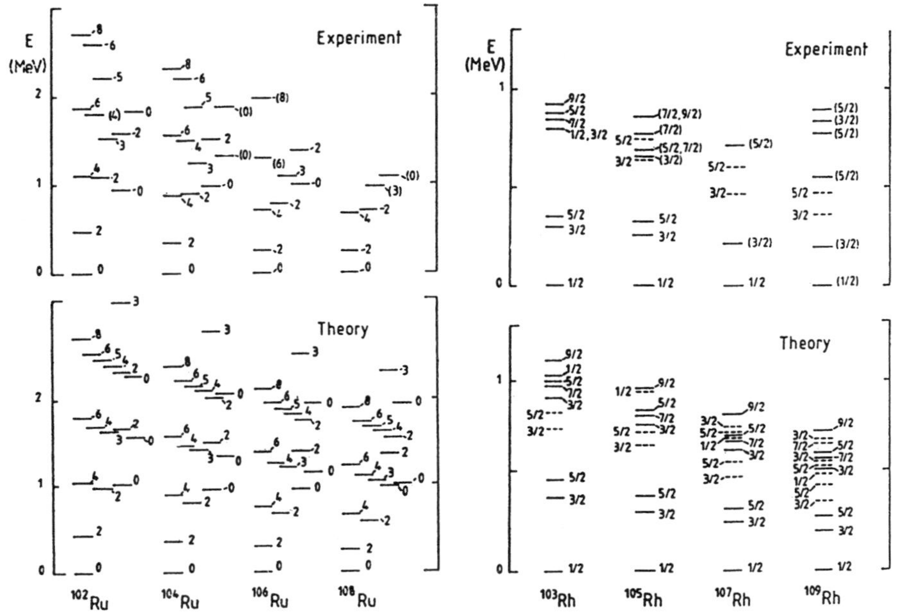

Figure 3. Experimental and calculated positive-parity states in $^{102-108}Ru$ and negative-parity states in $^{103-109}Rh$ [11].

odd-even and odd-odd nuclei. Having said that, it should be stated that the quartet dynamical susy of references [6–9] has the distinct advantage of immediately suggesting the form of the quartet's Hamiltonian and operators, while the general statement made above does not provide a general recipe. For some particular cases, however, this can be done in a straightforward way. In reference [11], for example, the $U(6/12)$ supersymmetry (without imposing one of the three dynamical IBM symmetries) was successfully tested for the Ru and Rh isotopes. In that case a combination of $U^{B+F}(5)$ and $SO^{B+F}(6)$ symmetries was shown to give an excellent description of the data, as shown in Figs. 3 and 4. The $U(6/12)$ case is simple because, using a pseudospin decomposition, there are isomorphic $U(6)$ algebras for the bosons and the fermion and any combination of the three dynamical IBM algebras can be considered [18]

$$\begin{aligned} &U(6/12) \quad j = 1/2, 3/2, 5/2 \\ &\tilde{l} = 0, 2 \qquad \tilde{s} = 1/2 \\ &G_l^{BF} \equiv G_l^B + G_{\tilde{l}}^F \ , \end{aligned} \tag{3}$$

and an arbitrary interaction expressed in terms of G^{BF} implies explicit correlations between the boson-boson and boson-fermion interactions [18].

An immediate consequence of this proposal is that it opens up the possibility of

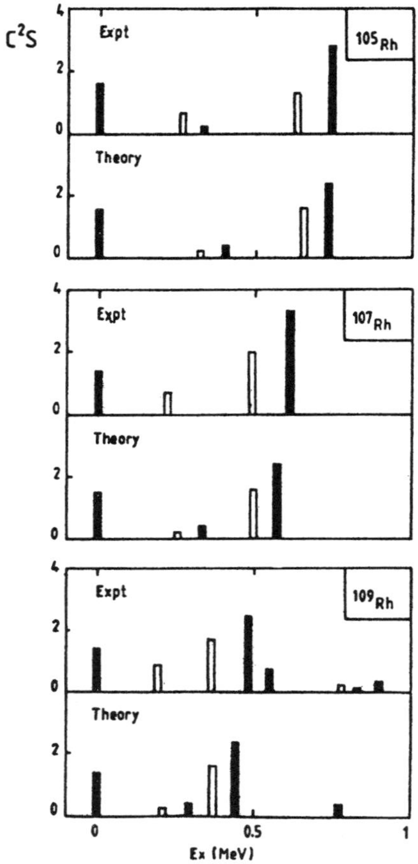

Figure 4. Experimental and calculated spectroscopic factors in Rh isotopes [11].

testing susy in other nuclear regions, since dynamical symmetries are very scarce and have severely limited the study of nuclear supersymmetry.

4 Generic susy

We have recently initiated a renewed search for supersymmetry in nuclei [12, 18]. We have yet to discover a general mechanism to generate all appropriate operators, but a set of general guiding rules are the following:

1) A single Hamiltonian should describe the members of a supersymmetric doublet or quartet.

2) The boson Hamiltonian, plus the single-particle orbits, should essentially determine the boson-fermion interaction and, in case of a quartet, also the fermion-fermion interaction.

Although the analysis is not concluded, some preliminary results for the W and Hf nuclei are quite encouraging [18]. The first calculation involves a mixture of $SU(3)$ and $SO(6)$ symmetries in $U(6/12)$. This calculation employs the Q-consistent formalism and a comparison between the experimental and calculated $BE(2)$ transitions and quadrupole moments is shown in tables 1 and 2. The agreement is very good, except for one transition in ^{183}W. We also show an example of generic susy in U(6/4). It corresponds to the supermultiplet composed of ^{174}Hf and ^{173}Hf. In this case the hamiltonian uses a combination of Casimir operators of the $U(5)$ and $SO(6)$ groups and of their subgroups. In figure 5 we compare the experimental and calculated level energies in these two nuclei [18].

One of our main interests is to apply the generic form of supersymmetry to the Pt-Au region and compare the results with the traditional scheme, particularly for the new transfer experiments [12]. In addition, we expect to find other examples of quartet supersymmetric behavior, once the constraints set by dynamical symmetry are dropped.

$B(E2)$ (e^2b^2) and Q (eb) in ^{182}W

$J_i^\pi \to J_f^\pi$	Exp.	Calc.
$2_1^+ \to 0_1^+$	0.839(18)	0.8422
$4_1^+ \to 2_1^+$	1.201(61)	1.1877
$6_1^+ \to 4_1^+$	1.225(135)	1.2777
$2_2^+ \to 0_1^+$	0.021(1)	0.0040
$2_2^+ \to 2_1^+$	0.041(1)	0.0072
$2_2^+ \to 4_1^+$	0.00021(1)	0.0006
$2_3^+ \to 0_1^+$	0.006(1)	0.0000
$2_3^+ \to 0_2^+$	1.225(368)	0.6840
$2_3^+ \to 2_1^+$	0.0039(5)	0.0001

Q	Exp.	Calc.
2_1^+	$-2.00^{+0.04}_{-0.08}$	-1.86
2_2^+	$1.94^{+0.10}_{-0.04}$	1.61

Table 1. Experimental and calculated reduced transition probabilities and quadrupole moments in ^{182}W [18].

$B(E2)$ (e^2b^2) in ^{183}W

$J_i^\pi \to J_f^\pi$	Exp.	Calc.
$\frac{3}{2}^-_1 \to \frac{1}{2}^-_1$	0.938(62)	0.603
$\frac{5}{2}^-_1 \to \frac{1}{2}^-_1$	0.68(4)	0.603
$\frac{5}{2}^-_1 \to \frac{3}{2}^-_1$	0.20(3)	0.173
$\frac{13}{2}^-_1 \to \frac{9}{2}^-_1$	1.1(3)	0.915
$\frac{17}{2}^-_1 \to \frac{13}{2}^-_1$	0.89(12)	0.925
$\frac{3}{2}^-_* \to \frac{1}{2}^-_1$	0.005(2)	0.000
$\frac{3}{2}^-_* \to \frac{3}{2}^-_1$	0.10(4)	0.010
$\frac{3}{2}^-_* \to \frac{5}{2}^-_1$	0.012(5)	0.023
$\frac{5}{2}^-_* \to \frac{1}{2}^-_1$	0.082(9)	0.011
$\frac{5}{2}^-_* \to \frac{3}{2}^-_1$	0.001(1)	0.001
$\frac{5}{2}^-_* \to \frac{5}{2}^-_1$	0.027(6)	0.004
$\frac{5}{2}^-_* \to \frac{7}{2}^-_1$	0.43(2)	0.001
$\frac{5}{2}^-_* \to \frac{3}{2}^-_2$	1.30(18)	0.860

Table 2. Experimental and calculated reduced transition probabilities in ^{183}W [18].

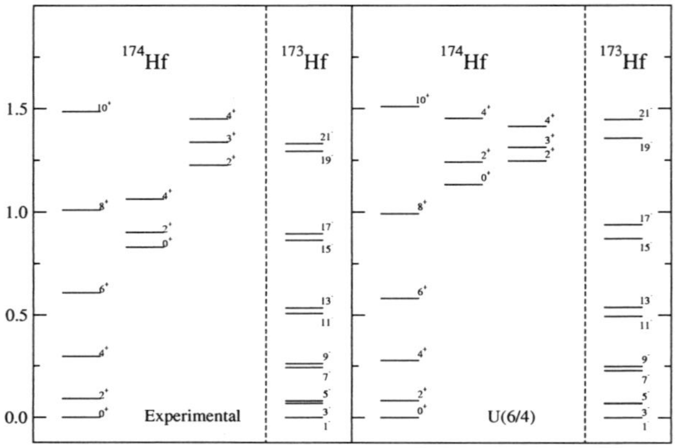

Figure 5. Experimental and calculated positive-parity states in ^{174}Hf and negative-parity states in ^{173}Hf [18]. The energy scale is in MeV.

We continue to search for a more general theoretical framework that can encompass the particular cases that we can solve at this point. We conclude by proposing that nuclear susy may be a more general phenomenon than was previously realized and that may yet play an important, unifying role in nuclei.

5 Dedication

We dedicate this paper to Franco Iachello, who has managed to uniquely combine the Platonic ideal of symmetry with the down-to-earth Aristotelic ability to recognize these patterns in Nature.

Acknowledgments

We are grateful to P. Van Isacker, G. Graw, J. Jolie, J. Arias and C. Alonso for their collaboration and much inspiration. This paper was supported in part by Conacyt, Mexico.

References

1. F. Iachello, *Phys. Rev. Lett.* **44**, 772 (1982).
2. F. Iachello and P. Van Isacker, *The Interacting Boson-Fermion Model* (Cambridge University Press, Cambridge, 1991).
3. A. Metz, J. Jolie, G. Graw, R. Hertenberger, J. Gröger, C. Günther, N. Warr and Y. Eisermann, *Phys. Rev. Lett.* **83**, 1542 (1999), and references therein.
4. A. Kostelecky and D.K. Campbell, Eds., *Supersymmetry in Physics* (North Holland, Amsterdam 1984);
   S. Weinberg, *The quantum theory of fields: Supersymmetry* (Cambridge University Press, Cambridge, 2000).

Figure 6. Detail of "The School of Athens" (Plato on the left and Aristoteles on the right), by Rafael.

5. A.B. Balantekin, I. Bars and F. Iachello, *Nucl. Phys.* A **370**, 284 (1981); A.B. Balantekin, I. Bars, R. Bijker and F. Iachello, *Phys. Rev.* C **27**, 1761 (1983); R. Bijker, Ph.D. Thesis, (1984); A. Mauthofer, K. Stelzer, Th.W. Elze, Th. Happ, J. Gerl, A. Frank and P. Van Isacker, *Phys. Rev.* C **39**, 1111 (1989).
6. P. Van Isacker, J. Jolie, K. Heyde and A. Frank, Phys. Rev. Lett. **54**, 653 (1985).
7. D.D. Warner, R.F. Casten and A. Frank, Phys. Lett. B **180**, 207 (1986).
8. A. Metz, Y. Eisermann, A. Gollwitzer, R. Hertenberger, B.D. Valnion, G. Graw and J. Jolie, Phys. Rev. C **61**, 064313 (2000).
9. A. Frank and P. Van Isacker, *Algebraic Methods in Molecular and Nuclear Structure Physics* (Wiley, New York, 1994).
10. R. Bijker, J. Barea and A. Frank, these proceedings.

11. A. Frank, P. Van Isacker and D.D. Warner, Phys. Lett. B **197**, 474 (1987).
12. G. Graw, private communication.
13. J. Barea, R. Bijker, A. Frank and G. Loyola, Phys. Rev. C **64**, 064313 (2001).
14. O. Scholten, Ph.D. Thesis (1980).
15. J. Barea, C.E. Alonso and J.M. Arias, Phys. Rev. C **65**, 34328 (2002).
16. I. Talmi, *Simple Models for Complex Nuclei* (Harwood Academic Publishers, Chur, 1993).
17. P. Navrátil and J. Dobes, Phys. Rev. C **37**, 2126 (1988).
18. J. Barea, Ph.D. Thesis (2002).

EVERYTHING YOU ALWAYS WANTED TO KNOW ABOUT SUSY, BUT WERE AFRAID TO ASK

R. BIJKER[1], J. BAREA[1] AND A. FRANK[1,2]

[1] *ICN-UNAM, AP 70-543, 04510 México, DF, México*

[2] *CCF-UNAM, AP 139-B, 62251 Cuernavaca, Morelos, México*

New experimental tests of nuclear supersymmetry are suggested. They involve the measurement of one- and two-nucleon transfer reactions between nuclei that belong to the same supermultiplet. These reactions provide a direct test of the 'fermionic' sector, i.e. of the operators that change a boson into a fermion or vice versa. We present some theoretical predictions for the supersymmetric quartet of nuclei: ^{194}Pt, ^{195}Pt, ^{195}Au and ^{196}Au.

1 Introduction

Dynamical supersymmetries were introduced [1] in nuclear physics in 1980 by Franco Iachello in the context of the Interacting Boson Model (IBM) [2] and its extensions. The spectroscopy of atomic nuclei is characterized by the interplay between collective (bosonic) and single-particle (fermionic) degrees of freedom. The IBM describes collective excitations in even-even nuclei in terms of a system of interacting monopole and quadrupole bosons with angular momentum $l = 0, 2$. The bosons are associated with the number of correlated proton and neutron pairs, and hence the number of bosons N is half the number of valence nucleons.

For odd-mass nuclei the IBM has been extended to include single-particle degrees of freedom [3]. The Interacting Boson-Fermion Model (IBFM) has as its building blocks N bosons with $l = 0, 2$ and $M = 1$ fermion with $j = j_1, j_2, \ldots$. The IBM and IBFM can be unified into a superalgebra $U(n/m)$, where $n = \sum_l (2l+1) = 6$ is the dimension of the boson space and $m = \sum_j (2j+1)$ of the fermion space [4]. In this framework, even-even and odd-mass nuclei form the members of a supermultiplet. The inclusion of the neutron-proton degree of freedom leads to supersymmetric quartets of nuclei consisting of an even-even, an odd-even, an even-odd and an odd-odd nucleus [5].

Supersymmetry (SUSY) distinguishes itself from other symmetries in that it includes, in addition to transformations among fermions or among bosons, also transformations between bosons and fermions. The spectroscopic properties of the nuclei that belong to the same supermultiplet, are linked and correlated by SUSY, i.e. they are described by the same form of the operators. Most tests of supersymmetry that have been discussed in the literature involve energies and transitions. These observables are described by the bosonic generators that transform bosons into bosons and fermions into fermions. Whereas the bosonic generators describe observables within a given nucleus, the fermionic generators that change a boson into a fermion or vice versa, describe the transitions between different nuclei of the same supermultiplet, such as observed in single-particle transfer reactions. Unlike for the bosonic sector, there are relatively few direct tests of the fermionic generators [6,7].

It is the purpose of this contribution to investigate one-nucleon transfer reactions in the context of nuclear supersymmetry, and to establish possible correlations between different transfer reactions. As an example, we consider the supersymmetric quartet of nuclei: ^{194}Pt, ^{195}Pt, ^{195}Au and ^{196}Au, whose energy spectra have been classified and described successfully in terms of the $U(6/12)_\nu \otimes U(6/4)_\pi$ supersymmetry [5,8].

2 The $U(6/12)_\nu \otimes U(6/4)_\pi$ supersymmetry

The mass region $A \sim 190$ has been a rich source of possible empirical evidence for the existence of (super)symmetries in nuclei. The even-even nucleus ^{196}Pt is the standard example of the $SO(6)$ limit of the IBM [9]. The odd-proton nuclei ^{191,193}Ir and ^{193,195}Au were suggested as examples of the $Spin(6)$ limit [1], in which the odd-proton is allowed to occupy the $\pi d_{3/2}$ orbit, whereas the pairs of nuclei ^{190}Os - ^{191}Ir, ^{192}Os - ^{193}Ir, ^{192}Pt - ^{193}Au and ^{194}Pt - ^{195}Au have been analyzed as examples of a $U(6/4)$ supersymmetry [4]. The odd-neutron nucleus ^{195}Pt, together with ^{194}Pt, were studied in terms of a $U(6/12)$ supersymmetry, in which the odd neutron occupies the $\nu p_{1/2}$, $\nu p_{3/2}$ and $\nu f_{5/2}$ orbits [10]. These ideas were later extended to the case where neutron and proton bosons are distinguished [5], predicting in this way a correlation among quartets of nuclei, consisting of an even-even, an odd-proton, an odd-neutron and an odd-odd nucleus. The best experimental example of such a quartet with $U(6/12)_\nu \otimes U(6/4)_\pi$ supersymmetry is provided by the nuclei ^{194}Pt, ^{195}Au, ^{195}Pt and ^{196}Au. The number of bosons N is taken to be the number of bosons in the odd-odd nucleus ^{196}Au: $N = N_\nu + N_\pi = 4 + 1 = 5$.

$$
\begin{array}{ccccc}
\text{even-odd} & & & & \text{odd-odd} \\
N_\nu + 1, N_\pi, j_\pi & & & & N_\nu, N_\pi, j_\nu, j_\pi \\
 & {}^{195}_{79}\mathrm{Au}_{116} & \leftrightarrow & {}^{196}_{79}\mathrm{Au}_{117} & \\
 & \updownarrow & & \updownarrow & \\
 & {}^{194}_{78}\mathrm{Pt}_{116} & \leftrightarrow & {}^{195}_{78}\mathrm{Pt}_{117} & \\
\text{even-even} & & & & \text{odd-even} \\
N_\nu + 1, N_\pi + 1 & & & & N_\nu, N_\pi + 1, j_\nu
\end{array} \tag{1}
$$

The supersymmetric classification of nuclear levels in the Pt and Au isotopes has been re-examined by taking advantage of the significant improvements in experimental capabilities developed in the last decade. High resolution transfer experiments with protons and polarized deuterons have led to strong evidence for the existence of supersymmetry (SUSY) in atomic nuclei. The experiments include high resolution transfer experiments to ^{196}Au at TU/LMU München [8,11], and in-beam gamma ray and conversion electron spectroscopy following the reactions ^{196}Pt$(d, 2n)$ and ^{196}Pt(p, n) at the cyclotrons of the PSI and Bonn [12]. These studies have achieved an improved classification of states in ^{195}Pt and ^{196}Au which give further support to the original ideas [10,13,5] and extend and refine previous experimental work [14,15,16] in this research area.

In a dynamical (super)symmetry, the Hamiltonian is expressed in terms of the Casimir invariants of the subgroups in a group chain. The relevant subgroup chain of $U(6/12)_\nu \otimes U(6/4)_\pi$ for the Pt and Au nuclei is given by [5]

$$\begin{aligned} U(6/12)_\nu \otimes U(6/4)_\pi &\supset U^{B_\nu}(6) \otimes U^{F_\nu}(12) \otimes U^{B_\pi}(6) \otimes U^{F_\pi}(4) \\ &\supset U^{B}(6) \otimes U^{F_\nu}(6) \otimes U^{F_\nu}(2) \otimes U^{F_\pi}(4) \\ &\supset U^{BF_\nu}(6) \otimes U^{F_\nu}(2) \otimes U^{F_\pi}(4) \\ &\supset SO^{BF_\nu}(6) \otimes U^{F_\nu}(2) \otimes SU^{F_\pi}(4) \\ &\supset Spin(6) \otimes U^{F_\nu}(2) \\ &\supset Spin(5) \otimes U^{F_\nu}(2) \\ &\supset Spin(3) \otimes SU^{F_\nu}(2) \\ &\supset SU(2) \ . \end{aligned} \tag{2}$$

In this case, the Hamiltonian

$$\begin{aligned} H = {} & \alpha\, C_{2U^{BF_\nu}(6)} + \beta\, C_{2SO^{BF_\nu}(6)} + \gamma\, C_{2Spin(6)} \\ & +\delta\, C_{2Spin(5)} + \epsilon\, C_{2Spin(3)} + \eta\, C_{2SU(2)} \ , \end{aligned} \tag{3}$$

describes simultaneously the excitation spectra of the quartet of nuclei. Here we have neglected terms that only contribute to binding energies. The coefficients α, β, γ, δ, ϵ and η have been determined in a simultaneous fit of the excitation energies of the four nuclei of Eq. (1) [12].

In a dynamical supersymmetry, closed expressions can be derived for energies, and selection rules and intensities for electromagnetic transitions and single-particle transfer reactions. While a simultaneous description and classification of these observables in terms of the $U(6/12)_\nu \otimes U(6/4)_\pi$ supersymmetry has been shown to be fulfilled to a good approximation for the quartet of nuclei ^{194}Pt, ^{195}Au, ^{195}Pt and ^{196}Au, there are important predictions still not fully verified by experiments. These tests involve the transfer reaction intensities among the supersymmetric partners. In the next section we concentrate on the latter and, in particular, on the one-proton transfer reactions $^{194}\text{Pt} \to {}^{195}\text{Au}$ and $^{195}\text{Pt} \to {}^{196}\text{Au}$.

3 One-nucleon transfer reactions

The single-particle transfer operator that is commonly used in the Interacting Boson-Fermion Model (IBFM), has been derived in the seniority scheme [17]. Although strictly speaking this derivation is only valid in the vibrational regime, it has been used for deformed nuclei as well. An alternative method is based on symmetry considerations. It consists in expressing the single-particle transfer operator in terms of tensor operators under the subgroups that appear in the group chain of a dynamical (super)symmetry [6,7,18]. The single-particle transfer between different members of the same supermultiplet provides an important test of supersymmetries, since it involves the transformation of a boson into a fermion or vice versa, but it conserves the total number of bosons plus fermions.

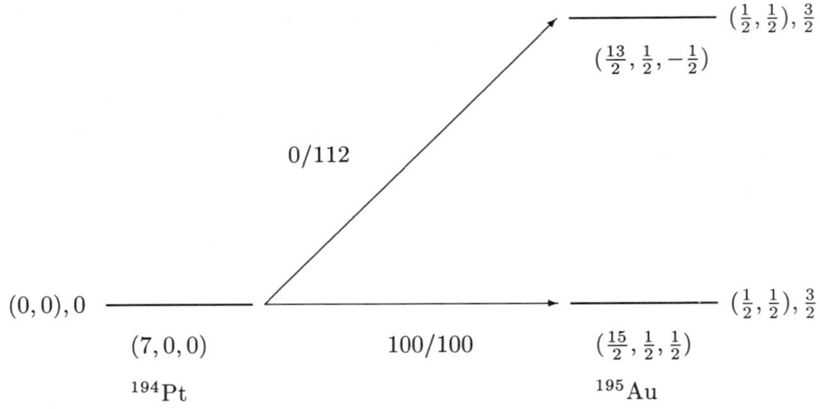

Figure 1. Allowed one-proton transfer reactions for $^{194}\text{Pt} \rightarrow {}^{195}\text{Au}$. The spectroscopic factors are normalized to 100 for the ground state to ground state transition for the operators T_1/T_2.

The operators that describe one-proton transfer reactions in the $U(6/12)_\nu \otimes U(6/4)_\pi$ supersymmetry are given by

$$T_{1,m}^{(\frac{1}{2},\frac{1}{2},-\frac{1}{2}),(\frac{1}{2},\frac{1}{2}),\frac{3}{2}} = -\sqrt{\frac{1}{6}}\left(\tilde{s}_\pi \times a^\dagger_{\pi,\frac{3}{2}}\right)^{(\frac{3}{2})}_m + \sqrt{\frac{5}{6}}\left(\tilde{d}_\pi \times a^\dagger_{\pi,\frac{3}{2}}\right)^{(\frac{3}{2})}_m ,$$

$$T_{2,m}^{(\frac{3}{2},\frac{1}{2},\frac{1}{2}),(\frac{1}{2},\frac{1}{2}),\frac{3}{2}} = \sqrt{\frac{5}{6}}\left(\tilde{s}_\pi \times a^\dagger_{\pi,\frac{3}{2}}\right)^{(\frac{3}{2})}_m + \sqrt{\frac{1}{6}}\left(\tilde{d}_\pi \times a^\dagger_{\pi,\frac{3}{2}}\right)^{(\frac{3}{2})}_m . \quad (4)$$

The operators T_1 and T_2 are, by construction, tensor operators under *Spin*(6), *Spin*(5) and *Spin*(3) [18]. The upper indices $(\sigma_1,\sigma_2,\sigma_3)$, (τ_1,τ_2), J specify the tensorial properties under *Spin*(6), *Spin*(5) and *Spin*(3). The use of tensor operators to describe single-particle transfer reactions in the supersymmetry scheme has the advantage of giving rise to selection rules and closed expressions for the spectroscopic factors.

Fig. 1 shows the allowed transitions for the transfer operators of Eq. (4) that describe the one-proton transfer from the ground state $|(N+2,0,0),(0,0),0\rangle$ of the even-even nucleus ^{194}Pt to the even-odd nucleus ^{195}Au belonging to the supermultiplet $[N_\nu+1\}_\nu \otimes [N_\pi+1\}_\pi$. The operators T_1 and T_2 have the same transformation character under *Spin*(5) and *Spin*(3), and therefore can only excite states with $(\tau_1,\tau_2)=(\frac{1}{2},\frac{1}{2})$ and $J=\frac{3}{2}$. However, they differ in their *Spin*(6) selection rules. Whereas T_1 can only excite the ground state of the even-odd nucleus with $(\sigma_1,\sigma_2,\sigma_3)=(N+\frac{3}{2},\frac{1}{2},\frac{1}{2})$, the operator T_2 also allows the transfer to an excited state with $(N+\frac{1}{2},\frac{1}{2},-\frac{1}{2})$. The ratio of the intensities is given by [18]

$$R_1 = \frac{I_{\text{gs}\rightarrow\text{exc}}}{I_{\text{gs}\rightarrow\text{gs}}} = 0 ,$$

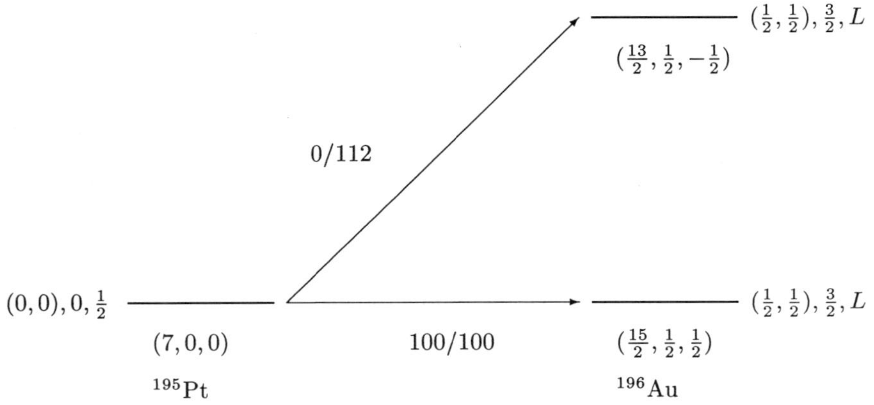

Figure 2. As Fig. 1, but for ^{195}Pt $\rightarrow$ ^{196}Au.

$$R_2 = \frac{I_{\text{gs}\rightarrow\text{exc}}}{I_{\text{gs}\rightarrow\text{gs}}} = \frac{9(N+1)(N+5)}{4(N+6)^2} , \tag{5}$$

for T_1 and T_2, respectively. In the case of the one-proton transfer ^{194}Pt $\rightarrow$ ^{195}Au, the second ratio is given by $R_2 = 1.12$ ($N = 5$).

The available experimental data from the proton stripping reactions ^{194}Pt$(\alpha, t)^{195}$Au and ^{194}Pt$(^3$He$, d)^{195}$Au [19] shows that the $J = 3/2$ ground state of ^{195}Au is excited strongly with $C^2S = 0.175$, whereas the first excited $J = 3/2$ state is excited weakly with $C^2S = 0.019$. In the SUSY scheme, the latter state is assigned as a member of the ground state band with $(\tau_1, \tau_2) = (5/2, 1/2)$. Therefore the one proton transfer to this state is forbidden by the *Spin*(5) selection rule of the tensor operators of Eq. (4). The relatively small strength to excited $J = 3/2$ states suggests that the operator T_1 of Eq. (4) can be used to describe the data.

In Fig. 2 we show the allowed transitions for the one-proton transfer from the ground state $|(N+2,0,0),(0,0),0,\frac{1}{2}\rangle$ of the odd-even nucleus ^{195}Pt to the odd-odd nucleus ^{196}Au. Also in this case, the operator T_1 only excites the ground state doublet of ^{196}Au with $(\sigma_1, \sigma_2, \sigma_3) = (N+\frac{3}{2}, \frac{1}{2}, \frac{1}{2})$, $(\tau_1, \tau_2) = (\frac{1}{2}, \frac{1}{2})$, $J = \frac{3}{2}$ and $L = J \pm \frac{1}{2}$, whereas T_2 also populates the excited state with $(N+\frac{1}{2}, \frac{1}{2}, -\frac{1}{2})$. The ratio of the intensities is the same as for the ^{194}Pt $\rightarrow$ ^{195}Au transfer reaction

$$\begin{aligned} R_1(^{195}\text{Pt} \rightarrow ^{196}\text{Au}) = R_1(^{194}\text{Pt} \rightarrow ^{195}\text{Au}) &= 0 , \\ R_2(^{195}\text{Pt} \rightarrow ^{196}\text{Au}) = R_2(^{194}\text{Pt} \rightarrow ^{195}\text{Au}) &= \frac{9(N+1)(N+5)}{4(N+6)^2} . \end{aligned} \tag{6}$$

This is direct consequence of the supersymmetry. Just as the energies and the electromagnetic transition rates of the supersymmetric quartet of nuclei were calculated with the same form of the Hamiltonian and the transition operator, here we have extended this idea to the one-proton transfer reactions. We find definite predictions for the spectroscopic factors of the ^{195}Pt $\rightarrow$ ^{196}Au transfer reactions, which can be tested experimentally. To the best of our knowledge, there are no data available for this reaction.

For the one-neutron transfer reactions there exists a similar situation. The available experimental data from the neutron stripping reactions ^{194}Pt$(d,p)^{195}$Pt [20] can be used to determine the appropriate form of the one-neutron transfer operator [7], which then can be used to predict the spectroscopic factors for the transfer reaction ^{195}Au $\rightarrow$ ^{196}Au. We believe that, as a consequence of the supersymmetry classification, a number of additional correlations exist for transfer reactions between different pairs of nuclei. This would be the first time that such relations are predicted for nuclear reactions which may provide a challenge and motivation for future experiments.

4 Summary and outlook

The recent measurements of the spectroscopic properties of the odd-odd nucleus ^{196}Au have rekindled the interest in nuclear supersymmetry. The available data on the spectroscopy of the quartet of nuclei ^{194}Pt, ^{195}Au, ^{195}Pt and ^{196}Au can, to a good approximation, be described in terms of the $U(6/4)_\pi \otimes U(6/12)_\nu$ supersymmetry. However, there is a still another important set of experiments which can further test the predictions of the supersymmetry scheme. These involve transfer reactions between nuclei belonging to the same supermultiplet, in particular between the even-odd (odd-even) and odd-odd members of the supersymmetric quartet. Theoretically, these transfers are described by the supersymmetric generators which change a boson into a fermion, or vice versa.

We investigated in some detail the example of proton transfer between the SUSY partners: ^{194}Pt $\rightarrow$ ^{195}Au and ^{195}Pt $\rightarrow$ ^{196}Au. The supersymmetry implies strong correlations for the spectroscopic factors of these two reactions which can be tested experimentally. A similar set of relations can be derived for the one-neutron transfer reactions ^{194}Pt $\leftrightarrow$ ^{195}Pt and ^{195}Au $\leftrightarrow$ ^{196}Au. Another interesting extension of supersymmetry concerns the recently measured two-nucleon transfer reaction ^{194}Pt$(\alpha,d)^{196}$Au [21], in which a neutron-proton pair is transferred to the target nucleus. This reaction presents a very sensitive test of the wave functions, since it provides a measure of the correlation within the transferred neutron-proton pair. Whether it is possible to describe this process by a transfer operator that is correlated by SUSY to that of the one-proton and one-neutron transfer reactions is an open question.

In conclusion, we emphasize the need for new experiments taking advantage of the new experimental capabilities [8,11,12] and suggest that particular attention be paid to one- and two-nucleon transfer reactions between the SUSY partners ^{194}Pt, ^{195}Au, ^{195}Pt and ^{196}Au, since such experiments provide the most stringent tests of nuclear supersymmetry. It remains to be seen whether the correlations predicted by SUSY are indeed verified by experiments.

Dedication

It is a great pleasure to dedicate this contribution to Franco Iachello on the occasion of the conference 'Symmetries in Nuclear Structure', held in his honor. Unfortunately it was not possible to attend the meeting, but I am grateful to the organizers for the invitation to write a contribution for the proceedings.

In the fall of 1977, I took a course on nuclear structure presented by a young Italian professor at the University of Groningen. The lectures were characterized by their clarity of presentation, a contagious enthusiasm, a link between the material presented in the course and ongoing research and, last but not least, the connection between theory and experiment. These ingredients have formed the basis and provided the motivation and inspiration for my own scientific career, first as a graduate student at the KVI in Groningen, and later as a research scientist. Over the years I have had the pleasure to collaborate with Franco on different subjects, such as supersymmetry, baryon spectroscopy and nuclear clusters which has resulted in 12 joint publications between 1983 and 2002.

I will not mention his other career as a racecar driver, nor comment on his uncanny likeliness to Woody Allen (with the exception of the title of this contribution). Finally, I wish Franco an equally productive and creative second half of his career. Congratulations, Franco!

Acknowledgments

We are grateful to Gerhard Graw for numerous stimulating discussions and for sharing the new data on the two-nucleon transfer reactions $^{198}\text{Hg}(\vec{d},\alpha)^{196}\text{Au}$ and $^{194}\text{Pt}(\alpha,d)^{196}\text{Au}$ prior to publication. The work presented in this contribution is motivated by the renewed experimental interest in this mass region. This work was supported in part by CONACyT.

References

1. F. Iachello, *Phys. Rev. Lett.* **44**, 772 (1980).
2. A. Arima and F. Iachello, *Phys. Rev. Lett.* **35**, 1069 (1975).
3. F. Iachello and O. Scholten, *Phys. Rev. Lett.* **43**, 679 (1979).
4. A.B. Balantekin, I. Bars and F. Iachello, *Phys. Rev. Lett.* **47**, 19 (1981); A.B. Balantekin, I. Bars and F. Iachello, *Nucl. Phys.* A **370**, 284 (1981).
5. P. van Isacker, J. Jolie, K. Heyde and A. Frank, *Phys. Rev. Lett.* **54**, 653 (1985).
6. F. Iachello and S. Kuyucak, *Ann. Phys. (N.Y.)* **136**, 19 (1981).
7. R. Bijker and F. Iachello, *Ann. Phys. (N.Y.)* **161**, 360 (1985).
8. A. Metz, J. Jolie, G. Graw, R. Hertenberger, J. Gröger, C. Günther, N. Warr and Y. Eisermann, *Phys. Rev. Lett.* **83**, 1542 (1999).
9. J.A. Cizewski, R.F. Casten, G.J. Smith, M.L. Stelts, W.R. Kane, H.G. Börner and W.F. Davidson, *Phys. Rev. Lett.* **40**, 167 (1978); A. Arima and F. Iachello, *Phys. Rev. Lett.* **40**, 385 (1978).
10. A.B. Balantekin, I. Bars, R. Bijker and F. Iachello, *Phys. Rev.* C **27**, 1761

(1983).
11. A. Metz, Y. Eisermann, A. Gollwitzer, R. Hertenberger, B.D. Valnion, G. Graw and J. Jolie, *Phys. Rev.* C **61**, 064313 (2000).
12. J. Gröger, J. Jolie, R. Krücken, C.W. Beausang, M. Caprio, R.F. Casten, J. Cederkall, J.R. Cooper, F. Corminboeuf, L. Genilloud, G. Graw, C. Günther, M. de Huu, A.I. Levon, A. Metz, J.R. Novak, N. Warr and T. Wendel, *Phys. Rev.* C **62**, 064304 (2000).
13. H.Z. Sun, A. Frank and P. van Isacker, *Phys. Rev.* C **27**, 2430 (1983); H.Z. Sun, A. Frank and P. van Isacker, *Ann. Phys. (N.Y.)* **157**, 183 (1984).
14. A. Mauthofer, K. Stelzer, J. Gerl, Th.W. Elze, Th. Happ, G. Eckert, T. Faestermann, A. Frank and P. van Isacker, *Phys. Rev.* C **34**, 1958 (1986).
15. J. Jolie, U. Mayerhofer, T. von Egidy, H. Hiller, J. Klora, H. Lindner and H. Trieb, *Phys. Rev.* C **43**, R16 (1991).
16. G. Rotbard, G. Berrier, M. Vergnes, S. Fortier, J. Kalifa, J.M. Maison, L. Rosier, J. Vernotte, P. van Isacker and J. Jolie, *Phys. Rev.* C **47**, 1921 (1993).
17. O. Scholten, *Prog. Part. Nucl. Phys.* **14**, 189 (1985).
18. J. Barea, R. Bijker, A. Frank and G. Loyola, *Phys. Rev.* C **64**, 064313 (2001).
19. M.L. Munger and R.J. Peterson, *Nucl. Phys.* A **303**, 199 (1978).
20. Y. Yamazaki and R.K. Sheline, *Phys. Rev.* C **14**, 531 (1976).
21. G. Graw, private communication.

SUPERSYMMETRY AND IDENTICAL BANDS

P. VON BRENTANO

Institut für Kernphysik, Universität zu Köln, 50937 Köln, Germany

Supersymmetry as applied to identical bands is discussed. A review of the work of the Koeln-Dubna group on this topic is given and examples in ^{171,172}Yb , ^{173,174}Hf and ^{195,194}Pt are discussed. The role of pseudo-spin in the supersymmetry is investigated. A recent precision lifetime measurement for identical bands in $^{171,172}Yb$ is discussed. Keywords: supersymmetry, identical bands, pseudo-spin

1 Introduction

Supersymmetry has been successfully transferred from particle physics to nuclear physics by Franco Iachello and coworkers[1,3,2,4,5,8,7,6,9,10,11]. Various attempts to obtain an extended classification, in which several neighboring nuclei including even, odd and odd–odd ones are described by the same Hamiltonian, have led to convincing results e.g.[12]. However, in nuclear structure theory primary attention has been paid to the investigation of models with Hamiltonians which exhibit dynamical supersymmetries. In a recent paper[13] it has been shown, however, that the concepts of identical bands[14,15,16] and pseudo-spin [18,19] allows one to discuss (partial) supersymmetry also for cases in which the even partner nucleus has no dynamical symmetry. This has also been discussed in the consistent Q formalism[21,22]. The concept of "Identical Bands" (IB) was suggested by a Strasbourgh-Liverpool collaboration led by Byrski [14], who found an agreement to better than 0.3% in the gamma energies of two superdeformed (sd) bands in ^{151}Tb and^{152}Dy. The role of pseudo-spin in identical bands was subsequently clarified by Nazarewicz et al. [15]. An explanation based on the U(6/12) supersymmetry was given in the frame of the Interacting Boson Model by Gelberg et al. [23] and in the frame of supersymmetric quantum mechanics by Amado, Bijker et al. [24]. Although the importance of the discovery of the IB was widely recognized the data on the Dy-Tb pair are still incomplete due to experimental difficulties. The group at Argonne has recently settled the spins of the identical superdeformed band in ^{152}Dy [17] but the spins in ^{151}Tb are still unknown. Also the 3/2, 7/2... signature partner band in ^{151}Tb is still missing. This shows that one has to rely on the spectroscopy of "Identical Bands" at normal deformation, where spins and parities are all known. In this case the moments of inertia of the even and odd nucleus system tend to differ by 10% or more as compared to 0.3% in the SD case. A survey is given by Baktash et al.[16]. We show in Fig. 1 a convincing example of the identical bands in ^{174}Hf and ^{173}Hf [13].

In the following I will review the Koeln-Dubna work[25,13,26,27] which focused on the supersymmetry connected with identical bands. In order to avoid equations or technical discussions a simple schematic model for supersymmetry will be discussed. This is done on the basis of the U(6/12) supersymmetry [1,7]which has a very transparent microscopic foundation. In U(6/12) respectively in U(15/30)[1,?] there are three (five) contributing shells in the odd nucleus and which have the spins 1/2, 3/2 and 5/2 (or 1/2, 3/2, 5/2, 7/2 and 9/2)

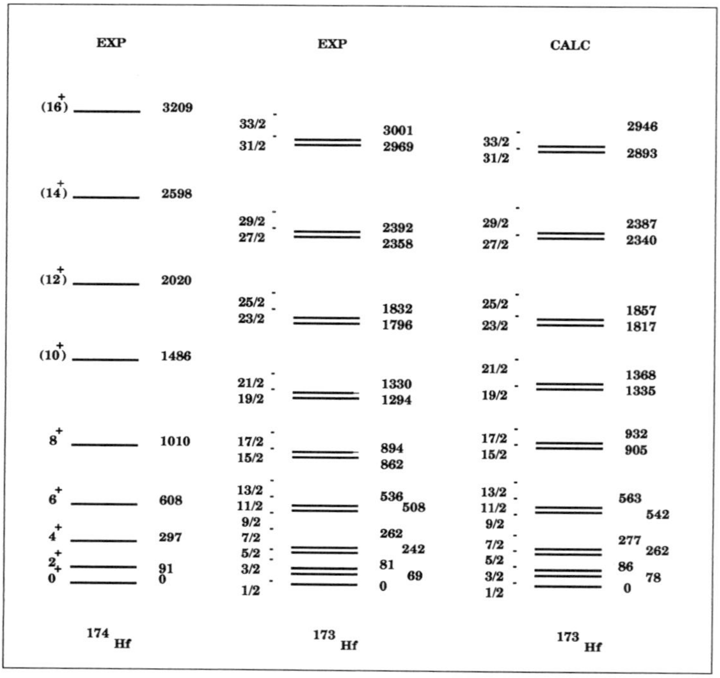

Fig.1
Comparison of the energies of SUSY partner bands in ^{174}Hf and ^{173}Hf. The energies labeled CALC are obtained by correcting for the change in the moment of inertia and by allowing a pseudo-spin orbit coupling. From Jolos et al.[13].

2 Pseudo-spin

In discussing the U(6/12) SUSY and the schematic model pseudo-spin plays an important role. Thus I make a few comments on pseudo-spin symmetry. This approximate symmetry is long known [18,19]. Recently it was shown, that pseudo-spin symmetry is actually based on a relativistic symmetry[20]. We remind briefly the pseudo-spin scheme: In Fig. 2 the s.p. neutron states around ^{208}Pb are shown. In the pseudo-spin scheme the 3p1/2, 2f5/2 and 3p3/2 levels are relabeled with the pseudo-spin quantum numbers $2\ \tilde{s}\ 1/2$, $2\ \tilde{d}\ 5/2$, $2\ \tilde{d}\ 3/2$. Whereas the angular momenta of the particle are l = 1 or l = 3 respectively, the pseudo angular momenta are $\tilde{l} = 0$ or $\tilde{l} = 2$. For simplicity it is assumed in the following that the pseudo-spin of the last odd particle is fully decoupled in energy. In this case to each state with spin I_e of the even nucleus there belongs a multiplet of states in the odd partner nucleus with the same excitation energy and with spins :

$$I^+_{odd} = I_e + 1/2 \ \&\ I^-_{odd} = \mid I_e - 1/2 \mid$$

respectively. In particular there is a doublet of states for $I_e \neq 0$ and a singlet for

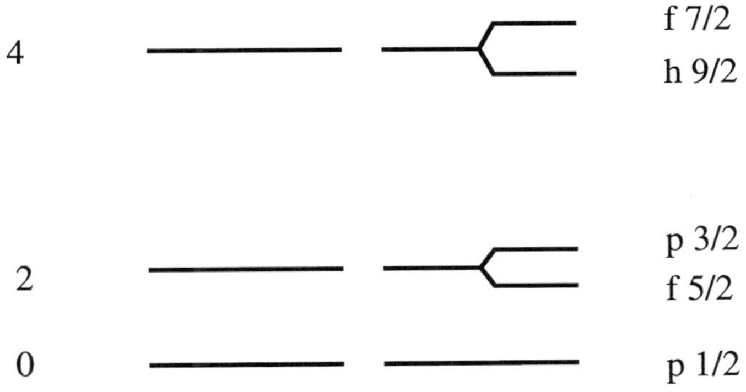

Fig. 2
The figure shows on the right side the spins of the neutron single particle energies in Pb and on the left side their pseudo angular momentum: $\tilde{l}$.

$I_e = 0$. An example of nearly decoupled pseudo-spin multiplets is found in very nice data by Jolie and Graw[32]. They measured a set of states which is presumably complete below 1 MeV. After this they were able to assign the observed levels to pseudo-spin multiplets. Indeed - as shown in Fig. 3 - each level in ^{195}Pt (below 0.8 MeV) comes either as a doublet or as a singlet if $I_e = 1/2$. This is to -my knowledge - the only case in which pseudo-spin multiplets in the odd nucleus are seen so clearly. So more data are needed. It is a beautiful prediction from the dynamical U(6/12) supersymmetry and from the pseudo-spin concept.

3 Supersymmetry with identical bands : A schematic model

I will now discuss a schematic model for supersymmetry (SUSY) with identical bands in the frame of U(6/12). The odd particle is in shells with spins 1/2, 3/2 and 5/2 only. The pseudo angular momenta $\tilde{l}$ of the odd particle have integer values and can be considered as bosonic excitations. They have the values $\tilde{l} = 0$ and $\tilde{l} = 2$ corresponding to the spins L = 0 and L = 2 of the s and d bosons. Thus it is "natural" following T. Otsuka [30] to describe the particles with pseudo angular momenta $\tilde{l} = 0$ and $\tilde{l} = 2$ by (pseudo angular momentum) quasi bosons with the spins $\tilde{l} = 0$: s^f and $\tilde{l} = 2$: d^f. Thus there are three kinds of bosons : neutron bosons b^n , proton bosons b^p, and quasi bosons b^f

The schematic model assumes :

- Equal energies $\varepsilon_d^n = \varepsilon_d^p = \varepsilon_d^f$ of the three kinds of bosons : b^n , b^p, b^f
- Equal interactions V_{ik} among the three kinds of bosons b^n , b^p, b^f .
- Fully decoupled pseudo-spin.

These assumptions imply identical bands in the following three bosonic systems : K_0, K_1 and K_2 defined in Fig. 3 and which have the same total boson numbers N_B : but with different numbers of bosons of the three kinds : b^n , b^p, b^f.

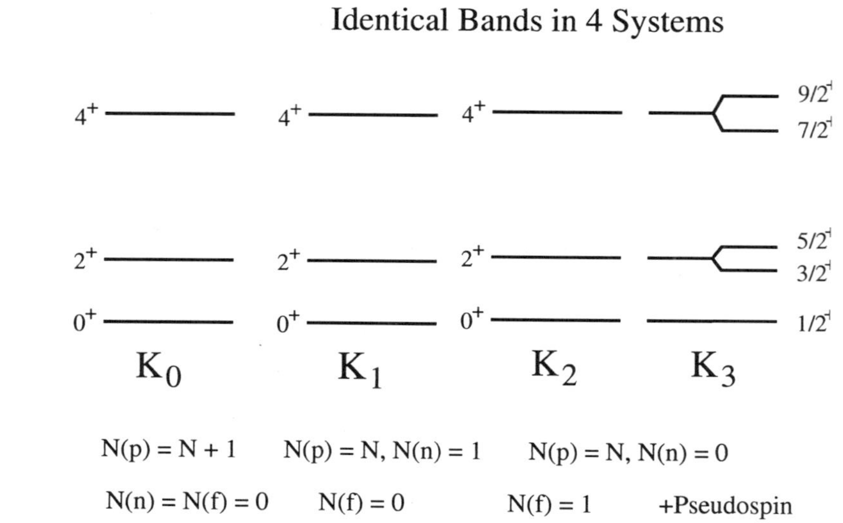

Fig. 3
The figure illustrates the "Identical Bands" suggested by the schematic model in four "nuclei" which have the same total number of "bosons" but differing numbers of bosons of various types as shown in the figure. The "nucleus" K_3 is obtained from K_2 by adding the pseudo-spin $\widetilde{s}$

In particular one finds :

- to every state in K_0 there corresponds a state in K_1 with the same energy. K_0 and K_1 are members of an F spin multiplet. They have good F-spin.
- to every state in K_1 there corresponds a state in K_2 with the same energy and vice versa
- to every state in K_2 there corresponds a pseudo-spin multiplet in K_3 with the same energy

In summary in the schematic model to each pseudo-spin multiplet of the odd system there belongs a unique state of the corresponding even system with the same energy. This is a somewhat surprising result. It means that there are as many states in the even "nucleus" K_1 as there are pseudo-spin multiplets in the "odd nucleus" K_3. One might think that the odd nucleus has many more states. One must remember, however, that also the "mixed symmetry" states in the "even nucleus" have to be considered.

4 Comparison of schematic model to data

Below the following predictions A)... D) of the schematic model are compared to data:

- A) to each level below about 0.8 MeV in ^{194}Pt there is a multiplet in ^{195}Pt with similar energy. We note that not only the yrast band but e.g. also the gamma band has a SUSY partner. (see Fig. 4 [32]) We note that the energies of the centroids of the doublets differ somewhat from the energy of the corresponding level in ^{194}Pt. Also there is some splitting in the doublets. Thus the schematic model is broken somewhat

- B) There are low lying multiplets in ^{195}Pt which do not correspond to low lying states in^{194}Pt, however. In the schematic model they might correspond to high lying mixed symmetry states [31]. An example is the (199 keV and 224 keV) doublet, which has a core state with $I^{\pi} = 1^{+}$.

- C) in the rare earth region there are identical bands with a vanishing projection $\widetilde{L_3} = \widetilde{\Lambda} = 0$ of the pseudo angular momentum on the symmetry axis. The $\widetilde{\Lambda} = 0$ band qualifies as super symmetrical partner band of the K = 0 ground band of the even partner nucleus. These were the bands considered in [13,26,27] . Examples are identical bands in^{171,172}Yb (Fig. 5) , ^{173,174}Hf (Fig. 1) and in ^{181,2}Pt [26]. These are convincing SUSY partner bands.

- D) the bands in ^{171,172}Yb , in ^{173,174}Hf and in ^{181,1822}Pt which have a non vanishing pseudo angular projection $\widetilde{L_3} = \widetilde{\Lambda} \neq 0$ have a pseudo-spin partner band which is separated in energy by a few hundred keV or more. Thus pseudo-spin is not decoupled for these bands. And the schematic model breaks down for these bands .

We note that A) and C) show that there are identical bands and spectra. B) indicates that there are levels in these nuclei which have no partners with the same energies. Thus the schematic SUSY survives only for a part of the levels as a partial symmetry[28,29]. We note, however, that these extra states are reproduced in the dynamical U(6/12) SUSY as this uses a different (more correct) interaction. Part of the reason for the failure is certainly that pseudo-spin is not fully decoupled i.e. the excitation energies in the pseudo-spin doublets are not equal. In ^{207}Pb one finds : ΔE (2f5/2 - 3p3/2) = 329keV. However the pseudo-spin splitting is still much smaller than the spin orbit splitting : ΔE (2f5/2 - 3p3/2) $\leq 0.36 * \Delta E$ (3p1/2 -3 p3/2)). An interesting point concerns the apparent validity of pseudo-spin in the $\widetilde{\Lambda} = 0$ bands in ^{171,172}Yb , ^{173,174}Hf and ^{181,182}Pt (see C) and the fact that pseudo-spin is not decoupled for the $\widetilde{\Lambda} \neq 0$ .The reason may be that the $\widetilde{\Lambda} = 0$ state has no pseudo-spin partner, as is shown e.g. in Fig. 2 for the 3 p 1/2 state, whereas for $\widetilde{\Lambda} \neq 0$ there are two pseudo-spin partner levels e.g. the 2f5/2 and 3p3/2 levels , which can interact. The small decoupling in the $\widetilde{\Lambda} = 0$ bands is also interesting and deserves a closer study.

5 Supersymmetry of identical bands in ^{171,172}Yb supported by lifetime data

There are 5 requirements for an K = 0 and an odd rotational band in neighbor nuclei to be super partners of an unbroken partial supersymmetry:

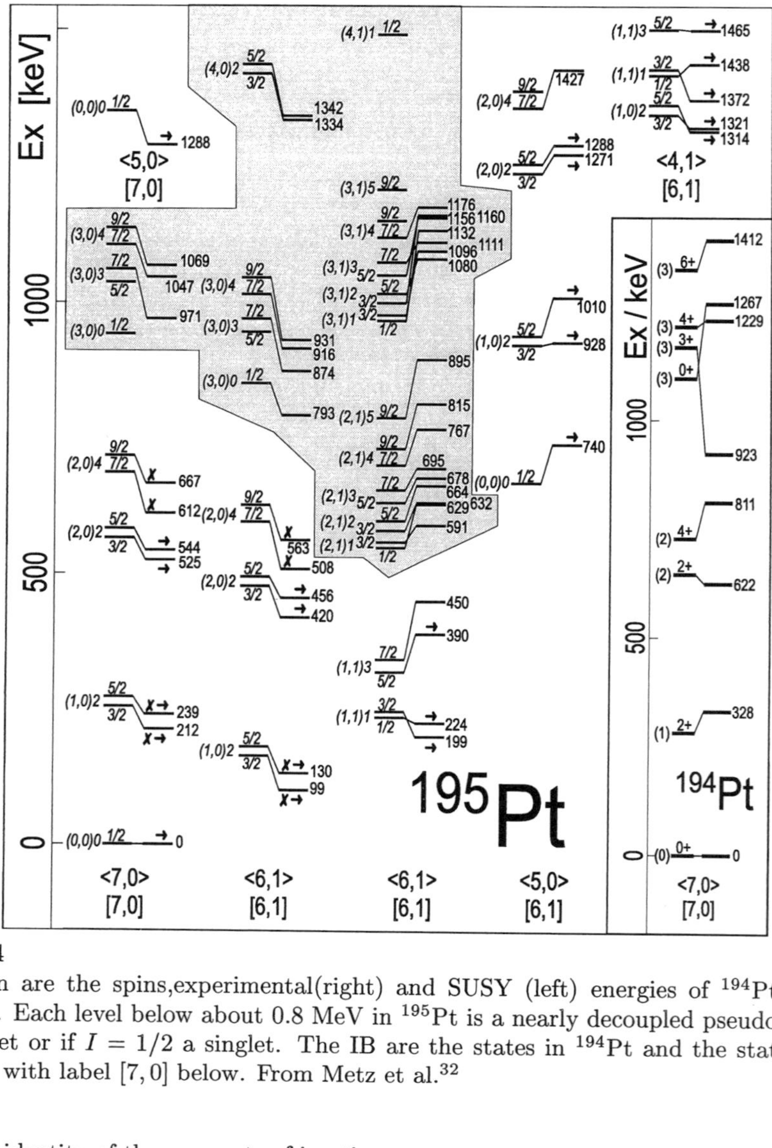

Fig. 4
Shown are the spins,experimental(right) and SUSY (left) energies of ^{194}Pt and ^{195}Pt. Each level below about 0.8 MeV in ^{195}Pt is a nearly decoupled pseudo-spin doublet or if $I = 1/2$ a singlet. The IB are the states in ^{194}Pt and the states in ^{195}Pt with label $[7,0]$ below. From Metz et al.[32]

1) identity of the moments of inertia,

2a) decoupling of the pseudo-spin.(close doublets in the odd nucleus),

2b) alternatively a special value a=1 of the decoupling parameter in the rotational energy formula of the $K = 1/2$ band,

3) an even pseudo angular momentum of the particle,

4) identity of the quadrupole transition moments, and

5) $K = 1/2$, $\tilde{\Lambda}$=0 in the odd band.

Tab.1
Ratios of transition quadrupole moments Q_t from SUSY partner bands in ^{171}Yb and^{172}Yb. The Q_t values are obtained from the measured B(E2) values by using the formulas for the quadrupole moments for the rigid rotor. From Klug, Dewald et al.[27]

	ΔI=2	$\Delta I = 2$	$\Delta I = 1$
I	$\frac{Q_t^{171}(I+1/2)}{Q_t^{172}(I)}$	$\frac{Q_t^{171}(I-1/2)}{Q_t^{172}(I)}$	$\frac{Q_t^{171}(I-1/2)}{Q_t^{172}(I)}$
8	1.02(6)	1.03(6)	1.21(10)
6	0.94(5)	0.94(5)	0.95(6)
4	0.96(2)	0.95(3)	0.97(2)

Some people like to remember these 5 conditions and these special values of special parameters. Others like to say there is a SUSY, that implies all.

At this point it might be appropriate to avoid a possible misunderstanding, namely that the $K = 1/2$, $\tilde{\Lambda}$=0 bands e.g, in ^{171}Yb could be obtained by a weak coupling of the spin of the core state I of ^{172}Yb to the pseudo-spin.

$$J = \mathrm{I} \oplus \tilde{s}$$

This is not true at all. Rather one couples the spins I of several core states strongly to the pseudo angular momentum $\tilde{l}$ of the particle and after this the pseudo-spin is weakly coupled.

$$J = \Sigma(I, \tilde{l})\,[\,(\,I\,\oplus\,\tilde{l}\,)\oplus\,\tilde{s}\,]$$

Thus neither I nor $\tilde{l}$ are not even approximately good quantum numbers. And although the ^{171,172}Yb pair looks like a weak coupling situation, it is actually a quite complicated coupling scheme: strong for $\tilde{l}$ and weak for $\tilde{s}$ [26].

Thus it is important to test the SUSY in more detail by establishing the identity of the quadrupole transitions. The pair ^{171,172}Yb is the only case in which one do Coulomb Excitation on both for both SUSY partners. The energies are shown in Fig. 5 agree with the SUSY. For these reasons Klug, Dewald et al. measured precision lifetimes by the recoil distance method following Coulomb excitation experiments using an EUROBALL CLUSTER detector at the FN-Tandem accelerator in Koeln [27]. From these data and previous data on branching and mixing ratios from Canberra, the ratios of the transition quadrupole moments for corresponding E2 transitions in ^{171}Yb and ^{172}Yb were obtained , which are given in the Table 1.

One notes that the values of the transition quadrupole moment Q_t for ^{171}Yb and ^{172}Yb are in equal within an error of a few percent[27]. This is one of the most accurate experimental comparisons of transition quadrupole moments for super-symmetry partner bands. This striking agreement supports strongly the proposed supersymmetry for ^{171}Yb and ^{172}Yb. Indeed all 5 of the requirements for a super-symmetry are fulfilled for the^{171}Yb and ^{172}Yb pair. On the contrary in the ^{171}Tm and ^{172}Yb pair only three conditions are fulfilled, whereas the conditions 3) and 5): an even pseudo angular momentum of the particle are violated. And indeed there is no supersymmetry in this case.

Summing up convincing examples of (partial) SUSY for energies and B(E2) values in identical bands are shown. An intuitive understanding of the U(6/12)

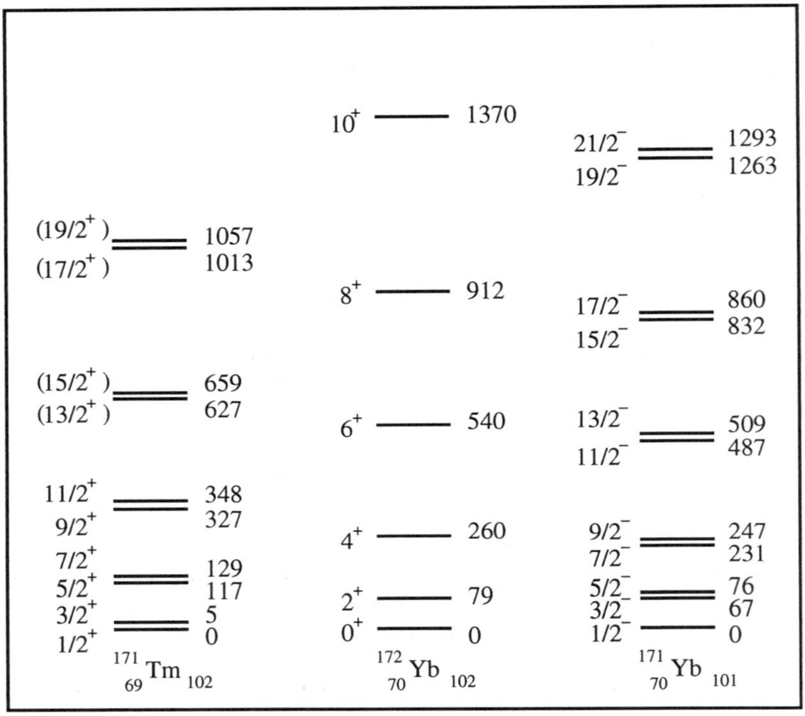

Fig. 5
The figure compares K= 1/2 bands in ^{171}Yb and ^{171}Tm with the K = 0 ground band in ^{172}Yb .The bands in Yb have $\widetilde{\Lambda} = 0$ and thus form a SUSY doublet . The band in ^{171}Tm has $\widetilde{\Lambda} = 1$ (doublet for groundstate) and is thus not the SUSY partner of ^{172}Yb, which has $\widetilde{\Lambda} = 0$. From Klug et al.[27]

SUSY in the frame of the pseudo-spin concept and the schematic model have been given. Limitations of supersymmetry and pseudo-spin symmetry restricting the SUSY to a partial SUSY have been discussed.

My particular thanks go to Franco Iachello, for his constant inspiration to Koeln work since many years. Thanks go also to A. Dewald, A. Gelberg , J. Jolie, R. V. Jolos and T. Otsuka for joint works published and in progress on supersymmetry which were used in this talk. Thanks for many discussions to R. F. Casten, F. Iachello, P. van Isacker and N. Pietralla. This work was partially supported by the DFG under the contract Br 799/10-2.

References

1. F. Iachello and P. Van Isacker, *The Interacting Boson–Fermion Model*, (Cambridge University Press, Cambridge, 1991).

2. A. Frank and P. Van Isacker, *Algebraic Methods in Molecular and Nuclear Physics*, (John Wiley & Sons, New York, 1994).
3. F. Iachello, Phys. Rev. Lett. 44, 772 (1980).
4. F. Iachello and S. Kuyucak, Ann. Phys. (N.Y.) **136**, 19 (1981).
5. A. B. Balantekin, I. Bars and F. Iachello, Nucl. Phys. **A370**, 284 (1981).
6. J. Jolie, P. Van Isacker, K. Heyde, and A. Frank, Phys. Rev. Lett. 55, 653 (1985).
7. P. Van Isacker, A. Frank and Hong-Zhou Sun, Ann. Phys.(N.Y.) 157 (1984)183
8. R. Bijker and F. Iachello, Ann. Phys. (N.Y.) **161**, 360 (1985).
9. D. D. Warner, R. F. Casten, and A. Frank, B, **180**, 207 (1986).
10. J. Vervier, P. Van Isacker, J. Jolie, V. K. B. Kota, and R. Bijker, Phys. Rev. C **32**, 1406 (1985).
11. P. Van Isacker, J. Jolie, K. Heyde, and A. Frank, **54**, 653 (1985).
12. A. Metz, J. Jolie, G. Graw, R. Hertenberger, J. Gröger, C. Günter, N. Warr, and Y. Eisermann, **83**, 1542 (1999).
13. R. V. Jolos and P. von Brentano, Phys. Rev. C **60**, 064318 (1999).
14. T. Byrski, F. A. Beck, D. Curien, C. Schuck et al., Phys.Rev.Lett. **64**, 1650 (1990).
15. W. Nazarewicz, P. J. Twin, P. Fallon, and J. D. Garrett, Phys.Rev.Lett., **64**, 1654 (1990).
16. C. Baktash, B. Haas, W. Nazarewicz, Annu.Rev.Nucl.Part.Sci., **45**, 485 (1995)R. Janssens, private communication
17. R. Janssens, private communication.
18. A.Arima, M.Harvey, and K.Shimizu, Phys.Lett.,**30 B**, 517 (1969).
19. K.T.Hecht and A.Adler Nucl.Physics A 137 129 (1969).
20. Ginocchio J N **78**,436 (1997)
21. D. D. Warner, R. F. Casten Phys. Rev. C **28**,1798, (1983).
22. A. Frank, P. Van Isacker, D. D. Warner, Phys. Lett**.**, **179**B, 474,(1987).
23. A. Gelberg, P. von Brentano, R. F. Casten, J.Phys. G **16**, L143 (1990).
24. R. D. Amado, R. Bijker, F. Cannata, and J. P. Dedonder, Phys. Rev. Lett., **67**, 2777 (1991).
25. R.V.Jolos, P. von Brentano, A. Gelberg, K.–H. Kim, and T.Otsuka, B, **430**, 1 (1998).
26. R. V. Jolos and P. von Brentano, Phys. Rev. C **6302**, 024304 (2001).
27. T. Klug, A. Dewald, R.V. Jolos , B. Saha, P. von Brentano and J.Jolie. Phys. Lett. B 525, 252 (2002).
28. Y. Alhassid and A. Leviatan, J. Phys. A **25**, L1265 (1992).
29. A. Leviatan, **77**, 818 (1996).
30. T.Otsuka, private communication
31. P. von Brentano and A. Gelberg, to be published.
32. A. Metz, Y. Eisermann, A. Gollwitzer, R. Hertenberger, G. Graw and J. Jolie, Phys. Rev. C 61 064313 (2000).

SEARCHING FOR BOSON-FERMION SYMMETRIES IN NEUTRON-RICH NUCLEI

J. A. CIZEWSKI, K. L. JONES, J. S. THOMAS
Rutgers University, New Brunswick, NJ 08903 USA
E-mail: cizewski@physics.rutgers.edu

D. W. BARDAYAN, J. C. BLACKMON, C. J. GROSS, F. LIANG, D. SHAPIRA, M. S. SMITH, D. W. STRACENER
Oak Ridge National Laboratory, Oak Ridge, TN 37831 USA

R. L. KOZUB, C. D. NESARAJA
Tennessee Technological University, Cookeville, TN 38505 USA

U. GREIFE, R. J. LIVESAY
Colorado School of Mines, Golden, CO 80401 USA

Z. MA
University of Tennessee, Knoxville, TN 37996 USA

Nuclear shell structure has been predicted to evolve as the N/Z ratio becomes much larger than characteristic of stable nuclei. To probe this expected evolution will require transfer reactions using beams of neutron-rich nuclei. The first measurements of a (d,p) reaction with a ^{82}Ge beam are reported. The prospects for future studies of single-particle transfer reactions on unstable neutron-rich species, and implications for future searches for boson-fermion symmetries and supersymmetries, are discussed.

1 Introduction

Many examples of boson-fermion symmetries and supersymmetries have been identified in nuclei when the boson cores take on the O(6) limit, the SU(3) limit, or the U(5) limit of the Interacting Boson Model. [1] The sensitive tests of boson-fermion symmetries, including supersymmetries, come from particle transfer reactions, which identify the single-particle nature of the nuclear excitations. The most compelling examples of boson-fermion symmetries have been identified in the A=196 nuclei, where the quartet of nuclei, ^{194}Pt, ^{195}Pt, ^{195}Au, and ^{196}Au have been identified as a supermultiplet. [2] For this supersymmetry the even-even core has O(6) symmetry, the odd neutrons occupy the pseudo-*sd* $j = 1/2, 3/2, 5/2$ $(p_{1/2}, p_{3/2}, f_{5/2})$ orbitals and the odd proton occupies the $j = 3/2$ $(d_{3/2})$ orbital. The realization of this symmetry relied on identifying the appropriate distribution of single-particle strengths for these configurations, as well as small single-particle strengths of neutron $j = 7/2$ or proton $j = 1/2$ strengths, which are outside of the symmetry.

Dynamical supersymmetric structures have also been discussed for superdeformed excitations in the A$\sim$150 and 190 regions [3,4] and normal deformed excitations in the rare-earth region, including recent studies [5] in the ^{172,171}Yb,^{171}Tm nuclei, where the boson core is near the SU(3) limit. In spherical nuclei, dynamical supersymmetries have been suggested for $j = 1/2$ or $j = 3/2$ particles coupled to an SU(5) core. [1]

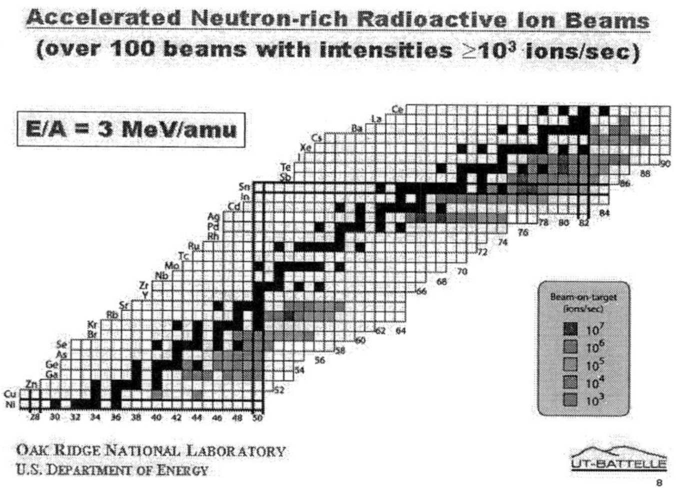

Figure 1: Predicted intensities (Ref. 7) of neutron-rich species accelerated to 3 MeV/A at HRIBF.

Most tests of boson-fermion symmetries or supersymmetries have been in stable nuclei, or nuclei near stability – a small fraction of the nuclei that are expected to exist. A particular fertile ground could be neutron-rich nuclei. However, in very neutron-rich nuclei the shell structure is expected to change because of the diffuse nature of the nuclear surface. [6] The future searches for boson-fermion symmetries will therefore rely even more heavily on single-particle transfer reactions: to identify the shell structure, as well as to identify the distribution of single-particle strengths in nuclei which are candidates for boson-fermion symmetries.

In stable nuclei the shell structure is that of a modified harmonic oscillator, with a strong spin-orbit interaction, with shell gaps at 50, 82, 126, for example. Boson-fermion symmetries in the 50-82 shell involve the normal parity, N=4 $d_{5/2}$, $g_{7/2}$, $s_{1/2}$, $d_{3/2}$ orbitals that can be considered a pseudo N=3 *fp* shell. If the spin-orbit interaction is reduced in neutron-rich nuclei, then the shell structure could involve the N=4 shell, either the full *sdg* shell, or maybe just *s* and *d* orbitals. For nuclei with a very diffuse surface, the shell structure could be weakened so that the single-particle levels would be equally spaced. [6] Depending upon the location of the Fermi surface, one might need the entire *sdg* space, or maybe just *sd* single-particle orbitals for the fermion space, when boson-fermion symmetries are considered. A specific prediction[6] is that $s_{1/2}$ orbitals are expected to come low in energy in weakly bound, neutron-rich nuclei. To determine the single-particle nature of excitations in neutron-rich nuclei will require particle transfer reactions with radioactive ion *beams* of nuclei far from stability.

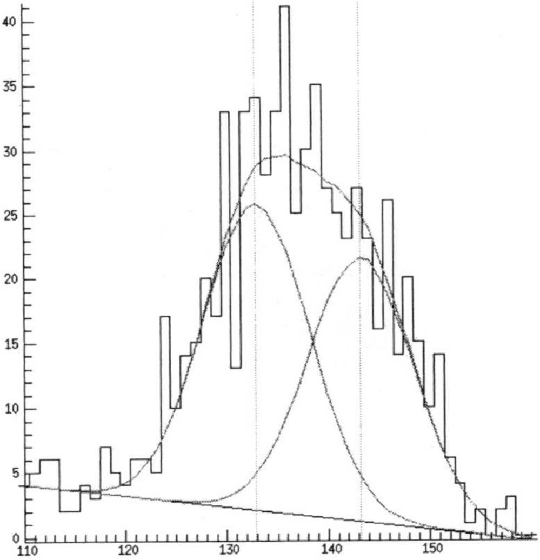

Figure 2: Partial Q-value spectrum of the d(^{82}Ge,p) reaction highlighting the group of states near the ground state.

2 Experimental Methods

The studies of transfer reactions with neutron-rich beams are enabled because of the plethora of such beams, summarized in Fig.1, that are available at the Holifield Radioactive Ion Beam Facility (HRIBF) at Oak Ridge National Laboratory. A UC target is bombarded with protons from the ORIC cyclotron. Negative ions or molecules of fission fragments are injected into the 25-MV Tandem Van de Graaff accelerator and delivered to the target area. Enhanced intensity beams of neutron-rich Sn and Ge isotopes have been enabled by injecting sulphides of these species, a method that reduces significantly the isobaric contamination.

The first direct transfer reaction measurements with a neutron-rich species involved an ^{82}Ge beam on a CD_2 target. The mixed beam of ^{82}Se/^{82}Ge bombarded a thin (~0.4mg/cm^2) deuterated polyethylene target. The heavy recoils were detected in an ion chamber with separate ΔE sections and in coincidence with the light recoils detected in the SIDAR array [8] of silicon strip detectors in a lampshade geometry. For this experiment SIDAR subtended 150 to 105 degrees in the laboratory, which corresponds to about 15-40 degrees in the center of mass for the (d,p) reaction with a 4 MeV/A ^{82}Ge beam. Typical intensities of 10^4 particles/s of ^{82}Ge were obtained on target.

Although the stable ^{82}Se contaminant was about 6 times stronger than the ^{82}Ge of interest, the species were readily separated in ΔE-E in the ion chamber. By injecting the sulfide into the tandem accelerator, the efficiency for extracting ^{82}Ge was enhanced and the relative intensities of the unstable ^{82}As and stable ^{82}Se were dramatically reduced.

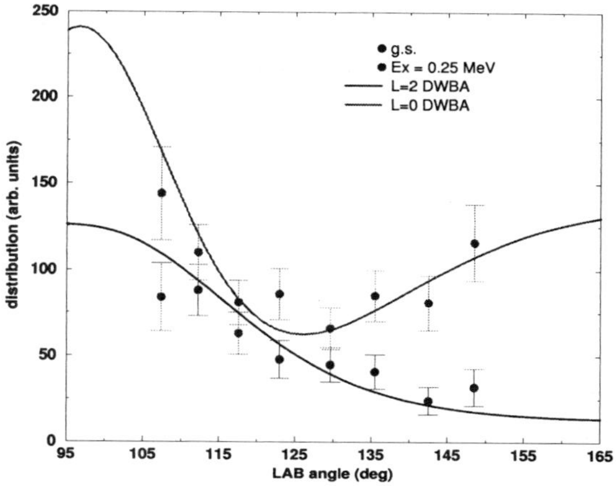

Figure 3: Preliminary angular distribution analysis of the two components of the Q-value spectrum in Fig.2 compared to DWBA predictions for $\ell = 0$ and $\ell = 2$ transfer in the lab frame.

A partial Q-value spectrum, associated with the lowest excited states in the N=51 isotone ^{83}Ge, is displayed in Fig. 2. The energy calibration and experimental response were obtained from the d(^{82}Se,p) data, since this reaction is well understood from previous measurements [9] in normal kinematics. The lowest group is about 450-keV wide, and can be resolved into two components: the ground state and an excited state at about 270 keV. The preliminary analysis of the angular distributions of these two components, shown in Fig. 3, is consistent with a ground state populated with ℓ=2, presumeably $d_{5/2}$ transfer, and a first excited $1/2^+$ state populated with ℓ=0.

3 Systematics of N=51 Isotones and Future Prospects

The systematics of the N=51 isotones are summarized in Fig. 4. All of the ground states are $5/2^+$. In Z=38,40, near the proton sub-shell closure, the $1/2^+$ state is above 1 MeV, as well as other excited states. Below Z=38, the $1/2^+$ becomes lower energy, as low as ~270 keV in ^{83}Ge. The $d_{3/2}$ excitations have not been identified in the neutron-rich ^{83}Ge and ^{85}Se isotones. In ^{85}Se the next excited states are around 1 MeV in excitation; experiments are planned to study the d(^{84}Se,p) reaction to probe the single-particle character of the excitations. In ^{83}Ge no additional states below 1 MeV have been populated, which implies that the $d_{3/2}$ strength is above this energy.

This first foray into the terra incognito of neutron-rich nuclei above N=50 can provide some insight into possible searches for boson-fermion symmetries in neutron rich nuclei with N>50. With low-lying $d_{5/2}$ and $s_{1/2}$ orbitals, and the $d_{3/2}$ orbital

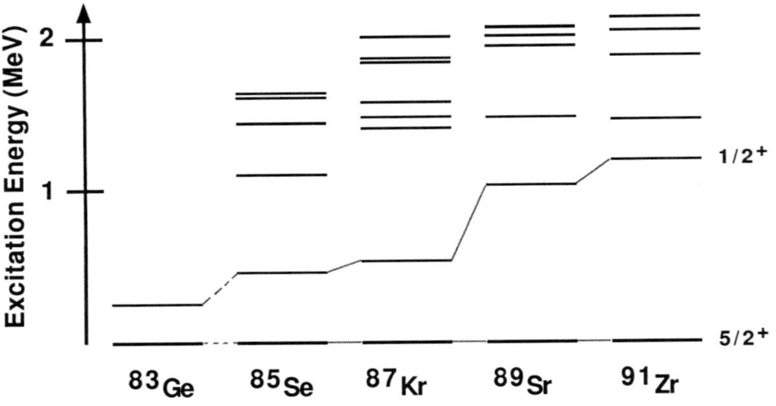

Figure 4: Systematics of the N=51 isotones. Data taken from Refs. 9,10 and present results.

much higher in energy, a fermion symmetry is unlikely. However, as even more neutron-rich species become available, the systematics suggest that the $s_{1/2}$ state could become the ground state, well separated from the $d_{5/2}$ orbital. So possible tests of a U(6/2) supersymmetry could be feasible, of course assuming that the collective structure of these exotic nuclei retain the boson symmetries.

In the near term the study of (d,p) reactions with neutron-rich unstable beams will continue to exploit the wealth of beams available at HRIBF. In particular, the study of the d(^{132}Sn,p) reaction is feasible since enriched ^{132}Sn beams of 10^5 particles/s at 4.5-MeV/A have been obtained. The prospects are also favorable to study open shell nuclei, such as the transitional ^{92,94}Sr species that are expected to be available with high intensity.

In the longer term, the exciting opportunities with the next generation of radioactive ion beam accelerators, including the Rare Isotope Accelerartor RIA, will enable the extension of the present studies to more neutron-rich N=50 isotones, as well as neutron-rich Ge isotopes.

Acknowledgements

This work was supported in part by the U.S. National Science Foundation and Department of Energy.

References

1. F. Iachello and P. Van Isacker, *The Interacting Boson-Fermion Model*, Cambridge University Press (1991), and references therein.
2. A. Metz et al., Phys. Rev. Lett. **83**, 1542 (1999); A. Metz et al., Phys. Rev. C **61**, 064313 (2000).
3. A. Gelberg, P. von Brentano, and R. F. Casten, J. Phys. **G16** L143, (1990).
4. J. A. Cizewski, in *Perspectives for the Interacting Boson Model*, R.F. Casten, A. Vitturi et al. (editors), World Scientific, (1994) p. 351.
5. P. von Brentano, these proceedings.
6. J. Dobaczewski, et al., Phys. Rev. C**53**, 2809 (1996).
7. http://www.phy.ornl.gov/hribf; D. Stracener, private communication.
8. D.W. Bardayan et al., Phys. Rev. C **62**, 055804 (2000).
9. J. P. Omtvedt, et al. Z. Phys. **A339**, 349 (1991).
10. Evaluated Nuclear Structure Data Files, http://www.nndc.bnl.gov.

SUPERSYMMETRY IN NUCLEAR CLUSTERIZATION

G. LÉVAI
Institute of Nuclear Research of the Hungarian Academy of Sciences (ATOMKI), Debrecen, Pf. 51, 4001 Hungary
E-mail: levai@atomki.hu

The α-cluster bands of the ^{20}Ne, ^{19}F and ^{18}F nuclei are analysed in terms of a U(4|12) superalgebra. These systems are interpreted as members of the 0, 1 and 2 fermion sectors of the supersymmetry scheme, and the correlations between their spectroscopic properties are discussed according to this assumption.

1 α-cluster Bands in Nuclei with Mass Number $A = 16$ to 20

α-cluster states are known to be present in the spectrum of many light nuclei. The simplest cluster systems of this type are core+α-type structures, and they usually account for the most well-known collective bands of nuclei in the $A = 16$ to 20 range. The best known example is the ^{20}Ne nucleus, the spectrum of which is known to contain several α-cluster bands [1] including the ground-state band $K^\pi = 0_1^+$, the 0^- band (sometimes interpreted as the parity doublet of the ground-state band [2]), the 0_4^+ band, which contains states with remarkably large α-width, and some other $K^\pi = 0^+$ bands [1]. The members of the ground-state band of ^{20}Ne are often interpreted as states with $Q = 8$ oscillator quanta assigned to the relative motion of the α-particle and the ^{16}O core [2,3]. There are a number of α-cluster bands in the spectrum of ^{19}F (and its less well-known mirror pair ^{19}Ne) [4]. These are the $K^\pi = \frac{1}{2}^+$ ground-state band, the lowest-lying negative-parity band with $K^\pi = \frac{1}{2}^-$ and some other bands lying above $E_x = 5$ MeV. Further well-known α-cluster bands of $A = 18$ nuclei are the $K^\pi = 1^+$ band based on the $T = 0$ state of ^{18}F at $E_x = 1.701$ MeV and a $K^\pi = 0^+$ band with $T = 1$ in the spectrum of ^{18}O at $E_x = 3.634$ MeV [3], which has isobaric analogues in the spectrum of ^{18}F and ^{18}Ne too. The $K^\pi = 0^+$ band of ^{16}O starting at $E_x = 6.049$ MeV can also be mentioned as another example for characteristic α-cluster bands in this mass range.

Not surprizingly, there were efforts to describe at least some of these α-cluster bands in a unified framework [5]. In a local potential model Buck et al. discussed the ground-state band of ^{20}Ne, the $K^\pi = \frac{1}{2}{}_1^-$ band of ^{19}F, the $K^\pi = 1^+$ $(T = 0)$ band of ^{18}F and the $K^\pi = 0^+$ $(T = 1)$ band of ^{18}O and interpreted them as bands with $Q = 8$ oscillator quanta in the relative motion of the clusters [3]. Introducing spin-orbit and other spin-dependent tensor forces for the description of the ^{18}F states, they got good agreement with the observed properties of these bands.

A more complete description of α-cluster states in a wider mass range was presented [6] in terms of the semimicroscopic algebraic cluster model (SACM) [7]. In this model the clusters are treated in terms of Elliott's SU(3) model [8], while the relative motion of the clusters is described in terms of the SU(3) basis of the vibron model [9]. The SACM basis is a symmetry-dictated truncation of the complete SU(3) shell-model basis of the whole nucleus and, therefore, is free from Pauli-forbidden states and from spurious centre-of-mass excitations. Furthermore, it also reflects the cluster character of states. The ^{16}O+α, ^{15}N+α, ^{14}C+α and ^{12}C+α systems were described using a standardized Hamiltonian [6], and the dependence of the model parameters on the mass number was extracted. As a test of the consistency of this systematic study the spectrum of the ^{17}O nucleus as a ^{13}C+α system was generated by using the *same* Hamiltonian with parameters *interpolated* from the parameters of the neighbouring nuclei. This systematic study confirmed that the parameters of the SACM Hamiltonian for core+α-type cluster systems vary smoothly in the $A = 16$ to 20 region [6].

Inspired by the success of this study we proposed a supersymmetry scheme based on the SACM, making use of the fact that both bosons and fermions appear naturally in it [10]. In particular, the bosonic sector is identified with the relative motion of the clusters, while the fermions are defined as *holes* on the p shell. The zero-fermion case then corresponds to a closed-shell core (^{16}O). The addition of fermions corresponds to a decrease in mass of the nucleus, while the addition of bosons is equivalent to an increase in the number of relative excitation quanta. A physical argument in support of this type of supersymmetry is that the typical energy of the fermionic and bosonic excitations is in the same range for these nuclei: The shell excitation quanta are $\hbar\omega \simeq 13$ MeV in this region and the typical nucleon separation energies are also of this order for most nuclei close to the valley of stability. In technical terms this supersymmetry scheme is rather similar to the ones based on the interacting boson and interacting boson–fermion models, which also establish connection between odd and even nuclei [11], but the physical picture associated with it (*e.g.* the nature of the bosons) is rather different.

2 The U(4|12) Superalgebra

The boson creation and annihilation operators $b^{\dagger(l,0)l}_{m_l}$ and $\tilde{b}^{(0,l)l}_{m_l}$ are those of the vibron model [9] (*i.e.* π and σ bosons for $l = 1$ and 0), where the superscript indicates the SU(3) tensor character (λ, μ). The generators of $\mathrm{U}_B(4)$ are constructed as [10]

$$B^{(\lambda,\mu)L}_{M_L}(l,l') = [b^{\dagger(l,0)} \times \tilde{b}^{(0,l')}]^{(\lambda,\mu)L}_{M_L} \ . \tag{1}$$

The eight $B^{(1,1)L}_{M_L}(1,1)$ operators (with $L = 1$ and 2) generate $\mathrm{SU}_B(3)$. The fermion operators $a^{\dagger(0,1)lst}_{m_l m_s m_t}$ and $\tilde{a}^{(1,0)lst}_{m_l m_s m_t}$, which create and annihilate a hole on the p shell, have SU(3) character $(\lambda, \mu) = (0,1)$ and (1,0), respectively, and carry orbital angular momentum $l = 1$, spin $s = 1/2$, and isospin $t = 1/2$. The bilinear products

$$A^{(\lambda,\mu)LST}_{M_L M_S M_T} = [a^{\dagger(0,1)\frac{1}{2}\frac{1}{2}} \times \tilde{a}^{(1,0)\frac{1}{2}\frac{1}{2}}]^{(\lambda,\mu)LST}_{M_L M_S M_T} \tag{2}$$

generate $\mathrm{U}_F(12)$. For $S = T = 0$ and $(\lambda, \mu) = (1,1)$ one gets the 8 generators of the orbital $\mathrm{SU}_F(3)$ algebra. For $(\lambda,\mu) = (0,0)$ the 16 generators of Wigner's $\mathrm{U}^{ST}_F(4)$ supermultiplet algebra are obtained. To embed the bosonic and fermionic algebras in a superalgebra, one has to define generators which create a fermion and annihilate a boson, or *vice versa*. The former type of operators can be written as

$$D^{(\lambda,\mu)Lst}_{M_L m_s m_t}(l') = [a^{\dagger(0,1)st} \times \tilde{b}^{(0,l')}]^{(\lambda,\mu)Lst}_{M_L m_s m_t} \ . \tag{3}$$

The relevant group chain contains that of the SACM for core+α cluster systems [6], embedded in U(4|12):

Table 1. α-cluster bands of A=20, 19 and 18 nuclei up to two major shell excitations. The bands are identified with the labels $n_\pi(\lambda,\mu)$, and also carry the T isospin quantum number.

	(^{16}O; T=0)+α	(^{15}N, ^{15}O; $T=\frac{1}{2}$) + α	(^{14}C, ^{14}N, ^{14}O; T=1)+α and (^{14}N; T=0)+α
$0\hbar\omega$	8(8,0)	7(6,0)	6(4,0)
$1\hbar\omega$	9(9,0)	8(7,0), 8(8,1)	7(5,0), 7(6,1)
$2\hbar\omega$	10(10,0)	9(8,0), 9(9,1)	8(6,0), 8(7,1), 8(8,2)

$$
\begin{aligned}
&\mathrm{U}(4|12) \supset \mathrm{U}_B(4) \times \mathrm{U}_F(12) \supset \mathrm{SU}_B(3) \times \mathrm{SU}_F(3) \times \mathrm{U}_F^{ST}(4) \\
&\supset \mathrm{SU}(3) \times \mathrm{SU}_F^S(2) \times \mathrm{SU}_F^T(2) \supset \mathrm{SO}(3) \times \mathrm{SU}_F^S(2) \times \mathrm{SU}_F^T(2) \supset \mathrm{Spin}(3) \times \mathrm{U}_F^T(1)\ . \quad (4)
\end{aligned}
$$

The associated quantum numbers are also those of the SACM [6], extended with the fermion and total particle numbers N_F and $\mathcal{N} = N_B + N_F$. The remaining algebras and the associated quantum numbers play the same role as in the SACM [6]: the $\mathrm{U}_B(4)$, $\mathrm{SU}_B(3)$, and $\mathrm{SU}_F(3)$ representation labels $N_B = n_\pi + n_\sigma$, n_π, and (λ_F, μ_F) denote the total boson number, the dipole boson number, and Elliott's SU(3) labels of the core. The SU(3), $\mathrm{U}_F^{ST}(4)$, $\mathrm{SU}_F^S(2)$, $\mathrm{SU}_F^T(2)$, SO(3), Spin(3), and $\mathrm{U}_F^T(1)$ labels (λ, μ), $[f_1, f_2, f_3, f_4]$, S, T, L, J, and M_T stand for the quantum numbers labelling the unified nucleus in terms of the Elliott's LS-coupled SU(3) shell model. As in the SACM, the SU(3) representations (λ, μ) are obtained from the SU(3) multiplication $(n_\pi, 0) \times (\lambda_F, \mu_F)$, keeping only those contained in the fully antisymmetric SU(3) model space of the unified nucleus. The number of the dipole bosons n_π, *i.e.* the number of harmonic oscillator quanta in the relative motion of the clusters, also determines the respective shell ($n\hbar\omega$) of the unified nucleus via $n = n_\pi - n_\pi^{\min}$: states with $n_\pi < n_\pi^{\min}$ are excluded due to the Pauli blocking between the nucleons of the two clusters [6] (*i.e.* the Wildermuth condition). For ^{20}Ne$\sim$ ^{16}O+α and ^{19}F$\sim$ ^{15}N+α $n_\pi^{\min} = 8$ and 7, respectively. Keeping only the essential quantum numbers, the basis states can be labelled as

$$
|\mathcal{N} N_B n_\pi, (\lambda_F, \mu_F); (\lambda, \mu)\chi LSJM_J TM_T\rangle\ . \quad (5)
$$

The total number of particles $\mathcal{N}$ is chosen by taking into account the physical relevance of N_F and N_B, the number of fermions and bosons. N_F should be at least the maximal number of holes on the p-shell, 12. With $\mathcal{N} = 12$, the maximal number of π bosons is also 12 ($N_B = n_\pi + n_\sigma$). The essential labels identifying α-cluster bands of $A = 20$, 19 and 18 nuclei are shown in Table 1.

3 $^{20}\mathrm{Ne} \sim {}^{16}\mathrm{O} + \alpha$ and $^{19}\mathrm{F} \sim {}^{15}\mathrm{N} + \alpha$ as Systems with $N_F = 0$ and 1

As the first test of cluster supersymmetry the the α-cluster states of ^{20}Ne and ^{19}F were analyzed in terms of the U(4|12) superalgebra [10]. 25 ^{20}Ne and 26 ^{19}F states were assigned to 5 and 6 bands, and they were fitted with the Hamiltonian

$$
\begin{aligned}
H = {}& \gamma_B n_\pi + \theta_B (-1)^{n_\pi} + \delta_B C^{(2)}(\mathrm{SU}_B(3)) \\
&+ \delta C^{(2)}(\mathrm{SU}(3)) + \beta L \cdot L + [\xi + (-1)^{n_\pi} \xi'] L \cdot S\ , \quad (6)
\end{aligned}
$$

which is diagonal in the basis (5). The first four terms of (6) set the bandhead energies, while the last three account for the relative position of states within a band. The γ_B parameter was not fitted but was kept at the value of an oscillator constant appropriate for $A = 20$ nuclei [6], $\gamma_B = 13.185$ MeV. In assigning the states to α-cluster bands, standard compilations [12] were used as well as other works that suggest further band assignments based on experimental arguments [1,4]. We found correlated behaviour between the ^{20}Ne and ^{19}F spectra in the sense that rather similar parameter sets were obtained in joint and separate fits of the two systems [10]. As a quantitative measure for the validity of the supersymmetry, we calculated [10] with the parameters of the joint fit $(\sum_i |E_{Th.i} - E_{Ex.i}|)/\sum_i E_{Ex.i} = 0.13$, which is comparable to the value 0.14 obtained for ^{190}Os and ^{191}Ir assuming U(6|4) supersymmetry [11]. Similar conclusions were drawn for E2 transitions too [10].

Since the SUSY generators of the type (3) connect states of neighbouring nuclei, the analysis of the C^2S spectroscopic factors observed in one-nucleon transfer reactions from the ground state of ^{20}Ne to various states of ^{19}F allow for a direct test of the supersymmetry construction. Calculating the matrix elements of the

$$D^{(\lambda,\mu)Jt}_{M_J m_t}(lLs) = \sum_{M_L m_s} \langle LM_L s m_s | JM_J \rangle D^{(\lambda,\mu)Lst}_{M_L m_s m_t}(l) \ . \tag{7}$$

SUSY generator we found that the allowed transitions correspond to strong C^2S values observed in the ^{20}Ne(t,α) reaction [12]. Furthermore, the relative intensity of two transitions was also reproduced with good accuracy [10]. In summary, we concluded that although there are few criteria by which the validity of the present supersymmetry can be tested, these all seemed to support the idea of correlating different cluster systems under this scheme.

4 Preliminary Results for ^{18}F $\sim$ ^{14}N $+ \alpha$ as a System with $N_F = 2$

The $N_F = 2$ sector of the model space consists of the α-cluster states of ^{18}O, ^{18}F and ^{18}Ne. The $U_F^{ST}(4)$ spin-isospin structure is more diverse in this case: the combination $(S,T) = (0,1)$ occurs for all three nuclei and corresponds to isospin triplets in their spectra, while the combination $(S,T) = (1,0)$ characterizes only ^{18}F, and leads to a rich angular momentum structure due to the the coupling of the orbital angular momentum L to spin 1. Obviously, this situation complicates the description of the $A = 18$ nuclei. First, the simple Hamiltonian (6) is not sufficient anymore for a joint fit of the nuclei, since it lacks isospin effects. Second, the presence of $S = 1$ structures requires the incorporation of tensorial couplings [3].

Here we do not consider the extension of the Hamiltonian (6), rather we analyse the spectroscopic information available for ^{18}F states and attempt their interpretation in terms of the U(4|12) SUSY scheme without trying to reproduce the actual energy eigenvalues. In the theoretical assignment of the individual states we make use of the experimentally established band structure, electromagnetic transition data [12] and our previous results concerning the α-cluster states of ^{18}O [13], which (might) have analogue $T = 1$ states in ^{18}F. Table 2 contains the essential experimental information on a number of ^{18}F states, including the band assignment (if any), levels to which the strongest E2 and M1 transitions occur and one-nucleon transfer data for the ^{19}F(p,d) reaction [12]. The electromagnetic transitions can have both isoscalar and isovector character, however, as expected [14], the dominant E2

Table 2. Spectroscopic information on ^{18}F levels [12]. Only those levels are shown, for which sufficient data exist for their theoretical band assignment. The indicated E2 and M1 transitions correspond to B(E2) and B(M1) values in excess of 3 and 0.01 W.u., respectively. The "w" character indicates known transitions weaker than these, while blank entries indicate missing data. All E2 transitions presented have isoscalar character, while the M1 transitions are isovector ones, except those indicated in italics. The C^2S values stand for data from the ^{19}F(p,d) reaction.

E_x			Experiment			Theory	
(MeV)	$J^\pi;T$	K^π	E2 to level	M1 to level	C^2S	$n_\pi(\lambda,\mu)KLS$	C^2S
0.0	$1^+;0$	0^+			0.65	6(4,0)001	
0.937	$3^+;0$	0^+	0.0		1.47	6(4,0)021	
1.121	$5^+;0$	0^+	0.937			6(4,0)041	0
4.360	$1^+;0$			3.062	0.04+0.13	6(4,0)021	
3.839	$2^+;0$			3.062	0.50	6(4,0)021	
6.311	$3^+;0$		3.724	3.062, 3.839, *4.116*, 4.964		6(4,0)041	0
1.042	$0^+;1$			0.0	0.27	6(4,0)000	
3.062	$2^+;1$		1.042	0.0, *0.937*	0.74	6(4,0)020	
4.652	$4^+;1$					6(4,0)040	0
1.081	$0^-;0$	0^-			0.38	7(5,0)011	
2.101	$2^-;0$	0^-	1.081		(0.03)	7(5,0)031	0
4.398	$4^-;0$	0^-	2.101			7(5,0)051	0
3.134	$1^-;0$	1^-		w	1.04	7(5,0)011	
4.226	$2^-;0$	(1^-)	w	w	(0.015)	7(5,0)011	
3.791	$3^-;0$	1^-				7(5,0)031	0
6.096	$4^-;0$	1^-	2.101	w		7(5,0)031	0
4.848	$5^-;0$	1^-				7(5,0)051	0
5.605	$1^-;0+1$			1.081	(0.82)	7(5,0)010	
6.240	$3^-;0+1$			2.101, 3.791, 4.226, 4.398		7(5,0)030	0
4.860	$1^-;0$			w	(0.11)	7(6,1)111	
5.786	$2^-;0$		1.081			7(6,1)111	
1.701	$1^+;0$	1^+		1.042	0.07	8(6,0)001	0
2.523	$2^+;0$	1^+			~ 0	8(6,0)021	0
3.358	$3^+;0$	1^+	1.701	w	~ 0	8(6,0)021	0
5.298	$4^+;0$	1^+	2.523, 3.358	4.652		8(6,0)041	0
6.567	$5^+;0$	1^+	3.358	*5.298*		8(6,0)041	0
9.58	$6^+;0$	1^+				8(6,0)061	0
11.22	$7^+;0$	1^+				8(6,0)061	0
3.724	$1^+;0$				0.015+0.22	8(6,0)021	0
4.116	$3^+;0$		w	3.062	$\simeq 0$	8(6,0)041	0
4.753	$0^+;1$				0.03+0.08	8(6,0)000	0
6.283	$2^+;1$		w	0.937, 1.701, 3.358, 3.724, 3.839, 4.116, 4.360		8(6,0)020	0
4.964	$2^+;1$				~ 0	8(8,2)220	0

and M1 transitions belong to the former and latter type, respectively.

In the theoretical interpretation of positive-parity $T=0$ states we assigned the members of the $K^\pi=0^+$ and $K^\pi=1^+$ bands [12] to the $J=L+1$ states of the $n_\pi(\lambda,\mu)=6(4,0)$ basis states and to the $J=L+1$ and $J=L$ states of the $8(6,0)$ states, respectively. We also assigned some further low-lying levels to the remaining combinations of J and L. We assigned the members of the negative-parity $T=0$

$K^\pi = 0^-$ and 1^- bands [12] to the model states $n_\pi(\lambda,\mu) = 7(5,0)$ with $J = L-1$, $J = L$ and $J = L+1$. In the interpretation of the $T = 1$ states we made use of the level assignment of the $^{18}O \sim {}^{14}C + \alpha$ system [13]. In this procedure we took into account the selection rules of the E2 and M1 operators [7], i.e. that they can connect states with $\Delta n_\pi = 0$, $|\Delta\lambda| \leq 1$ and $|\Delta\mu| \leq 1$. Most of the dominant transitions indicated in Table 2 fulfil this requirement.

Similarly to the case of ^{20}Ne and ^{19}F, a direct check of the validity of the supersymmetry scheme is possible by inspecting the C^2S spectroscopic factors obtained from the ^{19}F(p,d) reaction [12]. Calculating the matrix elements of the (7) operators (SUSY generators) with the ^{19}F ground state labelled by $n_\pi(\lambda,\mu)KL = 7(6,0)00$ and the general ^{18}F model state we find that the $6j$, $9j$ and SU(3) $9j$ symbols introduce strict selection rules for the L and (λ,μ) labels of the ^{18}F states. In particular, operator (7) with $(\lambda,\mu) = (0,2)$ character can generate transitions only to levels with $n_\pi(\lambda,\mu) = 6(4,0)$ and $L = 0, 2$, while that with $(\lambda,\mu) = (0,1)$ character can reach levels with $n_\pi(\lambda,\mu) = 7(5,0)$ and 7(6,1) and $L = 1$. As we can see from Table 2 these selection rules are respected in most cases.

This preliminary analysis seems to indicate that the U(4|12) cluster supersymmetry scheme remains valid in the $N_F = 2$ sector too, i.e. for $A = 18$ nuclei. Further work is necessary to generalise the formalism to the requirements of this domain, e.g. to incorporate isospin-dependence and tensorial terms in the Hamiltonian.

Acknowledgement

This work has been supported by by the OTKA grant No. T37502 (Hungary) and the MTA-CNRS (Hungarian-French) project No. 7878.

References

1. H.T. Richards, *Phys. Rev.* C **29**, 276 (1984). .
2. For reviews see *Prog. Theor. Phys. Suppl.* **68**, (1980).
3. B. Buck, H. Friedrich and A.A. Pilt, *Nucl. Phys.* A **290**, 204 (1977).
4. P. Descouvemont and D. Baye, *Nucl. Phys.* A **463**, 629 (1987); M. Dufour and P. Descouvemont, *Nucl. Phys.* A **672**, 153 (2000).
5. H. Abele and G. Staudt, *Phys. Rev.* C **47**, 742 (1993).
6. G. Lévai and J. Cseh, *Phys. Lett.* B **381**, 1 (1996).
7. J. Cseh, *Phys. Lett.* B **281**, 173 (1992); J. Cseh and G. Lévai, *Ann. Phys. (N.Y.)* **230**, 165 (1994).
8. J. P. Elliott, *Proc. Roy. Soc.* A **245**, 128 and 562 (1958).
9. A. Frank and P. Van Isacker, *Algebraic methods in molecular and nuclear structure physics* (Wiley Interscience, New York, 1994).
10. G. Lévai, J. Cseh and P. Van Isacker, *Eur. Phys. J.* A **12**, 305 (2001).
11. F. Iachello and P. Van Isacker, *The interacting boson-fermion model* (Cambridge University Press, Cambridge, 1991).
12. D.R. Tilley *et al*, *Nucl. Phys.* A **595**, 1 (1995); **636**, 249 (1998).
13. G. Lévai, J. Cseh and W. Scheid, *Phys. Rev.* C **46**, 548 (1992). .
14. R. D. Lawson, *Theory of the Nuclear Shell Model* (Clarendon, Oxford, 1980).

CLUSTER EFFECTS IN ALTERNATING PARITY AND SUPERDEFORMED BANDS OF MEDIUM MASS AND HEAVY NUCLEI

G. G. ADAMIAN[1,2], N. V. ANTONENKO[1,2], R. V. JOLOS[1,2],
YU. V. PALCHIKOV[1,2], W. SCHEID[2] AND T. M. SHNEIDMAN[1,2]
[1]*Joint Institute for Nuclear Research, 141980 Dubna, Russia,*
E-mail: jolos@thsun1.jinr.ru

[2]*Institut für Theoretische Physik der Justus-Liebig-Universität,*
35392 Giessen, Germany,
E-mail: Werner.Scheid@theo.physik.uni-giessen.de

A cluster model is applied to the description of the properties of the ground state alternating parity bands in actinides and some Ba, Ce, Nd and Sm isotopes and to the description of the superdeformed band in ^{60}Zn. The model is based on the assumption that cluster-type shapes are produced by a collective motion of the nuclear system in the mass asymmetry coordinate. The results of calculations of the parity splitting, electric multipole transition moments of the alternating parity bands and of the spectra, transitional quadrupole moment of the superdeformed band and transitions between superdeformed and ground bands in ^{60}Zn agree with the experimental data.

1 Introduction

During many years cluster states have been intensively investigated in light nuclei [1], especially in the even-even $N = Z$ nuclei with alpha-particle as a natural building block. Cluster states are also discussed in connection with the properties of heavy nuclei [2,3,4], where they are manifested in nuclear structure and reactions. The possibility of alpha-cluster type configurations in heavy nuclei has been taken into account in the algebraic model [5,6,7,8] by the introduction of the p-boson. It was clear from the beginning that the effects of clustering can be principally important for the description of the low lying collective negative parity states. Calculations for light nuclei [9] and heavy nuclei [2,3,10] have shown that configurations with large equilibrium deformations are strongly related to clustering, not necessary to the alpha clustering. Heavier clusters can also play an important role. As is well known, when the harmonic oscillator is deformed a degeneracy of the single particle states, which is presented at zero deformation, is lost at first, however, recreated again when the ratio of the frequencies of the axially symmetric harmonic oscillator becomes equal to the ratio of integer numbers. The clustering may be an important structural feature at these deformations because the magic numbers associated with the corresponding shell gaps are expressed as combinations of spherical magic numbers. This feature admits a description of the deformed harmonic oscillator in terms of a series of shifted overlapping spherical potentials [9]. In this description a deformation process can be viewed as a division of the original spherical potential into smaller potentials aligned along the deformation axis. The formation of different cluster configurations can be considered in this picture as a process of mass exchange between the potentials.

Cluster type configurations can be mirror symmetric or asymmetric ones. A direct consequence of the asymmetric cluster type structures is the presence of the negative parity rotational states with odd angular momenta together with the positive parity rotational states having even angular momenta. The negative parity states are shifted up with respect to the positive parity ones since there is a non-negligible penetration probability of the barrier separating configurations with the light cluster located to the left and to the right of the heavier cluster. Such bands are known not only in light but also in heavy nuclei: in actinides and some Ba, Ce, Nd and Sm isotopes. They are frequently considered as related to the octupole deformation. However, using the idea that the dynamics of the octupole deformation can be treated as a motion of mass between the clusters, we can apply a model based on the Hamiltonian with the mass asymmetry degree of freedom as the main collective variable to the description of the alternating parity bands in heavy and medium mass nuclei.

2 Model

2.1 Hamiltonian in the mass asymmetry coordinate

The dinuclear system consisting of a heavy cluster A_1 and a light cluster A_2 is used below to describe the properties of the alternating parity bands in actinides and some Ba, Ce, Nd and Sm isotopes [11,12] and of the superdeformed band in ^{60}Zn. The mass asymmetry coordinate η, defined as $\eta = (A_1 - A_2)/(A_1 + A_2)$ ($|\eta|$=1 if A_1=0 or A_2=0), which describes a partition of nucleons between the nuclei forming the dinuclear system, is used as the relevant collective variable [1]. The wave function in η can be thought as a superposition of different cluster-type configurations including the mononucleus configuration with $|\eta|$=1, the alpha-cluster configuration with $|\eta|$=1-8/A and the ^{8}Be-cluster configuration with $|\eta|$=1-16/A. The relative contribution of each cluster component to the total wave function is determined by the collective Hamiltonian

$$H = -\frac{\hbar^2}{2}\frac{d}{d\eta}\frac{1}{B(\eta)}\frac{d}{d\eta} + U(\eta, I), \tag{1}$$

where $B(\eta)$ is the inertia coefficient and $U(\eta, I)$ the potential.

2.2 Potential

The potential $U(\eta, I)$ in Eq. (1) is taken as a dinuclear potential energy for $|\eta| < 1$

$$U(\eta, I) = B_1(\eta) + B_2(\eta) - B_{CN} + V(R = R_m, \eta, I). \tag{2}$$

Here, the internuclear distance $R = R_m$ is the touching distance between the clusters and is set to be equal to the value corresponding to the minimum of the potential in R for a given η. The quantities B_1 and B_2 (which are negative) are the experimental binding energies of the clusters forming the dinuclear system at a given mass asymmetry η, and B_{CN} is the binding energy of the mononucleus. The quantity $V(R, \eta, I)$ in Eq. (2) is the nucleus-nucleus interaction potential. It is given as

$$V(R,\eta,I) = V_{coul}(R,\eta) + V_N(R,\eta) + V_{rot}(R,\eta,I) \tag{3}$$

with the Coulomb potential V_{coul}, the centrifugal potential V_{rot} and the nuclear interaction V_N. To calculate the potential energy for $I \neq 0$ we need the moment of inertia $\Im(\eta) = \Im(\eta, R_m)$. The calculations of $\Im(\eta)$ has been described in [11].

2.3 *Mass parameter*

The method of the calculation of the inertia coefficient $B(\eta)$ used in this paper is given in [13]. Our calculations show that $B(\eta)$ is a smooth function of the mass number A. As a consequence, we take nearly the same value of B=20 $\times 10^4 m_0$ fm^2 for almost all considered actinide nuclei with a variation of 10%. However for ^{222}Th and ^{220,222}Ra, we varied B in the range B=(10-20)$\times 10^4 m_0$ fm^2 to obtain the correct value of $E_0(I = 0)$. Using a smooth mass dependence of B [13] we get B=4.5$\times 10^4 m_0$ fm^2 in the Ba, Ce and Nd region. However, better results are obtained with B=3$\times 10^4 m_0$ fm^2.

In the case of lighter nuclei the scale of variation of η, which is proportional to A^{-1}, is larger than in heavy nuclei and the η-dependence of the inertia coefficient should be taken into account, especially for the description of the superdeformed band where the potential U has several minima which correspond to different cluster configurations. The value of the inertia coefficient B corresponding to a cluster configuration with the light cluster heavier than the alpha particle has been estimated in the following way. Rewriting the Schrödinger equation (1) with a constant B in a discrete form

$$-\frac{\hbar^2}{2B(\Delta\eta)^2}\left(\psi(\eta+\Delta\eta) + \psi(\eta-\Delta\eta) - 2\psi(\eta)\right) + U(\eta)\psi(\eta) = E\psi(\eta), \tag{4}$$

we obtain the following expression for the nondiagonal matrix element of H in a discrete basis

$$\langle\eta + \Delta\eta|H|\eta\rangle = -\frac{\hbar^2}{2B(\Delta\eta)^2}. \tag{5}$$

As it follows from the definition of η, for two–nucleon transfer $\Delta\eta$=4/A. The quantity $\langle\eta + \Delta\eta|H|\eta\rangle$ can be estimated as a matrix element of the pairing interaction. Then $|\langle\eta + \Delta\eta|H|\eta\rangle|$ takes a value between the pairing interaction constant G, which is estimated as $G = 25/A$ MeV, and the pairing gap Δ if there are pairing correlations. Substituting G instead of $\langle\eta+\Delta\eta|H|\eta\rangle$ we obtain the upper boundary for B

$$B \le \frac{\hbar^2 A^3}{2 \cdot 16 \cdot 25 MeV} = \frac{\hbar^2 A^3}{800 MeV}. \tag{6}$$

For ^{60}Zn it gives $B = 11165 m\cdot$ fm^2. This estimate can be used for orientation and only for the case of clusters where the pairing concept is well justified. The best description of the experimental data is obtained taking $B(\eta = \eta_{Be})$=7580 $m\cdot$ fm^2 which is close to the estimate in Eq. (6). The value of the inertia coefficient at the α– minimum determines the value of the zero point energy and is fixed by the

Table 1. Comparison of experimental (E_{exp}) and calculated (E_{calc}) energies of states of the ground state alternating parity bands in $^{220-226}$Ra and ^{242}Pu . Energies are given in keV. Experimental data are taken from [14,15,16].

	^{226}Ra		^{224}Ra		^{222}Ra		^{220}Ra		^{242}Pu	
I^π	E_{exp}	E_{calc}	E_{exp}	E_{calc}	E_{exp}	E_{calc}	E_{exp}	E_{calc}	E_{exp}	E_{calc}
1^-	254	254	216	193	242	224	413	385	781	778
2^+	68	68	85	85	111	96	179	125	45	45
3^-	322	327	291	282	317	324	474	509	832	843
4^+	212	206	251	253	302	287	410	375	147	146
5^-	447	455	433	434	474	486	635	709	927	958
6^+	417	414	480	482	550	550	688	688	306	304
7^-	627	635	641	642	703	728	873	962		1122
8^+	670	668	756	747	843	843	1001	1016	518	514
9^-	858	862	907	901	992	1014	1164	1252		1329
10^+	960	954	1069	1046	1173	1166	1343	1356	779	773
11^-	1134	1134	1222	1215	1331	1346	1496	1568		1578
12^+	1282	1270	1415	1378	1537	1525	1711	1706	1084	1077
13^-	1448	1449	1578	1558	1710	1722	1864	1904		1863
14^+	1629	1621	1789	1745	1933	1924	2106	2067	1431	1421
15^-	1797	1810	1970	1944	2125	2140	2263	2257		2181
16^+	1999	2003	2189	2153	2359	2366			1816	1800
17^-	2175	2220	2389	2372	2570	2602				2526
18^+									2236	2210
19^-										2894
20^+									2686	2646

experimental value of the ground state energy. We obtained $B(\eta = \eta_\alpha)$=790 $m\cdot$ fm^2 which is used in our calculations.

3 Results of calculations and discussion

3.1 Parity splitting

We first calculated the parity splitting for the isotopes of Ra, Th, U, Pu, Ba, Ce and Nd for different values of the angular momentum I. The results of our calculations [12] are partly shown in Tables 1 and 2.

As is seen from the Tables, the calculated results agree well with the experimental data. The largest deviations of the calculated values from the experimental ones are found at low spins in some of the considered nuclei. A good description of the experimental data, especially of the variation of the parity splitting with A at low I and of the value of the critical angular momentum at which the parity splitting disappears, means that the dependence of the potential energy on η and I for the considered nuclei is correctly described by the proposed cluster model. The used value of the inertia coefficient B is also important.

In the considered nuclei the ground state energy level lies near the top of the barrier in η, if exists, and the weight of the $\alpha-$cluster configuration (Fig. 1) estimated as that contribution to the norm of the wave function which is located at $|\eta| \leq \eta_\alpha$ is about 5×10^{-2} for ^{226}Ra, which is close to the calculated spectroscopic factor [17]. This means that our model is in quantitative agreement with the known

Table 2. Comparison of experimental (E_{exp}) and calculated (E_{calc}) energies of states of the ground state alternating parity bands in $^{144-148}$Ba and ^{146}Ce. Energies are given in keV. Experimental data are taken from [14,18].

	^{148}Ba		^{146}Ba		^{144}Ba		^{146}Ce	
I^π	E_{exp}	E_{calc}	E_{exp}	E_{calc}	E_{exp}	E_{calc}	E_{exp}	E_{calc}
1^-		623	739	664	759	607	925	776
2^+	142	124	181	143	199	157	259	195
3^-	775	771	821	818	838	763	961	956
4^+	423	400	514	469	530	505	668	614
5^-	963	1018	1025	1078	1039	1026	1183	1259
6^+	808	808	958	958	961	961	1171	1171
7^-	1256	1342	1349	1424	1355	1375	1550	1660
8^+	1265	1273	1483	1491	1471	1496	1737	1756
9^-	1645	1731	1778	1841	1772	1796	2019	2138
10^+	1768	1788	2052	2028	2044	2005	2552	2345
11^-	2117	2181	2293	2323	2278	2285	2562	2681
12^+	2304	2327	2632	2574	2667	2546	3013	2953
13^-			2877	2871	2863	2843	3163	3286
14^+			3193	3166	3321	3146		3603
15^-			3524	3489	3519	3473	3827	3954
16^+			3737	3823	3992	3815		
17^-					4242	4179		

$\alpha-$decay widths of the nuclei considered.

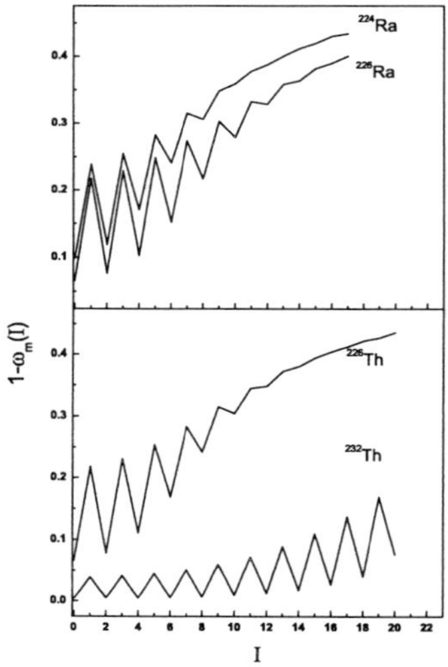

Figure 1. Calculated probability of the $\alpha-$cluster component in the wave function of the state with spin I of the alternating parity band for ^{224,226}Ra and ^{226,232}Th.

The spectra of those considered nuclei whose potential energy has a minimum at the alpha cluster configuration can be well approximated by the following analytical expression derived with the WKB approximation for the probability of the barrier penetration

$$E(I) = \frac{\hbar^2}{2J(I)} I[I+1], \quad \text{if} \quad I \quad \text{is} \quad \text{even},$$
$$E(I) = \frac{\hbar^2}{2J(I)} I[I+1] + \delta E(I), \quad \text{if} \quad I \quad \text{is} \quad \text{odd}. \tag{7}$$

Here, the parity splitting $\delta E(I)$ is given as

$$\delta E(I) = \frac{2E_1(I^\pi = 1^-)}{1 + \exp(b_0 \sqrt{B_0 I[I+1]})} \tag{8}$$

with

$$B_0 = \frac{\hbar^2}{2} \left(\frac{1}{\Im(\eta = 1)} - \frac{1}{\Im(\eta = \eta_\alpha)} \right). \tag{9}$$

The quantity $B_0 I(I+1)$ describes the change of the height of the barrier with spin I. The moment of inertia in Eq. (7) is given by the expression

$$J(I) = w_m(I)\Im(\eta = 1) + [1 - w_m(I)]\Im(\eta = \eta_\alpha) \tag{10}$$

containing a weight function $w_m(I)$

$$w_m(I) = \frac{w_m(I=0)}{1 + b_1 B_0 I[I+1]}, \tag{11}$$

which is the probability to find the mononucleus component in the wave function of the state with spin I of the ground state band. Since $w_m(I)$ decreases with increasing angular momentum, $J(I)$ increases with I in agreement with the experimental tendency. The quantity $w_\alpha(I) \equiv 1 - w_m(I)$ gives the corresponding probability of the α-cluster component. The constants $\Im(|\eta| = 1)$, $w_m(I = 0)$=0.93, $b_0 = \pi$ MeV$^{-1/2}$ and b_1=0.2 MeV^{-1} were obtained by fitting the experimental spectra for the nuclei considered.

These formulae clearly demonstrate that there are two important quantities which predetermine the description of the spectra of the alternating parity bands. These are $E_1(I^\pi = 1^-)$, which is determined by the depth of the α–minimum of the potential at $I = 0$ and by the value of the mass parameter $B(\eta)$, and B_0, which determines the angular momentum dependence of $\delta E(I)$ and $w_m(I)$, i.e. of $J(I)$.

3.2 $E\lambda$ transitions

With the wave functions obtained, we have calculated the reduced matrix elements of the electric multipole moments $Q(E1)$, $Q(E2)$ and $Q(E3)$. The effective charge for $E1$-transitions has been taken to be equal to $e_1^{eff} = e(1+\chi)$ with an average state-independent value of the $E1$ polarizability coefficient χ=–0.7 [19]. This renormalization takes a coupling of the mass–asymmetry mode into account to the giant dipole resonance in the dinuclear system. In the case of the quadrupole transitions we did not renormalize the charge $e_2^{eff} = e$ although an effective charge of 1.35 e describes the data for actinides better as it is seen from the results of calculations. For octupole transitions our cluster model Hamiltonian includes the octupole mode

Table 3. Calculated and experimental intrinsic multipole transition moments. The values of the dipole moment D_{10} are given for those values of the nuclear spin I for which there are experimental data. These values of I are given in the second column. The experimental data are taken from [14,15,16,20,21,22].

Nucleus	D_{10} (e fm) calc.	D_{10} (e fm) exp.	Q_{20} (e fm^2) calc.	Q_{20} (e fm^2) exp.	Q_{30} (e fm^3) calc.	Q_{30} (e fm^3) exp.
^{220}Ra	0.28 (I=7)	0.27	397	558	3167	
^{222}Ra	0.30 (I=7)	0.27	395	675	3064	
^{224}Ra	0.133 (I=3)	0.028	510	633	2889	
^{226}Ra	0.111 (I=1)	0.06–0.10	574	718	2611	2861
^{222}Th	0.29 (I=6)	0.38	397	548	3632	
^{224}Th	0.312 (I=10)	0.52	495		2985	
^{226}Th	0.223 (I=8)	0.30	561	830	2672	
^{228}Th	0.151 (I=8)	0.12	653	843	2255	
^{230}Th	0.054 (I=6)	0.04	666	899	1935	2144
^{232}Th	0.007 (I=1)		719	966	1616	1969
^{234}U	0.004 (I=1)		758	1035	1541	1895
^{236}U	0.004 (I=1)		786	1080	1433	1951
^{238}U	0.004 (I=1)		818	1102	1417	2041

responsible for the shape variation and deformation of the nuclear surface. This is the low–frequency collective octupole mode. However, high–frequency isovector as well as isoscalar octupole modes are not contained in the model Hamiltonian. For this reason the effect of the coupling of the low–frequency octupole mode to the high frequency mode should be taken into account by the octupole effective charge. The combined effect gives $\delta e_3^{(pol)} \approx (0.5 + 0.3\tau_z)e$ [19]. So, we have taken the effective charges to be equal to $e^{eff}_{3,proton} = 1.2e$ for protons and $e^{eff}_{3,neutron} = 0.8e$ for neutrons.

The part of the results of these calculations are listed in Table 3 and shown in Fig. 2. The obtained values are in agreement with the known experimental data for Q_λ^{exp}. Only in ^{224}Ra and ^{146}Ba (for $I = 7$) the calculated values of D_{10} are by factor of four larger than the experimental D_{10}. In Th the isotopic dependence of the dipole moment is well reproduced. Taking into account the collective character of our model and the absence of parameters to fit these data, the description of the experimental data is rather good. It should be also noted that the experimental data on the dipole moment have some uncertainties.

As example we present the angular momentum dependence of the reduced matrix elements of the electric dipole operator for ^{226}Ra in Fig. 2. The calculations qualitatively reproduce the angular momentum dependence of the experimental matrix elements. The same is true for the reduced matrix elements of the electric quadrupole and octupole operators.

The calculated angular momentum dependence of the intrinsic transition quadrupole moment shows an increase of the quadrupole moment with angular momentum in the transitional nucleus ^{226}Ra and a constant dependence in the well deformed isotope ^{238}U.

The calculated results for the E3–reduced matrix elements in ^{148}Nd exceed the experimental data for transitions to the ground state band [23]. This can be explained

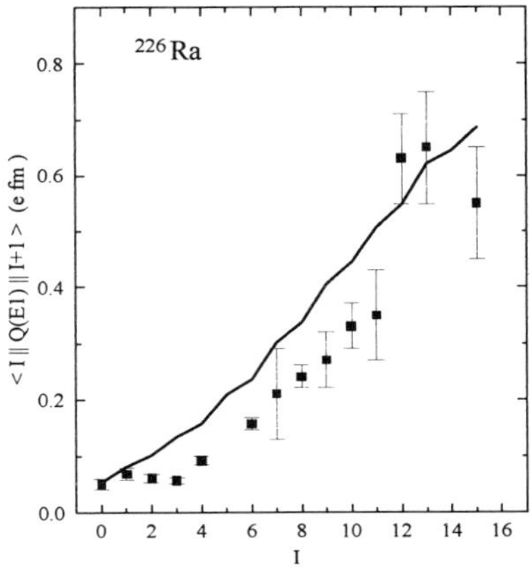

Figure 2. Angular momentum dependence of the calculated reduced matrix elements of the electric dipole operator (solid curve) in ^{226}Ra. The experimental data (squares) are taken from [21].

as follows. The experimental E3 matrix elements connecting the negative parity states of ^{148}Nd to the β band are unexpectedly large, about 70% of the matrix elements within the ground state band [24]. This shows a considerable partition of the E3 strength among the K=0 bands. In our model the β degree of freedom is absent and all the E3 strength is concentrated in the transitions to the ground state band. The nucleus ^{148}Nd is transitional in its collective properties between spherical and deformed nuclei and the β anharmonicity is quite large. This explains a partition of the E3 strength between the ground state and the β bands. The summed E3 strength for ^{148}Nd corresponds to the intrinsic transitional octupole moment of $\sim 2000e$ fm^3 (instead of $\sim 1500e$ fm^3 for the transitions in the ground state band) which agrees with the calculated value.

3.3 Superdeformed band of ^{60}Zn

The calculated spectra of the superdeformed band are described quite well including a variation of the experimental moment of inertia with spin. In our calculations this variation is reproduces because at the spin values around I=20 the superdeformed band is crossed with the ground state band, located mainly in the α–cluster minimum, and becomes the yrast one above I=20. In Fig. 3 the experimental upbending plot for the superdeformed band is compared with the results of calculations. The calculations qualitatively reproduce a deviation from the smooth dependence of the angular momentum on the γ–transition energy. However, the effect is overestimated.

We have calculated the branching ratios of the E2 ΔI=2 intensities of the γ–transitions. In the experiment for the 18^+_{sd}, 16^+_{sd} and 14^+_{sd} states only the decay into the superdeformed states has been observed. In our calculations we obtain for these transitions

$$\frac{I(18^+_{sd} \to 16^+_{gs})}{I(18^+_{sd} \to 16^+_{sd})} = 0.02, \quad \frac{I(16^+_{sd} \to 14^+_{gs})}{I(16^+_{sd} \to 14^+_{sd})} = 0.07, \quad \frac{I(14^+_{sd} \to 12^+_{gs})}{I(14^+_{sd} \to 12^+_{sd})} = 0.18. \quad (12)$$

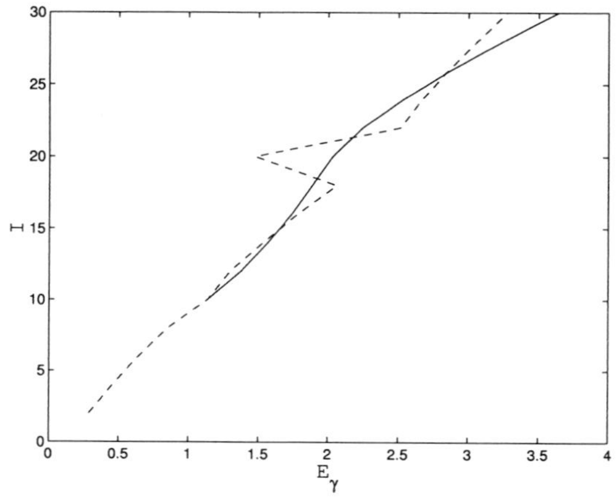

Figure 3. Spin as function of E_γ in the superdeformed band of ^{60}Zn. The results of calculations are shown by the dashed curve. The experimental data (solid curve) are taken from [25].

For the ratio $I(12^+_{sd} \to 10^+_{gs})/I(12^+_{sd} \to 10^+_{sd})$, where the lowest 10^+ state is considered as the 10^+_{gs} state, the experimental value is 0.54 [25] and the calculated one is 0.42. For the ratio $I(10^+_{sd} \to 8^+_{gs})/I(10^+_{sd} \to 8^+_{sd})$ the experimental value is 0.60 [25] and the calculated one 0.63. If the upper limit of $\sim$ 4% of the superdeformed band intensity is assumed for the unobserved $8^+_{sd} \to 6^+_{sd}$ transition, a lower limit of 0.01 W.u. was obtained for B(E2;$8^+_{sd} \to 6^+_{gs}$) in [25]. The calculated value of B(E2;$8^+_{sd} \to 6^+_{gs}$) is 0.19 W.u., which is large enough to explain the absence of the $8^+_{sd} \to 6^+_{sd}$ transition. The last two ratios in (12) are probably not small enough to explain the absence of the corresponding transitions between the states of the superdeformed and the ground state bands. However, it is quite possible that the strength of the corresponding transitions is fragmented over several states which are not presented in our model, but can be reproduced by shell model calculations.

The crucial evidence for asymmetric cluster structures is the existence of the alternating parity bands. Unfortunately, there is no experimental information about collective negative parity states in ^{60}Zn. The results of our calculations for the superdeformed band show that the parity splitting practically disappears. This fact contradicts the experimental observation in [25] where the negative parity states were not found. The absence of a population of the negative parity superdeformed states can be understood only if these negative parity states are shifted up by 0.5-2.0 MeV [26,27] with respect to the positive parity superdeformed states. This shift is not reproduced in our calculations. A reason could be a missing parity projection in the calculations of the potential $U(\eta, I)$. We mention also that our preliminary consideration indicates the possibility to consider the superdeformed band in ^{60}Zn as a mirror symmetric cluster configuration with two alpha particles on the opposite sides of ^{52}Fe like ^{24}Mg which is described as α-^{16}O-α. Thus, it is very interesting to look in more details in the experiment whether the negative parity superdeformed states exist or not.

Acknowledgments

This work was supported in part by DFG (Bonn), Volkswagen-Stiftung (Hannover) and RFBR (Moscow).

References

1. W. Greiner, J.Y. Park, and W. Scheid, *Nuclear Molecules*, (World Scientific, Singapore, 1995).
2. S. Cwiok, W. Nazarevicz, J.X. Saladin, W. Plociennik, and A. Johnson, *Phys. Lett.* B **322**, 304 (1994).
3. S. Aberg, L.-O. Jonsson, *Z. Phys.* A **349**, 205 (1994).
4. Yu.V. Pyatkov, V.V. Pashkevich, Yu.E. Penionzhkevich et al., *Nucl. Phys.* A **624**, 140 (1997).
5. F. Iachello and A.D. Jackson, *Phys. Lett.* B **108**, 151 (1982).
6. M. Gai, J.F. Ennis, M. Ruscev et al.,*Phys. Rev. Lett.* **51**, 646 (1983).
7. H. Daley and F. Iachello, *Phys. Lett.* B **131**, 281 (1983).
8. H. Daley and B. Barret, *Nucl. Phys.* A **449**, 256 (1986).
9. M. Freer, A.C. Merchant, *J. Phys.* G **23**, 261 (1997).
10. T.M. Shneidman, G.G. Adamian, N.V. Antonenko, S.P. Ivanova, and W. Scheid, *Nucl. Phys.* A **671**, 119 (2000).
11. T.M. Shneidman, G.G. Adamian, N.V. Antonenko, R.V. Jolos, and W. Scheid, *Phys. Lett.* B **526**, 322 (2002).
12. T.M. Shneidman, G.G. Adamian, N.V. Antonenko, R.V. Jolos, and W. Scheid, *Phys. Rev.* C **67**, 014313 (2003).
13. G.G. Adamian, N.V. Antonenko, and R.V. Jolos, *Nucl. Phys.* A **584**, 205 (1995).
14. http://www.nndc.bnl.gov/nndc/ensdf/
15. J.F.C. Cocks et al., *Nucl. Phys.* A **645**, 61 (1999).
16. I. Wiedenhover et al., *Phys. Rev. Lett.* **83**, 2143 (1999).
17. Yu.S. Zamiatin et al., *Phys. Part. Nucl.* **21**, 537 (1990).
18. W.R. Phillips et al., *Phys. Rev. Lett.* **57**, 3257 (1986).
19. A. Bohr and B. Mottelson, *Nuclear Structure v.II*, (Benjamin, New York, 1975).
20. P.A. Butler and W. Nazarewicz, *Rev. Mod. Phys.* **68**, 349 (1996).
21. H.J. Wollersheim et al., *Nucl. Phys.* A **556**, 261 (1993).
22. S. Raman et al., *At. Data Nucl. Data Tables* **36**, 1 (1987).
23. R. Ibbotson et al., *Phys. Rev. Lett.* **71**, 1990 (1993).
24. D. Cline, *Nucl. Phys.* A **557**, 615c (1993).
25. C.E. Svensson, D. Rudolph, C. Baktash et al., *Phys. Rev. Lett.* **82**, 3400 (1999).
26. C.E. Svensson, private communication.
27. J. Eberth, private communication.

THERMAL SIGNATURES OF PHASE TRANSITIONS IN FINITE NUCLEI

Y. ALHASSID
Center for Theoretical Physics, Sloane Physics Laboratory,
Yale University, New Haven, Connecticut 06520, U.S.A.
E-mail: yoram.alhassid@yale.edu

Phase transitions in nuclei are predicted in the mean-field approximation (e.g., Landau theory), but in the finite nucleus their singularities are smoothed out by the quantal and thermal fluctuations. We discuss thermal signatures of the shape and pairing transitions that are observed despite the large fluctuations.

1 Introduction

The fundamental quantity for calculating the thermal properties of a nucleus described by an Hamiltonian H at temperature T is the free energy $F(T) = -T \ln Z(T)$, where

$$Z(T) = \mathrm{Tr}\, e^{-H/T} \tag{1}$$

is the partition function. It is difficult to calculate (1) since H includes a two-body interaction. Because the interactions are strong, it is necessary to use non-perturbative methods such as the mean-field approximation.

In a mean-field approximation, we choose certain trial parameters σ to characterize the system, and express the free energy F at temperature T as a function of σ. The equilibrium configuration σ_{eq} is found by minimizing $F = F[T, \sigma]$ with respect to σ. Values of σ_{eq} with different symmetries can describe different phases of the system. We can then construct a phase diagram characterizing the equilibrium configuration as a function of temperature and/or any other control parameters. This approach is suitable for describing symmetry-breaking phase transitions. An example is the Landau theory, in which the variables σ play the role of the order parameters.[1]

However, in finite systems such as nuclei, thermal and quantal fluctuations around the equilibrium configuration are important and smooth out the singularities of the phase transition. At intermediate and high temperatures, it is often sufficient to consider the thermal fluctuations alone. The probability of finding specific values of σ (not necessarily the equilibrium values) is

$$P[\sigma] \propto e^{-F[T,\sigma]/T} \,, \tag{2}$$

and the partition function is calculated from

$$Z(T) = \int \mathcal{D}[\sigma] e^{-F[T,\sigma]/T} \tag{3}$$

where $D[\sigma]$ is the appropriate metric. Eq. (3) is known as the static-path approximation (SPA).[2] The mean-field theory is obtained when (3) is evaluated in the saddle-point approximation, leading to $\delta F/\delta\sigma = 0$. We note that the integration over σ restores the broken symmetry of the mean-field approximation.

At low temperatures, it is necessary to include quantal fluctuations, characterized by time-dependent fields $\sigma(\tau)$. This is expressed quantitatively in the Hubbard-Stratonovich transformation[3]

$$e^{-H/T} = \int \mathcal{D}[\sigma] G_\sigma U_\sigma \ , \tag{4}$$

in which the Gibbs ensemble $e^{-H/T}$ is written as a superposition of "ensembles" U_σ of non-interacting nucleons in fluctuating external fields $\sigma(\tau)$ (the imaginary time τ varies between 0 and $1/T$). The factor G_σ in (4) is a Gaussian weight in σ. Using (4), the canonical partition function can be written as in (3) but now $\sigma = \sigma(\tau)$ and

$$F[T,\sigma] = -T \ln \mathrm{Tr}\, U_\sigma - T \ln G_\sigma \ . \tag{5}$$

The Hubbard-Stratonovich transformation allows us to reproduce correlation effects through fluctuations of the mean field. The integrand in (4) is easily calculated, but integrating over the large number of σ fields is a difficult task. In the shell model Monte Carlo (SMMC) approach, this integration is done by Monte Carlo methods.[4] Some of the recent applications of the method are reviewed in Ref. [5].

The effect of fluctuations is to smooth out the singularities of the phase transitions. An interesting question is whether signatures of the phase transitions can still be observed despite the large fluctuations. In Sections 2 and 3 we address this question in the context of the shape and pairing transitions in nuclei, respectively.

2 Shape transitions

To illustrate the methods discussed in the introduction, we review the shape-transition theory of hot rotating nuclei.[6]

2.1 *Landau theory*

The order parameters of the nuclear shape transitions are the quadrupole deformation parameters $\sigma_{2\mu}$ in the laboratory frame. Alternatively, we can use the two intrinsic shape parameters β, γ, and the three Euler angles $\Omega \equiv (\phi, \theta, \psi)$ describing the orientation of the nuclear body-fixed frame with respect to the nuclear rotation axis $\hat{\omega}$.

The nuclear partition function at a given temperature and angular velocity ω is given by a generalization of (1)

$$Z(T,\omega) = \mathrm{Tr}\, e^{-(H - \boldsymbol{\omega}\cdot\mathbf{J})/T} \ , \tag{6}$$

where $\mathbf{J}$ is the total angular momentum of the nucleus. In the Landau theory, we expand the rotating-frame free energy in the order parameters. Since this free energy is a scalar, it depends only rotationally invariant combinations of the second rank tensor $\sigma_{2\mu}$ and the vector $\boldsymbol{\omega}$. To leading order in ω, we have[6]

$$F[T,\omega;\sigma_{2\mu}] = F(T,\omega=0;\beta,\gamma) - \frac{1}{2}\left(\hat{\omega}\cdot\mathcal{I}\cdot\hat{\omega}\right)\omega^2 \ , \tag{7}$$

where $F(T,\omega=0;\beta,\gamma)$ is the free energy in the absence of rotations and

$$\hat{\omega}\cdot\mathcal{I}\cdot\hat{\omega} = I_{x'x'}\sin^2\theta\cos^2\phi + I_{y'y'}\sin^2\theta\sin^2\phi + I_{z'z'}\cos^2\theta \tag{8}$$

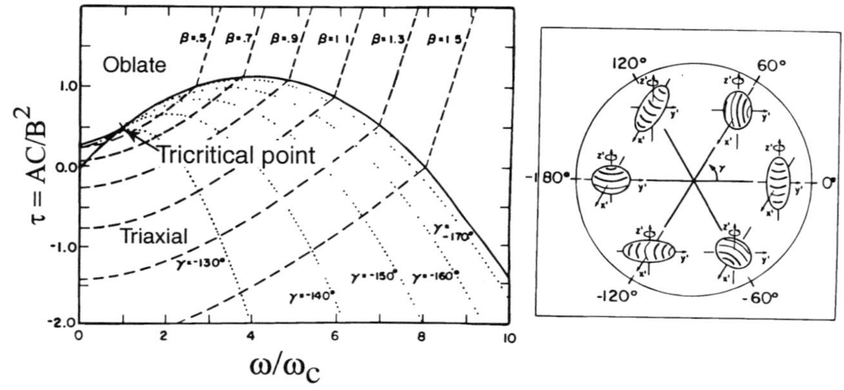

Figure 1. Right: shapes of a deformed nucleus rotating around a principal axis z'. Left: universal phase diagram of the Landau shape-transition theory in the τ-ω/ω_c plane (see text). The solid line separates the triaxial collective phase from the oblate non-collective phase. The tricritical point is indicated. From Refs. [6].

is the moment of inertia about the rotation axis $\hat{\omega}$, expressed in terms of the principal moments of inertia $I_{x'x'}, I_{y'y'}$ and $I_{z'z'}$. To fourth order in $\sigma_{2\mu}$

$$F(T, \omega = 0; \beta, \gamma) = F_0 + A\beta^2 - B\beta^3 \cos 3\gamma + C\beta^4 \tag{9}$$

where A, B, C are temperature-dependent constants, and global stability requires $C > 0$. Expanding the moment of inertia to second order in deformation, we find

$$\begin{aligned}
I_{z'z'}(T;\beta,\gamma) &= I_0 - 2R\beta\cos\gamma + 2I_1\beta^2 + 2D\beta^2\sin^2\gamma \ , \\
I_{x'x'}(T;\beta,\gamma) &= I_{z'z'}(T;\beta,\gamma - 2\pi/3) \ , \\
I_{y'y'}(T;\beta,\gamma) &= I_{z'z'}(T;\beta,\gamma + 2\pi/3) \ ,
\end{aligned} \tag{10}$$

where I_0, R, I_1 and D are temperature-dependent coefficients.

Minimizing the free energy with respect to the Euler angles Ω and the intrinsic shape β, γ leads to a universal phase-diagram in terms of two dimensionless parameters $\tau \equiv AC/B^2$ and a scaled angular velocity ω/ω_c (see Fig. 1). A transition line separates the triaxial collective phase from an oblate non-collective phase. The point

$$\tau_c = \frac{63}{128} \ ; \quad \omega_c = \frac{9}{16}\frac{B}{C}\left(\frac{B}{R}\right)^{1/2} \ , \tag{11}$$

is a tricritical point separating the first-order transitions ($\omega < \omega_c$) from the second-order transitions ($\omega > \omega_c$). In the absence of rotations ($\omega = 0$), a first-order transition from a deformed to a spherical shape occurs at $\tau = 1/4$. In the range $0 \le \tau \le 9/32$, both the deformed and spherical shapes coexist as local minima.

At a fixed temperature, it is also possible to change certain parameters ξ in the Hamiltonian. These parameters replace T as the control parameters. The corresponding Landau theory is identical to the one described above except that the parameters $A, B, \ldots$ are now functions of these control parameters ξ. The equilibrium shape is determined by τ and B. For $B > 0$ ($B < 0$), there is a first-order transition from a prolate (oblate) shape at $\tau < 1/4$ to a spherical shape at $\tau > 1/4$. For $B = 0$, the free energy in (9) describes a γ-soft nucleus (i.e., the free energy is independent of γ), and the transition from a deformed to a spherical shape (at $A = 0$) becomes second order. For $\tau < 1/4$ the nucleus is deformed, and at $B = 0$ there is a second order transition from a prolate ($B > 0$) to an oblate shape ($B < 0$).

Zero-temperature shape transitions were studied in the interacting boson model (IBM) as a function of the model's parameters (and for $\omega = 0$) in the limit of an infinite number of bosons.[7] This limit is equivalent to a mean-field (or classical) approximation where quantal fluctuations are ignored, and can therefore be described by the Landau theory of Section 2.1. Recently, such a Landau approach was implemented using two parameters of the IBM Hamiltonian as control parameters.[8] The Landau expansion is given by (9), where A, B, C are functions of these control parameters. The Casten triangle (for prolate nuclei) has a first-order transition line $\tau = 1/4$ (and $B > 0$) separating the deformed and spherical regions. This line ends at a second-order critical point[9] for which $B = 0$.

2.2 Thermal fluctuations

The probability of finding the nucleus in a state with deformation $\sigma_{2\mu}$ is $P[\sigma_{2\mu}] \propto e^{-F[T,\omega;\sigma_{2\mu}]/T}$ The thermal expectation value of an observable O is calculated by a thermal average over all quadrupole shapes $\sigma_{2\mu}$

$$\langle \mathcal{O} \rangle = \int D[\sigma] P[\sigma_{2\mu}] \langle \mathcal{O} \rangle_\sigma \,, \tag{12}$$

where $\langle \mathcal{O} \rangle_\sigma$ is the expectation value of the observable for a given quadrupole deformation σ and

$$D[\sigma] = \prod_\mu d\sigma_{2\mu} = \beta^4 |\sin 3\gamma| d\beta d\gamma d\Omega \,. \tag{13}$$

An observable that is sensitive to deformation is the giant dipole resonance (GDR). A theory of the GDR in hot rotating nuclei, based on the Landau plus shape fluctuation theory, was developed in Refs. [10] and was found to describe well the experimental data. While the measured GDR cross-sections are consistent with collective to non-collective shape transitions, the fluctuations dominate the GDR strength function around the transition temperature and the GDR cross section is less sensitive to shape transitions.

Another shape transition is the Jacobi transition from a non-collective oblate shape to a triaxial shape with major to minor axis ratio of $\approx 2:1$. This transition occurs as the spin of the nucleus increases above a certain value. Fig. 2 shows the GDR cross section in hot ^{45}Sc for increasing values of the spin. The calculated[11] GDR strength function (solid lines), based on the shape fluctuation theory and the actual free energy surfaces (in which a Jacobi transition occurs), describes well the experimental data.[12] However, if we use free energy surfaces in which no Jacobi transition occurs (i.e., the nucleus remains oblate at all spins), we find large

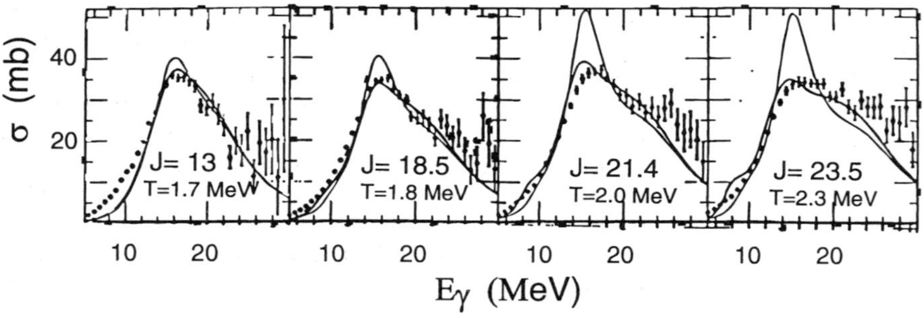

Figure 2. The GDR cross section in hot ^{45}Sc for several average spins (indicated in the figure in units of $\hbar$). The heavy solid lines are obtained in the thermal fluctuation theory of Section 2.2 using the actual free energy surfaces and are in agreement with the experiment. The light solid lines are obtained using free energy surfaces in which a Jacobi transition does not occur (i.e., assuming the nucleus remains oblate). From Ref. [12].

discrepancies with the data at the higher spin values (light solid lines). Thus the GDR strength function provides a clear signature of the Jacobi transition.

3 Pairing phase transition

Signatures of pairing in nuclei at $T = 0$ are well known and include odd-even mass differences and a gap to the first excited state in even-even nuclei. However, much less known are signatures of the pairing transition at finite temperature. In the BCS theory,[13] the heat capacity exhibits a discontinuity at the critical temperature T_c. However, in a finite system, the fluctuations smooth out this discontinuity. The pair correlation length in the nucleus is much larger than the size of the nucleus, and the expectation was that the fluctuations would wash out any signature of the pairing transition in the heat capacity. Recently, we performed realistic SMMC calculations of the heat capacity in iron isotopes that include both quantal and thermal fluctuations and found a signature of the pairing transition that is enhanced in the neutron-rich isotopes.[14] In Section 3.1 we briefly review the BCS theory and in Section 3.2 we discuss the results of the SMMC approach.

3.1 BCS theory

As an example we consider a simple pairing Hamiltonian

$$H = \sum_{k>0} \epsilon_k (a_k^\dagger a_k + a_{\bar{k}}^\dagger a_{\bar{k}}) - G P^\dagger P \, , \tag{14}$$

where k and $\bar{k}$ are time-reversed single-particle states and $P^\dagger = \sum_{k>0} a_k^\dagger a_{\bar{k}}^\dagger$ is a pair creation operator. In an axially-deformed nucleus, the angular momentum projection m on the symmetry axis is a good quantum number, and the time-reversed states are characterized by $\pm m$.

In the BCS theory, the free energy of the Hamiltonian (14) is calculated as a function of a complex order parameter ξ

$$F[\xi,\xi^*] = G|\xi|^2 - 2T\sum_{k>0} \ln\left[2\cosh(E_k/2T)\right] \;, \tag{15}$$

where $E_k = \sqrt{(\epsilon_k-\mu-G/2)^2 + G^2|\xi|^2}$ are the quasi-particle energies and μ is the chemical potential. Minimizing F with respect to ξ, we find the finite-temperature BCS equation

$$\frac{1}{G} = \sum_{k>0} \frac{\tanh(\beta E_k/2)}{2E_k} \;, \tag{16}$$

while the chemical potential is determined from the particle number N

$$N = \sum_{k>0}\left(1 - \frac{\epsilon_k - \mu - \frac{G}{2}}{E_k}\tanh\frac{\beta E_k}{2}\right) \;. \tag{17}$$

The solutions of Eqs. (16) and (17) determine the pairing gap $\Delta = G|\xi|$ and the chemical potential μ as a function of T and N. The pairing gap vanishes above a critical temperature T_c, but becomes non-zero below T_c.

3.2 Beyond the mean field

In the SPA, thermal fluctuations in the order parameter ξ are taken into account. In SMMC, both static and dynamic fluctuations of all fields are included. We have recently calculated the heat capacity in iron isotopes using the SMMC method in the complete $fpg_{9/2}$ shell.[14] The Hamiltonian includes monopole pairing plus quadrupole, octupole and hexadecupole interactions.[15] The pairing strength is determined from odd-even mass differences, while the strength of the multipole interaction is determined self-consistently and renormalized appropriately for the various multipoles.

In SMMC, we calculate the thermal energy as the expectation value of the Hamiltonian, and the heat capacity is found by a numerical derivative $C = dE/dT$. At low temperatures, this method leads to large statistical errors in C (even for a good-sign Hamiltonian). We have introduced a novel method that significantly reduces these statistical errors by taking into account correlated errors.[14]

Results are shown in Fig. 3 for ^{60}Fe (left) and ^{59}Fe (right). For the even-mass ^{60}Fe we observe a 'shoulder' in C around $T \sim 0.7$ MeV. This shoulder is correlated with the rapid decrease (with temperature) in the number of $J = 0$ neutron pairs $\langle\Delta^\dagger\Delta\rangle$ where $\Delta^\dagger = (j+1/2)^{-1/2}\sum_{j,m>0} a^\dagger_{jm}a^\dagger_{j\bar{m}}$ (see bottom panel of Fig. 3).

In the Fermi gas and BCS approximations we calculate the heat capacity from $C = TdS/dT$, where the entropy S is given by $S = -\sum_k[f_k \ln f_k + (1-f_k)\ln(1-f_k)]$ [f_k are the occupation numbers of the particles (Fermi gas) or the quasi-particles (BCS)]. In Fig. 3 we show both the BCS heat capacity (solid line for ^{60}Fe) and the Fermi gas heat capacity (dashed lines). In SMMC (symbols), we observe strong suppression of the BCS peak, but a 'bump' remains for ^{60}Fe around the neutron pairing transition temperature of ~ 0.7 MeV. No such signature of the pairing

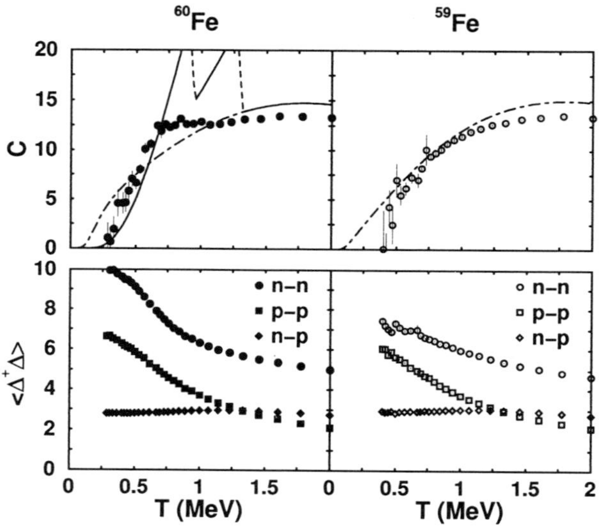

Figure 3. Top: heat capacity versus temperature for ^{60}Fe (left panel) and ^{59}Fe (right panel). The circles are the SMMC results, the dashed-dotted lines are the Fermi gas results and the solid line (left panel only) is the BCS heat capacity. Bottom: number of $J = 0$ n-n, p-p and n-p pairs versus temperature. From Ref. [14].

transition is observed for the odd-mass ^{59}Fe. Similar S-shape heat capacities were measured by the Oslo group in rare-earth nuclei[16] (see right panel of Fig. 5).

4 Extending the shell model theory to higher temperatures

The SMMC calculations described in Section 3.2 were restricted to a single major shell. At higher temperatures, contributions from other shells and possibly the continuum become important. For example, the calculations in the $fpg_{9/2}$ shell are limited to $T < 1.5 - 2$ MeV. We have recently introduced a method to calculate the partition function to higher temperatures.[17] In Section 4.1 we discuss the effects of truncations within an independent-particle model. In Section 4.2 we combine the independent-particle model calculations in the complete space (both the bound states and the continuum) with the correlated SMMC calculations in the truncated space.

4.1 Independent-particle model

In the independent-particle approximation, we can calculate the many-particle partition function in the complete space. The one-body mean-field potential is taken to be a Woods-Saxon central potential plus spin-orbit interaction.[18] Denoting the single-particle energies by ϵ_{nlj} (n is the principal quantum number, l is the orbital angular momentum, and j is the total spin) and the scattering phase shifts for $\epsilon > 0$

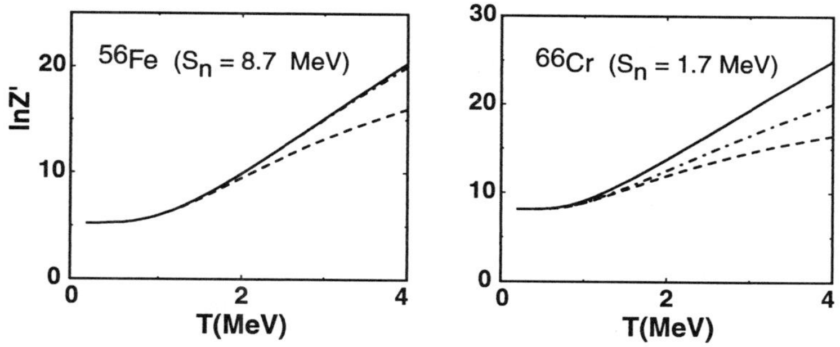

Figure 4. $\ln Z'$ versus temperature for ^{56}Fe (left) and ^{66}Cr (right) in the independent-particle model. Solid lines: bound states plus continuum, dotted-dashed lines: bound states only, dashed lines: $fpg_{9/2}$ shell. From Ref. [17].

by $\delta_{lj}(\epsilon)$, the many-particle grand-canonical partition function is given by

$$\ln Z_{sp}^{GC}(\beta,\mu) = \sum_{nlj}(2j+1)\ln[1+e^{-\beta(\epsilon_{nlj}-\mu)}] + \int_0^\infty d\epsilon\delta\rho(\epsilon)\ln[1+e^{-\beta(\epsilon-\mu)}] \ . \quad (18)$$

Here $\delta\rho(\epsilon) = \pi^{-1}\sum_{lj}(2j+1)d\delta_{lj}/d\epsilon$ is the continuum contribution to the single-particle level density obtained after the free-particle contribution is subtracted out, and μ the chemical potential determined from the number of particles $N = \partial \ln Z_{\rm sp}^{\rm GC}/\partial\alpha$.

To avoid the numerical derivative of the phase shift in (18), we integrate by parts and use Levinson's theorem to obtain

$$\ln Z_{sp}^{\rm GC}(\beta,\mu) = \sum_{lj}(2j+1)\left\{\sum_n \ln\left[\frac{1+e^{-\beta(\epsilon_{nlj}-\mu)}}{1+e^{\beta\mu}}\right] + \frac{\beta}{\pi}\int_0^\infty d\epsilon\ \delta_{lj}(\epsilon)f(\epsilon)\right\} \ , \quad (19)$$

where $f(\epsilon) = [1+e^{\beta(\epsilon-\mu)}]^{-1}$ is the Fermi-Dirac function. The canonical partition function is calculated in the saddle-point approximation

$$\ln Z \approx \ln Z^{\rm GC} - \beta\mu N - \frac{1}{2}\ln\left(2\pi\langle(\Delta N)^2\rangle\right) \ , \quad (20)$$

where $\langle(\Delta N)^2\rangle = \partial^2 \ln Z_{\rm sp}^{\rm GC}/\partial\alpha^2$ is the particle-number variance.

It is convenient to define an excitation partition function $Z' = Ze^{\beta E_0}$, where E_0 is the ground-state energy. In Fig. 4 we show $\ln Z'$ versus T in the independent-particle model for ^{56}Fe (left) and ^{66}Cr (right) using all bound states plus continuum (solid lines), bound states only (dotted-dashed lines) and $fpg_{9/2}$ shell only (dashed lines). We see that the truncation to the $fpg_{9/2}$ shell becomes problematic for $T > 1.5$ MeV. For the strongly bound ^{56}Fe, the contribution of the continuum is negligible (up to $T \sim 4$ MeV), but for the weakly-bound ^{66}Cr the continuum contribution is important already for $T > 1$ MeV.

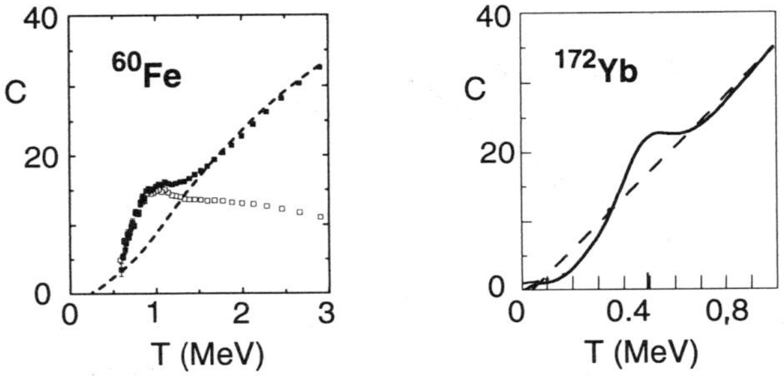

Figure 5. Heat capacity versus temperature. Left: the extended heat capacity of ^{60}Fe (solid squares) is compared with the heat capacity of the independent-particle model (dashed line). The open squares describe the SMMC heat capacity in the $fpg_{9/2}$ shell. Right (from Ref. [16]): the measured heat capacity of ^{172}Yb (solid line) is compared with the Fermi gas result (dashed line).

4.2 Interaction effects

Correlations are important at lower temperatures, while the size of the model space becomes important at higher temperatures. Assuming that the partition function is factorizable, we have combined the fully correlated partition in the truncated space with the independent-particle partition in the full space[17]

$$\ln Z_v = \ln Z_{v,tr} + \ln Z_{sp} - \ln Z_{sp,tr} \ . \tag{21}$$

Here Z_v and $Z_{v,tr}$ are the correlated partition functions in the full and truncated space, respectively. Similarly, Z_{sp} and $Z_{sp,tr}$ are the corresponding partition functions in the independent-particle model.

4.3 Extended heat capacity

Using the theory of Section 4.2, we have calculated the heat capacity up to $T \sim 4$ MeV. The results for ^{60}Fe are shown in the left panel of Fig. 5, where the extended heat capacity is seen to increase monotonically with T at high temperatures. The signature of the pairing transition is now much clearer than the one found in the truncated calculations of Fig. 3 and is in qualitative agreement with the S-shape heat capacity observed in rare-earth nuclei[16] (see right panel of Fig. 5).

5 Conclusion

In the finite nucleus, thermal and quantal fluctuations smooth out the singularities at the critical point of a phase transition. Despite the dominance of fluctuations, thermal signatures of the phase transition can sometimes be observed. In particular, we have discussed recent work on the the pairing transition and found a 'bump' in the heat capacity that is enhanced in neutron-rich isotopes. This signature is more clearly seen once our shell-model calculations are extended to higher temperatures.

Acknowledgments

It is a great pleasure to dedicate this article to Franco Iachello on the occasion of his 60th birthday and in appreciation of our fruitful collaborations. This work was supported in part by the Department of Energy grant DE-FG-0291-ER-40608. I would like to thank G.F. Bertsch, L. Fang and S. Liu for their collaboration on the work presented in Sections 3 and 4.

References

1. L.D. Landau and E.M. Liftshitz, *Statistical Physics*, vol. 5 of *Course of Theoretical Physics* (Butterworth-Heinemann, Boston, 1999).
2. Y. Alhassid and J. Zingman, *Phys. Rev.* C **30**, 684 (1984); B. Lauritzen, P. Arve, and G.F. Bertsch, *Phys. Rev. Lett.* **61**, 2835 (1988).
3. J. Hubbard, *Phys. Rev. Lett.* **3**, 77 (1959); R. L. Stratonovich, *Dokl. Akad. Nauk. S.S.S.R.* **115**, 1097 (1957).
4. G. H. Lang, C. W. Johnson, S. E. Koonin, and W. E. Ormand,*Phys. Rev.* C **48**, 1518 (1993); Y. Alhassid, D. J. Dean, S. E. Koonin, G. Lang, and W. E. Ormand, *Phys. Rev. Lett.* **72**, 63 (1994).
5. Y. Alhassid, *Int. J. Mod. Phys.* B **15**, 1447 (2001).
6. Y. Alhassid, S. Levit, and J. Zingman, *Phys. Rev. Lett.* **57**, 539 (1986); Y. Alhassid, J. Zingman, and S. Levit, *Nucl. Phys.* A **469**, 205 (1987).
7. J.N. Ginocchio and M.W. Kirson, *Phys. Rev. Lett.* **44**, 1744 (1980); A.E.L. Dieperink, O. Scholten, and F. Iachello, *Phys. Rev. Lett.* **44**, 1747 (1980).
8. J. Jolie, P. Cejnar, R.F. Casten, S. Heinze, A. Linnemann, and V. Werner, *Phys. Rev. Lett.* **89**, 182502 (2002).
9. F. Iachello, *Phys. Rev. Lett.* **85**, 3580 (2000); R.F. Casten and Z.V. Zamfir, *Phys. Rev. Lett.* **85**, 3584 (2000).
10. Y. Alhassid, B. Bush and S. Levit, *Phys. Rev. Lett.* **61**, 1926 (1988); Y. Alhassid and B. Bush, *Nucl. Phys.* A **509**, 461 (1990); Y. Alhassid, in *New Trends in Nuclear Collective Dynamics*, p. 41, Y. Abe, H. Horiuchi and K. Matsuyanagi, eds. (Springer Verlag, New York, 1992).
11. Y. Alhassid and N. Whelan, *Nucl. Phys. A* **565**, 427 (1993).
12. M. Kicinska-Habior, K.A. Snover, J.A. Behr, C.A. Gossett, Y. Alhassid and N. Whelan, *Phys. Lett.* B **308**, 225 (1993).
13. J. Bardeen, L.N. Cooper and J.R. Schrieffer, *Phys. Rev.* **108**, 1175 (1957).
14. S. Liu and Y. Alhassid *Phys. Rev. Lett.* **87**, 022501 (2001).
15. H. Nakada and Y. Alhassid, *Phys. Rev. Lett.* **79**, 2939 (1997).
16. A. Schiller, A. Bjerve, M. Guttormsen, M. Hjorth-Jensen, F. Ingebretsen, E. Melby, S. Messelt, J. Rekstad, S. Siem, S.W. Odegard, *Phys. Rev.* C **63**, 021306 (2001).
17. Y. Alhassid, G.F. Bertsch, and L. Fang, nucl-th/0303040 (2003).
18. A. Bohr and B. R. Mottelson, *Nuclear Structure*, vol. 1 (Benjamin, New York, 1969).

INTEGRABILITY AND QUANTUM PHASE TRANSITIONS IN INTERACTING BOSON MODELS

J. DUKELSKY[1], J. M. ARIAS[2], J. E. GARCIA-RAMOS[3] AND S. PITTEL[4]

[1] *Instituto de Estructura de la Materia, CSIC, Serrano 123, Madrid, Spain*

[2] *Departamento de Física Atómica, Molecular y Nuclear, Universidad de Sevilla, Spain*

[3] *Deparatmento de Física Aplicada, Universidad de Huelva, Spain*

[4] *Bartol Research Institute, University of Delaware, Newark, Delaware 19716, USA*

The exact solution of the boson pairing hamiltonian given by Richardson in the sixties is used to study the phenomena of level crossings and quantum phase transitions in the integrable regions of the *sd* and *sdg* interacting boson models.

One of the most fruitful features of the Interacting Boson Model (IBM)[1] of nuclei is the existence of three dynamical symmetry limits. Each represents a well defined nuclear phase, providing analytically the exact eigenstates of the system and offering a unique tool to deeply understand the physics involved.

A quantum system has a dynamical symmetry (DS) if the hamiltonian can be expressed as a function of the Casimir operators of a subgroup chain. A direct consequence of this definition is that a system exhibiting a DS is quantum integrable and analytically solvable. The concept of quantum integrability (QI), though sometimes associated with a dynamical symmetry, is more general, however. It can be stated as follows: A quantum system is integrable if there exists a complete set of hermitian, independent and commuting operators (constants of motion). Clearly, if a quantum system has a DS, the Casimir operators of its subgroups play the role of the constants of motion, fulfilling the definition of QI. But, by no means must a system have a DS to be quantum integrable.

Perhaps the most striking manifestation of quantum integrable systems is the absence of level repulsion in their energy spectra. This can be clearly seen in the IBM by looking at the transitions between the three dynamical symmetries. Using the Consistent-Q hamiltonian[2],

$$H = x n_d + \frac{x-1}{N} Q^{\chi} \cdot Q^{\chi} \tag{1}$$

where $n_d = \sum_m d_m^\dagger d_m$ and $Q_m^\chi = \left[d_m^\dagger s + (-)^m s^\dagger d_{-m}\right] + \chi \left[d^\dagger \tilde{d}\right]_m^2$, the transitions between the dynamical symmetries can be readily explored.

In Fig. 1A, we plot the 0^+ energy levels associated with the transition from $SU(3)$ ($x = 0$, $\chi = -\sqrt{7}/2$) to $O(6)$ ($x = 0$, $\chi = 0$) and with the transition from $O(6)$ to $\overline{SU(3)}$ ($x = 0$, $\chi = \sqrt{7}/2$). The results are shown as a function of χ and for $N = 10$ bosons. The level repulsion between pairs of energy levels is present everywhere, except at the $\chi = 0$ point where the $O(6)$ DS is realized. At this point, due to the absence of level repulsion, level crossings are allowed. In the inset we amplify the first of the two allowed level crossings.

In Fig. 1B we show the analogous results for the 0^+ states of a system of $N = 10$ bosons along the transition from $SU(3)$ ($x = 0$, $\chi = -\sqrt{7}/2$) to $U(5)$ ($x = 1$), as a function of the parameter x. The level repulsion phenomenon appears

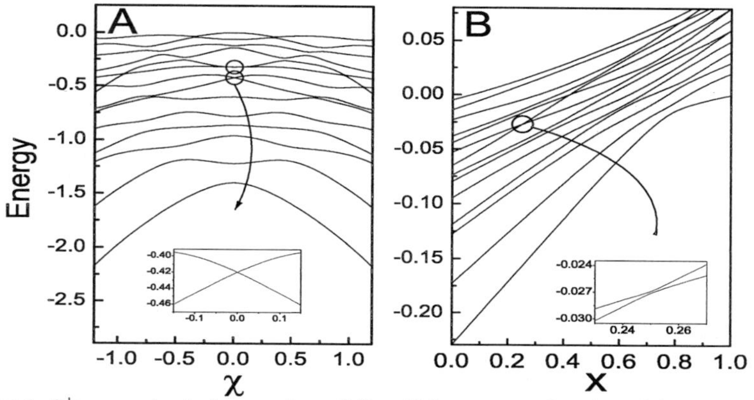

Figure 1. 0^+ energy levels for a system of $N = 10$ bosons as a function of the parameter χ for the $SU(3)$-$O(6)$-$\overline{SU(3)}$ transition (1A), and as a function of the parameter x for the $SU(3)$-$U(5)$ transition (1B)

any time two levels are close enough. The inset amplifies one of the closest two level approaches, showing that level repulsion prevents the crossing. The level repulsion between the ground state and the first excited 0^+ state leads in the thermodynamic limit to a non-analytic behavior of the ground state energy, defining a critical point characterized by a first-order phase transition[3].

An interesting question is what is the structure and origin of the phase transition from $SU(3)$ to $\overline{SU(3)}$. As noted earlier, this transition proceeds through the $O(6)$ DS. As also noted earlier, level crossings are permitted at that *critical symmetry* point. Thus, if the $SU(3)$ to $\overline{SU(3)}$ transition is indeed first-order, it is interesting to ask how this can come about in the absence of level repulsion at the critical point. Further study on this issue is clearly required.

The transition from $O(6)$ to $U(5)$ is described by the hamiltonian (1) by setting $\chi = 0$ and varying x from 0 to 1. Earlier investigation of the properties of the spectrum for this leg of the Casten triangle made clear that the system is integrable everywhere on the path, even though there is no global DS[4,5,6]. To facilitate discussion of this region of parameter space, it is useful to digress for a moment and to discuss a class of integrable boson pairing models. As we will see, the $O(6)$ to $U(5)$ leg of the Casten triangle falls within this class of models.

The models of interest are among three classes of integrable models for fermion and boson systems[7]. Though integrable, these models do not in general relate to a DS. The models we will be discussing[7] are based on the $SU(1,1)$ pair algebra

$$K_l^0 = \frac{1}{2}\sum_m \left(b_{lm}^\dagger b_{lm} + \frac{1}{2}\right), \quad K_l^+ = \sum_m b_{lm}^\dagger b_{l\overline{m}}^\dagger = \left(K_l^-\right)^\dagger \tag{2}$$

where the operators $b_{lm}^\dagger$ (b_{lm}) create (destroy) a boson in statelm. In terms of the generators (2), the constants of motion of the *rational* class are

$$R_l = K_l^0 + 2g \sum_{l'(\neq l)} \frac{1}{\eta_l - \eta_{l'}} \left[\frac{1}{2} \left(K_l^+ K_{l'}^- + K_l^- K_{l'}^+ \right) - K_l^0 K_{l'}^0 \right] \tag{3}$$

It can be readily confirmed that the operators R_l are hermitian, independent (none of them can be expressed as a function of the others), and mutually commute. Moreover, if the system has M single boson states l, there are as many operators (3) as quantum degrees of freedom, constituting a complete set of constants of motion. The pairing strength g and the set of M real numbers η_l are free parameters.

If a system is quantum integrable there should exist a unique basis of common eigenstates of the M operators R_l. It has been shown[7] that this complete set of eigenvectors can be formally written as a product of boson pairs acting on a subspace of unpaired boson states $|\nu\rangle \equiv |\nu_1, \nu_2, \cdots, \nu_M\rangle$

$$|\Psi\rangle = \prod_{\alpha=1}^{n} \left(\sum_{l=1}^{M} \frac{1}{2\eta_l - e_\alpha} K_l^+ \right) |\nu\rangle \tag{4}$$

where the n parameters e_α that apply to each eigenstate are particular solutions of the set of n coupled nonlinear equations

$$1 + g \sum_l \frac{2l + 2\nu_l + 1}{2\eta_l - e_\alpha} - 4g \sum_{\beta(\neq\alpha)} \frac{1}{e_\alpha - e_\beta} = 0 \tag{5}$$

The total number of bosons is $N = 2n + \sum_l \nu_l$. The eigenvalues r_l of the R_l can be found in Ref. [7].

Any hamiltonian constructed as a linear combination of the constants of motion (3), viz: $H = 2\sum_l \epsilon_l R_l(g, \eta)$ commutes with them, and thus is diagonal in the basis of eigenstates (4). The hamiltonian eigenvalues are the same linear combination $2\sum_l \epsilon_l r_l(g, \eta)$. In particular, the boson pairing hamiltonian that was solved by Richardson[8] is obtained by choosing the coefficients of the linear combination equal to the parameters inside the R_l operators, $epsilon_l = \eta_l$. We will use the following parameterization of the pairing hamiltonian:

$$H_P = \frac{x}{2} \sum_{lm} l b_{lm}^\dagger b_{lm} + \frac{1-x}{2N} \sum_{lml'm'} b_{lm}^\dagger b_{l\overline{m}}^\dagger b_{l'\overline{m}'} b_{l'm'} \tag{6}$$

and its eigenvalues are given by

$$E_P = \sum_l \eta_l \nu_l + \sum_{\alpha=1}^{M} e_\alpha \tag{7}$$

With the limitation to s and d bosons, the hamiltonian (6) describes a transition from a spherical vibrational $U(5)$ DI to the γ unstable $O(6)$ DI. In Fig. 2 we show the corresponding 0^+ states for $N = 10$ bosons as a function of x. In the limit of $x = 0$, we are in the exact $O(6)$ DS limit; for $x = 1$ we are in the $U(5)$ DS limit.

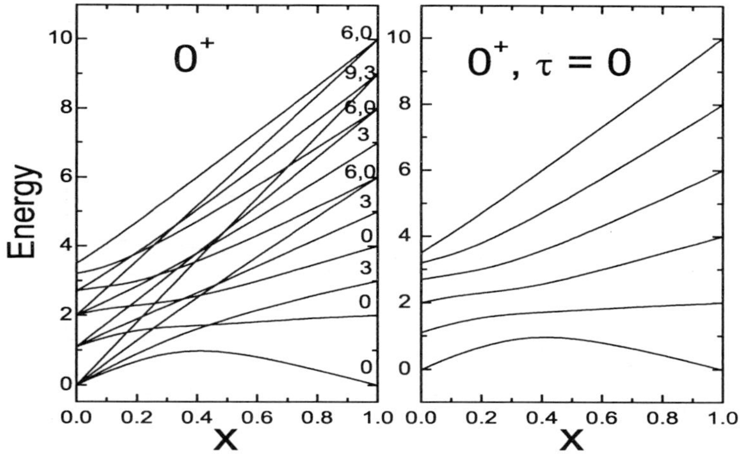

Figure 2. 0^+ energy levels for a system of $N = 10$ bosons as a function of the parameter x for the $O(3)$-$U(5)$ transition. The left panel shows all the 0^+ levels, while in the right panel only the $\tau = 0$ are displayed.

As previously discussed, the behavior of the levels as a function of x does not show any sign of level repulsion. Since the $O(5)$ Casimir operator commutes with the hamiltonian (6), the eigenstates can be labelled by the boson seniority quantum number τ.

In the left panel, we plot all the 0^+ levels, the Seniority quantum number τ is specified on the right vertical axis. There are several level crossings in the figure, but they all correspond to pairs of levels with different seniority quantum numbers. The 0^+ levels with $\tau = 0$ are displayed in the right panel. These levels evolve independently with the parameter x. The fact that the complete transitional region from $O(6)$ to $U(5)$ is quantum integrable has been been previously noted[6]. From the six quantum degrees of freedom of the $U(6)$ dynamical group, four of them are taken into account by the Casimir operators of the $O(5)$ subgroup chain, the fifth is the conserved number of bosons N and the sixth is any linear combination of the Casimir operators of $O(6)$ and $U(5)$, e.g. the hamiltonian, which by construction commutes with all the other constants of motion. We arrive to the same conclusion from the constants of motion given in (3). In an *sd* space there are two constants of motion, R_s and R_d. The sum gives the boson number N and any other combination defines a hamiltonian interpolating between $O(6)$ and $U(5)$. The states with unpaired bosons are completely classified by the $O(5)$ subgroup.

A non-analytic point in the thermodynamic limit due to level repulsion is precluded in the transition from $O(6)$ to $U(5)$. The only source of non-analyticity within an integrable region is a level crossing. However, as we see in Fig. 2, states with $\tau = 0$ do not cross. The exact solvability of the model implies that the energy is an analytic function of the parameter x and only phase transitions of order greater than one are permitted. We may wonder, therefore, under what conditions are crossings between $\nu = 0$ states possible. It can be shown that crossings between

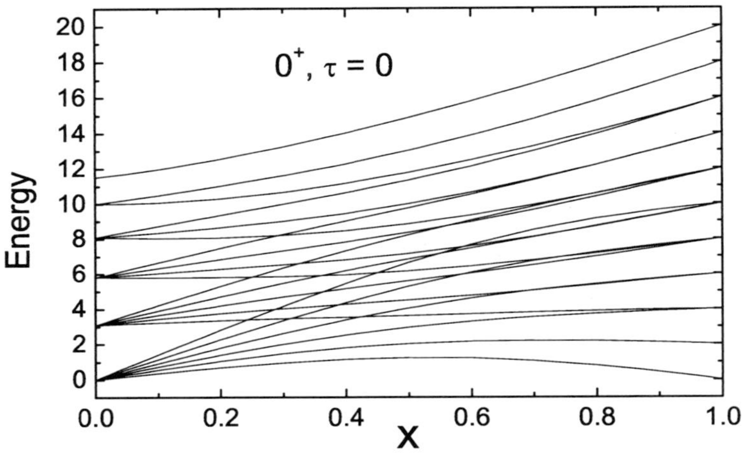

Figure 3. sdg $0^+, \tau = 0$ energy levels for a system of $N = 10$ bosons as a function of the parameter x for the $O(15)$-$U(14)$ transition.

states with the same set of quantum numbers can take place when there are at least two parameter dependent constants of motion. The simplest example is a boson model with sdg bosons. In this case, there are three independent constants of motion, R_s, R_d and R_g, but factorizing out the boson number N as the sum of the three, we are left with two that parameter dependent. An analysis based on the structure of the $U(15)$ dynamical group of the sdg IBM leads to the same conclusion. The pairing hamiltonian (6) is a linear combination of the Casimir operators of two subgroups, those associated with $U(14)$ and $O(15)$. Both subgroups constitute a DS of the $U(15)$ dynamical group and both have the group $O(14)$ as a common subgroup. In Fig. 3 we show the 0^+, $\tau = 0$ states for $N = 10$ bosons as a function of the parameter x for this model. As expected, there are no signs of level repulsion, but there are level crossings. However, there are no crossings with the ground state, which would be interpreted as a first order phase transition even for a finite system.

In previous works[9,10], we showed that the rational class of integral models with repulsive pairing exhibit a quantum phase transition (QPT) to a symmetry broken phase with a macroscopic occupation of the two lowest single-boson states. This led us to conclude that repulsive pairing between bosons is a new and robust mechanism for enhancing sd dominance[10] in interacting boson models of nuclei.

Within the Landau theory, a second-order phase transition is related to a continuous change of the order parameter from 0 in the disordered phase to a non-zero value in the ordered phase. The order parameter is not unique, as its definition depends on the particular problem. In the Ehrenfest approach, the order of the transition is related to the order of the first discontinuous derivative of the energy.

The advantage of an exactly solvable model is that we can reconcile these two approaches to phase transitions, by defining the order parameter as the derivative of the hamiltonian (6) with respect to x. Making use of the Helmann-Feyman

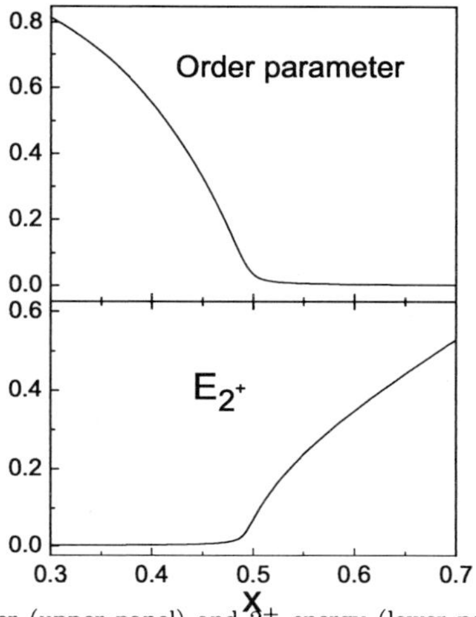

Figure 4. Order parameter (upper panel) and 2^+ energy (lower panel) as a function of x for a system of $N = 1000$ bosons in the $O(6)$-$U(5)$ transition.

theorem, the order parameter is given by

$$O = \frac{\partial H}{\partial x} = n_d - \frac{1}{N}\left(d^\dagger \cdot d^\dagger - s^\dagger s^\dagger\right)(d \cdot d - ss) \quad \rightarrow \quad \langle O \rangle = \frac{\partial E}{\partial x}$$

In Fig. 4 we show the order parameter and the energy of the first excited state with angular momentum 2^+, for a system of $N = 1000$ bosons and in the vicinity of the phase transition. Though a more detailed study, which includes finite size corrections in the thermodynamic (large N) limit, we can readily show that the phase transition is indeed of second order in the *sd* IBM space as well as in any larger space. The present analysis based on the exact solution for very large systems confirms recent results[11] based on the phenomenological Landau theory on the characteristics of this phase transition.

References

1. F. Iachello and A. Arima, *The Interacting Boson Model of Nuclei* (Cambridge University Press, Cambridge, UK. 1987).
2. R. F. Casten and D. D. Warner, *Rev. Mod. Phys.* **60**, 389 (1988).
3. N.V. Zamfir em et. al., *Phys. Rev.* C **66**, 021304 (2002).
4. Y. Alhassid and N. Whelan, *Phys. Rev. Lett.* **67**, 816 (1991).
5. N. Fhelan and Y. Alhassid, *Nucl. Phys.* A **556**, 42 (1993).
6. P. Cejnar and J. Jolie, *Phys. Rev.* E **58**, 387 (1998).
7. J. Dukelsky, C. Esebbag and P. Schuck, *Phys. Rev. Lett.* **87**, 066403 (2001).
8. R. W. Richardson, *J. Math. Phys.* **9**, 1327 (1968).
9. J. Dukelsky and P. Schuck, *Phys. Rev. Lett.* **86**, 4207 (2001).
10. J. Dukelsky and S. Pittel, *Phys. Rev. Lett.* **86**, 4791 (2001).
11. J. Jolie *et. al.*, *Phys. Rev. Lett.* **86**, 4791 (2001).

PHASE TRANSITIONAL BEHAVIOR IN SPHERICAL-DEFORMED TRANSITIONS REGIONS

R.F. CASTEN

Wright Nuclear Structure Laboratory, Yale University, New Haven, CT 06520, USA
E-mail: rick@riviera.physics.yale.edu

The recent proposal and discovery of examples of critical point symmetries [in particular X(5) in ^{152}Sm] has spurred the search for other empirical manifestations of these symmetries and given new understanding of how deformation arises in nuclei. This talk will review the characteristic signatures for X(5) and the resulting evidence for X(5) symmetry in nuclei near N = 90. New experiments on nuclei in this region will be mentioned. The different behavior of nuclei as a function of neutron number (using Sm and Ba as examples) will be discussed in terms of the phase structure of the nuclear structure symmetry triangle.

1 Introduction

It is an honor to contribute this paper in honor of Franco Iachello. Through his seminal achievements in a symmetry-based understanding of nuclear structure, he has transformed the entire field, brought new paradigms into the arsenal of structural benchmarks, and shown the power of simple group theory-based analytic solutions to the nuclear many-body problem. With the development of the IBA, he and Akito Arima provided an extraordinarily elegant model that gives not only the dynamical symmetries that characterize special nuclei but which also allows simple and remarkably accurate descriptions of nuclei spanning a wide range of collective structures. His recent development of critical point symmetries provides comparably simple paradigms for the most difficult of all nuclei, those in shape transitional regions, whose structural evolution is controlled by the competition of different configurations. This legacy of achievements is unique in the history of nuclear structure physics. Personally, I want to thank Franco, not only for his astonishing physics insights, but also for his friendship and the inspiration he has provided for more than a quarter century.

The physics focus of this paper is on the phase transitional nuclei referred to above, including a discussion of the new critical point symmetries, especially X(5), and the data supporting their manifestation in nuclei. Phase transitional behavior in nuclei is an ideal venue to illustrate two of the major themes and challenges facing modern science. These challenges may be phrased in the following way:

- Understanding how complex many-body objects can be constructed from simple ingredients—elementary constituents, basic forces, quantum mechanics and statistics, and a few conservation laws.

  In nuclei, the constituents at our level of focus are the nucleons, and the primary forces are the strong and Coulomb forces.

- Understanding how these complex many-body systems can display such astonishing regularity.

It is remarkable how often it happens that complex many-body systems display simple phenomenology. For nuclei, we exhibit this by two examples. First is the beautiful regularity of rotational spectra illustrated by the uniform spacing of γ-ray energies of superdeformed bands, as illustrated in the right part of the spectrum in Fig. 1. The rotational paradigm accounts for this behavior by the extraordinarily simple ansatz of the symmetric rotor.

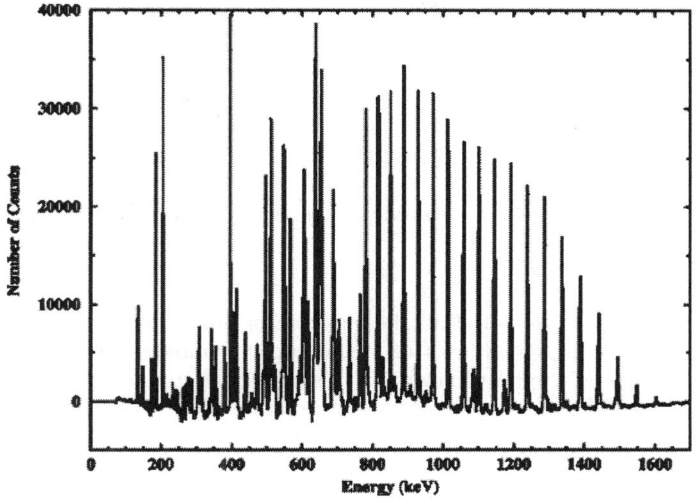

Figure 1. Spectrum of γ-rays in ^{150}Gd showing on the right the beautiful cascade of transitions in a superdeformed band. From ref. [1].

The second example is the regularity of quasi-band spectra over very large segments of the nuclear chart. This is illustrated in Fig. 2, which plots the energy of the first 4^+ state in collective even-even nuclei against the energy of the first 2^+ state[2]. While these nuclei span a large range of structures from near-harmonic vibrator ($R_{4/2} \equiv E(4_1^+)/E(2_1^+) \sim 2.0$) to well–deformed rotor, the data lie on a compact bilinear trajectory with slopes (least squares fits) of 2.00 and 3.33. The data along the former segment clearly satisfy the anharmonic vibrator (AHV) equation $E(4_1^+) = 2E(2_1^+) + \epsilon_4$. The remarkable empirical result is that ϵ_4 is nearly constant ($\sim$ 167 keV) for all collective nuclei from A $\sim$ 100 - 200. Why these complex many-body systems display such profound simplicities is still not understood.

2 Symmetries and Nuclear Structural Evolution

Attempting to understand such regularities often entails the application of ideas founded in the concept of symmetries, and, in particular, dynamical symmetries. The SU(3) scheme of Elliott[3] for s-d shell nuclei was one of the first applications

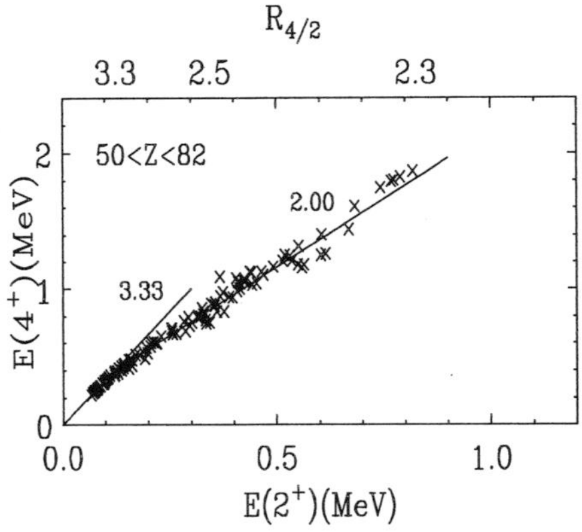

Figure 2. $E(4_1^+)$ against $E(2_1^+)$ for Z = 50 - 82. The slopes on the two linear segments are least squares fitted values.

of a symmetry perspective to the detailed spectra of atomic nuclei. His approach also illustrates the complementarity of macroscopic (symmetry) and microscopic approaches. Of course, the best known use of symmetries to understand the structure of nuclei is the Interacting Boson Approximation (IBA) model of Arima and Iachello[4], in which the s-d boson ansatz leads to the group structure of U(6) and the three group chains of dynamical symmetries known as U(5) [vibrator], SU(3) [symmetric rotor], and O(6) [γ-unstable deformed nucleus]. These symmetries are indicated at the vertices of the symmetry triangle on the left of Fig. 3.

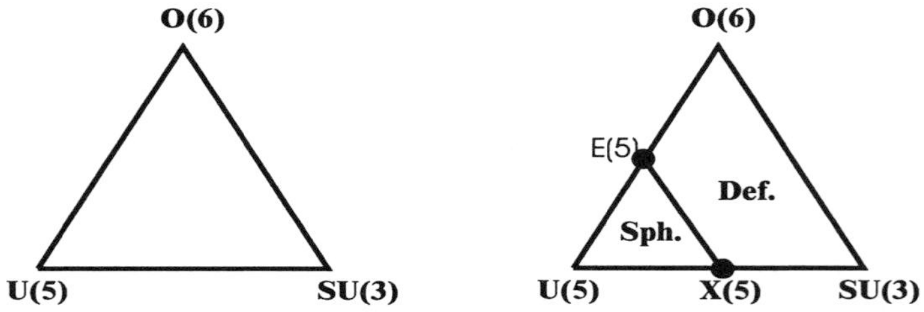

Figure 3. Left: Symmetry triangle of the IBA. Right: Same but with the two critical point symmetries and the line of first order phase transitions connecting them.

Nuclear structure evolves as a function of nucleon number. In particular, the competition between the valence pairing interaction and the valence proton-neutron interaction largely determines this evolution and the onset of collectivity and deformation. The pairing interaction is typically about 1 MeV, while the p-n interaction is around 200 keV in heavy nuclei. Therefore, one might expect deformation to develop when the number of valence p-n interactions is about 5 times the number of pairing interactions. Though this is just a crude guide, it works remarkably well and has been codified in the P-Factor, defined as

$$P = \frac{N_p N_n}{N_p + N_n} \tag{1}$$

In nearly all regions of medium and heavy mass nuclei, the transition from spherical to deformed nuclei occurs near P $\sim$ 5 or slightly lower.

These shape transition regions often exhibit behavior resembling phase transitions in macroscopic systems, despite the finite number of nucleons and the much smaller number of valence nucleons. It is in the concept of phase transitions that the two scientific challenges we introduced earlier come together in a dramatic way. Phase transitions, especially first order phase transitions which are, by far, the most common in nuclei, arise due to the competition between different degrees of freedom. Generally, these represent spherical and deformed shapes, although with varying degrees of axial asymmetry (softness). The microscopic many-body treatment of such systems is therefore extremely difficult and, for this reason, such nuclei are often the preferred (albeit challenging) testing ground for nuclear models.

A complementary approach from the standpoint of macroscopic symmetries, which could produce the spectra of phase transitional nuclei analytically, has been lacking until very recently. However, Iachello's recent development of the concept of critical point symmetries[7,8] changes this situation in an important way. Iachello's idea is sketched in Fig. 4 which illustrates the competition of spherical and deformed equilibrium configurations as nucleon number varies. The curves labelled 1 - 4 on the left show schematically the successive energy surfaces for successively larger numbers of valence nucleons. The point where the spherical and deformed minima are degenerate corresponds to the critical point, or the point of the phase transition. If the barrier between the two minima is small, as seems to happen and as Iachello assumes, this "critical" energy surface has a nearly flat bottom and a steep rise at some value of β. The primary assumption behind the concept of critical point symmetries is to replace curve 3 by the square well approximation on the right in Fig. 4. With this assumption, if the potential is flat in the γ degree of freedom, one obtains E(5), and if the potential has a minimum at $\gamma = 0^0$ (or 60^0), one has X(5). We will focus on the latter here.

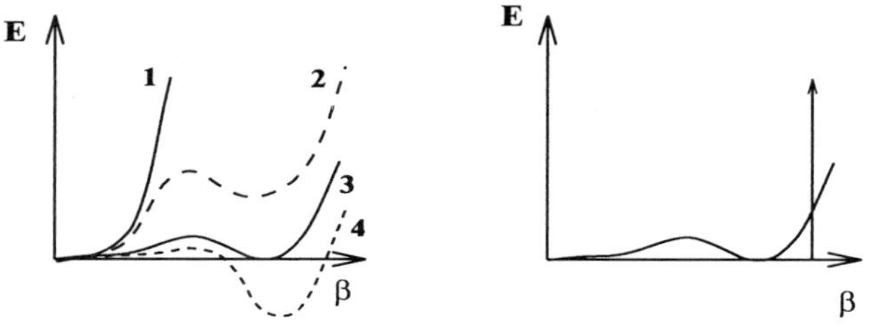

Figure 4. Left: Schematic illustration of the energy surfaces in a sequence of nuclei spanning a transition region. Curve 3 corresponds to the phase transition. Right: Curve 3 and the approximation to it embodied in the concept of critical point symmetries.

In both cases, the solutions are Bessel functions and the energies are given by the zeroes of these functions. For E(5) the order of the Bessel function is half integer, for X(5) they are irrational numbers. All predictions are analytic and parameter free (except for scale). Empirical evidence for nuclei resembling both E(5) and X(5) has been found[9–13]. These critical point symmetries are indicated along the U(5)–O(6) and U(5)–SU(3) legs of the symmetry triangle on the right in Fig. 3. The use of large solid symbols for E(5) and X(5) is meant to distinguish these geometrical constructs, which are solutions to a differential equation, from the algebraic dynamical symmetries of the IBA. Together, these paradigms provide benchmarks both for limiting ideals of structure and regions of rapid structural change.

The predictions for X(5) are shown on the left in Fig. 5. In the A∼150 region, ^{152}Sm was the first nucleus to be identified as near to X(5). Its yrast and yrare levels are shown for comparison on the right in the figure. The agreement is excellent for the yrast energies and B(E2) values, for the energy of the 0_2^+ state, the B(E2) values within the 0_2^+-based sequence, and the relative inter-sequence B(E2) values. The disagreement for the scale of inter-sequence B(E2) values is probably due to the fact that these B(E2) values vanish in the SU(3) limit and that ^{152}Sm is slightly to the rotor side of the exact point of the phase transition. The disagreement for the energy spacings in the 0_2^+-based levels is a well known problem that plagues almost all treatments of the N = 90 nuclei.

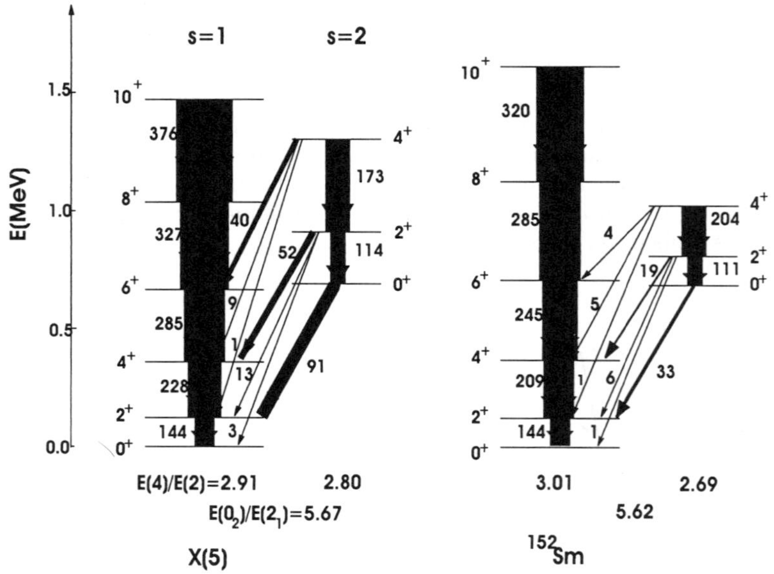

Figure 5. Comparison of X(5) with ^{152}Sm. From ref. 10.

There are three particular parameter-free signatures of X(5) that are satisfied in ^{152}Sm, namely the $R_{4/2}$ values for the yrast and 0_2^+-based levels, the energy ratio $E(0_2^+)/E(2_1^+)$ which is 5.67 in X(5) and 5.62 in ^{152}Sm, and the near degeneracy of the 0_2^+ and 6_1^+ levels. The predictions of X(5) are hardly trivial and differ from both the vibrator and rotor, as shown on the left in Fig. 6. It is clear that X(5) provides a very good account of the data.

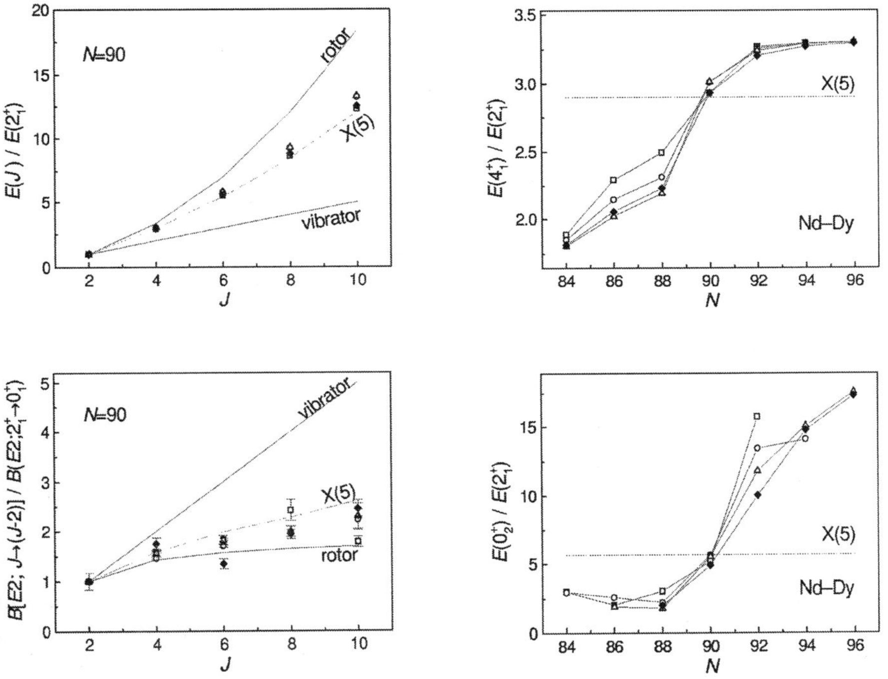

Figure 6. Comparison of X(5) predictions with data in the N = 90 region as a function of angular momentum (left) and neutron number (right). The different symbols (and the lines connecting them on the right), correspond to different isotopic chains.

Figure 6 shows another important feature of the A = 150 region, namely that all the elements from Nd - Dy exhibit remarkably similar behavior both as a function of N and angular momentum. Moreover, the signatures shown agree with X(5) only for N = 90, highlighting the (literally) critical role of these isotones. The similarity of the N = 90 isotones is vividly shown in Fig. 7 where relative level schemes are shown (2_1^+ energies are normalized for ease of comparison). The three signatures of X(5) mentioned earlier are summarized below the level schemes.

This result brings up the issue of where to search for X(5)-like nuclei in this and other mass regions. Here, we return to the P-factor and recall that the spherical-

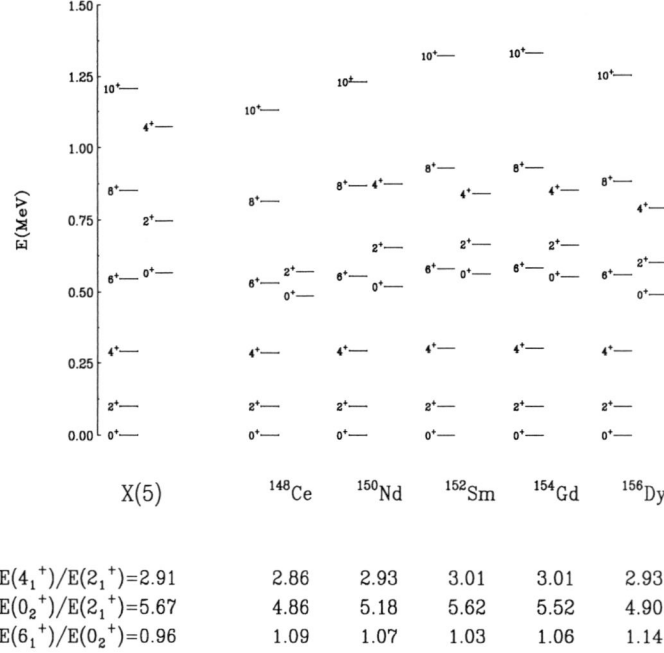

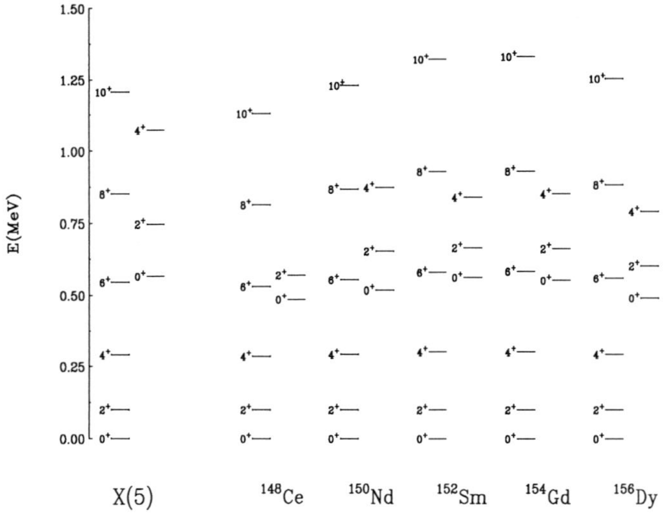

$E(4_1^+)/E(2_1^+)=2.91$	2.86	2.93	3.01	3.01	2.93
$E(0_2^+)/E(2_1^+)=5.67$	4.86	5.18	5.62	5.52	4.90
$E(6_1^+)/E(0_2^+)=0.96$	1.09	1.07	1.03	1.06	1.14

Figure 7. X(5) (left column) and the N = 90 isotones (right). The 2^+ energies are normalized to each other. Below the level schemes are the values of the three most characteristic signatures of X(5) and their empirical values in these nuclei.

deformed shape transition occurs when P $\sim$ 5. Figure 8 shows a broader view of the rare earth region and the locus of P $\sim$ 5. Each box gives the empirical $R_{4/2}$ value. The correlation of P $\sim$ 5 and the X(5) value $R_{4/2} \sim 2.9$ is quite extensive. In particular, Fig. 8 suggests that ^{162}Yb is a candidate for X(5). We have recently carried out experiments at WNSL on this nucleus that revise earlier results and confirm this candidacy (pending still further experiments now underway). Specifically, the previously existing first excited 0^+ state was found not to exist so that the (now) 0_2^+ occurs near the energy predicted in X(5).

We now take a wider perspective and inspect the locus of P $\sim$ 5 values over large regions of heavy nuclei. This is shown in Fig. 9 which exhibits some interesting features. First of all, the possible candidates for first order phase transitional behavior and X(5) symmetry are numerous. Secondly, many lie far off the valley of stability and will be accessible only with exotic beam facilities such as RIA. Thirdly, $28 \leq$ N, Z ≤ 50 region is interesting in that there is only a single cluster of contiguous nuclei near the N = Z nuclides ^{80}Zr and ^{76}Sr. The former nucleus is intriguing since 40 is sometimes a magic number but the enhanced p-n interaction near N = Z favors collectivity and ^{80}Zr has long been considered a deformed nucleus following the initial experiments by Lister *et al*[14]. However, its $R_{4/2}$ value is only

78			2.30		2.26	2.44	2.51	2.70	2.68	Pt
76				2.62	2.66	2.74	2.93	3.02	3.09	Os
74			2.68	2.82	2.95	3.07	3.15	3.22	3.24	W
72	2.31	2.56	2.79	2.97	3.11	3.19	3.25	3.27	3.28	Hf
70	2.33	2.63	2.93	3.12	3.23	3.27	3.29	3.31	3.31	Yb
68	2.32	2.74	3.10	3.23	3.28	3.29	3.31	3.31	3.31	Er
66	2.23	2.93	3.21	3.27	3.29	3.30	3.31			Dy
64	2.19	3.02	3.24	3.29	3.30	3.30				Gd
62	2.32	3.00	3.25	3.29	3.30	3.30				Sm
60	2.49	2.93	3.26	3.29	3.32					Nd
58	2.59	2.86	3.15							Ce
56	2.66	2.84	2.99							Ba
Z/N	88	90	92	94	96	98	100	102	104	

Figure 8. Rare earth nuclei with empirical $R_{4/2}$ values and the locus of P $\sim$ 5 values.

2.86 and therefore structural features different from the symmetric rotor must come into play. One possibility is axial asymmetry, which lowers $R_{4/2}$ from 3.33 for $\gamma = 0^0$ to 2.5 for $\gamma = 30^0$. Alternately, ^{80}Zr could conceivably be a candidate for X(5) where $R_{4/2} = 2.91$. Only time and more data will tell.

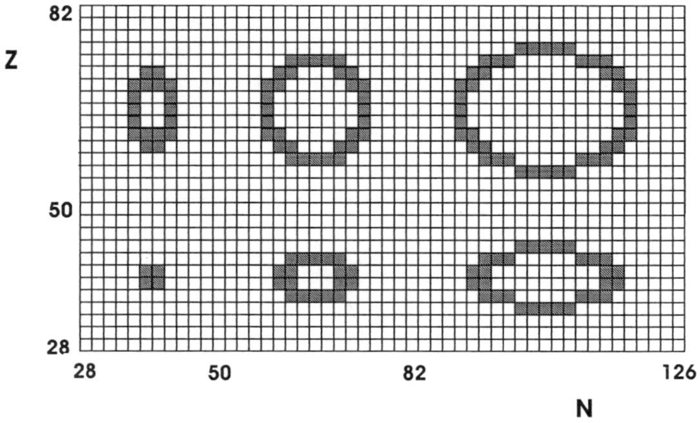

Figure 9. Similar to Fig. 8 for a broader region.

Lastly, we return to the symmetry triangle on the right of Fig. 3. Linking

E(5) and X(5) is a line (rigorously, a region of finite width) of first order phase transitions, terminating at an isolated point of second order phase transition at E(5). Nuclei to the lower left of this line are spherical, to the upper right deformed.

This implies that, contrary to prior general perceptions, nearly all spherical-deformed transition regions must be first order and involve a crossing of equilibrium configurations as in Fig. 4 (curve 3) since one cannot go from the spherical to the deformed phase without crossing the line. Thus, with one exception, the phase diagram in Fig. 3 (right) does not admit the idea of a gradual increase in the value of the deformation at the energy minimum. The one exception corresponds to completely γ-soft nuclei for which the phase transition passes through E(5). (Of course, for internal trajectories in the triangle passing close to E(5), the spherical-deformed phase transition would be formally first order but, in such nuclei with large $< \gamma >$ values, not easily distinguished from a second order phase transition.) Figure 3 (right) thus suggests that first order phase transitional behavior (muted by finite nucleon number effects) is therefore the rule not the exception in spherical–deformed transition regions.

With these ideas in mind, let us consider the data in Fig. 10, showing the evolution of $R_{4/2}$ values with neutron number in Sm and Ba with N $>$ 82. The steep rise in $R_{4/2}$ for Sm is one of the key pieces of data that originally inspired the phase transitional interpretation. However, we now see that this, in itself, cannot be a real signature for first order phase transitional behavior since Ba must also traverse a line of first order phase transition. While phase transitions are rapid with respect to a control parameter they are not necessarily rapid with respect to any given observable, nor, in the nuclear case, is neutron number the appropriate control parameter. Rather, a model Hamiltonian variable that specifies the location in the symmetry triangle (specifically, the "radial" distance from the U(5) vertex) is a proper control parameter. Figure 10 shows that the relation between such a control parameter and neutron number, that is, its rate of change with neutron number, need not be the same for different elements. Alternately, the Ba trajectory may pass along an internal path in the symmetry triangle, while Sm closely follows the bottom leg.

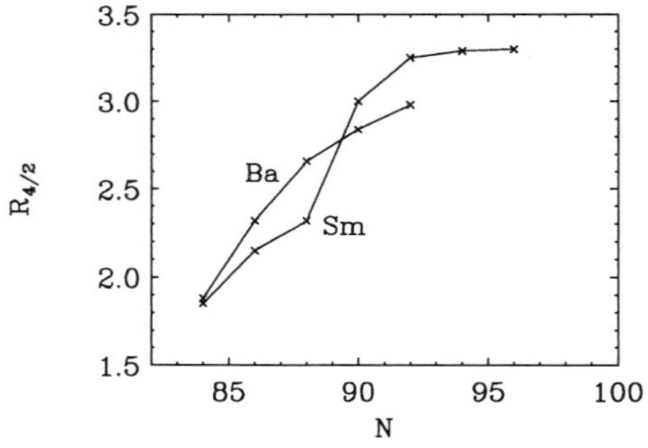

Figure 10. Comparison of $R_{4/2}$ values in Sm and Ba.

3 Conclusions

We have discussed some of the empirical features of phase transition-like behavior in nuclei, some of the key observables, the concept of critical point symmetries [specifically E(5) and X(5)] and the data showing their manifestations in actual nuclei. Study of the nuclear phase diagram[15] reveals new insights into spherical-deformed transition regions, and the ubiquity of first order phase transitional behavior. Finally, the use of the P-factor provides a guide to possible candidates for other critical point nuclei. All this work owes a tremendous debt to the stunning ideas and achievements of Franco Iachello.

Acknowledgments

Once again, I would like to thank Franco Iachello for his magnificent contributions to nuclear physics and his personal friendship. I also thank my collaborators, especially N.V. Zamfir, P. von Brentano, J. Jolie, M.A. Caprio, and E.A. McCutchan. Work supported by US DOE grant no. DE-FG02-91ER-40609.

References

1. C.W. Beausang, private communication from S. Erturk.
2. R.F. Casten, N.V. Zamfir, and D.S. Brenner, Phys. Rev. Lett. **71**, 227 (1993).
3. J.P. Elliott, Proc. Roy. Soc. **A245**, 128 (1958).
4. F. Iachello and A. Arima, *The Interacting Boson Model* (Cambridge University Press, Cambridge, England, 1987).
5. P. Federman and S. Pittel, Phys. Lett. **69B**, 385 (1977).
6. R.F. Casten, D.S. Brenner, and P.E. Haustein, Phys. Rev. Lett. **58**, 658 (1987).
7. F. Iachello, Phys. Rev. Lett. **85**, 3580 (2000).
8. F. Iachello, Phys. Rev. Lett. **87**, 052502 (2001).
9. R.F. Casten and N.V. Zamfir, Phys. Rev. Lett. **85**, 3584 (2000).
10. R.F. Casten and N.V. Zamfir, Phys. Rev. Lett. **87**, 052503 (2001).
11. R. Krücken *et al.*, Phys. Rev. Lett. **88**, 232501 (2002).
12. P.G. Bizzeti and A.M. Bizzeti-Sona, Phys. Rev. C **66**, 031301 (2002).
13. A. Frank, C.E. Alonso, and J.M. Arias, Phys. Rev. C **65**, 014301 (2001).
14. C.J. Lister *et al.*, Phys. Rev. Lett. **59**, 1270 (1987).
15. J. Jolie, these Proceedings.

CRITICAL POINT NUCLEI IN THE INTERACTING BOSON MODEL

N. V. ZAMFIR, E.A. MCCUTCHAN AND R.F. CASTEN
WNSL, Yale University, New Haven, Connecticut 06520-8124, USA
E-mail: victor.zamfir@yale.edu

The nuclear phase/shape transition in the framework of the Interacting Boson Model [1] is analyzed and predictions of observables along trajectories crossing the transition region are presented. The model calculations are compared with the evolution of basic observables and a mapping of the symmetry triangle in terms of a new proposed set of polar coordinates is realized for the N=82-104 region.

1 Introduction

The nature of shape/phase transitions in low energy nuclear spectra has been recently the subject of many experimental and theoretical investigations. It has been known for a long time [2,3] that the spherical-deformed transition regions in the Interacting Boson Approximation (IBA) Model from U(5) to SU(3) and from U(5) to O(6) behave as first and second order phase transitions, respectively. We will discuss these phase/shape transition regions in detail and will compare different observables predicted by the IBA along these paths with the available data.

2 Phase/shape transition in low energy nuclear spectra

In discussing phase transitions in low energy nuclear spectra it is convenient to write the IBA Hamiltonian in the following form[4,5]:

$$H(\zeta) = c[(1-\zeta)\hat{n}_d - \frac{\zeta}{4N_B}\hat{Q}^\chi \cdot \hat{Q}^\chi] \tag{1}$$

where N_B is the total number of bosons, $\hat{n}_d = d^\dagger \cdot \tilde{d}$ and $\hat{Q}^\chi = (s^\dagger \tilde{d} + d^\dagger s) + \chi(d^\dagger \tilde{d})^{(2)}$. This Hamiltonian contains 2 parameters, ζ and χ (c is only a scaling factor) and the parameter space is usually illustrated by a triangle [6] with its three symmetries at the vertices: ζ=0, any χ for U(5), ζ=1, $\chi = -\sqrt{7}/2$ for SU(3), and ζ=1, $\chi = 0$ for O(6). It is possible to provide a quantitative description of the parameter space by representing each set of parameters (ζ, χ) by polar coordinates (ρ, θ):

$$\rho = \frac{\zeta\sqrt{3}}{\sqrt{3}cos\theta_\chi - sin\theta_\chi}$$
$$\theta = \frac{\pi}{3} + \theta_\chi \tag{2}$$

where $\theta_\chi = \frac{2}{\sqrt{7}}\chi\frac{\pi}{3}$

Figure 1 (left) represents the IBA symmetry triangle showing the definition of these polar coordinates and the three dynamical symmetries in terms of the Hamiltonian parametrization in eq. (1).

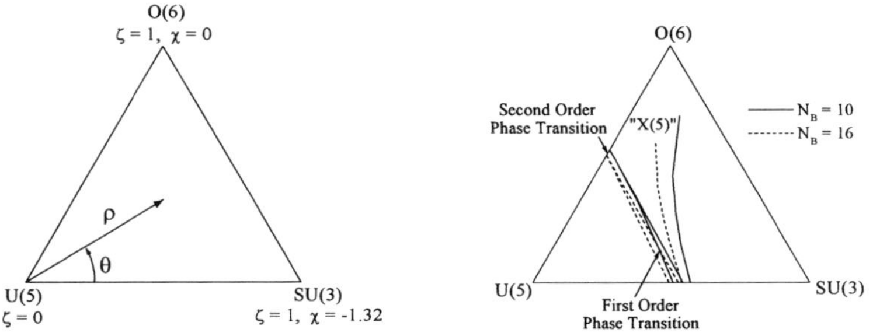

Figure 1. The IBA symmetry triangle. Left: definition of the polar coordinates ρ, θ. Right: the phase transition region and the locus of the X(5) type spectra [i.e. those IBA parameters which produce $R_{4/2} = 2.91$ and $E(0_2^+)/E(2_1^+) \sim 6$] for N_B=10 (continuous lines) and N_B=16 (dashed lines).

The total energy corresponding to the IBM Hamiltonian can be obtained using the intrinsic state formalism and is expressed in terms of the intrinsic shape variables β, γ [7,8]:

$$E(\beta, \gamma; \zeta, \chi, N_B) = A\frac{\beta^2}{1+\beta^2} + B\frac{\beta^4}{(1+\beta^2)^2} + C\frac{\beta^3 cos3\gamma}{(1+\beta^2)^2}. \tag{3}$$

where

$$\begin{aligned} A &= \frac{4N_B - 8N_B\zeta - \zeta\chi^2 + 8\zeta}{\zeta} \\ B &= (N_B - 1)(4 - \frac{2}{7}\chi^2) \\ C &= 4\sqrt{\frac{2}{7}}\chi. \end{aligned} \tag{4}$$

The study of this functional form has shown that there is a phase/shape transition as a function of the control parameters ζ and χ. For fixed χ a phase transition occurs in the ground state energy at a critical value of the parameter $\zeta = \zeta_{crit}$: for $\chi \neq 0$ there is a first-order phase transition (the first derivative of E_{min}, $\partial E_{min}/\partial\zeta$, is discontinuous) and for $\chi = 0$ there is a second-order phase transition (second derivative $\partial^2 E/\partial\beta^2$ is discontinuous).

In the first order phase transition there is a region where two minima, spherical and deformed, occur in the total energy. This is a region of phase/shape coexistence [9]. Figure 2 illustrates the evolution of the total energy (for γ=0) as a function of ζ for $\chi = -\sqrt{7}/2$ [U(5)-SU(3) transition]and N_B=10. The phase/shape coexistence region starts, with increasing ζ, where the deformed minimum develops (ζ=0.507) in addition to the spherical one and ends where the spherical minimum disappears (ζ=0.542). The critical point, where the first derivative is discontinuous, is $\zeta_{crit} = 16N_B/(34N_B - 27)$=0.512.

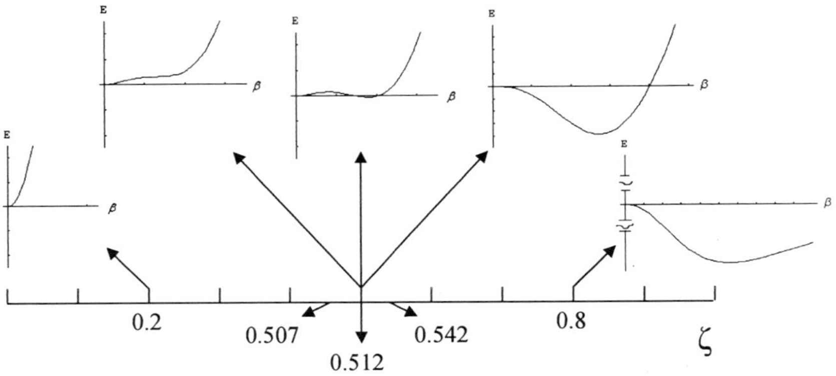

Figure 2. The evolution of total energy along the U(5)-SU(3) transition ($\chi = -\sqrt{7}/2$) for N_B=10 with emphasize on the phase transition region.

The range of ζ corresponding to the region of coexistence becomes smaller for smaller $|\chi|$ and for χ=0 [U(5)-O(6) transition] converges to one point, the critical point of the second order phase transition. This critical point is given by $\zeta = N_B/(2N_B - 2)$. In figure 1(right) the region of phase coexistence is shown for N_B=10 and 16.

3 The evolution of nuclear observables in the phase/shape transition region

We next explore how key observables behave for different χ values across the $\zeta = 0 \rightarrow 1$ trajectory in the triangle. Two characteristic observables that provide insight into nuclear structure, the energy ratios $R_{4/2}$ and $E(3_1^+)/E(0_2^+)$, are shown in figure 3. For $\zeta \rightarrow 0$ [U(5)] $R_{4/2}$=2.0 and $E(3_1^+)/E(0_2^+)$=1.5 since 2_1^+ is 1-phonon state, 4_1^+, 0_2^+ are 2-phonon states, and 3_1^+ is a 3-phonon state. For finite ζ, the predictions diverge for different χ (and N_B) values. For ζ=1, $R_{4/2}$ varies from 2.5 for χ=0 [O(6)] to 3.33 for χ=-1.32 [SU(3)]. The $E(3_1^+)/E(0_2^+)$ ratio shows a very interesting behavior: for large χ it presents a maximum which is located exactly in the phase transition region. In figure 3 the shaded areas show the predicted values for the two observables in the phase transition region. The observation of empirical values of these quantities within the range corresponding to the shaded areas could be a signature of phase transitional behavior although one has to determine, by a detailed fit, the appropriate (ζ, χ) values for a given nucleus. A simple signature could constitute the maximum in the evolution of the $E(3_1^+)/E(0_2^+)$ ratio along an isotopic chain.

The phase/shape transition region is characterized by a pronounced β softness, as can be seen in figure 2. This characteristic led to the discovery of new symmetries, E(5)[10](for γ-unstable nuclei) and X(5) [11](for axially symmetric nuclei), which

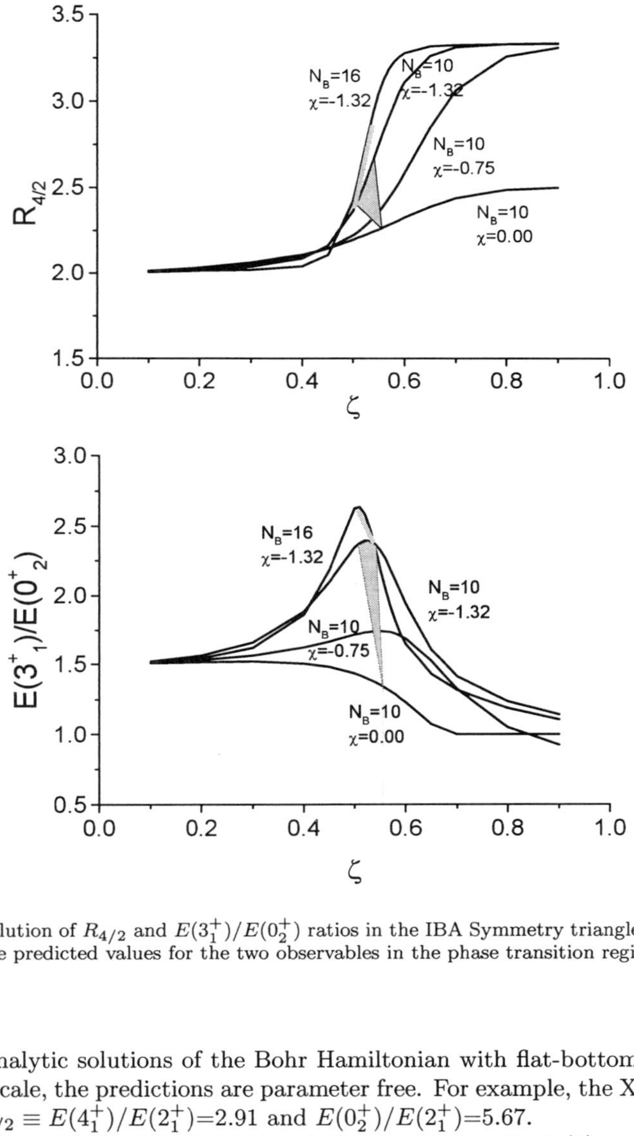

Figure 3. Evolution of $R_{4/2}$ and $E(3_1^+)/E(0_2^+)$ ratios in the IBA Symmetry triangle. The shaded areas show the predicted values for the two observables in the phase transition region.

consist of analytic solutions of the Bohr Hamiltonian with flat-bottom potentials. Except for scale, the predictions are parameter free. For example, the X(5) solution predicts $R_{4/2} \equiv E(4_1^+)/E(2_1^+)$=2.91 and $E(0_2^+)/E(2_1^+)$=5.67.

The IBA calculations produce very similar spectra to the X(5) predictions for parameters which are not exactly in the phase/shape transition but very close to it. In figure 1(right) are shown the locus of the IBA parameters, which produce exactly the X(5) value for $R_{4/2}$, i.e. 2.91, and within a reasonable deviation the other characteristic ratio, $E(0_2^+)/E(2_1^+)$. For example for N_B=16 it varies between 5.0 and 7.0 for χ between -1.32 and -0.4, respectively. For χ=-0.9 and $\zeta = 0.59$ the IBA predicts an energy spectrum very close to the X(5) solution, but, as can be

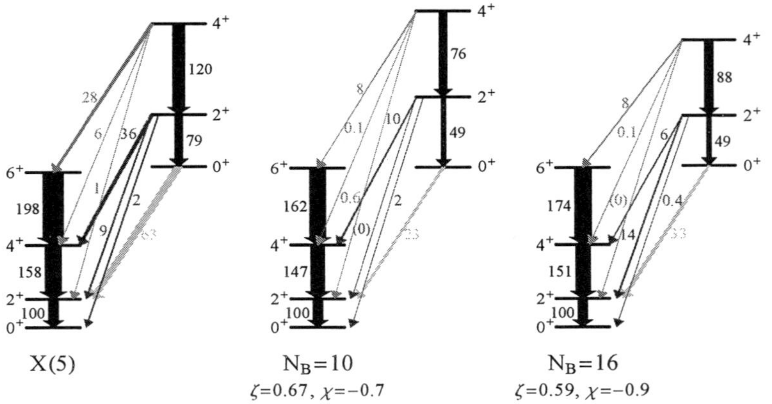

Figure 4. Comparison of the IBA results for (N_B=10, χ=-0.7, ζ=0.67) and (N_B=16, χ=-0.9, ζ=0.59) with the X(5) predictions.

seen in figure 4, the transition probabilities are poorly reproduced. The agreement for in-band transitions seems to increase with increasing N_B but the disagreement for cross-over transitions remains quite large.

4 Mapping the symmetry triangle

A well known example of an U(5)-SU(3) transition is the N∼ 90 region. In fact Sm isotopes were well described in the framework of the IBA [13] and $^{152}Sm_{90}$ was the first empirical example of a nucleus very close to the X(5) solution [12]. The Nd, Gd, and Dy isotopes have a very similar evolution and they can be described by the same set of parameters (very close to the parameters in ref. [13]): $\zeta = 0.23+0.085N_\nu$, (N_ν is the number of neutron bosons) and $\chi = -\sqrt{7}/2$ [5].

The empirical evolution of basic observables $R_{4/2} \equiv E(4_1^+)/E(2_1^+)$, the band-head of the quasi-gamma band, $E(2_\gamma^+)$, and $E(0_2^+)$ is compared with the IBA results in Figure 5.

The $R_{4/2}$ ratio evolve from ∼2.0, characteristic for the U(5) symmetry, to ∼3.33, characteristic for SU(3) symmetry, with a sharp rise at N=90. The energies of the intrinsic excitations 0_2^+ and 2_γ^+ have a minimum also at N=90. This point corresponds in the IBA to a calculation very near the critical point of the phase/shape U(5)-SU(3) transition.

The calculations performed with this simple parametrization reproduce the entire transition quite well. However, the exact description of the evolution of the empirical observables and the correspondence between the data and the IBA predictions in terms of ζ and χ can be established only by means of a detailed fit.

A fit for the Gd, Dy, Er, Yb, and Hf isotopes with 82< N < 104 was performed considering the basic observables, energies of the 2_1^+, 4_1^+, 2_γ^+, and 0_2^+ states and available transition probabilities. In figure 6 is presented the comparison of the

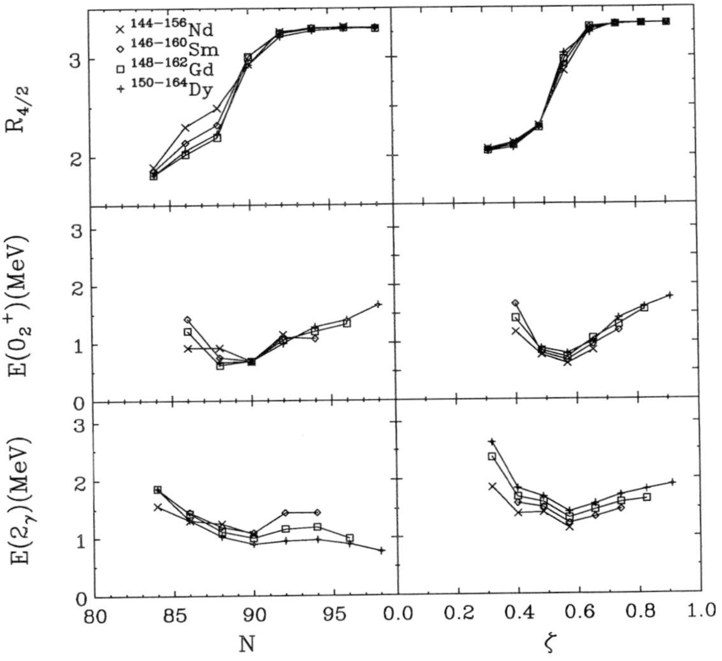

Figure 5. The evolution of basic empirical observables in the Nd,Sm,Gd, and Dy isotopic chains with N>82 compared with the IBA results.

empirical energies of these states with the IBA results. The agreement is impressive, including the description of the 0_2^+ states which were poorly described in previous fits [14]. The other excited states are reproduced quite well. [The main exceptions are the $2^+_{K=0}$ - 0_2^+ relative energies for N$\sim$90 nuclei which are larger in the calculations than the empirical values.] As an example in figure 7 is shown the comparison of the experimental level scheme with the IBA results for three isotopes of Dy. The total energy obtained using eqs. (3,4) is also shown in each case. The minimum energy is shifted with increasing N from β=0 to finite β with ^{156}Dy very close to the critical point. In fact, this nucleus is considered a very good X(5) candidate [15].

The trajectories in the symmetry triangle corresponding to the fit parameters for these isotopic chains are shown in figure 8 in terms of the polar coordinates defined in eq. (2). All isotopic chains are, as it is expected, transitional from U(5) for N just above 82, toward the SU(3)-O(6) leg (large ζ) with increasing N. The Sm isotopic chains correspond to a U(5)-SU(3) transition (which corresponds to rigid γ) and the higher Z chains show an increasing γ-softness for low N ($\leq$ 90). For higher N the Dy and Er isotopes show an increasing γ-softness [closer to the U(5)-O(6) leg with increasing N] while the Yb and Hf isotopes with N> 90 present an increased γ-rigidity. An interesting feature is that the Dy-Hf nuclei with N=92 are reproduced with very similar ρ, θ (ζ, χ) sets of parameters.

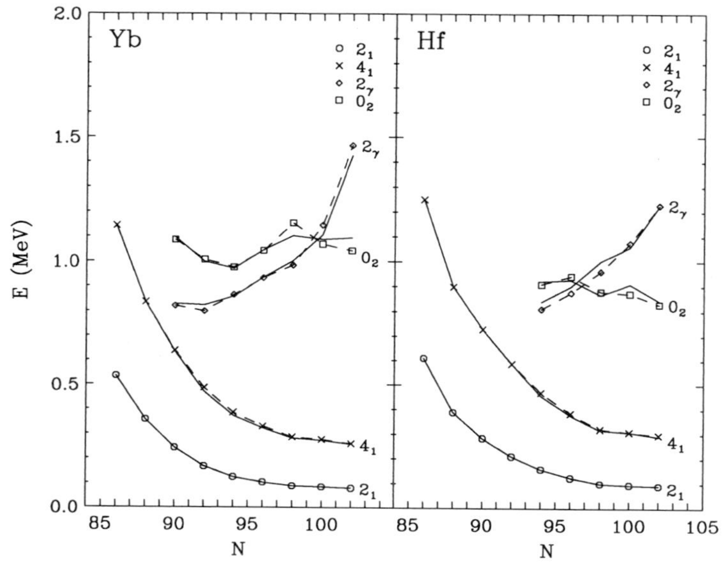

Figure 6. Comparison between empirical level energies (symbols connected by dashed lines) and the IBA calculations (connected by solid lines) for Yb and Hf isotopes.

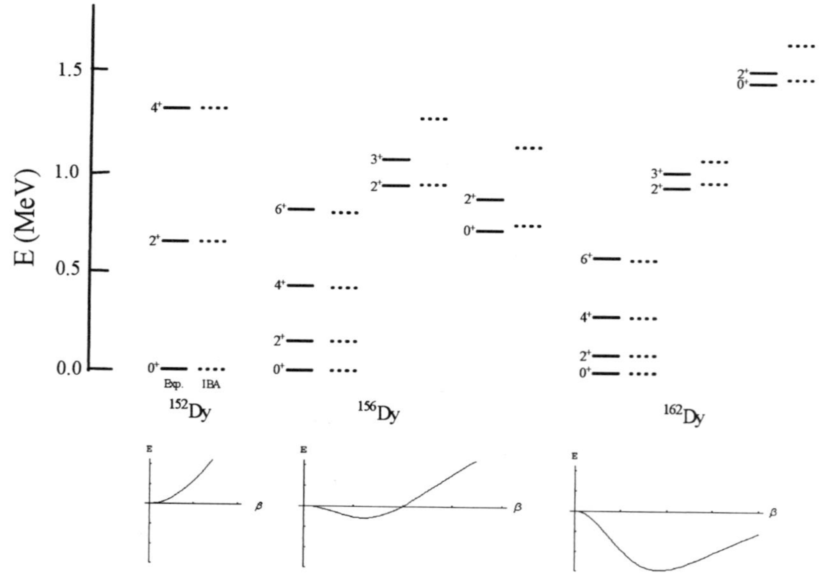

Figure 7. Comparison between empirical level energies and the IBA calculations for three Dy isotopes. The total energy is shown in each case.

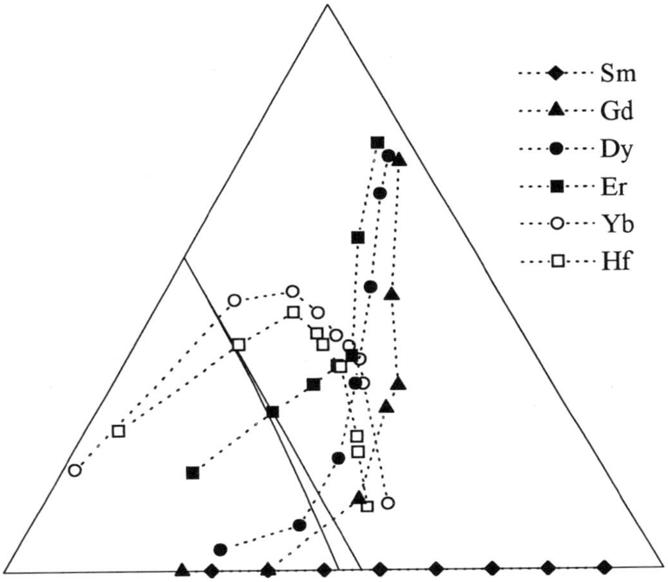

Figure 8. The trajectories corresponding to the IBA fits for $^{144-156}Nd_{84-96}$, $^{146-160}Sm_{84-98}$, $^{148-162}Gd_{84-98}$, $^{152-164}Dy_{86-98}$, $^{152-164}Er_{84-96}$, $^{156-170}Yb_{86-100}$, and $^{158-170}Hf_{84-98}$ isotopic chains in the IBA Symmetry triangle.

5 Conclusion

The IBA phase transition predictions are discussed. The model calculations reproduce rather well the evolution of basic observables in the N=82-104 region and a mapping of the symmetry triangle is realized for this mass region. Based on the phenomenological relation between the *continuous* model order parameters (ζ,χ) and the *discrete* nucleon numbers (N,Z), it is possible to identify critical phase/transition behavior in finite systems like atomic nuclei.

Acknowledgments

A crucial role in the foundation of the concept of phase transitions in low energy nuclear spectra has been played by Franco Iachello. It is a pleasure to dedicate this paper to him on the occasion of this conference in his honor. Useful discussions with Peter von Brentano and Gustavo Fernandes are acknowledged. This work was supported by U.S. DOE Grants No. DE-FG02-91ER-40609 and DE-FG02-88ER40417.

References

1. F. Iachello and A. Arima, *The Interacting Boson Model* (Cambridge University Press, Cambridge, England, 1987).
2. A.E.L. Dieperink, O. Scholten, and F. Iachello, *Phys. Rev. Lett.* **44**, 1747 (1980); *Nucl. Phys.* **A346**, 125 (1980).
3. D.H. Feng, R. Gilmore, and S.R. Deans *Phys. Rev.* **C 23**, 1254 (1981).
4. V. Werner, P. von Brentano, R.F. Casten and J. Jolie, *Phys. Lett. B* **527**, 55 (2002).
5. N.V. Zamfir, P. von Brentano, R.F. Casten, and J. Jolie, *Phys. Rev. C* **66**, 021304(R) (2002).
6. R.F. Casten and D.D. Warner, in *Progress in Particle and Nuclear Physics*, vol. 9, ed. D. Wilkinson (Pergamon, Oxford, 1983), p. 311.
7. J.N. Ginocchio and M.W. Kirson, *Nucl. Phys.* **A350**, 31 (1980).
8. P. Cejnar, W.T. Chou, N.V. Zamfir, and R.F. Casten, *Phys. Rev. C* **64**, 054305 (2001).
9. F. Iachello, N.V. Zamfir, and R.F. Casten, *Phys. Rev. Lett.* **81**, 1191 (1998).
10. F. Iachello, *Phys. Rev. Lett.* **85**, 3580 (2000).
11. F. Iachello, *Phys. Rev. Lett.* **87**, 052502 (2001).
12. R.F. Casten and N.V. Zamfir, *Phys. Rev. Lett.* **87**, 052503 (2001).
13. O Scholten, F. Iachello, and A. Arima, *Ann. Phys. (NY)* **115**, 325 (1978).
14. W.T. Chou, N.V. Zamfir, and R.F. Casten, *Phys. Rev.* **C56**, 829 (1997).
15. M.A. Caprio et al., *Phys. Rev.* **C 66**, 054310, 2002.

CRITICAL POINTS IN NUCLEI AND INTERACTING BOSON MODEL INTRINSIC STATES

JOSEPH N. GINOCCHIO

Theoretical Division, Los Alamos National Laboratory, Los Alamos, NM, 87545, U.S.A.
E-mail: gino@lanl.gov

A. LEVIATAN

Racah Institute of Physics, The Hebrew University, Jerusalem 91904, Israel
E-mail: ami@vms.huji.ac.il

We consider properties of critical points in the interacting boson model, corresponding to flat-bottomed potentials as encountered in a second-order phase transition between spherical and deformed γ-unstable nuclei. We show that intrinsic states with an effective β-deformation reproduce the dynamics of the underlying non-rigid shapes. The effective deformation can be determined from the the global minimum of the energy surface after projection onto the appropriate symmetry. States of fixed N and good $O(5)$ symmetry projected from these intrinsic states provide good analytic estimates to the exact eigenstates, energies and quadrupole transition rates at the critical point.

"*During these moments of abstraction he seemed more intimately absolved, in the sense of being linked anew with the universe.*
Giuseppe de Lampedusa, "The Leopard"

1 Introduction

In the days that one of us (JNG) was a graduate student, group theory was considered almost a dirty word in the nuclear physics community even though Wigner had introduced spin-isospin symmetry (SU(4)), Elliott had exploited the symmetry of the harmonic oscillator to link collective motion and the shell model (SU(3)) and the symmetry of the hadrons had been discovered (SU(3) again). Francesco Iachello changed that attitude and brought group theory front and center in nuclear physics and in other fields of physics.

Franco Iachello is a descendant of a noble Sicilian family similar to that portrayed in Giuseppe de Lampedusa's classic Italian novel, "The Leopard". Set in Sicily in 1860 at the time of the campaign for the unification of Italy, the hero, the Prince, struggles with how to keep the old while embracing the new. The Prince often took solace from the turmoil of the real world by studying astronomy and mathematics much the same as Franco has by his significant contributions to physics and group theory.

2 Critical Points in the Geometric Collective Model

Recently Franco has been studying two critical points associated with shape phase transitions in nuclei within the geometric framework of the collective model for infinite square well potentials [1,2] . This model involves a Bohr Hamiltonian which

Table 1. Excitation energies (normalized to the energy of the first excited state) and B(E2) values (in units of $B(E2;\, 2^+_{1,1} \to 0^+_{1,0}) = 1$) for the E(5) critical point [1], for several N=5 calculations and for the experimental data of ^{134}Ba [15]. The finite-N calculations involve the exact diagonalization of the critical IBM Hamiltonian (H_{cri}) [Eq. (17)], τ-projected states for H_{cri} [Eqs. (9),(12) with $y = 0.314$], the $U(5)$ limit [ϵn_d] and the $O(6)$ limit [$(A/4)(N-\sigma)(N+\sigma+4) + B\tau(\tau+3)$].

	E(5)	exact N=5	τ-projection N=5	U(5) N=5	O(6) N=5	^{134}Ba exp
$E(0^+_{1,0})$	0	0	0	0	0	0
$E(2^+_{1,1})$	1	1	1	1	1	1
$E(L^+_{1,2})$	2.20	2.195	2.19	2	2.5	2.32
$E(L^+_{1,3})$	3.59	3.55	3.535	3	4.5	3.66
$E(0^+_{2,0})$	3.03	3.68	3.71	2	$1.5\frac{A}{B}$	3.57
$B(E2;\, 4^+_{1,2} \to 2^+_{1,1})$	1.68	1.38	1.35	1.6	1.27	1.56(18)
$B(E2;\, 6^+_{1,3} \to 4^+_{1,2})$	2.21	1.40	1.38	1.8	1.22	
$B(E2;\, 0^+_{2,0} \to 2^+_{1,1})$	0.86	0.51	0.43	1.6	0	0.42(12)

describes the dynamics of a macroscopic quadrupole shape via a differential equation in the intrinsic quadrupole shape variables beta and gamma. In the current contribution we shall discuss the critical point (CP) for a second order shape phase transition between spherical and deformed γ-unstable nuclei, which Franco called E(5) [1]. An empirical example of such a critical point has been found in ^{134}Ba [3,4] and possibly in ^{104}Ru [5], ^{102}Pd [6] and ^{108}Pd [7].

In the geometric approach the E(5) eigenfunctions [1] are proportional to Bessel functions of order $\tau + \frac{3}{2}$ and the corresponding eigenvalues are proportional to $(x_{\xi,\tau})^2$. Hamiltonians that are γ- unstable have an $O(5)$ symmetry and τ is the $O(5)$ quantum number. Furthermore

$x_{\xi,\tau}$ is the ξ-th root of these Bessel functions. A portion of an E(5)-like spectrum is shown in Fig. 1. It consists of states, $L^+_{\xi,\tau}$, arranged in major families labeled by $\xi = 1, 2, \ldots$ and $O(5)$ τ-multiplets ($\tau = 0, 1, \ldots$) within each family. The angular momenta L for each τ-multiplet are obtained by the usual $O(5) \supset O(3)$ reduction [8]. The E(5) CP leads to analytic parameter-free predictions for energy ratios and $B(E2)$ ratios which persist when carried over to a finite-depth potential [9]. As seen in Table 1, the E(5) predicted values are in-between the values expected of a spherical vibrator [$U(5)$] and a deformed γ-unstable rotor [$O(6)$].

3 Wave Function Ansatz for the γ-Unstable Limit of the Interacting Boson Model

The Interacting Boson Model [10] (IBM) was one of the great achievements of Akito Arima and Franco Iachello. This model describes low-lying quadrupole collective states in nuclei in terms of a system of N monopole (s) and quadrupole (d) bosons representing valence nucleon pairs. The IBM Hamiltonian relevant to the critical point of the phase transition between spherical and γ-unstable deformed nuclei has $O(5)$ symmetry [11]. Its energy surface, obtained by the method of coherent states [11,12], is γ-independent and exhibits a flat-bottomed behavior in β which resembles the infinite square-well potential used to derive the E(5) CP in the geometric

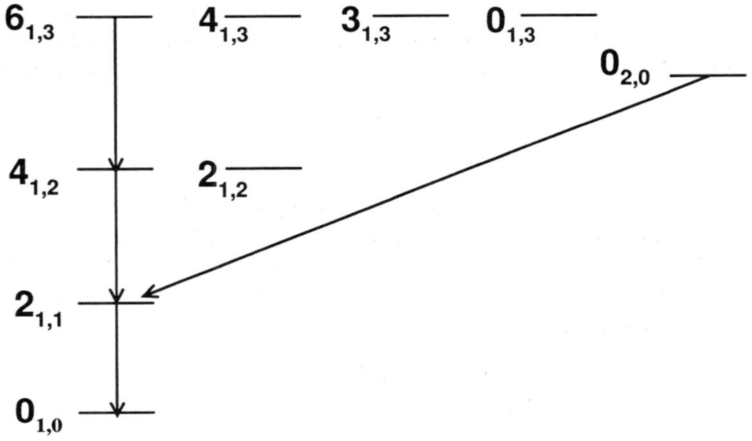

Figure 1. An E(5)-like spectrum of states labeled by $L_{\xi,\tau}$. Shown are the transitions whose B(E2) values are given in Eq. (12). The E2 rates for other $\Delta\tau = 1$ transitions (not shown) are governed by O(5) symmetry. Specifically, $B(E2;\, L^+_{1,2} \longrightarrow 2^+_{1,1})$ for $L = 4, 2$ are in the ratio 1 : 1 respectively, $B(E2;\, L^+_{1,3} \longrightarrow 4^+_{1,2})$ for $L = 6, 4, 3$ and $B(E2;\, L^+_{1,3} \longrightarrow 2^+_{1,2})$ for $L = 4, 3, 0$ are in the ratios 1 : 10/21 : 2/7 : 11/21 : 5/7 : 1, respectively. Taken from ref.[13].

approach [1]. Calculations with finite N values (N=5 for ^{134}Ba) have found that this critical IBM Hamiltonian can replicate numerically the E(5) CP and its analytic predictions [3,5,6,7]. In the present contribution, based on recent work [13], we examine the properties and conditions that enable features of E(5) CP to occur in a finite system described by the interacting boson model. For that purpose we propose wave functions of a particular analytic form, which can simulate accurately the exact IBM eigenstates at the critical point. These wave functions with fixed N and good $O(5)$ symmetry are used to derive accurate estimates for energies and quadrupole rates at the critical point without invoking large-N approximations. The proposed wave functions can be obtained by projection from intrinsic states with an effective β-deformation.

In the IBM, the γ-unstable transition region is modeled by the Hamiltonian

$$H = \epsilon\, \hat{n}_d + \frac{1}{4} A \left[d^\dagger \cdot d^\dagger - (s^\dagger)^2 \right] [\, H.c. \,] \tag{1}$$

with ϵ and A positive parameters. Here $\hat{n}_d$ is the d-boson number operator, $H.c.$ stands for Hermitian conjugate and the dot implies a scalar product. In the $U(5)$ limit ($A = 0$), the spectrum of H is harmonic, $\epsilon\, n_d$, with $n_d = 0, 1, 2 \ldots N$. The eigenstates are classified according to the chain $U(6) \supset U(5) \supset O(5) \supset O(3)$

with quantum numbers $|\, N, n_d, \tau, L\rangle$ (for $\tau \geq 6$ an additional multiplicity index is required for complete classification). These states can be organized into sets characterized by $n_d = \tau + 2k$. States in the lowest-energy set ($k = 0$) satisfy

$$\begin{aligned} P_0 \,|\, N, n_d = \tau, \tau, L\rangle &= 0 \\ P_0^\dagger &= d^\dagger \cdot d^\dagger \ . \end{aligned} \tag{2}$$

Other sets ($k > 0$) are generated by $|\, N, n_d, \tau, L\rangle \propto (P_0^\dagger)^k |\, N - 2k, n_d = \tau, \tau, L\rangle$. In the $O(6)$ limit ($\epsilon = 0$), the spectrum is $\frac{1}{4}A(N - \sigma)(N + \sigma + 4)$ with $\sigma = N, N-2, N-4, \ldots 0$ or 1. The eigenstates are classified according to the chain $U(6) \supset O(6) \supset O(5) \supset O(3)$ with quantum numbers $|\, N, \sigma, \tau, L\rangle$. The ground band has $\sigma = N$ and its members satisfy

$$\begin{aligned} P_1 \,|\, N, \sigma = N, \tau, L\rangle &= 0 \\ P_1^\dagger &= [\, d^\dagger \cdot d^\dagger - (s^\dagger)^2 \,] \ . \end{aligned} \tag{3}$$

The remaining bands with $\sigma = N - 2k$ are generated by $|\, N, \sigma, \tau, L\rangle \propto (P_1^\dagger))^k \,|\, N - 2k, \sigma, \tau, L\rangle$. These results suggest that in-between the $U(5)$ and $O(6)$ limits, we consider a ground band ($\xi = 1$) for the Hamiltonian (1) determined by the condition

$$\begin{aligned} P_y \,|\, \xi = 1; y, N, \tau, L\rangle &= 0 \ , \\ P_y^\dagger &= \left[\, d^\dagger \cdot d^\dagger - y\,(s^\dagger)^2 \,\right] . \end{aligned} \tag{4}$$

In the $U(5)$ basis these states are

$$|\, \xi = 1; y, N, \tau, L\rangle = \sum_{n_d} \frac{1}{2} \left[1 + (-1)^{n_d - \tau} \right] \xi_{n_d,\tau} |\, N, n_d, \tau, L\rangle \ , \tag{5}$$

and the n_d summation covers the range $\tau \leq n_d \leq N$. The coefficients $\xi_{n_d,\tau}$ have the explicit form

$$\begin{aligned} \xi_{n_d,\tau} &= \left[\frac{(N-\tau)!\,(2\tau+3)!!}{(N-n_d)!\,(n_d-\tau)!!\,(n_d+\tau+3)!!} \right]^{1/2} y^{(n_d-\tau)/2}\, \xi_{\tau,\tau} \ , \\ (\xi_{\tau,\tau})^2 &= \frac{2(N-\tau+1)}{(2\tau+3)!!}\, y^{2\tau+3} \left[G^{(\tau+1)}_{N+1-\tau}(y) \right]^{-1} \ , \\ G^{(n)}_\alpha(y) &= 2y^{2n+1} \sum_p \binom{\alpha}{2p+1} y^{2p} \frac{(2p+1)!!}{(2p+2n+1)!!} \ . \end{aligned} \tag{6}$$

$G^{(n)}_\alpha(y)$ is an odd function of y, $G^{(n)}_\alpha(-y) = -G^{(n)}_\alpha(y)$, and satisfies the follwing recursion relation

$$\begin{aligned} G^{(n)}_\alpha(y) &= \frac{1}{\alpha+2}\, G^{(n-1)}_{\alpha+2}(y) - \frac{1}{\alpha+1}\, G^{(n-1)}_{\alpha+1}(y) \qquad n \geq 1 \\ G^{(0)}_\alpha(y) &= (1+y)^\alpha - (1-y)^\alpha \ . \end{aligned} \tag{7}$$

Furthermore, $G^{(n)}_\alpha(y) = \pm \frac{2^{\alpha+n}\,\alpha(\alpha+n-1)!}{(\alpha+2n)!}$ for $y = \pm 1$ and $G^{(n)}_\alpha(y) \sim \frac{2\alpha}{(2n+1)!!}\, y^{2n+1}$ for $y \to 0$.

Members of the first excited band ($\xi = 2$) have approximate wave functions of the form

$$
\begin{aligned}
|\,\xi = 2; y, N, \tau, L\rangle &= \mathcal{N}_\beta\, P_y^\dagger\,|\,\xi = 1; y, N-2, \tau, L\rangle \\
\mathcal{N}_\beta &= \left[2(2N + y^2 + 1) + 4(y^2-1)S_{1,\tau}^{(N-2)}\right]^{-1/2} ,
\end{aligned} \tag{8}
$$

where $S_{1,\tau}^{(N)}$ is defined in Eq. (10) below.

The states of Eqs. (5) and (8) have fixed N, L and good $O(5)$ symmetry τ. Henceforth, for reasons to be explained below, they will be referred to as τ-projected states. Diagonal matrix elements of the Hamiltonian (1) in these states, denoted by $E_{\xi,\tau} = \langle \xi; y, N, \tau | H | \xi; y, N, \tau\rangle$, can be evaluated in closed form

$$
\begin{aligned}
E_{\xi=1,\tau} &= \epsilon\left[N - S_{1,\tau}^{(N)}\right] + \frac{1}{4}A\,(1-y)^2\,S_{2,\tau}^{(N)} \\
E_{\xi=2,\tau} &= \epsilon\left\{N - 2\mathcal{N}_\beta^2\left[2y^2 + (2N + 7y^2 - 1)\,S_{1,\tau}^{(N-2)} + 2(y^2-1)\,S_{2,\tau}^{(N-2)}\right]\right\} \\
&+ \frac{1}{4}A\,2\mathcal{N}_\beta^2\left\{2(y-1)^2\,(y^2-1)\,S_{3,\tau}^{(N-2)} + (y-1)^2\,(2N + y^2 - 8y + 5)\,S_{2,\tau}^{(N-2)}\right. \\
&\left. + 16(y-1)(N+y)\,S_{1,\tau}^{(N-2)} + 2\left[(2N+y)(2N+y+2)+1\right]\right\} .
\end{aligned} \tag{9}
$$

The $E_{\xi,\tau}$ are independent of L since H is an $O(5)$ scalar. Their expressions involve the quantities $S_{k,\tau}^{(N)} = \langle \xi = 1; y, N, \tau | (s^\dagger)^k s^k | \xi = 1; y, N, \tau\rangle$ which are given by

$$
S_{1,\tau}^{(N)} = (N - \tau + 1)\frac{G_{N-\tau}^{(\tau+1)}(y)}{G_{N-\tau+1}^{(\tau+1)}(y)} \tag{10}
$$

with $S_{2,\tau}^{(N)} = S_{1,\tau}^{(N)} S_{1,\tau}^{(N-1)}$ and $S_{3,\tau}^{(N)} = S_{1,\tau}^{(N)} S_{1,\tau}^{(N-1)} S_{1,\tau}^{(N-2)}$. Non-diagonal matrix elements of the Hamiltonian H between τ-projected states in different ξ-bands, $H_{1,2;\tau} = \langle \xi = 2; y, N, \tau | H | \xi = 1; y, N, \tau\rangle$, can be evaluated as well

$$
\begin{aligned}
&H_{1,2;\tau} = \\
&2\mathcal{N}_\beta\left[S_{2,\tau}^{(N)}\right]^{1/2}\left\{\epsilon\, y + \frac{1}{4}A(y-1)\left[(2N + y + 1) + 2(y-1)\,S_{1,\tau}^{(N-2)}\right]\right\} .
\end{aligned} \tag{11}
$$

By techniques similar to that employed in the $O(6)$ limit of the IBM [8], explicit expressions of quadrupole rates can be derived for transitions between the τ-projected states. For the relevant IBM quadrupole operator, $T(E2) = d^\dagger s + s^\dagger \tilde{d}$, these transitions are subject to the $O(5)$ selection rule $\Delta\tau = \pm 1$, and, as explained in the caption of Fig. 1, it is sufficient to focus on the B(E2) values

$$
\begin{aligned}
&B(E2;\, \xi = 1;\, \tau+1, L = 2\tau+2 \longrightarrow \xi = 1,\, \tau, L = 2\tau) = \\
&\quad \frac{(\tau+1)}{(2\tau+5)(N-\tau+1)}\left(S_{1,\tau}^{(N)}\right)^2 \frac{G_{N-\tau+1}^{(\tau+1)}(y)}{G_{N-\tau}^{(\tau+2)}(y)}\left[y + (N-\tau)\frac{G_{N-\tau-1}^{(\tau+2)}(y)}{G_{N-\tau}^{(\tau+1)}(y)}\right]^2 , \\
&B(E2;\, \xi = 2,\, \tau, L = 2\tau \longrightarrow \xi = 1,\, \tau+1, L = 2\tau+2) = \\
&\quad \frac{(\tau+1)(4\tau+5)}{(4\tau+1)(2\tau+5)} 4\,\mathcal{N}_\beta^2\, y^2\,(y-1)^2\,(N-\tau)\,\frac{G_{N-\tau-1}^{(\tau+1)}(y)}{G_{N-\tau}^{(\tau+2)}(y)} .
\end{aligned} \tag{12}
$$

4 Intrinsic State of the γ-Unstable Wavefunction Ansatz

The states in Eqs. (5) and Eq. (8) can be obtained by $O(5)$ projection from the IBM intrinsic states for the ground band

$$
\begin{aligned}
|\,c; N\rangle &= (N!)^{-1/2}(b_c^\dagger)^N\,|0\,\rangle \\
b_c^\dagger &= (1+\beta^2)^{-1/2}\left[\beta\,\cos\gamma\, d_0^\dagger + \beta\,\sin\gamma\,\frac{1}{\sqrt{2}}\left(d_2^\dagger + d_{-2}^\dagger\right) + s^\dagger\right]
\end{aligned} \tag{13}
$$

and for the β band respectively

$$|\,\beta; N\rangle = \mathcal{N}_\beta P_y^\dagger |\,c; N-2\rangle \ , \tag{14}$$

provided $y = \beta^2$. The expressions in Eqs. (5)-(12) depend on the so far unspecified parameter y. Normally, the equilibrium value of β, and hence y, is chosen as the global minimum of the intrinsic energy surface

determined from the expectation value of H in the intrinsic state (13). This is a standard procedure for a Hamiltonian describing nuclei with rigid shapes, for which the global minimum is deep and well-localized. Such is the case for the Hamiltonian of Eq. (1) whose intrinsic energy surface has the form

$$
\begin{aligned}
E(\beta) &= E_0 + N(N-1)\,\beta^2(1+\beta^2)^{-2}\,\left[\,a + c\,\beta^2\,\right] \\
a &= \bar{\epsilon} - A \ \ , \ \ c = \bar{\epsilon} \ \ , \ \ \bar{\epsilon} = \epsilon/(N-1)
\end{aligned} \tag{15}
$$

with $E_0 = \frac{1}{4}AN(N-1)$ a constant. The topology of the energy surface is such that

$$
\begin{array}{ll}
a > 0 & \text{global spherical minimum at } \beta = 0 \\
a < 0 & \text{global deformed minimum at } \beta = a/(a-2c)\ .
\end{array} \tag{16}
$$

When $a \neq 0$ the intrinsic energy surface behaves quadratically ($\sim \beta^2$) near the single minimum, and the standard procedure of determining the equilibrium value of β from the global minimum of $E(\beta)$ is applicable. However, near the critical point of the phase transition an alternative procedure is required.

5 γ-unstable Wave function Ansatz at the Critical Point

The IBM Hamiltonian, H_{cri}, at the critical point of the $U(5)-O(6)$ phase transition corresponds to a special choice of parameters in the Hamiltonian of Eq. (1)

$$H_{cri}: \quad \epsilon = (N-1)A \ , \tag{17}$$

for which $a = 0$ in Eq. (15), and the corresponding energy surface reduces to

$$E(\beta) = E_0 + A\,N(N-1)\,\beta^4(1+\beta^2)^{-2}\ . \tag{18}$$

In this case, the intrinsic energy surface $E(\beta)$, shown in Fig. (2a), has a flat behavior ($\sim \beta^4$) for small β, an inflection point at $\beta = 1$ and approaches a constant for large β. The global minimum at $\beta = 0$ is not well-localized and $E(\beta)$ exhibits considerable instability in β, resembling a square-well potential for $0 \leq \beta \leq 1$. Under such circumstances fluctuations in β are large and play a significant role in the dynamics. Some of their effect can be taken into account by introducing into the intrinsic states of Eqs. (13) and (14) an effective β-deformation. The

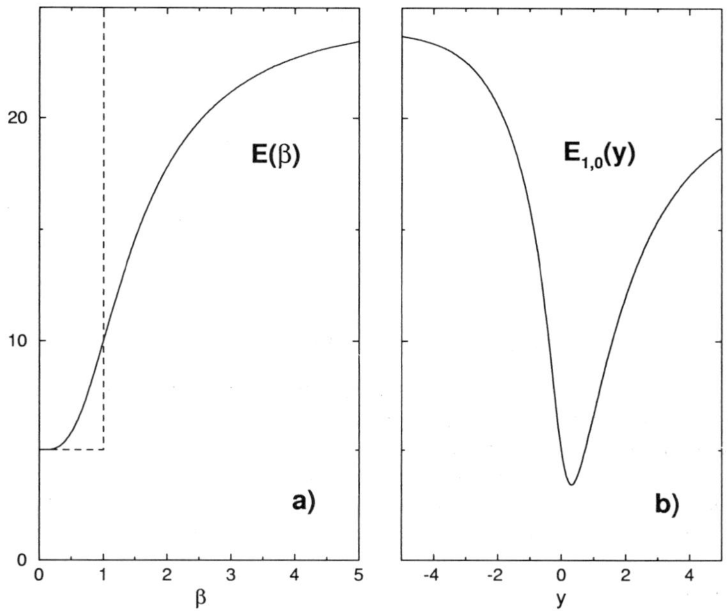

Figure 2. Energy surfaces of the critical IBM Hamiltonian H_{cri} (17) with $N = 5$ and $A = 1$. (a) Intrinsic energy surface $E(\beta)$, Eq. (18) [solid line], and its approximation by a square-well potential [dashed line]. (b) $O(5)$ projected energy surface $E_{\xi=1,\tau=0}(y)$, Eq. (9). The global minimum is at $y = 0.314$. Taken from ref.[13].

effective deformation is expected to be in the range $0 < y = \beta^2 < 1$, in-between the respective $U(5)$ and $O(6)$ value of β. This will enable a reproduction of E(5) characteristic signatures which are in-between these limits (see Table I). In contrast to $E(\beta)$ of Eq. (18), we see from Fig. (2b) that the $O(5)$ projected energy surface $E_{\xi=1,\tau=0}(y)$ of H_{cri} (Eqs. (9) and (17) with $N = 5$), does have a stable minimum at a certain value of y, which we interpret as an effective β-deformation. This procedure, based on variation after projection, is in the spirit of [14] in which it is shown that in finite boson systems, a γ-unstable $O(6)$ state can be generated from a rigid triaxial intrinsic state with an effective γ-deformation of 30°. In the present case the γ-instability is treated exactly by means of $O(5)$ symmetry, while the β-instability is treated by means of an effective deformation. The appropriate value of y can be used to evaluate the band-mixing, $\eta_\tau(y) = \frac{|H_{1,2;\tau}|}{E_{2,\tau} - E_{1,\tau}}$. A small value of η_τ will ensure that the τ-projected states of Eqs. (5), (8) form a good representation of the actual eigenstates of H_{cri}, and turn the expressions of Eqs. (9), (12) into meaningful estimates for energies and quadrupole transition rates at the critical point.

Table 2. $U(5)$ decomposition (in %) of the $L^+_{\xi,\tau}$ states for $N = 5$. The calculated values are obtained from the τ-projected states, Eqs. (5), (8) with $y = 0.314$. The exact values are obtained from numerical diagonalization of the critical IBM Hamiltonian H_{cri}, Eq. (17).

	$0^+_{1,0}$		$2^+_{1,1}$		$L^+_{1,2}$		$L^+_{1,3}$		$0^+_{2,0}$	
n_d	calc	exact	calc	exact	calc	exact	calc	exact	calc	exact
0	83.2	83.4							15.8	16.4
1			92.2	90.2						
2	16.4	16.2			96.8	95.2			70.9	76.2
3			7.8	9.7			99.1	98.4		
4	0.4	0.4			3.2	4.8			13.3	7.4
5			0.0	0.1			0.9	1.6		

6 Comparison with E(5) and Experiment

To test the suggested procedure we compare in Table 2 the $U(5)$ decomposition of exact eigenstates obtained from numerical diagonalization of H_{cri} for $N = 5$ with that calculated from the τ-projected states with $y = 0.314$ [the global minimum of $E_{1,0}(y)$]. As can be seen, the latter provide a good approximation to the exact eigenstates (the corresponding band-mixing is $\eta_\tau = 0.12, 3.53, 4.14, 3.05\%$ for $\tau = 0, 1, 2, 3$). This agreement in the structure of wave functions is translated also into an agreement in energies and B(E2) values as shown in Table 1. The results of Table 1 and 2 clearly demonstrate the ability of the suggested procedure to provide analytic and accurate estimates to the exact finite-N calculations of the critical IBM Hamiltonian, which in-turn agree with the experimental data in ^{134}Ba and captures the essential features of the E(5) critical point.

7 Large N Limit

In the large N limit, using Stirling's Formula in Eq. (6), we obtain

$$G^{(n)}_\alpha(y) \to 2x \left(\frac{x}{N}\right)^{2n} \sum_p \frac{x^{2p}}{(2p)!!(2p+2n+1)!!} , \tag{19}$$

where $x = Ny$. Therefore,

$$S^{(N)}_{k,\tau}(y) \to \mathcal{S}_{k,\tau}(x) , \tag{20}$$

that is, $S^{(N)}_{k,\tau}(y)$ becomes a function of x only. This suggests that the energy spectrum depends only on x and not N and y separately as N goes to infinity, except for possibly an overall scale. To test this in Fig. 3 we plot the energy ratio at the critical point

$$R = \frac{E_{\xi=1,\tau=2} - E_{\xi=1,\tau=0}}{E_{\xi=1,\tau=1} - E_{\xi=1,\tau=0}} \tag{21}$$

for several values of N as a function of x. Indeed as N increases, the curves asymptote to one universal curve. This means that the quadrupole beta deformation of the geometrical model is proportional to $\sqrt{N}\beta$. This observation may be useful in relating the predictions of E(5) CP to the IBM.

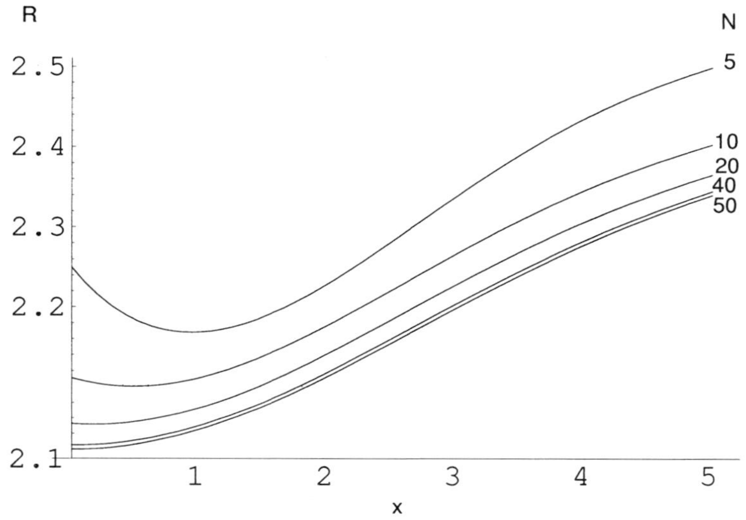

Figure 3. The energy ratio R at the critical point defined in Eq. (21) as a function of $x = Ny$, for different values of N. Note that the vertical axis is displaced from zero.

8 Summary and Future Outlook

In this contribution we have examined properties of a critical point in a finite system[13]. We have focused on the E(5) critical point relevant to a second-order shape phase transition between spherical and deformed γ-unstable nuclei. At the critical point of such a phase transition the intrinsic energy surface is flat and there is no stable minimum value of the deformation. However, for a finite system, we have shown that there is an effective deformation which can describe the dynamics at the critical point. The effective deformation is determined by minimizing the energy surface after projection onto the appropriate symmetries. States of finite N and good $O(5)$ symmetry, projected from intrinsic states with this effective deformation simulate accurately the exact eigenstates, and can be used to derive analytic estimates for energies and quadrupole transition rates at the critical point.

In the future we shall explore in depth the N-dependence as well as the large-N limit and its relationship to the geometric E(5) critical point. We shall also explore the first-order critical point in the phase transition from a spherical vibrator and an axially symmetric deformed nucleus within the IBM.

Acknowledgments

This work was supported in part by the Israel Science Foundation (A.L.) and in part by the U.S. Department of Energy under contract W-7405-ENG-36 (J.N.G).

References

1. F. Iachello, Phys. Rev. Lett. **85**, 3580 (2000).
2. F. Iachello, Phys. Rev. Lett. **87**, 052502 (2001).
3. R.F. Casten and N.V. Zamfir, Phys. Rev. Lett. **85**, 3584 (2000).
4. J.M. Arias, Phys. Rev. C **63**, 034308 (2001).
5. A. Frank, C.E. Alonso and J.M. Arias, Phys. Rev. C **65**, 014301 (2002).
6. N.V. Zamfir, *et al.*, Phys. Rev. C **65**, 044325 (2002).
7. Da-li Zhang and Yu-xin Liu, Phys. Rev. C **65**, 057301 (2002).
8. A. Arima and F. Iachello, Ann. Phys. (N.Y.) **123**, 468 (1979).
9. M.A. Caprio, Phys. Rev. C **65**, 031304 (2002).
10. F. Iachello and A. Arima, *The Interacting Boson Model* (Cambridge University Press, Cambridge, 1987).
11. A.E.L. Dieperink, O. Scholten and F. Iachello, Phys. Rev. Lett. **44**, 1747 (1980).
12. J.N. Ginocchio and M.W. Kirson, Phys. Rev. Lett. **44**, 1744 (1980); J.N. Ginocchio and M.W. Kirson, Nucl. Phys. A **350**, 31 (1980).
13. A. Leviatan and J.N. Ginocchio, submitted to Phys. Rev. Lett. (2003).
14. T. Otsuka and M. Sugita, Phys. Rev. Lett. **59**, 1541 (1987).
15. Yu. V. Sergeenkov, Nucl. Data Sheets **71**, 557 (1994).

THE CRITICAL POINT SYMMETRY E(5) AND THE IBM

J.M. ARIAS AND C.E. ALONSO

Departamento de Física Atómica, Molecular y Nuclear, Facultad de Física
Universidad de Sevilla, Apartado 1065, 41080 Sevilla, Spain
E-mail: pepe@nucle.us.es

A. VITTURI

Dipartimento di Fisica Galileo Galilei Via Marzolo 8
35131 Padova - Italy

J.E. GARCíA-RAMOS

Departamento de Física Aplicada, Universidad de Huelva
21071 Huelva, Spain

J. DUKELSKY

Instituto de Estructura de la Materia, CSIC, Serrano 123
28006 Madrid, Spain

A. FRANK

Instituto de Ciencias Nucleares, UNAM, Circuito Exterior C.U.
México D.F., 04510 México

The relation of the recently proposed E(5) critical point symmetry with the interacting boson model is investigated. It is established that the large N limit of IBM at the critical point in the transition from U(5) to O(6) dynamical symmetries is given by a β^4 potential rather than by an infinite square well as in E(5).

1 Introduction

The study of phase transitions is one of the most exciting topics in Physics. Recently a new critical point symmetry has been proposed by Iachello[1]. This symmetry applies when a quantal system undergoes transitions between traditional dynamic symmetries. In Ref.[1] the particular case of the Bohr Hamiltonian[2] in Nuclear Physics is worked out. In this case, for the situation in which the potential energy surface is γ-unstable and the dependence in the β degree of freedom can be modeled by a five-dimensional infinite well, the so called E(5) symmetry appears. This situation is expected to be realized in actual nuclei when they undergo a transition from spherical to deformed γ-unstable shapes. The E(5) symmetry is obtained by solving the Bohr differential equation with a particular potential and then used in connection with the IBM. The predictions of the E(5) symmetry are supposed to be valid in the large N limit, but appropriate calculations with the IBM should provide predictions for finite N as stated in Ref.[3]. In IBM the spherical and the deformed γ-unstable phases correspond to the U(5) and O(6) dynamical symmetries respectively. In this paper we show that the large N limit of IBM for the critical point in the transition from U(5) to O(6) is not a square well in five dimensions but a β^4 potential. With this in mind, still the IBM calculations provide a tool for including corrections due to the finite number of bosons.

The paper is structured as follows. In Sect. 2 a brief review of the E(5) symmetry is presented. In Sect. 3 the coherent state formalism used in order to obtain the energy surfaces from an IBM hamiltonian is summarized and worked out for two schematic hamiltonians modeling the transition from U(5) to O(6) dynamical limits. For one of these hamiltonians an exactly solvable model is worked out in Sect. 4 so as to obtain solutions in the large N limit. Finally, we present our conclusions in Sect. 5.

2 The E(5) symmetry

Consider the Bohr Hamiltonian

$$H = -\frac{\hbar^2}{2B}\left[\frac{1}{\beta^4}\frac{\partial}{\partial\beta}\beta^4\frac{\partial}{\partial\beta} + \frac{1}{\beta^2\sin 3\gamma}\frac{\partial}{\partial\gamma}\sin 3\gamma\frac{\partial}{\partial\gamma} - \frac{1}{4\beta^2}\sum_\kappa \frac{Q_\kappa^2}{\sin^2(\gamma - \frac{2}{3}\pi\kappa)}\right] + V(\beta,\gamma) \;, \quad (1)$$

where β, γ are the shape variables and the Q_κ's are the components of the angular momentum written in terms of Euler angles. In cases in which the potential depends only on β, $V(\beta,\gamma) = U(\beta)$, the hamiltonian is separable and if we set

$$\Psi(\beta,\gamma,\theta_i) = f(\beta)\Phi(\gamma,\theta_i) \quad (2)$$

where θ_i stands for the three Euler angles, the Schrödinger equation can be split in two equations,

$$\left[-\frac{1}{\sin 3\gamma}\frac{\partial}{\partial\gamma}\sin 3\gamma\frac{\partial}{\partial\gamma} + \frac{1}{4}\sum_\kappa \frac{Q_\kappa^2}{\sin^2(\gamma - \frac{2}{3}\pi\kappa)}\right]\Phi(\gamma,\theta_i) = \tau(\tau+3)\Phi(\gamma,\theta_i) \;; \quad (3)$$

where $\tau = 0, 1, 2, \ldots$ and

$$\left[-\frac{\hbar^2}{2B}\left(\frac{1}{\beta^4}\frac{\partial}{\partial\beta}\beta^4\frac{\partial}{\partial\beta} - \frac{\tau(\tau+3)}{\beta^2}\right) + U(\beta)\right] f(\beta) = Ef(\beta) \;. \quad (4)$$

If $U(\beta)$ can be modeled as a five dimensional infinite well, the problem is exactly solvable and the corresponding symmetry is called E(5). The solutions of the Schrödinger equations in β and (γ, θ_i) with the appropriate boundary conditions are known[1]. The wavefunctions on β are,

$$f_{\xi,\tau}(\beta) = C_{\xi,\tau}\beta^{-\frac{3}{2}} J_{\tau+\frac{3}{2}}\left(\frac{x_{\xi,\tau}}{\beta_w}\beta\right) \;, \quad (5)$$

where τ is the label associated to the O(5) algebra, $x_{\xi,\tau}$ is the ξth zero of the Bessel function $J_{\tau+\frac{3}{2}}(x)$, $C_{\xi,\tau}$ are normalization constants, and β_w is the range of the potential in the β variable. In Table 1 and in Fig. 1 the main E(5) results for excitation energies (relative to the excitation of the first excited state) and B(E2)'s are reproduced from Ref.[1],

The solutions of the (γ, θ_i) part were studied in Ref.[4] and tabulated in Ref.[5], where $\Phi_{\tau,L,M}(\gamma,\theta_i)$ are written in terms of $\mathcal{D}$-functions as,

$$\Phi_{\tau,L,M}(\gamma,\theta_i) = \sum_\rho g_{\tau,L,\rho}(\gamma)\mathcal{D}^{(L)}_{M\rho}(\theta_i) \;. \quad (6)$$

Table 1. Excitation energies for the E(5) symmetry relative to the excitation energy of the first excited state.

	$\xi = 1$	$\xi = 2$	$\xi = 3$	$\xi = 4$
$\tau = 0$	0.00	3.03	7.58	13.64
$\tau = 1$	1.00	4.80	10.11	16.93
$\tau = 2$	2.20	6.78	12.86	20.44
$\tau = 3$	3.59	8.97	15.81	24.16

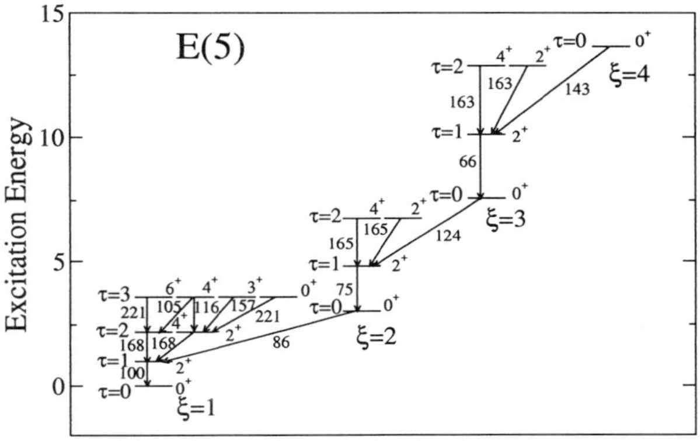

Figure 1. Schematic spectrum in the E(5) symmetry as taken from Ref.[1]. The B(E2) values are relative to the transition $2_{1,1} \to 0_{1,0}$ whose value is taken as 100.

The intrinsic functions $g_{\tau,L,\rho}(\gamma)$ are explicitly given in Ref.[5] and only even values of ρ appear in the sum.

3 Energy surfaces in IBM

The geometrical interpretation of the abstract IBM hamiltonian can be obtained by introducing a coherent state[6,7,8] which allows to associate to it a geometrical shape in terms of the deformation variables (β, γ). The basic idea of this formalism is to consider that the pure quadrupole states are globally described by a boson condensate of the form

$$|g; N, \beta, \gamma\rangle = \frac{1}{\sqrt{N!}} (\Gamma_g^\dagger)^N |0\rangle \quad , \tag{7}$$

where the basic boson is given by

$$\Gamma_g^\dagger = \frac{1}{\sqrt{1+\beta^2}} \left[s^\dagger + \beta \cos\gamma d_0^\dagger + \frac{1}{\sqrt{2}} \beta \sin\gamma (d_2^\dagger + d_{-2}^\dagger) \right] \quad , \tag{8}$$

which depends on the β and γ shape variables. The energy surface is defined as

$$E_N(\beta, \gamma) = \langle g; N, \beta, \gamma | H | g; N, \beta, \gamma \rangle \quad , \tag{9}$$

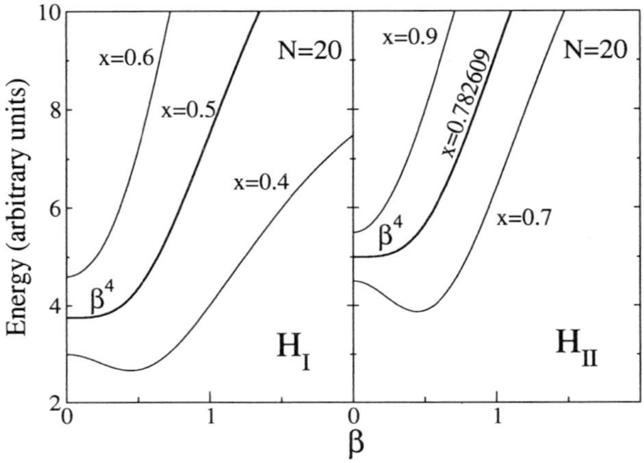

Figure 2. Representation of the energy surfaces as functions of the shape parameter β obtained for two schematic hamiltonians, Eq. (10) (left panel) and Eq. (17) (right panel). In each case three values of the order parameter are presented, one at the critical value, one above and one below that value. Curves are arbitrarely displaced in energy.

where H is the IBM hamiltonian. Here we are interested in the case in which the hamiltonian can undergo a transition from U(5) to O(6) and, consequently, the corresponding potential energy surfaces are γ-independent.

3.1 Schematic transitional hamiltonians

In this subsection we propose two schematic transitional hamiltonians that in terms of a single parameter produce the transition from U(5) (spherical) to O(6) (deformed γ-unstable) dynamical symmetries. The first one is

$$H_I = x\hat{n}_d + \frac{1-x}{N-1} P^\dagger P \ , \tag{10}$$

where

$$\hat{n}_d = \sum_\mu d^\dagger_\mu d_\mu \ , \tag{11}$$

$$P^\dagger = \frac{1}{2}(d^\dagger \cdot d^\dagger - s^\dagger s^\dagger) \ . \tag{12}$$

(13)

The corresponding energy surface is

$$E_I(N,\beta) = N\left[x\frac{\beta^2}{1+\beta^2} + \frac{1-x}{4}\left(\frac{1-\beta^2}{1+\beta^2}\right)^2\right] . \tag{14}$$

The condition to localize the critical point

$$\left(d^2E_I(N,\beta)/d\beta^2\right)_{\beta=0}=0 \tag{15}$$

gives the critical point $x_c^I=0.5$. For this critical value of x the expansion of the energy surface around $\beta=0$ is

$$E_I(N,\beta)\approx N\left[\frac{1}{8}+\frac{1}{2}\beta^4+\ldots\right] \tag{16}$$

In Fig. 2 we represent as an example the energy surfaces for the hamiltonian (10) with three elections for the order parameter x: one at the critical point, one above that value and one below it. For $x>x_c$ an equilibrium spherical shape is obtained, while for $x<x_c$ the equilibrium shape is deformed. The value x_c gives a β^4 surface which is flat close to $\beta=0$.

The second schematic hamiltonian we propose is

$$H_{II}=x\hat{n}_d-\frac{1-x}{N}Q\cdot Q\ , \tag{17}$$

where

$$Q=(s^\dagger\times\tilde{d}+d^\dagger\times\tilde{s})^{(2)}\ . \tag{18}$$

The corresponding energy surface is

$$E_{II}(N,\beta)=-(5+\beta^2)\ \frac{1-x}{1+\beta^2}+N\ x\ \frac{\beta^2}{1+\beta^2}-4(N-1)(1-x)\ \frac{\beta^2}{(1+\beta^2)^2} \tag{19}$$

Condition (15) gives in this case the critical point $x_c^{II}=\frac{4N-8}{5N-8}$ that in the large N limit gives 4/5. The expansion of the energy surface for x_c^{II} around $\beta=0$ is in this case

$$E_{II}(N,\beta)\approx\frac{N}{5N-8}\left[-5+4(N-1)\beta^4+\ldots\right] \tag{20}$$

In Fig. 2 the corresponding energy surfaces are plotted in the right panel. Same comments as in the preceding case are in order. Thus, we conclude that in the transition from spherical systems to γ-unstable deformed ones the critical point in IBM should be β^4 rather that an infinite square well. The question is then how different are the E(5) predictions from those obtained with a β^4 potential? In order to investigate this point we have solved numerically the Bohr hamiltonian for a potential β^4. The results for energies are presented in Table 2 and in Fig. 3. Here we keep the label ξ used in the E(5) case.

Particularly interesting are the energy ratios given in Table 3 (the notation is $L_{\xi,\tau}$) which have been used in recent works to identify possible nuclei as critical. In this table the E(5) and β^4 values are shown for comparison.

Table 2. Excitation energies for a β^4 potential relative to the excitation of the first excited state.

	$\xi = 1$	$\xi = 2$	$\xi = 3$	$\xi = 4$
$\tau = 0$	0.00	2.39	5.15	8.20
$\tau = 1$	1.00	3.63	6.56	9.75
$\tau = 2$	2.09	4.92	8.01	11.34
$\tau = 3$	3.27	6.26	9.50	12.95

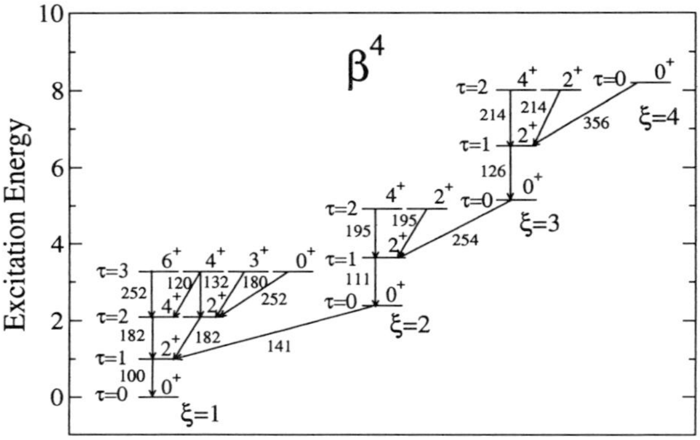

Figure 3. Same as Fig. 1 but for a β^4 potential.

Table 3. Energy ratios in the E(5) symmetry and for the β^4 potential.

	$E_{4_{1,2}}/E_{2_{1,1}}$	$E_{0_{2,0}}/E_{2_{1,1}}$	$E_{0_{1,3}}/E_{2_{1,1}}$	$E_{0_{2,0}}/E_{0_{1,3}}$
E(5)	2.20	3.03	3.59	0.84
β^4	2.09	2.39	3.27	0.73

Apart from energies, B(E2) transition probabilities can be calculated using the quadrupole operator

$$T_{\mu}^{(E2)} = t\ \beta\ \left[\mathcal{D}_{\mu 0}^{(2)}(\theta_i)\cos\gamma + \frac{1}{\sqrt{2}}\left(\mathcal{D}_{\mu 2}^{(2)}(\theta_i) + \mathcal{D}_{\mu -2}^{(2)}(\theta_i)\right)\sin\gamma\right] \quad , \qquad (21)$$

where t is a scale factor. In Table 4 two important B(E2) ratios are given for E(5) and β^4 cases. In Fig. 3 the β^4 B(E2) values are given normalized to the $B(E2; 2_{1,1} \rightarrow 0_{1,0})$ value which is taken as 100. Comparing Figs. 1 and 3 and Tables 1-4 we can observe important differences between E(5) and β^4 potentials. In order to see which is the actual large N limit of IBM we have performed calculations with the IBM codes for hamiltonians H_I (Eq. 10) and H_{II} (Eq. 17) at the critical point for different number of bosons. These codes allow to manage a small number of bosons, typically 20. In Fig. 4 the results of these calculations are shown with full line for Eq. (10) and with dashed line for Eq. (17). The values for E(5) and β^4 are shown as dotted lines as references. The last two panels labeled with R_1 and R_2 refer to the B(E2) ratios presented in Table 4.

Table 4. Selected B(E2) transition rate ratios in the E(5) symmetry and for the β^4 potential.

	$R_1 = \frac{B(E2;4_{1,2}\rightarrow 2_{1,1})}{B(E2;2_{1,1}\rightarrow 0_{1,0})}$	$R_2 = \frac{B(E2;0_{2,0}\rightarrow 2_{1,1})}{B(E2;2_{1,1}\rightarrow 0_{1,0})}$
E(5)	1.68	0.86
β^4	1.82	1.41

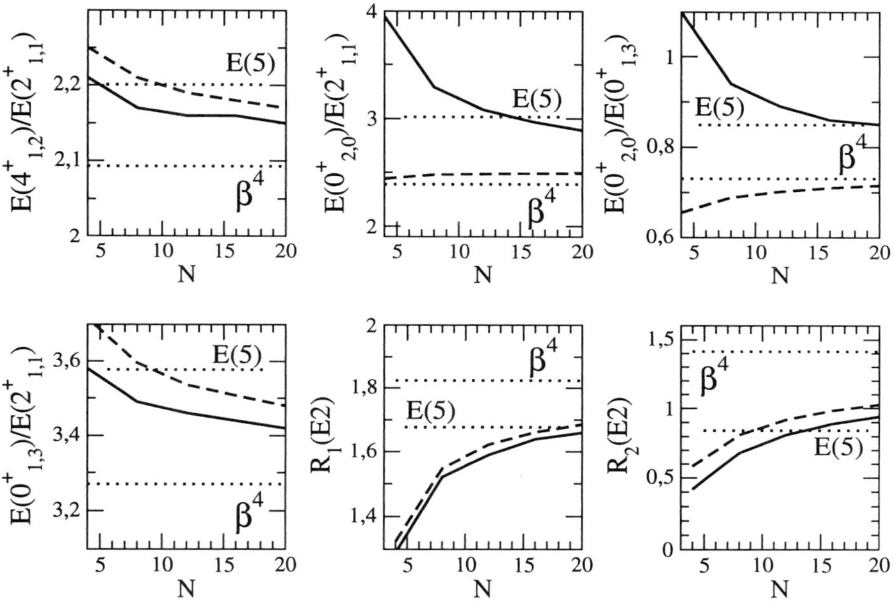

Figure 4. Variation with the number of bosons (up to $N = 20$) of selected energy and B(E2) ratios for IBM calculations performed at the critical points of hamiltonian (10) (full line) and (17) (dashed line). The corresponding E(5) and β^4 values are marked with dotted lines.

From Fig. 4 it is difficult to conclude whether E(5) or β^4 is the large N limit of the corresponding IBM hamiltonian. It is needed to perform calculations with larger values of N.

4 An exactly solvable model

Fortunately, Dukelsky et al.[9] have recovered an exactly solvable model for pairing proposed by Richardson in the 60's[10]. We follow here Ref.[9] to solve the schematic hamiltonian (10) in order to investigate the transition between U(5) and O(6) IBM limits using the Exactly Solvable Model (ESM). A similar procedure can be used for the case of hamiltonian (17).

Consider the hamiltonian (10), with the definitions

$$P_d^\dagger = d^\dagger \cdot d^\dagger , \tag{22}$$

$$P_s^\dagger = s^\dagger \cdot s^\dagger , \tag{23}$$

it can be rewritten as

$$H_I = x n_d + \frac{1-x}{4(N-1)} \left(P_d^\dagger P_d + P_s^\dagger P_s - P_d^\dagger P_s - P_s^\dagger P_d \right) . \tag{24}$$

The $SU(1,1)$ Casimir operators for s and d bosons are

$$-\frac{1}{2}\left(K_s^+ K_s^- + K_s^- K_s^+\right) + \left(K_s^0\right)^2 = \left(\frac{\nu_s}{2} + \frac{\Omega_s}{4}\right)\left(\frac{\nu_s}{2} + \frac{\Omega_s}{4} - 1\right) , \tag{25}$$

$$-\frac{1}{2}\left(K_d^+ K_d^- + K_d^- K_d^+\right) + \left(K_d^0\right)^2 = \left(\frac{\nu_d}{2} + \frac{\Omega_d}{4}\right)\left(\frac{\nu_d}{2} + \frac{\Omega_d}{4} - 1\right) , \tag{26}$$

with

$$K_s^0 = \frac{1}{4} + \frac{1}{2} n_s \quad , \quad K_d^0 = \frac{5}{4} + \frac{1}{2} n_d , \tag{27}$$

$$K_s^+ = \frac{1}{2} P_s^+ \quad , \quad K_d^+ = -\frac{1}{2} P_d^+ . \tag{28}$$

Substitution of these on Eqs. (25) and (26) and the use of the $SU(1,1)$ commutation relations allows us to write

$$P_s^\dagger P_s - n_s^2 + n_s + \nu_s \left(\nu_s - 1\right) = 0 , \tag{29}$$

$$P_d^\dagger P_d - n_d^2 - 3 n_d + \nu_d \left(\nu_d + 3\right) = 0 . \tag{30}$$

The relevant constant of motion is

$$R_d = K_d^0 - \frac{g}{\eta_s - \eta_d}\left(K_s^+ K_d^- + K_d^+ K_s^-\right) + \frac{2g}{\eta_s - \eta_d} K_s^0 K_d^0 . \tag{31}$$

Taking $\eta_s = 0$, and $\eta_d = 1$ we get

$$R_d = \frac{5}{4} + \frac{1}{2} n_d - \frac{g}{4}\left(P_d^\dagger P_s + P_s^\dagger P_d\right) - \frac{g}{2}\left(n_s + \frac{1}{2}\right)\left(n_d + \frac{5}{2}\right) . \tag{32}$$

Adding Eqs. (29) and (30) to R_d in order to have $P^\dagger P$ we obtain

$$R_d = \frac{5}{4} + \frac{1}{2} n_d + \frac{g}{4}\left(4 P^\dagger P - N^2 - 4N - \frac{5}{2} + \nu_s\left(\nu_s - 1\right) + \nu_d\left(\nu_d + 3\right)\right) . \tag{33}$$

Then the hamiltonian (10) can be written as

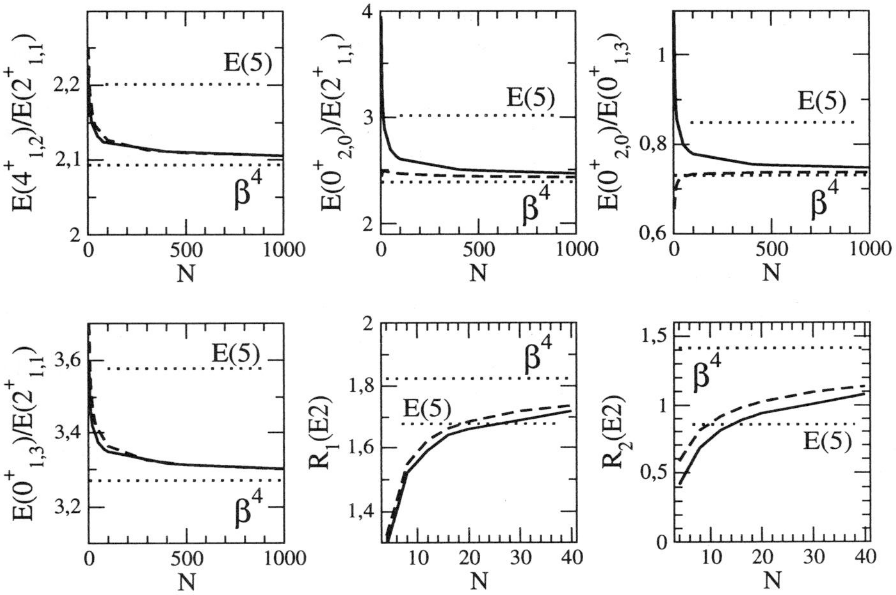

Figure 5. Same as Fig 4 but here the number of bosons runs up to 1000 in the energy ratios and up to 40 in the B(E2) ratios.

$$H_I = 2xR_d - \frac{5x}{2} + \frac{g}{2}x\left(N^2 + 4N + \frac{5}{2} - \nu_s(\nu_s - 1) - \nu_d(\nu_d + 3)\right) , \tag{34}$$

with

$$g = \frac{(1-x)}{2x(N-1)} . \tag{35}$$

The Richardson equations to be solved are

$$1 - \frac{g(2\nu_s + 1)}{e_\alpha} + \frac{g(2\nu_d + 5)}{2 - e_\alpha} - 4g\sum_{\beta(\neq\alpha)} \frac{1}{e_\alpha - e_\beta} = 0 , \tag{36}$$

and the eigenvalues of R_d are

$$r_d = \left(\frac{\nu_d}{2} + \frac{5}{4}\right)\left[1 - g\left(\nu_s + \frac{1}{2}\right) - 4g\sum_\alpha \frac{1}{2 - e_\alpha}\right] . \tag{37}$$

We have solved these equations for the hamiltonian (10) (and the corresponding ones for the hamiltonian (17)) and have been able to calculate up to $N = 1000$. In

Fig. 5 we present the results of these calculations for energy ratios up to $N = 1000$ and B(E2) ratios up to $N = 40$ together with the corresponding values for the E(5) symmetry and the β^4 potential.

Now it is clear that the large N limit for the IBM hamiltonian studied is β^4. Both hamiltonians Eq. (10) and Eq. (17) converge to the same results in the large N limit, although the corresponding corrections for finite N are quite different.

5 Conclusions

In this paper we have shown that the large N limit of the IBM hamiltonian at the critical point in the transition from U(5) (spherical) to O(6) (deformed γ-unstable) is represented by a β^4 potential rather than by an infinite square well as in the E(5) critical point symmetry. Results presented at this meeting by Ginocchio and Leviatan[11] also point to the same conclusion. Analysis of the IBM energy surface followed by an IBM calculation as presented in Ref.[12] can provide with the appropriate finite N corrections and so bring the opportunity of identifying nuclei at the critical points.

Acknowledgements

This work was supported in part by the Spanish DGICYT under project number BFM2002-03315, by the Mexican CONACyT and by an INFN-DGICYT agreement. The authors acknowledge gratefully continuous collaboration with Prof. F. Iachello and dedicate this work to him on the occasion of his 60th birthday.

References

1. F. Iachello, *Phys. Rev. Lett.* **85**, 3580 (2000).
2. A. Bohr and B. Mottelson, *Nuclear Structure, vol II*, Benjamin, Reading, Mass. 1975.
3. R. F. Casten and N. V. Zamfir, *Phys. Rev. Lett.* **85**, 3584 (2000).
4. L. Wilets and M. Jean, *Phys. Rev.* **102**, 788 (1956).
5. D. Bès, *Nucl. Phys.* **10**, 373 (1959).
6. J. N. Ginocchio and M. W. Kirson, *Nucl. Phys.* **A350**, 31 (1980).
7. A. E. L. Dieperink, O. Scholten and F. Iachello, *Phys. Rev. Lett.* **44**, 1747 (1980).
8. A. Bohr and B. Mottelson, *Phys. Scripta* **22**, 468 (1980).
9. J. Dukelsky, C. Esebbag and P. Schuck, *Phys. Rev. Lett.* **87**, 066403 (2001).
10. R.W. Richardson, *Phys. Lett.* **3**, 277 (1963). R.W. Richardson and N. Sherman, *Nucl. Phys.* **52**, 221 (1964). R.W. Richardson, *J. Math. Phys. (N.Y.)* **9**, 1327 (1968).
11. J. N. Ginocchio and A. Leviatan, *contribution to these proceedings.*
12. A. Frank, C.E. Alonso, and J.M. Arias, *Phys. Rev.* **C65**, 014301 (2001).

FINITE WELL SOLUTIONS FOR THE E(5) AND X(5) HAMILTONIANS

M. A. CAPRIO
Wright Nuclear Structure Laboratory, Yale University,
New Haven, Connecticut 06520-8124, USA
E-mail: mark.caprio@yale.edu

The solutions of the E(5) and X(5) Hamiltonians for finite well depth are described, and the effects of finite depth on observables are discussed.

1 Introduction

Intriguing observations regarding nuclei in the spherical-deformed shape transitional regions have motivated new approaches to their theoretical description. The evolution of observables across the transitional regions exhibits behaviors characteristic of first and second order thermodynamic phase transitions. The nuclei near the critical points of these transitions have historically been among the most difficult to understand. However, a new family of models recently proposed by Iachello, based upon the dynamical symmetries exhibited by square-well potentials, provide a simple description of such nuclei. These models — E(5) for the spherical to deformed γ-soft transition[1] and X(5) for the spherical to axially symmetric rotor transition[2] — yield analytic solutions with essentially parameter free predictions.

The E(5) and X(5) models are based upon solution of the Bohr geometric Hamiltonian[3] for quadrupole deformation potentials which are infinite square wells with respect to β and are either γ-soft [E(5)] or γ-stabilized [X(5)]. The eigenfunctions for these Hamiltonians may be expressed analytically in terms of Bessel functions, and the energy eigenvalues are simply related to the zeroes of the Bessel functions. However, boson algebraic models suggest that a square well of finite depth may be more applicable to nuclei of finite boson number.[1] In this contribution, the solutions of the E(5) and X(5) Hamiltonians for finite well depth are described, and the effects of finite well depth on observables are discussed.

2 Finite well: γ-soft case

The five-dimensional finite square well potential,

$$V(\beta) = \begin{cases} V_0 & \beta \leq \beta_w \\ 0 & \beta > \beta_w, \end{cases} \tag{1}$$

is, like the infinite-depth E(5) potential,[1] independent of γ. Consequently, the Hamiltonian is separable,[4,5] and the eigenfunctions are of the form $f(\beta)\Phi(\gamma,\underline{\theta})$, where β and γ are the shape coordinates and θ_i are the Euler angles. The solutions for the "angular" (γ,$\underline{\theta}$) wave functions[6] are common to all γ-soft problems, while the dependence upon the potential $V(\beta)$ is isolated in the "radial" (β) wave function.

The equation for $f(\beta)$ is

$$\left[\frac{\hbar^2}{2B}\left(-\frac{1}{\beta^4}\frac{\partial}{\partial\beta}\beta^4\frac{\partial}{\partial\beta}+\frac{\tau(\tau+3)}{\beta^2}\right)+V(\beta)\right]f(\beta)=Ef(\beta), \tag{2}$$

where the separation constant τ assumes the values $\tau = 0, 1, \ldots$. The notation is simplified if an overall factor of $\hbar^2/(2B)$ is extracted from the Hamiltonian, leaving the equation in terms of the reduced eigenvalue $\varepsilon \equiv \frac{2B}{\hbar^2}E$ and the reduced potential $v(\beta) \equiv \frac{2B}{\hbar^2}V(\beta)$, of depth $v_0 \equiv \frac{2B}{\hbar^2}V_0$. Bound state solutions can only occur with eigenvalues in the range $v_0<\varepsilon<0$. Each solution of the β equation results in a multiplet of solutions to the full problem, degenerate with respect to angular momentum according to the γ-soft τ multiplet structure.

The finite well potential is piecewise constant as a function of β. Within a region of constant potential, the radial equation (2) reduces[1] to the Bessel equation. In the interior of the well ($\beta<\beta_w$), where the difference $\varepsilon - v_0$ is positive, the solutions involve the ordinary spherical Bessel functions of integer order, while in the classically forbidden region exterior to the well ($\beta>\beta_w$), where the difference $\varepsilon - v_0$ is negative, the solutions involve the modified spherical Bessel functions of integer order. The wave function in β for the ξth solution with separation constant τ is

$$f_{\xi,\tau}(\beta)=\begin{cases} A_{\xi,\tau}\beta^{-1}j_{\tau+1}[(\varepsilon_{\xi,\tau}-v_0)^{1/2}\beta] & \beta\le\beta_w \\ B_{\xi,\tau}\beta^{-1}k_{\tau+1}[(-\varepsilon_{\xi,\tau})^{1/2}\beta] & \beta>\beta_w. \end{cases} \tag{3}$$

The eigenvalues for the finite well are determined by the requirement that $f(\beta)$ be continuous and smooth at the matching point $\beta=\beta_w$. Since the spherical Bessel functions of integer order may be expressed in terms of trigonometric and exponential functions, the matching condition assumes the form of a transcendental equation, which is solved numerically for the allowed values of ε. Once an eigenvalue $\varepsilon_{\xi,\tau}$ is found, the coefficients $A_{\xi,\tau}$ and $B_{\xi,\tau}$ follow from the continuity condition at $\beta=\beta_w$ and the normalization condition $\int_0^\infty \beta^4 d\beta |f(\beta)|^2=1$. Further details of the solution procedure may be found in Ref. 7.

The finite square well of dimensions β_w and v_0 is characterized by a "well size" parameter

$$x_0 \equiv (-v_0)^{1/2}\beta_w. \tag{4}$$

The infinite E(5) well is obtained in the limit $x_0\to\infty$. The spectrum of eigenvalues depends upon β_w and v_0 only through the combination x_0,[7] except for an overall normalization factor on the energy scale. For a given value of x_0, the numerical solution procedure need only be carried out once, at some "reference" choice of the well width and depth (*e.g.*, $\beta_w=1$ and thus $v_0=-x_0^2$). The solution for any other well with the same value of x_0 but arbitrary width β'_w (and thus depth $v_0=-x_0^2/\beta_w'^2$) can be deduced by a simple rescaling of all energies,

$$\varepsilon' = \beta_w'^{-2}\varepsilon. \tag{5}$$

This rescaling leads, by (3), to a dilation of all wave functions,

$$f'(\beta) = \beta_w'^{-5/2} f(\beta/\beta'_w). \tag{6}$$

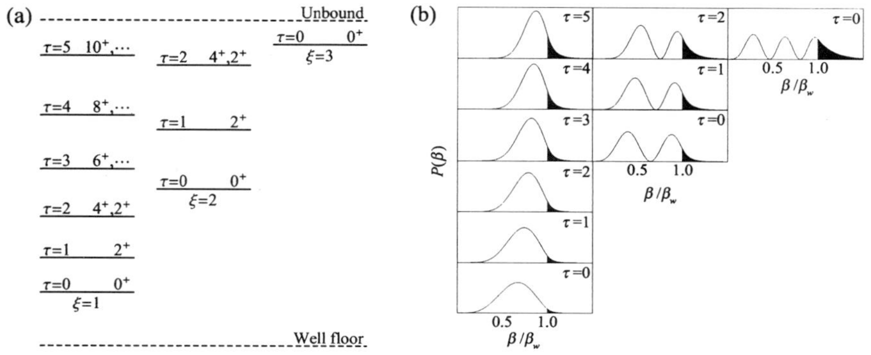

Figure 1. Bound states of the $x_0=10$ γ-soft well: (a) eigenvalues and (b) probability density functions $P(\beta)\equiv\beta^4|f(\beta)|^2$. The shaded areas under the probability density functions indicate penetration into the classically forbidden region of β values ($\beta>\beta_w$). (Figure adapted from Ref. 7.)

All ratios of energies are unaffected by the rescaling, and thus are the same for all wells with the same value of x_0, as are ratios of matrix elements of electromagnetic transition operators $\propto\beta^m$.

Finite depth for the potential well has several consequences, many of which are naturally to be expected. The level energies and radial probability distributions for the case $x_0=10$ are shown for illustration in Fig. 1. There are only a finite number of bound states. The wave functions penetrate the classically forbidden region ($\beta>\beta_w$), and for the highest-lying states a substantial portion of the probability distribution with respect to β lies outside β_w. The eigenvalues are lowered relative to those for an infinite E(5) well of the same width, as shown in Fig. 2 (inset). This is a straightforward consequence of the finite well depth: the wave functions are given the freedom to spread into the region $\beta>\beta_w$, which is analogous in its effect to a widening of the well, causing the energies to "settle" lower. Similarly, $B(E2)$ strengths are larger than for the infinite well, due to the larger β extent of the wave functions.

However, an examination of the solutions reveals that the level energies are nearly *uniformly* lowered by the same factor for all levels in the well, and so energy *ratios* are essentially unchanged. This is illustrated in Fig. 2 (main panel), where, for various well sizes, the excitation energies are shown normalized to that of the first excited state. The enhancement of $B(E2)$ strengths is likewise nearly uniform for all transitions, as shown in Fig. 3. The uniform reduction of all energies and enhancement of all transition matrix elements does not serve as a useful identifying feature of finite well depth, since the same effects are obtained for the infinite E(5) well with an increase in β_w.

Only the very highest levels, just short of being unbound, show appreciable deviations from the E(5) normalized energies and $B(E2)$ strengths. The third $\tau=0$ state for $x_0=10$, just below the top of the well, demonstrates these effects nicely, exhibiting a lowered energy (Fig. 2) and enhanced $E2$ decay strength (Fig. 3).

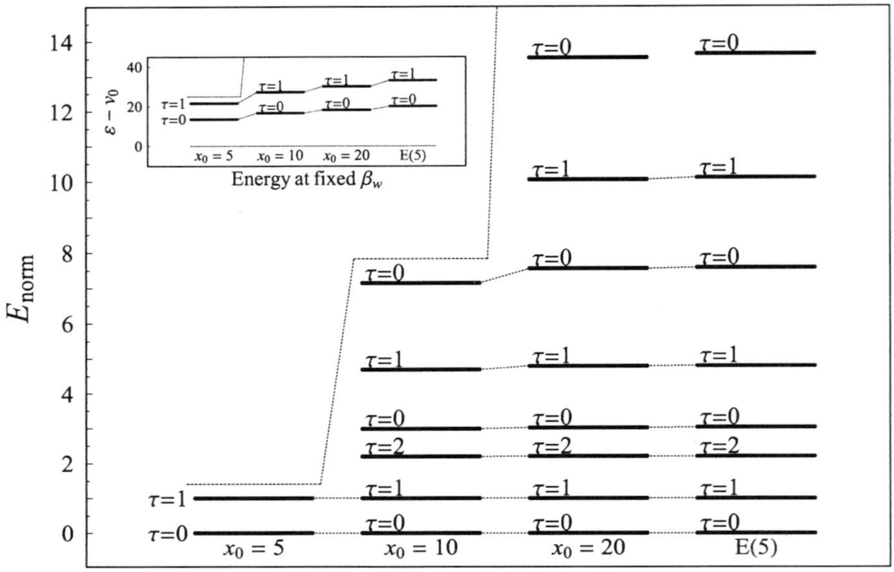

Figure 2. Evolution of level energies as a function of well size parameter x_0 for the γ-soft well. In the inset, the absolute energies relative to the floor of the well, $\varepsilon - v_0$, are shown for fixed well width (β_w=1). In the main graph, excitation energies are taken normalized to that of the first τ=1 state. The upper dashed line indicates the energy at which the system becomes unbound (the top of the well). (Figure adapted from Ref. 7.)

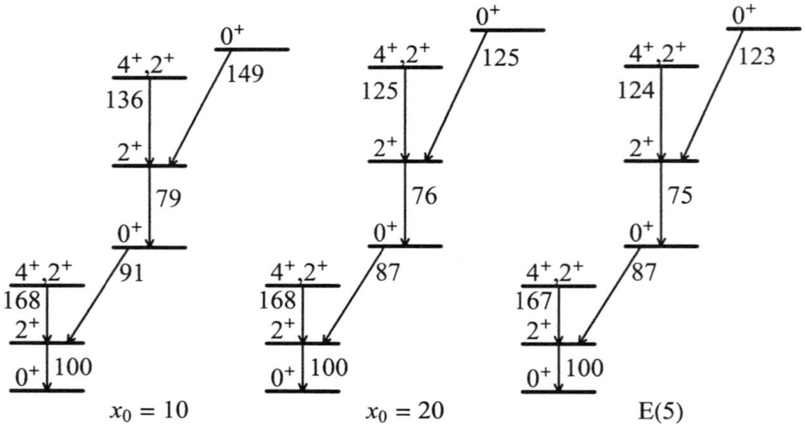

Figure 3. Evolution of $B(E2)$ strengths as a function of well size parameter x_0 for the γ-soft well. Values are calculated using $\mathfrak{M}(E2) \propto \beta$ and are normalized to $B(E2; 2_1^+ \rightarrow 0_1^+) = 100$. (Figure adapted from Ref. 7.)

3 Finite well: γ-stabilized case

In the X(5) model,[2] a square well potential in β is combined with a term $V_\gamma(\gamma)$ providing stabilization about γ=0. An approximate separation of the wave function into radial, γ, and rotational factors, $f(\beta)\eta(\gamma)\phi(\underline{\theta})$, occurs. To the extent that this separation of variables holds, the details of the potential V_γ are irrelevant to the calculation of the properties of states involving only rotational and β excitations. A band structure analogous to that of the rigid rotor arises, with K=0 for all bands involving only radial excitations. However, the energies differ substantially from those of a rigidly-deformed rotor, and the intrinsic state is different for each member of a band.

Let us now consider the case of a finite depth square well. The radial equation for $f(\beta)$ is again equivalent to the Bessel equation both inside and outside the well, but now yields solutions involving the spherical Bessel functions of order

$$n = \left[\frac{L(L+1)}{3} + \frac{9}{4}\right]^{1/2} - \frac{1}{2}, \tag{7}$$

where the separation constant L is the angular momentum quantum number. Only for the special values L=0,14,54,... do the wave functions involve the spherical Bessel functions of integer order encountered in the E(5) solution. The radial wave function for the sth radial excitation of angular momentum L is

$$f_{s,L}(\beta) = \begin{cases} A_{s,L}\beta^{-1} j_n[(\varepsilon_{s,L} - v_0)^{1/2}\beta] & \beta \le \beta_w \\ B_{s,L}\beta^{-1} k_n[(-\varepsilon_{s,L})^{1/2}\beta] & \beta > \beta_w. \end{cases} \tag{8}$$

The matching condition

$$\eta\frac{j_{n-1}[x_0\eta]}{j_n[x_0\eta]} = -(1-\eta^2)^{1/2}\frac{k_{n-1}[x_0(1-\eta^2)^{1/2}]}{k_n[x_0(1-\eta^2)^{1/2}]}, \tag{9}$$

where $\eta \equiv (1-\varepsilon/v_0)^{1/2}$, cannot be reexpressed as a transcendental equation involving trigonometric and exponential functions, as in the E(5) case, but must instead be solved directly in this form.

The solutions for the γ-stabilized well of finite depth exhibit properties similar to those described for the γ-soft well in the preceding section. Energies are lowered and electromagnetic transition strengths enhanced for smaller depths at fixed width, but these effects serve primarily to produce only an overall change in normalization. Ratios of energies or transition strengths are essentially unaffected, except for those involving the levels immediately below the top of the well. The level energies for the x_0=10 well are shown alongside the X(5) infinite well predictions in Fig. 4.

4 Conclusions

The results found for the finite-depth square well potential, in both the γ-soft [E(5)] and γ-stabilized [X(5)] cases, demonstrate that the eigenvalue spectrum is highly robust with respect to well depth. There are few clear signatures of finite well depth. Those which are present consist of moderate modifications to energies or transition strengths for high-lying levels, but such levels are typically the most subject to

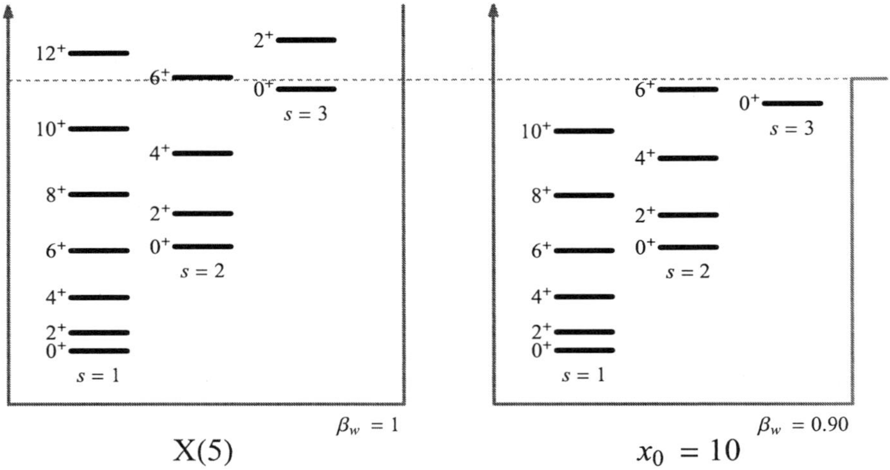

Figure 4. Level eigenvalues for the γ-stabilized x_0=10 well, alongside those for the lowest-energy states of the X(5) infinite well. At fixed well width, introduction of finite depth lowers the level energies, but a simple contraction of the well (from β_w=1 to $\beta_w\approx$0.90) almost entirely offsets this effect. Only the levels just below the top of the well — the $6^+_{s=2}$ and $0^+_{s=3}$ levels — show appreciable deviations from the X(5) energies. The horizontal dashed line indicates the energy of the top of the well, to facilitate comparison.

contamination from degrees of freedom outside the collective model framework and also the least accessible experimentally. Although realistic potentials are expected to be of finite depth, these results suggest that the infinite depth of the E(5) or X(5) potentials is not a limitation in their application to actual nuclei.

Acknowledgments

Discussions with F. Iachello, R. F. Casten, and N. V. Zamfir are gratefully acknowledged. This work was supported by the US DOE under grant DE-FG02-91ER-40609.

References

1. F. Iachello, *Phys. Rev. Lett.* **85**, 3580 (2000).
2. F. Iachello, *Phys. Rev. Lett.* **87**, 052502 (2001).
3. A. Bohr and B. R. Mottelson, *Nuclear Deformations*, Vol. 2 of *Nuclear Structure* (World Scientific, Singapore, 1998).
4. L. Wilets and M. Jean, *Phys. Rev.* **102**, 788 (1956).
5. G. Rakavy, *Nucl. Phys.* **4**, 289 (1957).
6. D. R. Bès, *Nucl. Phys.* **10**, 373 (1959).
7. M. A. Caprio, *Phys. Rev.* C **65**, 031304(R) (2002).

ANALYTICAL SOLUTIONS OF BOHR COLLECTIVE HAMILTONIAN WITH γ–INSTABILITY

L.FORTUNATO AND A.VITTURI

Dipartimento di Fisica "G.Galilei" and INFN
via Marzolo,8 I-35131 Padova - ITALY

The main aim of this contribution is to discuss the analytical solution of the collective Bohr equation with a Coulomb-like and a Kratzer-like γ–unstable potentials in quadrupole deformation space. Eigenvalues and eigenfunctions are given in closed form and transition rates are calculated for the two cases. The corresponding SO(2,1)×SO(5) algebraic structure is discussed. A few remarks concerning the possibility to study an approximate solution in the γ–stable case with these potential are treated briefly.

1 Introduction

It has been recently observed by Iachello [1] that a new class of dynamic symmetries can be associated with the critical point of the shape phase transition. The first example has received the name of E(5) symmetry and it is related to the phase transition between the harmonic oscillator and the γ-unstable case (that correspond to the U(5) and O(6) limits in the IBM). Later, the new group X(5) has been associated [2] with the phase transition between harmonic oscillator and axially symmetric rotor (U(5) and SU(3), respectively). Experimental confirmation of the actual occurrence of such situations was found for instance in the examination of the level scheme of ^{134}Ba and ^{94}Ru for E(5) and ^{152}Sm for X(5) [3,4,5].

In the former case of the transition from spherical to γ–unstable nuclei, the E(5) description assumes, for the transition potential in the β variable, an infinite square-well potential [1], a case that has been generalized by Caprio with the introduction of a finite square well [6]. Other choices are possible, like the Davidson potential studied by Elliott [7] and later by Rowe [8], that also generates an analytic vibration-rotation spectrum. The work of Rowe displays the same SO(2,1)×SO(5) algebraic structure and has many points in common with ours.

The content of the following is the analytic solution, still within the condition of γ–instability, of another class of potentials, namely the analogous of the Coulomb $(-A/\beta)$ and Kratzer $(-A/\beta + B/\beta^2)$ potentials in the β variable.

The Bohr hamiltonian for a γ–unstable potential (hence dependent only on β) may be separated and the second order differential equation in β may be recast in its standard form with a simple substitution. This equation, without discussing the details, reads:

$$\chi''(\beta) + \left\{\epsilon - u(\beta) - \frac{(\tau+3/2)^2}{\beta^2} + \frac{1}{4\beta^2}\right\}\chi(\beta) \;=\; 0. \qquad (1)$$

This equation is transformed in the Whittaker differential equation when the reduced potential $u(\beta)$ takes two particular forms: a Coulomb-like form $u(\beta) = -A/\beta$ and a Kratzer-like one $u(\beta) = -A/\beta + B/\beta^2$. Eigensolutions of this particular cases are the Whittaker's functions (that may be further reduced to expressions involving Laguerre polynomials) and the spectra acquire simple regular patterns. In the case of Coulomb-like case the spectrum is akin to the well-know spectrum of the hydrogen atom, albeit in a space of dimension 5 and with different meanings of quantum numbers and constants:

$$\varepsilon_{\tau,\xi} = \frac{A^2/4}{(\tau+\xi+2)^2} \tag{2}$$

where the τ quantum number is associated with the SO(5) invariance and ξ labels different bands. The latter comes from the fact that the hypergeometric function hidden in the Whittaker's function should be a finite series and hence the first argument must be a negative integer. This condition fixes the spectrum that we show, in reduced units (independent on the overall constant A), in fig.1.

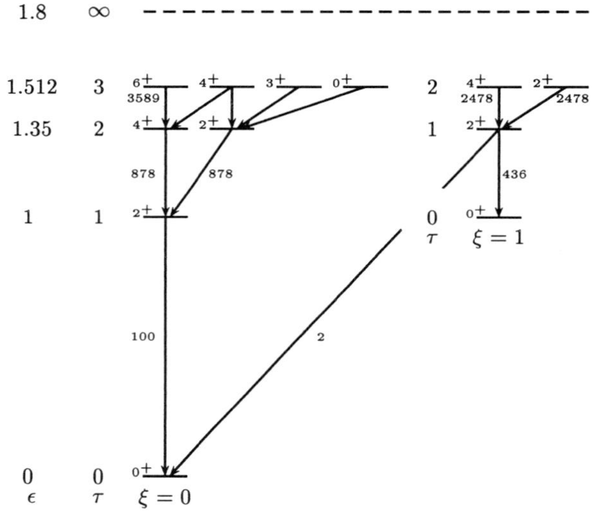

Figure 1.

This spectrum displays very peculiar features: there is a threshold at 1.8 that corresponds to an infinite quantum number and there is degeneracy of the τ-th member of the ξ-th band with the $(\tau-1)$-th state of the $(\xi+1)$-th band.

This degeneracy is removed if one introduces in the Bohr hamiltonian the Kratzer-like potential. Playing with the parameters one can move from a spherical situation $(B \to 0)$ to deformed one. An example of parameterization is shown in fig. 2 where the horizontal scale is expressed in the adimensional variable β_0/β, while the vertical scale is in units of the depth of the pocket $(\mathcal{D})$. The minimum corresponds to the value β_0, as one can easily see rewriting the Kratzer potential as:

$$u_K(\beta) = -2\mathcal{D}\Big(\frac{\beta_0}{\beta} - \frac{1}{2}\frac{\beta_0^2}{\beta^2}\Big) \tag{3}$$

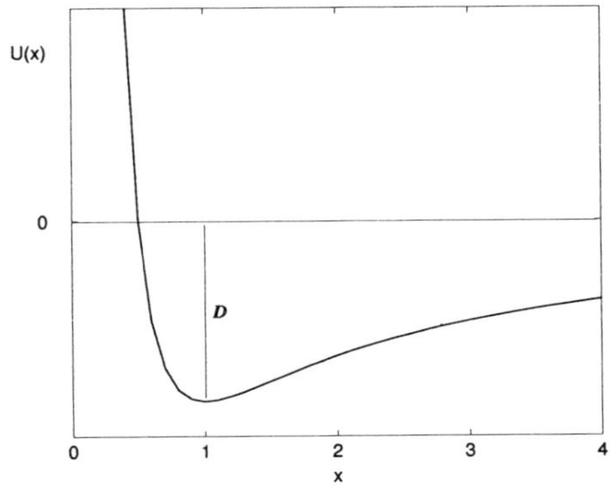

Figure 2.

Whenever one of the two parameters $\{\beta_0, \mathcal{D}\}$ is fixed two limits may be evidenced: when the second parameter is very small we regain the Coulomb-like spectrum, while when the parameter is very big we obtain a typical $\beta-$rigid, $\gamma-$unstable O(6) spectrum that displays only one band with a $\tau(\tau+3)$ spacing rule. This is shown, for instance in the fig. 3, where the behaviour of the spectra is studied fixing one of the two parameters ($\beta_0 = 0.5$) and increasing $\mathcal{D}$. The first two bands ($\xi = 0, 1$) are displayed with their lowest states ($\tau = 0, 1, 2, ..$). The various sub-states are not displayed for the sake of simplicity. In fig. 4 is instead displayed the case in which $\mathcal{D}$ is fixed to 10 and β_0 is varied. The behaviour at the two extremes is similar.

We have also calculated electromagnetic transition rates for the most important transitions in the various case. It is worth noticing that in the Coulomb-like limit of the Kratzer case we regain not only the energy levels, but also the B(E2) values, as can be deduced from the left parts of figures 3 and 4.

From an algebraic point of view, the hamiltonian may be recast as a linear combination of the infinitesimal generators of the SU(1,1) group

$$\hat{Z}_1 = 4\beta\Big(p^2 + \frac{B}{\beta^2}\Big) \qquad \hat{Z}_2 = \beta \qquad \hat{Z}_3 = 2(\hat{\vec{a}}\cdot\hat{\vec{p}} - i) \tag{4}$$

which have the commutation relations:

$$[\hat{Z}_1, \hat{Z}_2] = -4i\hat{Z}_3 \quad [\hat{Z}_3, \hat{Z}_2] = -2i\hat{Z}_2 \quad [\hat{Z}_3, \hat{Z}_1] = 2i\hat{Z}_1. \tag{5}$$

Thus the hamiltonian reads
$$\beta\mathcal{H} = \hat{Z}_1/4 - A. \tag{6}$$

that can be diagonalized in the basis that we discuss below.
There are four non-compact isomorphic Lie algebras su(1,1) $\sim$ so(2,1) $\sim$ sl(2,R)$\sim$ sp(2,R) that can be thus associated with this problem. The chain of subalgebras that gives the labels of the set of orthonormal states $\{| \xi\tau\alpha LM\rangle\}$ is explicitly given as [9]:

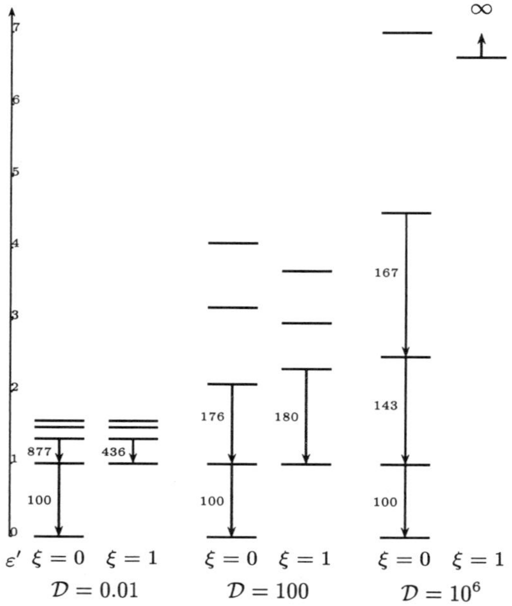

Figure 3.

$$\begin{array}{ccccccc} SU(1,1) \times SO(5) & \supset & U(1) \times SO(3) & \supset & SO(2) \\ \lambda \qquad\qquad \tau & & \alpha \quad \xi \qquad L & & M \end{array} \tag{7}$$

where λ is an SU(1,1) lowest weight and α indexes the SO(3) multiplicity.

The possibility of analytic solutions in the presence of a Kratzer potential is not limited to the case of $\gamma-$instability. Following Iachello[2] one can assume a harmonic dependence on the γ variable, as an approximation to the true potential in the region around $\gamma \simeq 0^o$. This potential is expected to describe an axially symmetric deformed rotor. An approximate separation of variables for the Bohr hamiltonian can still be adopted for potentials of the type $u(\beta) + v(\gamma)$. The total spectrum is thus composed of two parts: $\epsilon \simeq \epsilon_\beta + \epsilon_\gamma$. Inserting our choice of the potentials in the hamiltonian we find, for the $\beta-$part of the spectrum, a general structure that resembles the one we have presented for the $\gamma-$unstable case.
The spectrum for the Coulomb-like potential in β plus an harmonic oscillator in γ is easily found to be

$$\epsilon^{(\beta)}_{v,L} = \frac{A^2/4}{\left(\sqrt{\frac{9}{4} + \frac{L(L+1)}{3}} + \frac{1}{2} + v\right)^2}. \tag{8}$$

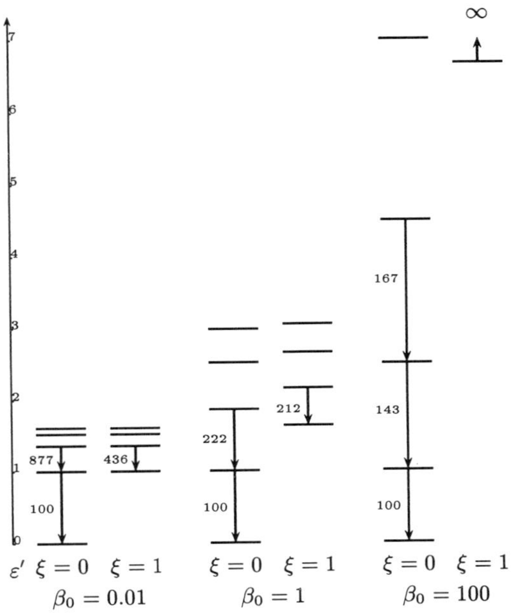

Figure 4.

This spectrum (fig. 5) looks rather similar to the one studied in the case of γ−instability. Nevertheless it is worth mentioning that the ratio of the ($v = 0, L = 4$) to the ($v = 0, L = 2$) energies is 1.718, while the ratio of the ($v = 1, L = 0$) (beginning of the second band) to the ($v = 0, L = 2$) energies is 1.423. Both this values may be very interesting for a comparison with experimental data. Other peculiarities are the presence of the threshold, corresponding to an infinite L quantum number, that is located at 2.5616 and the absence of the degeneracy pattern found in the γ−unstable case.

We expect that the corresponding Kratzer-like rotor will display remarkable features, having a spectrum that depends on one parameter. The two limiting cases are from one side the Coulomb-like spectrum showed above and from the other side a typical $I(I + 1)$ spectrum. We will discuss these issues in detail elsewhere.

Acknowledgments

This paper is a contribution to the Erice Conference “Symmetries in Nuclear Structure” that was also the occasion to celebrate Francesco Iachello’s 60th birthday. The conference testifies the vivid interest of the scientific community for the study of the fundamental symmetries showed not only in the realm of nuclei, but also in

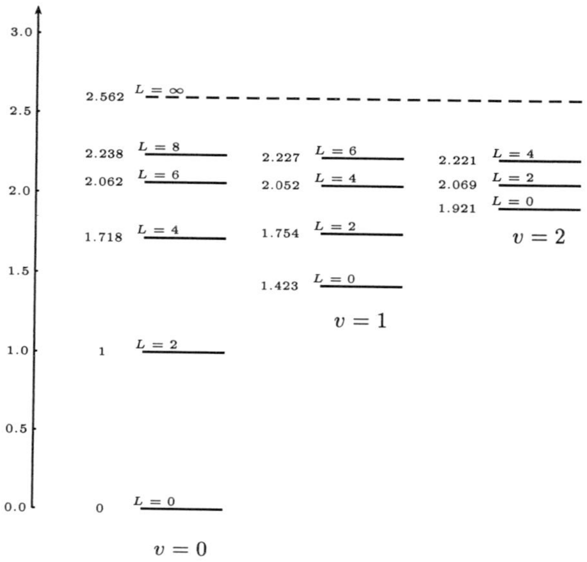

Figure 5. β−spectrum of the Coulomb-like rotor ($n_\gamma = 0$). The vertical scale is expressed in units of the energy of the ($v = 0, L = 1$) state.

the realm of molecules and subnuclear particles, and, more generally, in the whole aspects of natural systems. In all these fields Francesco Iachello has been one of the leading scientists, a promoter of novel developments and a source of inspiration for other researchers.

We wish also to express our gratitude to Prof. D.J.Rowe, whose suggestions were fundamental to complete this work.

References

1. F.Iachello, *Phys. Rev. Lett.* **85**, 3580 (2000).
2. F.Iachello, *Phys. Rev. Lett.* **87**, 052502 (2001).
3. R.F.Casten and N.V.Zamfir, *Phys. Rev. Lett.* **85**, 3584 (2000); R.F.Casten and N.V.Zamfir, *Phys. Rev. Lett.* **87**, 052503 (2001).
4. A.Frank, C.A.Alonso, and J.M.Arias, *Phys. Rev. C* **65**, 0143801 (2001).
5. P.G.Bizzeti and A.M.Bizzeti-Sona, *Phys. Rev. C* **66**, 031301 (2002).
6. M.A.Caprio, *Phys. Rev. C* **65**, 031304 (2002).
7. J.P.Elliott *et al.*, *Phys. Lett.* **169B**, 309 (1986).
8. D.J.Rowe, private communications.
   D.J.Rowe and C.Bahri, *J. Phys.* **A10**, 4947 (1998).
9. D.J.Rowe, *Prog. Part. Nucl. Phys.* **37**, 265 (1996).

PHASE TRANSITIONS AND CRITICAL POINTS IN THE INTERACTING BOSON MODEL

J.E. GARCÍA-RAMOS

Departamento de Física Aplicada, Universidad de Huelva, 21071 Huelva, Spain
E-mail: jegramos@nucle.us.es

J.M. ARIAS

Departamento de Física Atómica, Molecular y Nuclear, Universidad de Sevilla, 41080 Sevilla, Spain

J. BAREA

Departamento de Física Atómica, Molecular y Nuclear, Universidad de Sevilla, 41080 Sevilla, Spain

Instituto de Ciencias Nucleares, UNAM, 04510 México, DF, México

A. FRANK

Instituto de Ciencias Nucleares, UNAM, 04510 México, DF, México

We present a general analysis of phase-shape transitions in the framework of the Interacting Boson Model (IBM) using the catastrophe theory. The application to the the rare earth region shows that ^{148}Nd and ^{150}Sm are critical nuclei.

1 Introduction

Macroscopic phase transitions since long time ago are well known and have been studied for a long time.[1] However, the study of quantum (microscopic) phase transitions is of great interest. In this context, there has recently been in Nuclear Physics a renewed interest in the study of nuclear shape-phase transitions. New classes of symmetries that apply to systems localized at the critical points have been proposed. In particular, the "critical symmetry" $E(5)$[2] has been suggested to describe the critical point in the phase transition from spherical to γ-unstable shapes, while $X(5)$[3] is proposed to describe systems lying at the critical point in the transition from spherical to axially deformed shapes.

The Interacting Boson Model (IBM)[4] provides a natural way of studying nuclear shape-phase transitions in atomic nuclei. Although the corresponding analysis was carried out two decades ago, in most of the cases only schematic Hamiltonians were used. In this Hamiltonian the transition from one phase to the other is governed by a single parameter.[5] The main purpose of this paper is to check how the use of a more general Hamiltonian affects the predictions on phase transitions. We use a global approach, following Refs.[6,7], that allows to locate the different nuclei in an isotopic chain with respect to the phase transition point.

In section 2 we present the IBM Hamiltonian and the results of the fits made for the selected isotope chains. In section 3, the IBM energy surfaces are presented and the location of the critical points in the shape transition region stated for each isotope chain using catastrophe theory.[8] Finally, in section 4 our main results are summarized.

2 IBM general description

The most general (including up to two–body terms) IBM Hamiltonian, using the multipolar form, can be written as

$$\hat{H} = \mathcal{A}\hat{N} + \mathcal{B}\frac{\hat{N}(\hat{N}-1)}{2} + \varepsilon_d \hat{n}_d + \kappa_0 \hat{P}^\dagger \hat{P} + \kappa_1 \hat{L}\cdot\hat{L} + \kappa_2 \hat{Q}\cdot\hat{Q} + \kappa_3 \hat{T}_3\cdot\hat{T}_3 + \kappa_4 \hat{T}_4\cdot\hat{T}_4 \qquad (1)$$

where $\hat{N}$, and $\hat{n}_d$ are the total boson number operator, and the d boson number operator, respectively and

$$\hat{P}^\dagger = \frac{1}{2}\left(d^\dagger\cdot d^\dagger - s^\dagger\cdot s^\dagger\right), \qquad \hat{L} = \sqrt{10}(d^\dagger\times\tilde{d})^{(1)},$$
$$\hat{Q} = (s^\dagger\times\tilde{d} + d^\dagger\times\tilde{s})^{(2)} - \frac{\sqrt{7}}{2}(d^\dagger\times\tilde{d})^{(2)},$$
$$\hat{T}_3 = (d^\dagger\times\tilde{d})^{(3)}, \qquad \hat{T}_4 = (d^\dagger\times\tilde{d})^{(4)}. \qquad (2)$$

The electromagnetic transitions can also be analyzed in the framework of the IBM. In particular, in this work we will focus on $E2$ transitions. The most general $E2$ transition operator including up to one body terms can be written as

$$\hat{T}^{E2}_M = e_{eff}\left[(s^\dagger\times\tilde{d} + d^\dagger\times\tilde{s})^{(2)}_M + \chi(d^\dagger\times\tilde{d})^{(2)}_M\right], \qquad (3)$$

where e_{eff} is the boson effective charge and χ is a structure parameter.

Two-neutron separation energies (S_{2n}) are also studied in the present work. This observable is defined as the difference in binding energy between an even-even isotope and the preceding even-even one,

$$S_{2n} = BE(N) - BE(N-1), \qquad (4)$$

where N corresponds to the total number of valence bosons.

Using the Hamiltonian (1) and the $E2$ transition operator (3) we have studied the isotopic chains $^{144-154}_{60}$Nd, $^{146-160}_{62}$Sm, $^{148-162}_{64}$Gd, and $^{150-166}_{66}$Dy. For each chain the parameters in (1) and (3) are obtained from a simultaneous fit to the experimental energy spectra, $B(E2)$ transition rates and two neutron separation energies (S_{2n}). All parameters are fixed for the description of a given isotope chain but the parameter ε_d that is allowed to vary slightly from isotope to isotope. In tables 1 and 2 we summarize the parameters obtained for the Hamiltonian. Regarding the $E2$ transition operator, the values of the effective charges (in $e\cdot b$) and χ are $(e_{eff},\chi) = (0.119,-1.43)$, $(0.119,-1.69)$, $(0.110,-1.77)$ and $(0.103,-1.60)$ for Nd, Sm, Gd, and Dy, respectively. For a detailed description of the fitting procedure the reader is referred to Ref.[9].

Due to its special relevance in finding phase transitional regions, we compare in figure 1 the experimental and calculated S_{2n} values.[5] First order phase transitions are related to the appearance of a kink in the S_{2n} values. It can be observed in the figure that the calculation matches the experimentally observed behavior.

Table 1. Parameters in the Hamiltonian (1), excepting ε_d (see table 2).

	$\mathcal{A}$ (MeV)	$\mathcal{B}$ (MeV)	κ_0 (keV)	κ_1 (keV)	κ_2(keV)	κ_3 (keV)	κ_4 (keV)
Nd	-16.50	0.51	83.753	-13.928	-17.151	-101.27	-187.57
Sm	-17.82	0.46	53.209	-11.267	-14.674	-31.769	-131.24
Gd	-22.17	0.76	45.207	-7.932	-13.129	-35.224	-156.24
Dy	-24.66	0.80	38.651	-6.416	-13.638	-59.165	-163.05

Table 2. Values of ε_d in the Hamiltonian (1) (in keV) for each isotopic chain as a function of the neutron number.

	Neutron Number								
	84	86	88	90	92	94	96	98	100
Nd	1686.3	1606.7	1645.4	1602.9	1536.1	1595.9			
Sm	1427.3	1393.5	1289.3	1210.8	1158.6	1192.5	1312.2	1452.0	
Gd	1479.3	1508.7	1409.0	1300.4	1221.5	1174.4	1162.0	1176.5	
Dy	1558.8	1607.6	1562.4	1503.9	1461.0	1427.7	1413.4	1409.2	1443.1

3 Energy surfaces and phase transitions

The study of phase transitions in the IBM requires the use of the so called intrinsic-state formalism[5,10,11] in order to introduce an order parameter. This formalism is very useful to discuss phase transitions in finite systems because it provides a description of the behavior of a macroscopic system up to $1/N$ effects. To define the intrinsic, or coherent, state it is assumed that the dynamical behavior of the system can be described in terms of independent bosons ("dressed bosons") moving in an average field. The ground state of the system is a condensate, $|c\rangle$, of bosons occupying the lowest–energy phonon state, $\Gamma_c^\dagger$,

$$|c\rangle = \frac{1}{\sqrt{N!}}(\Gamma_c^\dagger)^N|0\rangle, \; \Gamma_c^\dagger = \frac{1}{\sqrt{1+\beta^2}}\Big(s^\dagger + \beta\cos\gamma\, d_0^\dagger + \frac{1}{\sqrt{2}}\beta\sin\gamma\,(d_2^\dagger + d_{-2}^\dagger)\Big) \quad (5)$$

where β and γ are variational parameters related with the shape variables in the geometrical collective model. The expectation value of the Hamiltonian in the intrinsic state (5) provides the energy surface of the system, $E(N,\beta,\gamma) = \langle c|\hat{H}|c\rangle$. The energy surface in terms of the parameters of the Hamiltonian (1) and the shape variables can be readily obtained,

$$\langle c|\hat{H}|c\rangle = \frac{N\beta^2}{(1+\beta^2)}\Big(\varepsilon_d + 6\,\kappa_1 - \frac{9}{4}\kappa_2 + \frac{7}{5}\kappa_3 + \frac{9}{5}\kappa_4\Big) + \frac{N(N-1)}{(1+\beta^2)^2}\Big[\frac{\kappa_0}{4}$$
$$+ \beta^2(-\frac{\kappa_0}{2} + 4\,\kappa_2) + 2\sqrt{2}\,\beta^3\,\kappa_2\,\cos(3\,\gamma) + \beta^4(\frac{\kappa_0}{4} + \frac{\kappa_2}{2} + \frac{18}{35}\kappa_4)\Big], \quad (6)$$

where the terms which do not depend on β and/or γ (corresponding to $\mathcal{A}$ and $\mathcal{B}$ in Eq. (1)) have not been included.

The equilibrium values of the variational parameters β and γ are obtained by minimization of the ground state energy $\langle c|\hat{H}|c\rangle$ and are taken as the order param-

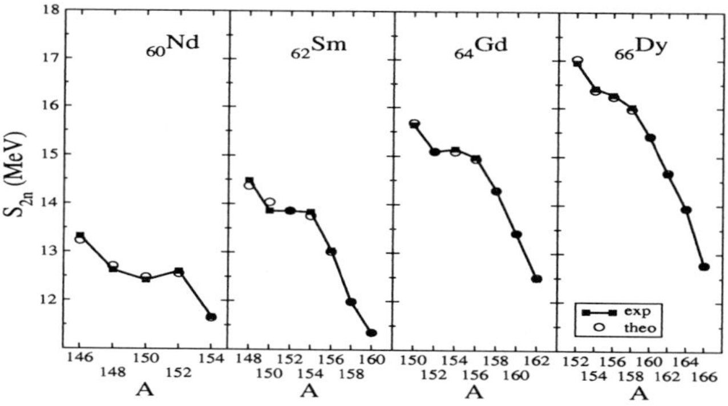

Figure 1. S_{2n} values for Nd, Sm, Gd, and Dy isotopes.

eters. Due to the properties of the energy surface the only relevant order parameter is β.

The classification of phase transitions that we use in this paper, that is the one followed traditionally in the IBM, is the Ehrenfest classification.[1] In this context, the origin of a phase transition resides in the way the energy surface (their minima positions) is changing as a function of the control parameter which, in this work, is a combination of Hamiltonian parameters. First order phase transitions appear when there exists a discontinuity in the first derivative of the energy with respect to the control parameter. This discontinuity appears when two degenerate minima exist in the energy surface (for two values of the order parameter β). Second order phase transitions appear when the second derivative of the energy with respect to the control parameter displays a discontinuity. This happens when the energy surface presents a single minimum for $\beta = 0$ and the surface satisfies the condition $\left(\frac{d^2E}{d\beta^2}\right)_{\beta=0} = 0$.

For the study of phase transitions in IBM within the framework of catastrophe theory[12,8] we already have the basic ingredients: the Hamiltonian of the system, Eq. (1), and the intrinsic state, Eq. (5). With them, we have generated the corresponding energy surface, given in Eq. (6), in terms of the Hamiltonian parameters and the shape variables. The objective is to find the values of the Hamiltonian parameters that correspond to critical points, *i.e.* the points where a phase transition is developing. In principle this analysis involves the 6 parameters in the Hamiltonian, but a first simplification occurs since the energy surface only depends on 5 parameters (see Eq. (6)). Fortunately, the catastrophe theory allows to reduce to two the number of relevant (or essential) parameters. We refer the reader to Ref.[12] for a detailed description of the application of this theory to the IBM case. Following this scheme one can find that the catastrophe germ of the IBM is β^4 ("cusp" germ) and the essential parameters for the Hamiltonian (1) are

$$r_1 = \frac{\tilde{\varepsilon} - (N-1)(\kappa_0 - 4\kappa_2)}{\tilde{\varepsilon} + (N-1)(\kappa_0 - 3\kappa_2 + \frac{36}{35}\kappa_4)}, \quad r_2 = -\frac{4\sqrt{2}\kappa_2(N-1)}{\tilde{\varepsilon} + (N-1)(\kappa_0 - 3\kappa_2 + \frac{36}{35}\kappa_4)}. \qquad (7)$$

The basic point is to translate every set of Hamiltonian parameters to the plane

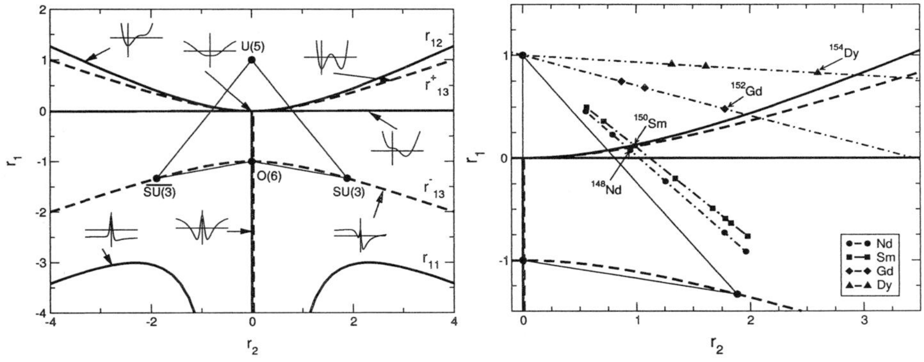

Figure 2. Left: Separatrix plane with a positive energy scale. Right: Representation of isotopes in the separatrix plane. Full and dashed lines correspond to the bifurcation and Maxwell sets, respectivelly.

formed by the essential parameters r_1 and r_2 (see figure 2). This plane is divided into several sectors by the bifurcation sets, that form the geometrical place in the parameter space where $\frac{d^2E}{d\beta^2} = 0$ for a critical value of β, and the Maxwell sets, the geometrical place in the space of parameters where two or more critical points are degenerate[8]. Both sets form the separatrix of the system, in this case of the IBM. In Ref.[12] the IBM bifurcation (r_2 axis, $r_2 = 0$ and $r_1 < 0$ semi-axis, r_{11}, and r_{12}) and Maxwell (negative r_1 semi-axis, r^+_{13}, and r^-_{13}) sets were obtained. They are all indicated in Fig. 2 (left). In this figure not all the lines correspond to phase transitions, but only r^+_{13}, that is related to a first order phase transition, and the point ($r_1 = 0$, $r_2 = 0$) where a second order phase transition takes place. The separatrix for $r_1 > 0$ is associated to minima while for $r_1 < 0$ is associated to maxima (except the negative r_1 semi-axis). In order to clarify the figure on the separatrix, the energy surfaces corresponding to each set are plotted as insets. The half plane with $r_2 > 0$ corresponds to prolate nuclei, while the one with $r_2 < 0$ corresponds to oblate nuclei. Note that expressions (7) are only valid for prolate nuclei, but can be readily obtained for the oblate case. On figure 2 the IBM symmetry limits and the correspondence with Casten's triangle[4] are also represented.

In figure 2 (right) the location of the different isotopes in an area close to the Maxwell set r^+_{13} are plotted. The main feature we find is that some nuclei are close to this Maxwell set: the closest are ^{148}Nd (boson number $N = 8$) and ^{150}Sm (boson number $N = 9$) and not far away lies ^{152}Gd (boson number $N = 10$). For Dy there is no isotope close to the critical point. According to our calculations, the transition from spherical to deformed for Dy occurs between $N = 11$ and $N = 12$.

In conclusion, from this global analysis we find that ^{148}Nd, ^{150}Sm are close to criticality. These isotopes are quite close but do not exactly coincide with previously proposed critical nuclei ^{150}Nd and ^{152}Sm[13,14], where the basic criterion was the closeness of their low-lying excitation spectra and transition intensities to the $X(5)$ values.

4 Conclusions

In this paper we have analyzed four chains of isotopes in the rare-earth region using the more general, up to two–body terms, IBM Hamiltonian. The Hamiltonian parameters for each isotopic chain were fixed through a global fit to several observables (excitation energies, E2 transitions and two-neutron separation energies). In these chains nuclei evolve from spherical to deformed shapes. We have performed an analysis of the corresponding shape transitions to look for possible nuclei at or close to a critical point using catastrophe theory. In this region we find that ^{148}Nd and ^{150}Sm are very close to the critical point. It seems that in this mass region criticality is associated with a number of neutrons equal to 88. The critical point symmetry $X(5)$ and the formalism presented here to identify a critical nucleus provide different, but close, results.

Acknowledgments

This work was supported in part by the Spanish DGICYT under projects number FPA2000-1592-C03-02 and BFM2002-03315 and by CONACYT (México).

References

1. H.E. Standley, "Introduction to phase transitions and critical phenomena", Oxford University Press, Oxford (1971).
2. F. Iachello, Phys. Rev. Lett. **85**, 3580, (2000).
3. F. Iachello, Phys. Rev. Lett. **87**, 052502, (2001).
4. F. Iachello and A. Arima. "The interacting boson model". Cambridge University Press, Cambridge, (1987).
5. A.E.L. Dieperink and O. Scholten, Nucl. Phys. A **346**, 125, (1980).
6. O. Castaños, A. Frank and P. Federman, Phys. Lett. **88B**, 203, (1979).
7. A. Gómez, O. Castaños, and A. Frank, Nucl. Phys. A **589**, 267, (1995). A. Gómez, O. Castaños, A. Frank, C.E. Alonso, and J.M. Arias, Nucl. Phys. A **594**, 483, (1995).
8. R. Gilmore. *"Catastrophe theory for scientists and engineers"*. Wiley, New York, (1981).
9. J.E. García-Ramos, J.M. Arias, J. Barea and A. Frank, submitted to Phys. Rev. C .
10. J.N. Ginocchio and M.W. Kirson, Nucl. Phys. A **350**, 31, (1980).
11. A.E.L. Dieperink, O. Scholten, and F. Iachello, Phys. Rev. Lett. **44**, 1747, (1980).
12. E. López-Moreno and O. Castaños, Phys. Rev. C **54**, 2374, (1996).
13. R.F. Casten and N.V. Zamfir, Phys. Rev. Lett. **87**, 052503, (2001).
14. R. Krücken et al, Phys. Rev. Lett. **88**, 232501, (2002)

TEST OF THE EMPIRICAL REALIZATION OF THE X(5) SYMMETRY IN ^{150}ND AND ^{104}MO

R. KRÜCKEN

Physik Department E12, Technische Universität München
and Maier-Leibnitz-Laboratory
85748 Garching, Germany

Lifetimes of excited states were measured in the stable nucleus ^{150}Nd and the neutron-rich unstable nuclei ^{144}Ba and ^{104,106}Mo isotopes by means of the recoil distance method. In ^{150}Nd excited states were populated via Coulomb excitation while the excited states in the neutron-rich Mo and Ba isotopes were populated in the spontaneous fission of ^{252}Cf. For ^{150}Nd and ^{104}Mo energies and B(E2) values are compared to the predictions for the X(5) critical point symmetry discovered by Iachello. While ^{150}Nd is found to be a good example for the empirical realization of the X(5) symmetry, the B(E2) values in ^{104}Mo clearly follow a rotational behavior. In ^{144}Ba the dipole moment of the octupole deformed band was measured for the first time directly and was found to be in agreement with theoretical predictions while the quadrupole deformation of this band seems to be significantly higher than that of the ground state band.

1 Introduction

Phase transitions play a very important role in many areas of physics and it is of general interest to understand the critical point behavior of a system undergoing a phase transition. Recently, a major breakthrough was made in the understanding of the critical point behavior of nuclei undergoing a phase/shape-transition from spherical to deformed shapes. In two Letters Iachello introduced new critical point symmetries [1,2] that allowed parameter free - except for scale - predictions of the excitation energies and electric quadrupole (E2) strengths of electromagnetic transitions for nuclei at the critical point of a phase/shape-transition.

^{152}Sm was established to be the first empirical realization for the X(5) symmetry [3]. The neighboring N=90 isotones are also good candidates for X(5) nuclei and in this contribution it will be shown that excitation energies and transition matrix elements in ^{150}Nd also agree very well with the X(5) predictions [4]. In both nuclei the level energies and B(E2) values were found to agree well with the parameter free (except for two scale factors) analytical predictions of the X(5) symmetry. An overview of work on other N=90 isotones can be found in Refs. [5,6].

Recently, ^{104}Mo was suggested to be another candidate for the empirical realization of the X(5) critical point symmetry [7,8]. The energies of the yrast band indeed follow closely the X(5) predictions. This is also true for the energies of the $n_\gamma = 1$ band in ^{104}Mo. However, the B(E2) values in ^{104}Mo, which are also very sensitive to the X(5) character of a nucleus, were not known with sufficient accuracy to draw any final conclusions. In this contribution it will be shown that the B(E2) values of the ground state band in ^{104}Mo are in better agreement with a rotational description and do not follow the X(5) predictions [9].

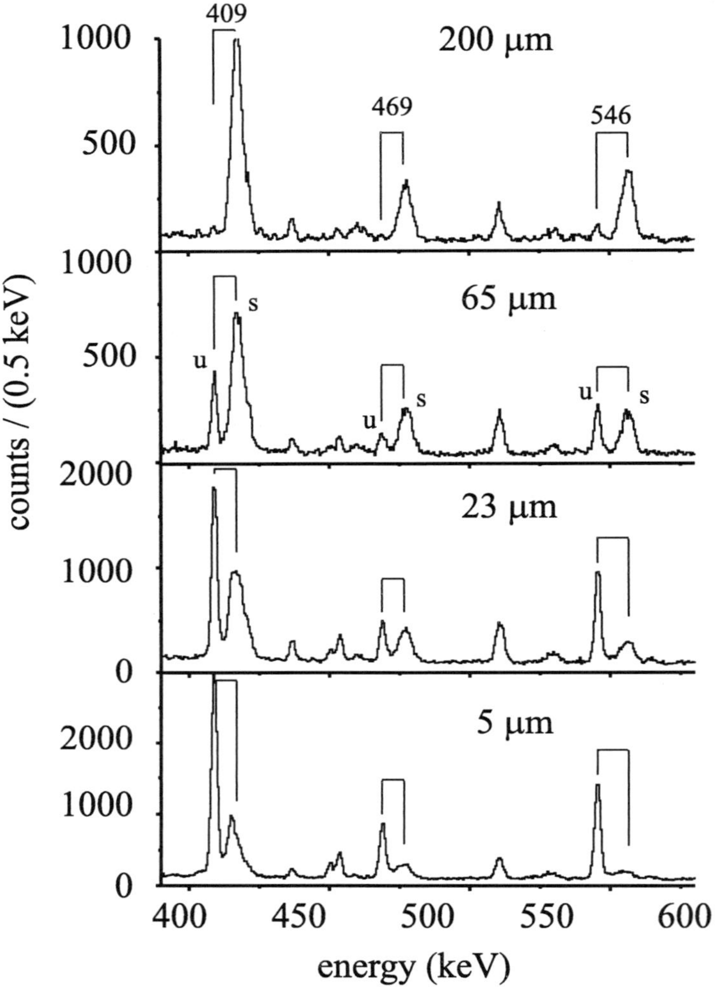

Figure 1. Spectra for transition in ^{150}Nd taken at different target-to-stopper distances.

2 Experimental Details

2.1 Coulex RDM in ^{150}Nd

The lifetimes of excited states in ^{150}Nd were measured by means of the recoil distance method (RDM) after being Coulomb excited by a 132 MeV ^{32}S beam delivered by the ESTU Tandem accelerator of the A.W. Wright Nuclear Structure Laboratory (WNSL) at Yale University. The beam was incident on a 1 mg/cm^2

enriched ^{150}Nd layer evaporated onto a 1.5 mg/cm^2 Ta foil used for mechanical support, which faced the incoming beam. The deexcitation γ-rays were detected by the Ge detectors of the SPEEDY array [10], and recorded when coincident with backward scattered beam particles detected by an array of photo cells covering an angular range from 153° to 171°. The forward recoiling nuclei had an average velocity of $v = 0.025(1)c$ and were stopped in a 9.3 mg/cm^2 thick Nb stopper foil. Target and stopper foil were mounted inside the New Yale Plunger Device (NYPD) [11]. Data were collected for 20 distances between 5μm and 800μm for 4 to 8 hours each, with the longer runs performed at the shortest distances to be most sensitive to shorter lifetimes.

For each distance, spectra of γ-rays detected at 138.5 ° and 41.5° were produced under the condition that the energies of the coincident backscattered projectiles were in the energy range corresponding to scattering on ^{150}Nd (see Fig. 1). This eliminated the γ-ray background from Ta Coulomb excitation. Lifetimes were deduced by means of the Differential Decay Curve Method (DDCM) [14]. Details of the analysis can be found in [4].

2.2 Fission RDM

Lifetimes in neutron-rich nuclei were measured by employing the recoil distance method (RDM) in conjunction with a ^{252}Cf source. A thin $\approx 70\mu$Ci ^{252}Cf source, placed on a stretched 1 mg/cm^2 Ni foil, was mounted in the New Yale Plunger Device (NYPD)[11]. Fission fragments flying within a cone of $\pm 20°$ along the axis of the plunger were slowed in the Ni foil and after flying a variable distance d were stopped in a stretched 10 mg/cm^2 Au stopper foil, while the complementary fragments were detected in a matrix of nine 1 cm^2 photo cells. Gamma rays in coincidence with the registered fission fragments were detected by the 100 large volume Compton-suppressed Ge detectors of the Gammasphere array [12]. Events with one detected fission fragment and at least three suppressed gamma-rays were recorded in list-mode format on magnetic tape.

Data was taken at 22 source-to-stopper distances between 9μm and 7000μm for about 1 day per distance. The data was presorted and stored in a BLUE database [13]. Spectra were created by gating on the energy of the complementary fission fragment and energies of two transitions in the decay cascade of the neutron-rich nucleus of interest. The gamma-ray gating condition was such, that at least one gate was placed on a Doppler-shifted component of a transition above the level of interest. For the second gate we required either a Doppler-shifted energy of a second transition above the level of interest or an unshifted or shifted component of a transition below the level of interest. Sample spectra are shown in Fig. 2 for the $4^+ \rightarrow 2^+$ transition in ^{104}Mo. Lifetimes of excited states in strongly populated fission fragments were deduced by means of the Differential Decay Curve Method (DDCM) [14]. Details of the analysis can be found in Ref. [9].

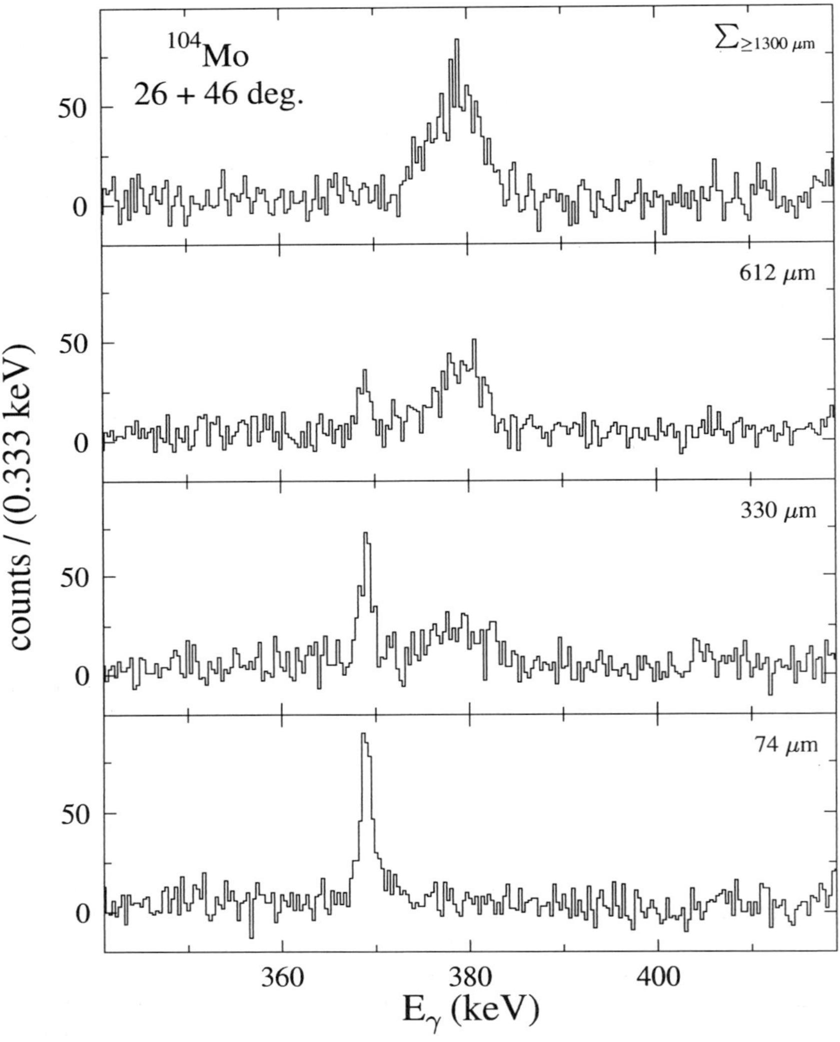

Figure 2. Background subtracted double gated coincidence spectra for the $4^+ \to 2^+$ transition in ^{104}Mo taken at different source-to-stopper distances.

3 Results and Discussion

3.1 ^{150}*Nd*

Using the present values of the branching ratios and (except for the 2_1^+ level) level lifetimes we have determined reduced transition matrix elements B(E2). For

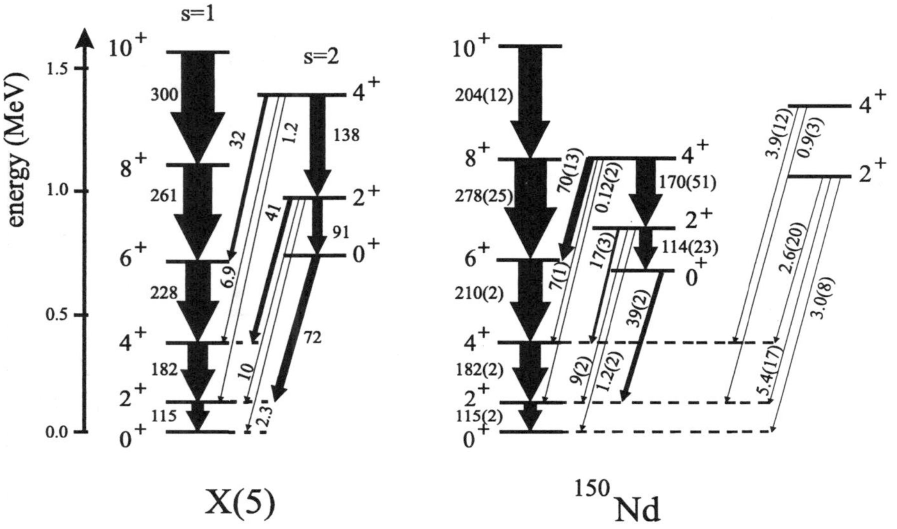

Figure 3. Partial experimental level scheme of ^{150}Nd (right) and the normalized predictions of the X(5) symmetry (left). The thickness of arrows indicates the B(E2) values, which are also given next to the arrows in Weisskopf units. The transition strengths for the X(5) predictions are normalized to the experimental best value of the $2_1^+ \rightarrow 0_1^+$ transition. At the far right of the experimental level scheme, data on the 2_3^+ and 4_3^+ levels, thought to be part of the $n_\gamma=1$ strucutre, are shown.

ΔI=0 transitions we have assumed pure quadrupole character if no information on multipole mixing ratios was available. The right of Figure 3 shows a partial level scheme of ^{150}Nd. The numbers on the transition arrows are the B(E2) values given in Weisskopf units. The left of the same figure shows the $s = 1$ and $s = 2$ levels of the X(5) description. The energy scale was normalized to the energy of the 2_1^+ level. B(E2) values are shown in W.u., normalized to the experimental B(E2; $2_1^+ \rightarrow 0_1^+$) value. The 2_3^+ and 4_3^+ levels are not included in the framework of the current X(5) predictions. These two states are most likely part of the excitation in the γ-degree of freedom, for which there is no analytical prediction in the current X(5) predictions.

The overall agreement is remarkable between the experimental energies and B(E2) values and those predicted by the X(5) symmetry. As already pointed out in Ref. [3] the energy spacing of the yrast (s=1) states is in nearly perfect agreement with X(5), in fact better than for ^{152}Sm. Also the energy of the 0_2^+ state is almost exactly predicted. Note that the energy ratio E(0_2^+)/E(2_1^+) is fixed in X(5) and cannot be adjusted to fit the data. Taken together, our results have established ^{150}Nd as a clear example of the empirical realization of a critical point nucleus for this phase transition from spherical to axially deformed nuclei.

The good overall agreement of the yrast B(E2) values for ^{150}Nd is illustrated in Fig. 4 where the normalized B(E2) values for X(5) and ^{150}Nd are shown as a

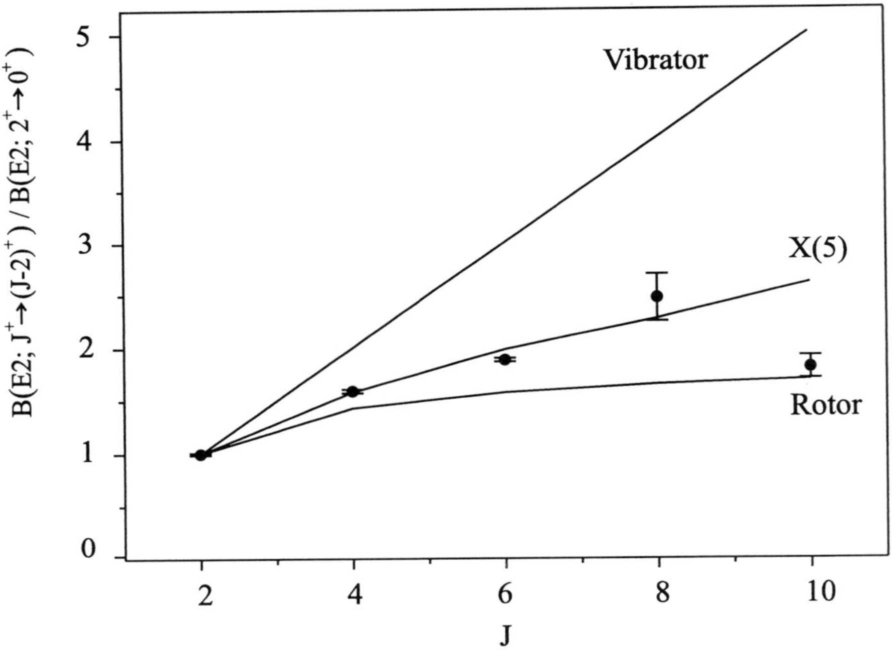

Figure 4. Normalized B(E2) values for the yrast transitions in ^{150}Nd in comparison to the predictions for the X(5) critical point symmetry as well as for the geometric paradigms of harmonic vibrator and axially symmetric rotor.

function of spin. Apart from the $10^+ \rightarrow 8^+$ transition, the values for ^{150}Nd are remarkably close to the X(5) predictions and show a significant deviation from the rotor values.

Regarding the cross-over transitions from s=2 to s=1 states, we see for most transitions less than a factor of two difference between the experimental and predicted B(E2) values. Considering the simplicity of the calculations and the lack of free parameters of the model, this is a remarkable level of agreement. It is important to note that many observables change most rapidly in the vicinity of the critical point of the phase transition and therefore small deviations in structure from X(5) are amplified by these sensitive cross-over B(E2) values. The fact that the average deviation of the cross-over B(E2) values of X(5) from the data is less than in ^{152}Sm is consistent with the fact that the yrast energies and B(E2) values are also closer to X(5) in ^{150}Nd. We also note here an interesting difference with ^{152}Sm where there is an almost constant factor between the predictions and the data for these cross-over B(E2) values, while in ^{150}Nd, there are more fluctuations present. Additionally, for the 203 keV $4_2^+ \rightarrow 1_1^-$ E1 transition, with a 13% gamma branching ratio, our lifetime data leads to a B(E1) value of 0.0011(3) W.u., which suggests some two-phonon octupole component in the wave function of the s=2

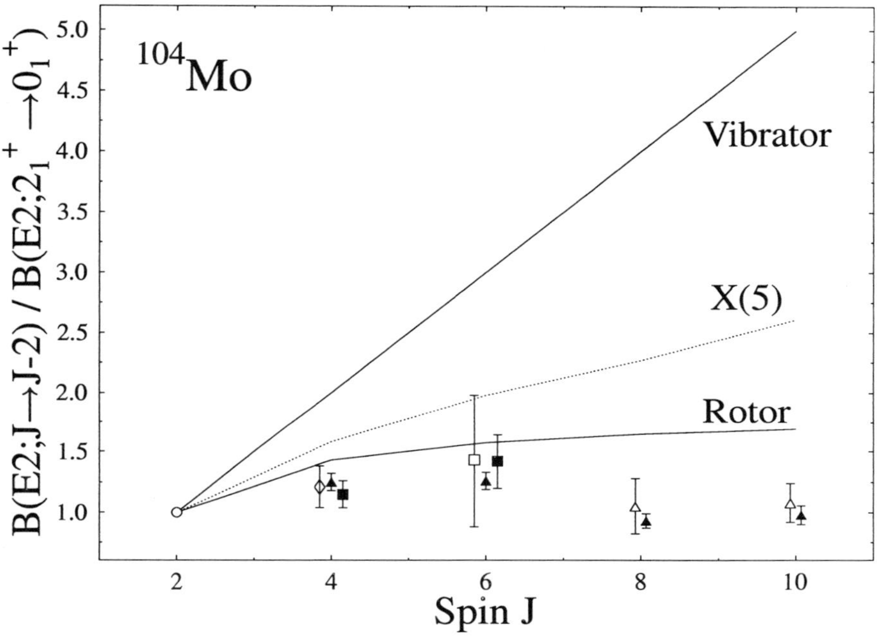

Figure 5. Comparison of the relative experimental B(E2)-values of the yrast transitions in ^{104}Mo from this work (filled squares), Refs. [16,17,18] (open diamond), Ref. [19] (open triangles), Ref. [20] (open square), and Ref. [21] (closed triangles) with the predictions for the vibrational and rotational limits and for the X(5) symmetry. Normalization is done using the combined value of Refs. [16,17,21] for the $2^+ \rightarrow 0^+$-transition (open circle).

states[15]. This particular degree of freedom is not included in the X(5) picture.

3.2 ^{104,106}*Mo*

Lifetimes have been obtained for the $4_1{}^+$ and $6_1{}^+$ levels in ^{104}Mo and ^{106}Mo. We have deduced reduced transition matrix elements B(E2), which are shown for ^{104}Mo in Fig.5 relative to the average B(E2) for the $2^+ \rightarrow 0^+$-transition. Previously published B(E2) values [16,17,19,20,21] are also shown.

Both Refs. [7,8] suggest ^{104}Mo as a X(5) candidate, but only quote the lifetimes (or equivalently BE(2)-values) of the $2_1{}^+$ and $4_1{}^+$ state. As the first excited state is used for normalization, effectively only one data point was used in these references, with statistics insufficient to draw conclusions. With the values from Refs. [19,21] together with the newly measured lifetimes, it is apparent that the data favors a rotational rather than an X(5) interpretation for the yrast cascade of ^{104}Mo (see Fig. 5). However, it is also clear that additional lifetimes for the K=2, and K=4 bands as well as the level sequence built on the first excited 0^+ state would give further insight. The new values for $4_1{}^+$ and $6_1{}^+$ lifetimes in ^{104}Mo from the present work are in agreement with those from previous measurements and are close to the predictions for a rotational behavior of the yrast states.

In ^{106}Mo we were for the first time able to measure the lifetime of the 6^+ yrast level with $\tau_{6^+} = 6.1(19)$ ps). The quadrupole moment for the 4^+ level in ^{106}Mo from this work ($Q_{4^+} = 382(28)$ efm^2) is significantly lower than the one quoted in Ref. [22] ($Q_{4^+} = 468(34)$ efm^2) but is consistent with both quadrupole moments for the 2^+ ($Q_{2^+} = 360(4)$ efm^2) and 6^+ states ($Q_{6^+} = 339(50)$ efm^2) . The constant quadrupole moment shows the rotational character of the yrast cascade in ^{106}Mo, which also has the same dynamical moment of inertia as the rotational bands built on the one- and two-gamma phonon vibrational states reported in Ref.[23].

For both nuclei there appears to be a drop in the quadrupole moments for the 8^+ and 10^+ levels. This has been interpreted in Ref. [19] as an indication for triaxiality at higher rotational frequency. It may also be possible that the adopted value for the 2_1^+ levels is to high.

3.3 ^{144}*Ba*

By measuring lifetimes of the 4^+ (τ=74(4)ps), 6^+ (τ=21(2)ps), and 7^- (τ=5.7(10)ps) levels in ^{144}Ba we were able to independently determine the quadrupole moments for the positive and negative parity bands. While our results for the quadrupole moments in the ground state band ($Q_{4^+} = 308(9)$ efm^2, $Q_{6^+} = 287(14)$ efm^2)is close to the result previously obtained from the lifetime of the 2^+ level ($Q_{2^+} = 333(7)$ efm^2), we find a significantly larger quadrupole moment for the negative parity band of $Q_{7^-} = 500(60)$ efm^2 (see left panel of Fig. 6). This surprising result was not, as far as we are aware, predicted in any theoretical description of ^{144}Ba.

At the same time we were for the first time able to directly measure the absolute B(E1) strength of a transition between negative and positive parity states in ^{144}Ba. The transition dipole moment we have extracted from the B(E1) value (D_t = 0.115(11) efm) agrees well with the predictions by Butler and Nazarewicz[24] of D_0= 0.12 efm for this dipole moment using the shell-correction method based on the reflection-asymmetric Woods-Saxon model (see right panel of Fig. 6). However, no significant difference of the quadrupole deformation is predicted in Ref.[24] between the positive and the negative parity band.

3.4 Summary

In summary, results from lifetime measurements were presented for excited states in ^{150}Nd, ^{104,106}Mo and ^{144}Ba using the recoil distance method. In the stable nucleus ^{150}Nd we have populated the excited states via Coulomb excitation while the excited states in the β-unstable neutron-rich Mo and Ba isotopes were populated in the spontaneous fission of ^{252}Cf.

Energies and B(E2) values of ^{150}Nd agree very well with the predictions of the critical point symmetry X(5). In ^{104}Mo the situation is more complicated. The energies of the g.s. band and the n_γ=1 band agree well with the X(5) predictions, while the B(E2) values in the ground state band favor a rotational interpretation over the X(5) predictions. Thus one may conclude that one has to be cautious in

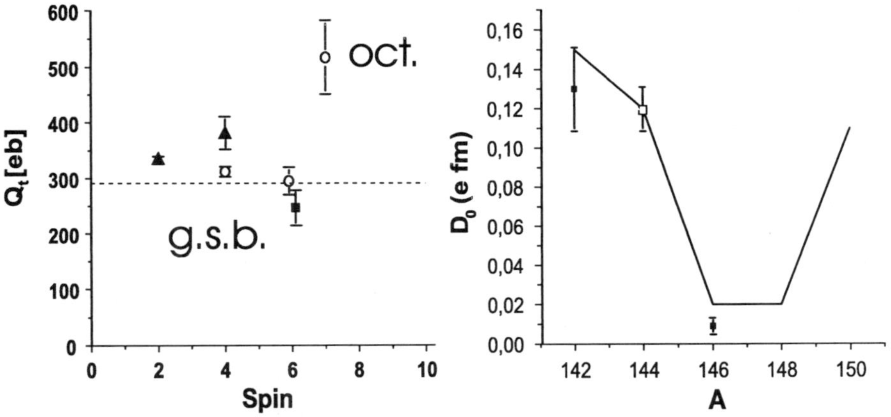

Figure 6. *Left:* Quadrupole moments of the g.s. band and the octupole band in ^{144}Ba. Open circles denote results from this work while filled triangles and squares are from Refs.,[16,17,18,25] and [20], respectively. The dashed line represents the Möller-Nix prediction. *Right:* Experimental dipole moments in the octupole bands of the Ba isotopes. The open square is from this work, while theoretical predictions and the other experimental data points are taken from Ref.[24]. Previously obtained dipole moments were obtained from branching ratios and the assumption of similar quadrupole moment in g.s. band and octupole band.

the interpretation of transitional nuclei in general and the comparison to the X(5) picture in particular. The use of a large number of experimental observables has to be stressed for the interpretation of transitional nuclei.

B(E2) values were determined for the ground state band and the octupole band in ^{144}Ba. Additionally, the B(E1) value for the $7^- \rightarrow 6^+$ transition was used for the first time to directly determine the dipole moment of the octupole deformed ^{144}Ba from experiment. The quadrupole moment of the g.s. band was found to be in agreement with previous measurements. For the octupole band a significantly larger quadrupole moment was found while the dipole moment was found to be in agreement with theoretical predictions. The large quadrupole moment of the octuple band remains a puzzle.

Acknowledgments

This work has been mostly performed during my time at Yale University and I would like to thank everybody who has contributed to this work. In particular I would like to thanks B. Albanna, C. Hutter, R.F. Casten and N.V. Zamfir. I would like to particularly thank F. Iachello for very valuable discussions and for the inspiration of this work. This work was supported by the U.S. DOE and the German DFG and BMBF.

References

1. F. Iachello, *Phys. Rev. Lett.* **85**, 3560 (2000).
2. F. Iachello, *Phys. Rev. Lett.* **87**, 052502 (2001).
3. R.F. Casten, N.V. Zamfir, *Phys. Rev. Lett.* **87**, 052503 (2001).
4. R. Krücken *et al.*, *Phys. Rev. Lett.* **88**, 232501 (2002).
5. M.A. Caprio *et al.*, *Phys. Rev.* C **66**, 054310 (2002).
6. A. Dewald, these proceedings.
7. P. G. Bizzeti and A. M. Bizzeti-Sona, *Phys. Rev.* C **66**, 031301(R) (2002).
8. D.S. Brenner in *Mapping the Triangle: Int. Conf. Nuclear Structure*, eds. A. Aprahamian, J.A. Cizewski, S. Pittel, N.V. Zamfir, AIP **638**, 223 (2002). *Phys. Rev.* C **66**, 031301(R) (2002).
9. C. Hutter *et al.*, *Phys. Rev. C*, in press.
10. R. Krücken in *Proc. of the International Symposium on Advances in Nuclear Physics*, eds. D. Poenaru and S. Stoica (World Scientific, Singapore, 2000), p. 336.
11. R. Krücken, J. Res. Natl. Inst. Stand. Technol. **105**, 53 (2000)
12. I.Y. Lee, *Nucl. Phys.* A **520**, 641c (1990).
13. M. Cromaz *et al.*, *Nucl. Instrum. Methods* A **462**, 519 (2001).
14. A. Dewald, S. Harissopulos, and P. von Brentano, *Z. Phys.* A **334**, 163 (1989).
15. P.A. Butler *et al.*, Acta Phys.Pol. **B24**, 117 (1993).
16. R. C. Jared and H. Nifenecker and S. G. Thomson in *Proc. 3rd IAEA Symp. Phys. Chem. Fission*, Vienna, 211 (1973).
17. E. Cheifetz, H.A. Selic, A. Wolf, R. Chechik, J.B. Wilhelmy, in *spectroscopy of fission products*, ed. T. von Egidy (IOP, 1980), p. 193.
18. G. Mamane and E. Cheifetz and E. Dafni and A. Zemel and J. B. Wilhelmy, *Nucl. Phys.* A **454**, 213 (1986).
19. A.G. Smith *et al.*, *Phys. Rev. Lett.* **77**, 1711 (1996).
20. R. Krücken *et al.*, *Phys. Rev.* C **64**, 017305 (2001).
21. A. G. Smith, *et al.*, *J. Phys.* G **77**, 1711 (1996). *J. Phys.* G **28**, 2307 (2002).
22. J. H. Hamilton *et al.*, Prog. Part. Nucl. Phys. **35**, 635 (1995).
23. A. Guessous *et al.*,*Phys. Rev. Lett.* **75**, 2280 (1995).
24. P.A. Butler and W. Nazarewicz, *Nucl. Phys.* A **533**, 249 (1991).
25. H.Mach *et al.*, *Phys. Rev.* C **41**, R2469 (1990).

THE RICH STRUCTURES OF A VERY SIMPLE HAMILTONIAN

J. JOLIE

Institute for Nuclear Physics, University of Cologne,
Zulpicherstr. 77, D-50937 Cologne, Germany

Within the Interacting Boson Model a very simple hamiltonian can be constructed by combining a vibrational one body term with a quadrupole-quadrupole interaction. This hamiltonian which essentially depends on two control parameters, has four dynamical symmetries, three first order phase transitions and an isolated second order phase transition. The shapes can be related to Landau theory of phase transitions. Extensions of the simple hamiltonian to more complicated situations are proposed and studied.

1 Introduction

The combination of group theoretical techniques with simple models containing the essential physics (degrees of freedom) is the trademark of F. Iachellos scientific work. This is exemplified in his three monographs [1,2,3] as in his numerous articles. One of the most successful of these models is the Interacting Boson Model (IBM) [1] of which we will use here a very simple, but physically relevant, hamiltonian. It consists of only two interactions: one which gives rise to harmonic vibrations and one which causes nuclear deformation. The structures this very simple hamiltonian generates are extremely rich as we will show. Then we will extend this simple hamiltonian to more complex systems.

2 A very simple hamiltonian

The hamiltonian which encorporates the two opposite forces, one driving to spherical shapes and one driving to deformation, can be constructed in the s,d IBM as follows:

$$\hat{H} = a(\eta \hat{n}_d - \frac{(1-\eta)}{N} \hat{Q}_\chi \hat{Q}_\chi) \tag{1}$$

whereby a is a general energy scaling factor, which will be taken equal to 100 keV, and N the number of s and d bosons. The quadrupole operator is given by:

$$\hat{Q} = (s^\dagger \tilde{d} + d^\dagger s)^{(2)} + \chi (d^\dagger \tilde{d})^{(2)}. \tag{2}$$

This parameterisation has been used to study in the context of the IBM chaotic and regular behavior [4], catastrophe theory [5] and quantum phase transitions [6]. Important for the structures associated with the hamiltonian (1) are the two independent dimensionless parameters η, describing the transition between spherical and deformed, and χ, describing the transition between prolate, γ-soft, and oblate deformation. In the hamiltonian (1) an explicit N dependence is introduced which is needed for scaling when N $\to \infty$ [4]. In order to visualise this simple parametrisation the Casten triangle is used [7] in which one axis is formed by η and the other one by χ. The great interest of the simple hamiltonian (1) lies in the fact that most nuclei can be located on or close to one of the legs of the triangle and that one can visualise the general six dimensional IBM hamiltonian in a two dimensional image. Moreover the triangle is constructed such that the three corners are formed by the three dynamical limits of the IBM: U(5) ($\eta = 1$), O(6) ($\eta = 0$, $\chi = 0$) and SU(3)($\eta = 0$,$\chi = -\frac{\sqrt{7}}{2}$).

It is instructive to rewrite the hamiltonian (1) in a form that reflects the group structure of the IBM yielding:

$$\hat{H} = (\eta + \frac{2(1-\eta)}{7N}(\chi^2 + \frac{\chi\sqrt{7}}{2}))\hat{C}_1[U(5)] + \frac{2(1-\eta)}{7N}(\chi^2 + \frac{\chi\sqrt{7}}{2})\hat{C}_2[U(5)] \quad (3)$$
$$+\frac{(1-\eta)}{N}(1 + \frac{2\chi}{\sqrt{7}})\hat{C}_2[O(6)] + \frac{(\eta-1)\chi}{N\sqrt{7}}\hat{C}_2[SU(3)]$$
$$+\frac{(\eta-1)}{N}(1 + \frac{3\chi}{\sqrt{7}} + \frac{2\chi^2}{7})\hat{C}_2[O(5)] + \frac{(1-\eta)}{14N}(\chi^2 + 2\sqrt{7}\chi)\hat{C}_2[O(3)].$$

One notices that while the number of parameters is reduced to two, the hamiltonian contains the whole rich group structure of the IBM except that the Casimir operators associated with U(6) does not appear. In the case that η equals one the hamiltonian reduces to $\hat{C}_1[U(5)]$ as expected. However, in the case that η equals zero which represents a transition inbetween O(6) and SU(3) the hamiltonian still contains contributions of the type $\hat{C}_1[U(5)] + \hat{C}_2[U(5)]$. This somewhat confusing situation can be avoided when one considers one of the two additional limits of the IBM that generally are not discussed [9,10], namely the $\overline{SU(3)}$ limit. This $SU(3)$-like limit with $\chi = +\sqrt{7}/2$ has a clear physical interpretation as it corresponds to an oblate rotor, if one consistently uses positive effective charges in the electromagnetic transition operator. Then one obtains [9]:

$$2[\hat{C}_1[U(5)] + \hat{C}_2[U(5)]] = \quad (4)$$
$$\hat{C}_2[\overline{SU(3)}] - 4\hat{C}_2[O(6)] + 6\hat{C}_2[O(5)] + \hat{C}_2[SU(3)] - 2\hat{C}_2[SO(3)],$$

and the ambiguity is removed when η is zero. Located on the $\eta = 0$ side are $SU(3)$ and $\overline{\mathrm{SU(3)}}$ at the $\chi = -\sqrt{7}/2$ and $\chi = +\sqrt{7}/2$ vertices and O(6) at $\chi = 0$ [11]. The halves of the triangle with positive and negative χ are related by a $\chi \leftrightarrow -\chi$ parameter symmetry[8].

3 Nuclear shapes and Landau theory

The groundstate of a given atomic nucleus has a well defined deformation. Whenever this deformation changes in a series of nuclei, one can speak of a phase transition provided criteria like the Ehrenfest classification are fulfilled. In contrast to conventional phase transition the shape phase transitions are not caused by changes of pressure or temperature nor by external forces, but they are governed by the changing single particle orbitals and their occupancy by protons and neutrons.

Shape phase transitions came again to the forefront of nuclear structure physics when Iachello developed new symmetries that describe atomic nuclei at the critical points[12,13]. These symmetries, called X(5) and E(5), are obtained within the framework of the collective model[14] under some simplifying approximations. Remarkably, the parameter-free predictions provided by the new symmetries are closely realised in some nuclei, such as ^{152}Sm and ^{134}Ba, respectively[15,16]. In a study of the shape phase transitions of the extended Casten triangle we have derived the structure shown in Fig (1) [11]. The groundstate can show three phases each separated by first order phase transitions. The phases are: spherical (I), prolate deformed (II) and oblate deformed (III). At the triple point where the three phase coexist a second order phase transition occurs. A similar structure was also found using catastrophe theory [5].

The general phase diagram can be described using the Landau theory of phase transitions[17,18], when using a geometric interpretation of the Hamiltonian (1) [19,20].

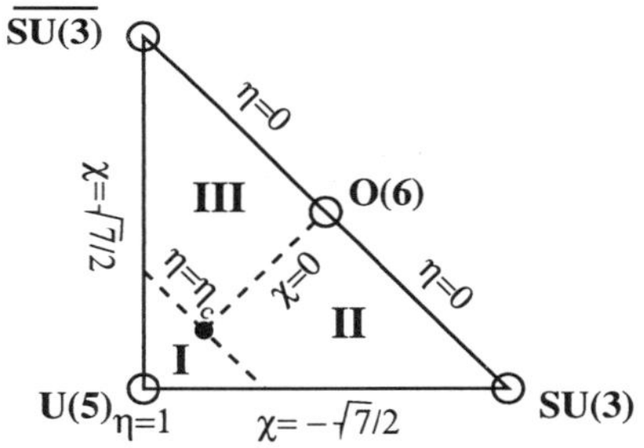

Figure 1. The extended Casten triangle[11] and its different phases. The solid dot in the center represents the second-order transition at the nuclear triple point between spherical nuclei (Phase I) and deformed nuclei with prolate (Phase II) and oblate (Phase III) forms. The dashed lines correspond to first-order phase transitions.

The latter can be derived by the method of Gilmore[21] using the s, d-boson condensate states[22]. The energy functional $E(N, \eta, \chi; \beta, \gamma)$ becomes:

$$E(N, \eta, \chi; \beta, \gamma) = \frac{1}{(1+\beta^2)^2}\Bigg[\left\{N\eta - (1-\eta)(4N + \chi^2 - 8)\right\}\beta^2$$

$$+4(N-1)(1-\eta)\sqrt{\frac{2}{7}}\chi\,\beta^3\cos 3\gamma + \left\{N\eta - (1-\eta)\left(\frac{2N+5}{7}\chi^2 - 4\right)\right\}\beta^4\Bigg] . \quad (5)$$

with the standard deformation parameters β and γ. The energy functional encodes the phase-transitional phenomena within the Landau theory of phase transitions [17,18]. Instead of the thermodynamic potential $\Phi(P, T; \xi)$ that depends on external parameters (pressure P and temperature T) and the order parameter ξ, we have $E(N, \eta, \chi; \beta, \gamma)$ depending on the external parameters N, η and χ, and on the order parameters β and γ. The task is to minimize the functional by varying β and γ for each N, η and χ—the optimal values being denoted β_0 and γ_0. Due to the $\cos(3\gamma)$ dependence Eq. (2) yields either $\gamma_0 = 0$ (for negative χ), or $\pi/3$ (for positive χ). The latter case can be equivalently described by a substitution $\gamma_0 \to 0$ and $\beta_0 \to -\beta_0$ which allows us to omit the parameter γ from further considerations and to distinguish prolate ($\beta_0 > 0$) and oblate ($\beta_0 < 0$) deformations, respectively, while $\beta_0 = 0$ corresponds to the spherical symmetry. Such the energy functional takes the form:

$$E(N,\eta,\chi;\beta,\gamma) = E_0(\eta) + A(N,\eta,\chi)\,\beta^2 + B(N,\eta,\chi)\,\beta^3 + C(N,\eta,\chi)\,\beta^4 + \dots \tag{6}$$

where, clearly, $B(N,\eta,\chi)$ is generally nonzero. Explicitly the first order phase transitions at the separatrix, where both phases have the same energy, obey when neglecting the higher order terms:

$$A(N,\eta,\chi) = \frac{B(N,\eta,\chi)^2}{4C(N,\eta,\chi)} \tag{7}$$

leading to the critical value η_c:

$$\eta_c(N,\chi) = \frac{4+2\chi^2/7}{5+2\chi^2/7} + \mathcal{O}\left(\frac{1}{N}\right) , \tag{8}$$

for the spherical deformed transition. The prolate-oblate transition is obtained for $A(N,\eta,\chi) < 0$ and:

$$B(N,\eta,\chi) = 0 \tag{9}$$

yielding the critical value $\chi_c = 0$. If both conditions are simultaneuosly fulfilled the triple point or isolated second order phase transition is obtained.

In analysing the behavior of these function for phase transitions Landau noted that they occur on one hand between a more symmetric ($\beta_0 = 0$) and less symmetric ($\beta_0 \neq 0$) phases, whereby the latter two phases occur with minima having different signs (in our case either prolate, $\beta_0 > 0$, or oblate minima, $\beta_0 < 0$) who are given by:

$$\beta_0(N,\eta,\chi) = \pm\frac{3|B| + \sqrt{9B^2 - 32AC}}{4C}, \tag{10}$$

with the plus (minus) sign corresponding to a negative (positive) B coeficient and thus in our case a negative (positive) χ value. Note that the parameter symmetry is built into Landaus theory.

General signatures of nuclear shapes are the excitation energy ratio of the first excited 4^+ and 2^+ states, $R_{4/2}$, the quadrupole moment $Q(2_1^+)$ and the $B(E2;2_2^+ \to 2_1^+)$ value using the quadrupole operator (2) as transition operator. They are shown in Figure 2, using an effective charge of 0.15 e.b, for nuclei ranging from ^{180}Hf to ^{200}Hg having boson numbers N from 14 to 4. The location of the prolate-oblate phase transition is clearly observed [23]. The signatures for $^{198,200}Hg$ are surprising. While they are not very well described quantitatively by the one-parameter hamiltonian, they qualitatively reveal unexpected features since they do not resemble a vibrational or shell model structure. Such structures would have $R_{4/2}$ around or below 2. Instead a slight increase of deformation indicates a deviation from the U(5)-O(6) line towards $\overline{\mathrm{SU(3)}}$.

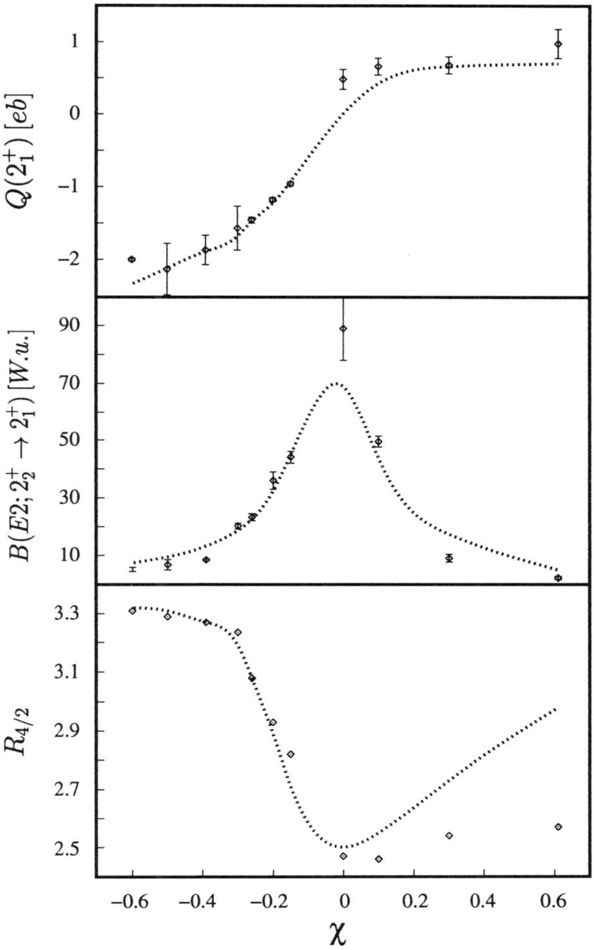

Figure 2. The signatures $R_{4/2}$, $Q(2_1^+)$ and $B(E2; 2_2^+ \rightarrow 2_1^+)$ as a function of the control parameter χ in the Hf-Hg mass region. Shown are the experimental values and the theoretical ones obtained using $\eta = 0$. Negative χ values correspond to prolate nuclei, positive to oblate.

The phase structure we have discussed here has also been observed in other systems. Notably the isotropic-nematic transition as described by the Landau-de Gennes theory shows a large similarity in the uniaxial case [24].

4 Phase transitions in more complex systems

We now turn to possible extensions of the simple model to more complex systems. Here we would like to stay as close as possible to the idea of constructing an hamiltonian that has essentially two competing interactions but has more degrees of freedom. In order to achieve this we will heavily rely on symmetry concepts.

The first extension is to boson systems with the neutron-proton degree of freedom. Here a simple extension can be obtained by considering the hamiltonian:

$$\hat{H} = a(\eta(\hat{n}_{d\nu} + \hat{n}_{d\pi}) - \frac{(1-\eta)}{N_\nu + N_\pi}(\hat{Q}_\nu + \hat{Q}_\pi)_{\chi_\nu,\chi_\pi}.(\hat{Q}_\nu + \hat{Q}_\pi)_{\chi_\nu,\chi_\pi}) \tag{11}$$

where the quadrupole operators are given by equation (2) for the protons and neutrons, respectively. The advantage of this parametrisation is that it also includes the normal extended Casten triangle as a subset of states when $\chi_\nu = \chi_\pi$ in which case we have a F-spin symmetric hamiltonian [25]. In the general case with $\chi_\nu \neq \chi_\pi$ a richer structure is obtained, because the hamiltonian now depends on three control parameters η, χ_ν and χ_π for given boson numbers. It can be represented as a pyramid containing six dynamical symmetries. The new symmetries correspond to Bijker and Dieperinks $SU^*(3)$ symmetry[26] in which proton and neutrons have respectively prolate and oblate deformation producing triaxial shapes [27].

Another extension of the hamiltonian (1) to a more complex system concerns odd-A nuclei. In order to achieve this in a straightforward way we will need to restrict the possible single particle orbits the odd nucleon can occupy and the boson-fermion interaction. This can be done when relying on the U(6/12) supersymmetric scheme [28], because supersymmetry automatically generates a boson-fermion hamiltonian from the boson hamiltonian (1). In U(6/12) the odd fermion can occupy $j = 1/2, 3/2$ and $5/2$ orbitals and all dynamical symmetries of the IBM-1 are included. Using the consistent-Q formalism for odd-A nuclei introduced in ref [29] one obtains the hamiltonian:

$$\hat{H} = a(\eta\hat{C}_1[U^{B+F}(5)] - \frac{(1-\eta)}{N}(\hat{Q}^{B+F}_\chi.\hat{Q}^{B+F}_\chi)) \tag{12}$$

with $\hat{C}^{B+F}_1[U^{B+F}(5)]$ the first order Casimir operator of $U^{B+F}(5)$ and $\hat{Q}^{B+F}_\chi$ the quadrupole operator of the form given by eq. (2) but using generators of $U^{B+F}(6)$. It is worth noting that this approach is still supersymmetric even when there are no dynamical symmetries present [30].

The use of the U(6/12) supersymmetric scheme allows thus to extend the Casten triangle to odd-A nuclei. While this is sufficient for a theoretical study the application to real nuclei is limited to those nuclei where $j = 1/2, 3/2$ and $5/2$ are important, e.g. the A=70-80, the Rh-Ag and the Pt-Os-Hg mass regions. Here we discuss the application to the prolate-oblate phase transition. Figure 5 compares the absolute energies of the 0^+ states of a ten bosons system as a function of χ when $\eta = 0$ to the ones of $\frac{1}{2}$ states in the ten bosons plus one fermion system. In the even-even system the phase transition is associated with the kink of the lowest energy at O(6) [11] and the sequence of energies is symmetric around O(6), although

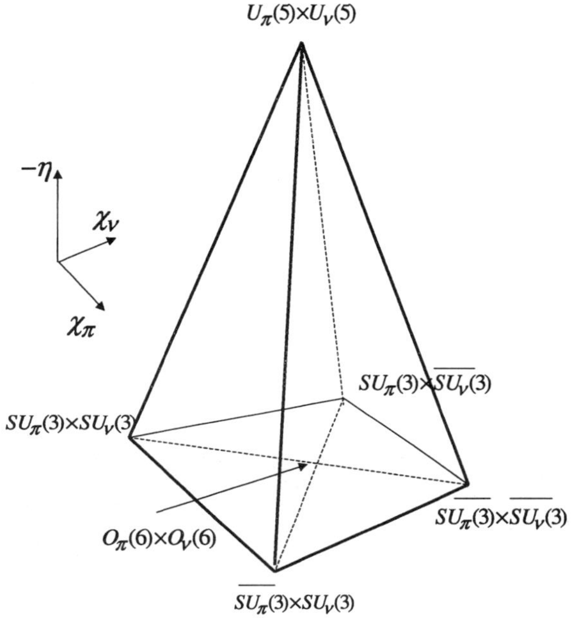

Figure 3. The dynamical symmetries of the hamiltonian (11).

the wavefunctions are different (see ref [11]). A quite similar behavior is found for the odd-A case.

5 Conclusion

We have reviewed a very simple IBM-1 hamiltonian that contains competing forces driving to spherical or deformed shapes. Like the Ising model [31], it turns out to generate a very rich structure with its two essential parameters. We have explained, using Landau theory of second-order phase transitions, why the IBM exhibits an isolated second-order phase transition and three continuous lines of first-order phase transitions. We stress that, although we have used the IBM, all these results are quite general for any collective model where the potential can be expanded as in Eq.(6). Moreover, it shows how general theoretical frameworks like Landau theory are universal [32].

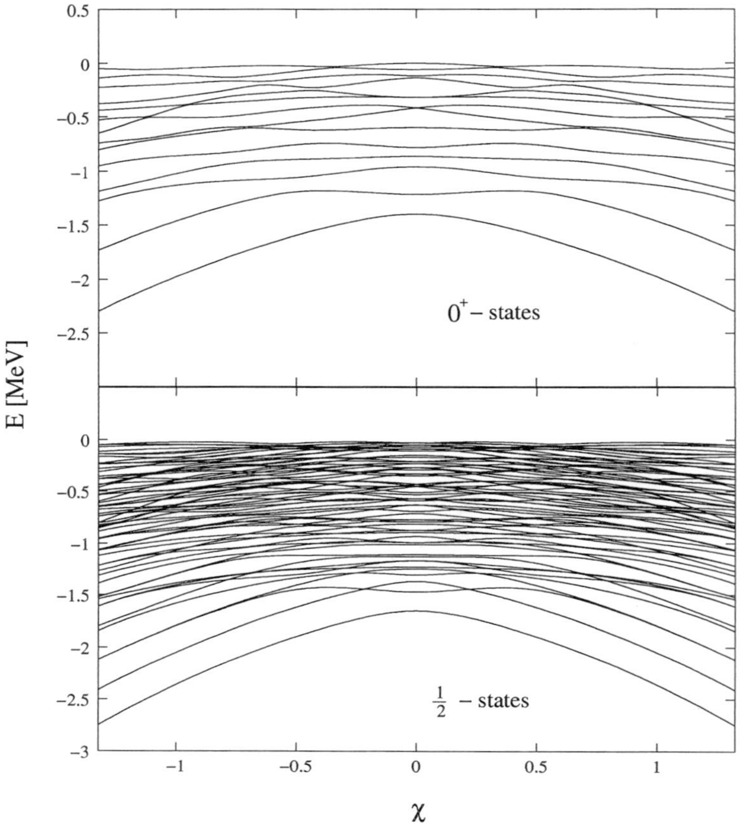

Figure 4. The absolute energies of the 0^+ states of a even-even nucleus (top) and of the corresponding $\frac{1}{2}$ states in the odd-A nucleus (bottom) as a function of χ when $\eta = 0$ and $N = 10$

This study allows the definition of a nuclear triple point. Its application to the Hf-Hg nuclei provides a new perspective on the evolution of nuclear structure in this mass region. We proposed two extensions of the simple hamiltonian: the introduction of the neutron-proton degree of freedom and the extension to odd-A nuclei. Both rely on the use of symmetries, F-spin symmetry and supersymmetry, as a way to keep the essential physics in these more complicated systems.

I would like to acknowledge P. Cejnar, R.F. Casten, S. Heinze, A. Linnemann, P. von Brentano, and V. Werner with whom part of the presented work was done. Last but not least, I would like to thank F. Iachello for his inspiring research and permanent interest in our work. This work was supported by the BMBF under

Project number 06OK958.

References

1. F. Iachello and A. Arima, *The Interacting Boson Model* (Cambridge University Press, Cambridge, England, 1987).
2. F. Iachello and P. Van Isacker, *The Interacting Boson-Fermion Model* (Cambridge University Press, Cambridge, England, 1991).
3. F. Iachello and R.D Levine, *Algebraic Theory of Molecules* (Oxford University Press, England, 1995).
4. N. Whelan, and Y. Alhassid, Nucl. Phys. **A556**, 42 (1993)
5. E. Lopez-Moreno and O. Castanos, Phys. Rev. C**54**, 2374 (1996).
6. P. Cejnar and J. Jolie, Phys. Rev. E **61**, 6237 (2000).
7. R.F. Casten in *Interacting Bose-Fermi Systems in Nuclei*, ed. F. Iachello Plenum, p 1 (1981).
8. A.M. Shirokov, N.A. Smirnova, Yu.F. Smirnov, Phys. Lett. **B434**, 237 (1998).
9. P. Cejnar and J. Jolie. Phys. Lett **B420**, 241 (1998).
10. J. Jolie and P. Cejnar, J. Phys. G **25**, 843 (1999).
11. J. Jolie, R.F. Casten, P. von Brentano, and V. Werner, Phys. Rev. Lett. **87**, 162501 (2001).
12. F. Iachello, Phys. Rev. Lett. **85**, 3580 (2000).
13. F. Iachello, Phys. Rev. Lett. **87**, 052502 (2001).
14. A. Bohr and B.R. Mottelson, Nuclear Structure (Benjamin, New York, 1975).
15. R.F. Casten and N.V. Zamfir, Phys. Rev. Lett. **85**, 3584 (2000).
16. R.F. Casten and N.V. Zamfir, Phys. Rev. Lett. (2001), **87**, 052503 (2001).
17. L. Landau, Phys. Z. Sowjet. **11**, 26 (1937); **11**, 545 (1937); reprinted in *Collected Papers of L.D. Landau* (Pergamon, Oxford, 1965), p. 193.
18. L.D. Landau and E.M. Lifshitz, *Statistical Physics, Part 1*, volume V of the *Course of Theoretical Physics* (Butterworth-Heinemann, Oxford, 2001).
19. J. Jolie, P. Cejnar, R.F. Casten, S. Heinze, A. Linnemann, V. Werner Phys. Rev. Lett. **89**, 182502 (2002).
20. P. Cejnar, S. Heinze, J. Jolie, subm. to Phys. Rev. C (2003).
21. R. Gilmore, J. Math. Phys. **20**, 891 (1979).
22. J.N. Ginocchio and M.W. Kirson, Nucl. Phys. **A350**, 31 (1980).
23. J. Jolie, A. Linnemann, subm to Phys. Rev. Lett. (2003).
24. E.F. Gramsbergen, L. Longa, W.H. de Jeu, Phys. Rep. **135**, 195 (1986).
25. P. Van Isacker, K.L.G. Heyde, J. Jolie, A. Sevrin, Ann. of Phys. (N.Y.) **171**, 253 (1986).
26. A.E.L. Dieperink and R. Bijker, Phys. Lett. **116B**, 77 (1982).
27. A. Sevrin, K.L.G. Heyde, J. Jolie, Phys. Rev. C **36**, 2621 (1987).
28. P. Van Isacker, A. Frank, H.Z. Sun, Ann. of Phys. (N.Y.) **157**, 183 (1984).
29. D.D. Warner, P. Van Isacker, J. Jolie, A.M. Bruce, Phys. Rev. Lett. **54**, 1365 (1985).
30. A. Frank, P. Van Isacker, D.D. Warner, Phys. Lett. **197B**, 474 (1987).
31. E. Ising, Z. Phys. **31**, 253 (1925).
32. D.D. Warner, Nature (London) **420**, 614 (2002).

THE EXCITED 0^+ STATES IN ^{162}YB AND THE CRITICAL POINT PHASE/SHAPE TRANSITION

E.A. MCCUTCHAN, N. V. ZAMFIR AND R.F. CASTEN

WNSL, Yale University, New Haven, Connecticut 06520-8124, USA

E-mail: elizabeth.ricard-mccutchan@yale.edu

Excited, non-yrast states in ^{162}Yb were populated through β decay and studied through off-beam γ-ray spectroscopy. New coincidence data provided evidence for the elimination of a previously reported low-lying 0^+ state. The revised level scheme of ^{162}Yb is compared to the predictions of the X(5) critical point model.

1 Introduction

The introduction of new theoretical approaches[1,2] to describe the spherical-deformed shape transitional region has spurred interest in the search for empirical examples in nuclei. Iachello has developed[1] the critical point model, X(5), which describes the behavior at the critical point of the transition between spherical and axially deformed nuclei with analytic solutions that are essentially parameter free (except for scale). Recent experiments have found several empirical realizations of the X(5) model predictions, mainly in the $N = 90$ isotonic chain. ^{152}Sm and ^{150}Nd have been studied [3,4,5] extensively and are the closest manifestations of the X(5) critical point model, while studies[6,7] on ^{154}Gd and ^{156}Dy have shown that these nuclei are also good candidates.

We will discuss the motivation for moving away from N=90 in search of additional candidates for X(5). The results of a recent experiment on ^{162}Yb will be presented with a focus on the low lying excited 0^+ states and a comparison will be made to the predictions of the X(5) critical point model.

2 The Search for X(5) Nuclei: The P Factor

The X(5) model predicts an yrast band structure intermediate between the vibrator and symmetric rotor with $R_{4/2}\equiv E(4_1^+)/E(2_1^+) = 2.91$. Additional characteristic signatures include the energy ratio $E(0_2^+)/E(2_1^+)=5.67$ and a near degeneracy of the 0_2^+ and 6_1^+ levels. In the search for possible candidates for the X(5) symmetry, the simplest observable to identify is the $R_{4/2}$ ratio. The empirical $R_{4/2}$ values for nuclei in the mass 150 region are shown in Fig. 1 with the stable nuclei highlighted in dark grey for reference. Those nuclei with $R_{4/2}$ close to the characteristic X(5) prediction are outlined in bold.

Another useful quantity that can be indicative of the onset of deformation is the P factor[8], given by

$$P = \frac{N_p N_n}{N_p + N_n} \tag{1}$$

where N_p and N_n are the numbers of valence protons and neutrons, respectively[9]. Substantial collectivity occurs when the total strength of valence p-n interactions

78			2.30		2.26	2.44	2.51	2.70	2.68	2.56	2.53	2.49	2.48	2.48	Pt
76				2.62	2.66	2.74	2.93	3.02	3.09	3.15	3.20	3.17	3.08	2.93	Os
74			2.68	2.82	2.95	3.07	3.15	3.22	3.24	3.26	3.29	3.27	3.24	3.09	W
72	2.31	2.56	2.79	2.97	3.11	3.19	3.25	3.27	3.28	3.29	3.31	3.30	3.26		Hf
70	2.33	2.63	2.93	3.12	3.23	3.27	3.29	3.31	3.31	3.31					Yb
68	2.32	2.74	3.10	3.23	3.28	3.29	3.31	3.31	3.31						Er
66	2.23	2.93	3.21	3.27	3.29	3.30	3.31								Dy
64	2.19	3.02	3.24	3.29	3.30	3.30									Gd
62	2.32	3.00	3.25	3.29	3.30	3.30									Sm
60	2.49	2.93	3.26	3.29	3.32										Nd
58	2.59	2.86	3.15												Ce
56	2.66	2.84	2.99												Ba
Z/N	88	90	92	94	96	98	100	102	104	106	108	110	112	114	

P ~ 5

$$P=\frac{N_N N_p}{N_N+N_p}$$

Figure 1. Empirical $R_{4/2}$≡E(4_1^+)/E(2_1^+) ratios for nuclei in the $54 < Z < 80$, $86 < N < 116$ region. Highlighted in dark grey are the stable nuclei. Those nuclei with $R_{4/2}$ values similar to the X(5) prediction are outlined in bold. The curve intersects those nuclei with P near 5, pointing to possible candidates for the critical point symmetry X(5).

(each on the order of 200 keV) become comparable to that from the pairing interaction (on the order of 1 MeV), or roughly 5 p-n interactions for each pairing interaction. This corresponds to P∼5. Of course, P values near 5 can also be encountered in other types of transition regions (e.g. γ-soft to axial rotor) which do not involve phase transitional behavior. Hence P values serve only as a guide to possible candidates for X(5). Included in Fig. 1 is a curve which traces out the contour P∼5. In fact, the P∼5 curve intersects all of the $N = 90$, X(5) candidates along with other nuclei with $R_{4/2}$ ratios close to that predicted by the X(5) model. The combination of $R_{4/2}$ values and the P factor provides the motivation for the current study of ^{162}Yb: by continuing along the P∼5 curve away from the $N = 90$ nuclei, the next nucleus with a $R_{4/2}$ value similar to characteristic X(5) prediction is ^{162}Yb.

While the $R_{4/2}$ ratio is in excellent agreement with the X(5) predictions, the existing data[10] showed that the other two characteristic signatures involving the first excited 0^+ state deviated dramatically, with E(0_2^+)/E(2_1^+)=3.63 and E(6_1^+)/E(0_2^+)=1.53. These differences can be attributed to the placement of a very low-lying first excited 0^+ state in the reported level scheme[10]. Further up in energy, however, there is a second excited 0^+ state whose energy agrees well with the predictions of X(5), with E(0_3^+)/E(2_1^+)=6.03 and E(6_1^+)/E(0_3^+)=0.96. The characteristic ratios for the different possible level schemes along with a comparison with the X(5) predictions are summarized in Fig. 2. The placement of the low-lying 0^+ state, at 605 keV, is based on an unpublished β decay study[11] where the level scheme was constructed using singles γ-ray data. In the present work, a similar

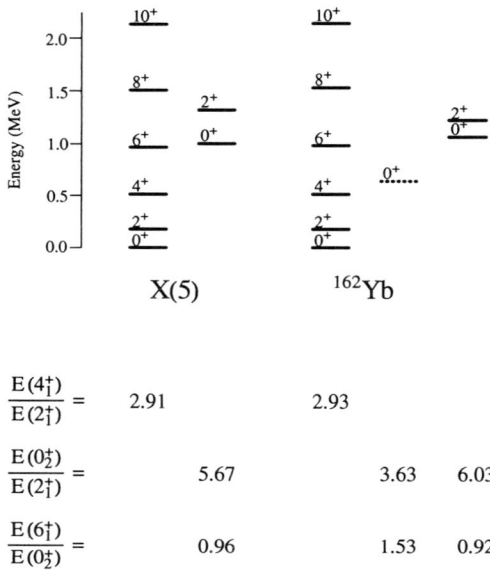

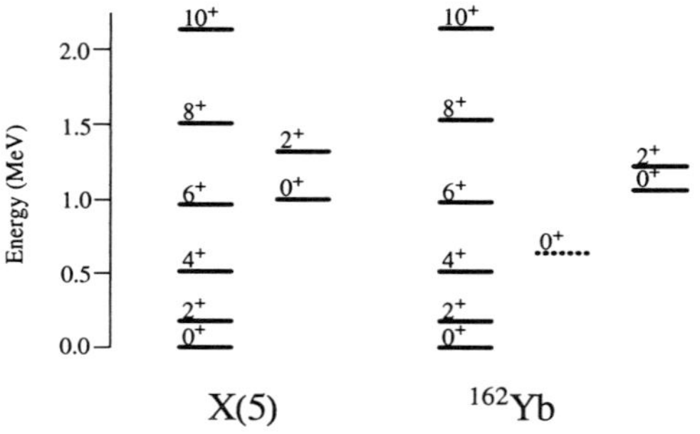

	X(5)		^{162}Yb		
$\frac{E(4_1^+)}{E(2_1^+)}$ =	2.91		2.93		
$\frac{E(0_2^+)}{E(2_1^+)}$ =		5.67		3.63	6.03
$\frac{E(6_1^+)}{E(0_2^+)}$ =		0.96		1.53	0.92

Figure 2. Low lying level scheme of ^{162}Yb and for the critical point symmetry $X(5)$ along with values for the characteristic signatures of $X(5)$. The two sets of empirical values involving the 0_2^+ state refer to the two candidates for the lowest excited 0^+ level.

experiment was repeated at WNSL in order to assess the identification of excited 0^+ states using γ-ray coincidence data.

3 Off-beam γ-ray spectroscopy of ^{162}Yb

The low-lying non-yrast states of ^{162}Yb were populated in β^+/ϵ decay and studied through γ-ray coincidence spectroscopy at the Yale Moving Tape Collector[12]. The parent ^{162}Lu nuclei were produced through the ^{147}Sm(^{19}F, 4n) reaction using a 95 MeV ^{19}F beam provided by the Yale ESTU tandem accelerator. The recoil product nuclei were deposited onto a 16-mm-wide aluminized Kapton tape. A 3-mm diameter gold plug, located between the target and tape, stopped the primary beam before reaching the tape. In contrast, fusion evaporation products, which are primarily emitted at larger angles, bypassed the plug and reached the tape with ~80% geometrical acceptance. Figure 3 (right) illustrates the angular distribution of products along with the range of acceptance. To enhance the yield of ^{162}Yb the tape was advanced at intervals corresponding to the half life of the parent($T_{1/2}$ = 1.2 min., τ_{tape}=2.0 min.), transporting the collected activity to a low background counting area which consisted of 3 clovers and 1 LEPS detector. A schematic of the experimental setup is given in Fig. 3 (left).

The coincidence data in the present experiment allowed for considerable modifi-

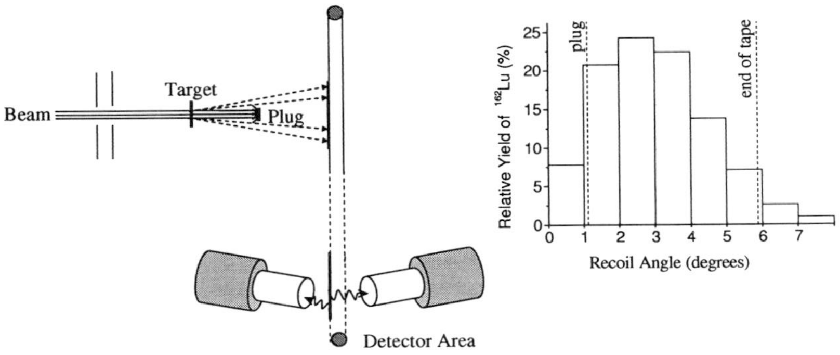

Figure 3. The Yale Moving Tape Collector. Left: schematic diagram of the target and detector areas. Right: yield of reaction products as a function of angle. The range of acceptance of recoil products onto the tape is given by the dashed lines.

cations to the existing level scheme of ^{162}Yb. We focus now on the evidence for the first excited 0^+ state at 605 keV since its placement is a key factor in the structural interpretation of this nucleus. The first excited 0^+ state at 605 keV is based on two feeding transitions, 1031 keV and 525 keV, as illustrated in Fig. 4 (left). The authors of ref. 11 observed no γ rays depopulating the level, thereby assuming that it decayed entirely by E0 transitions.

The present data show that there is no evidence for the level at 605 keV. The 2_1^+ to 0_1^+ transition of 166 keV was found to be strongly coincident with the 1031 keV transition as shown in Fig. 5 (left). A new level at 1198 keV is identified, based on the strong coincidence of the 1031 keV transition with only the 166 keV, 2_1^+ to 0_1^+ transition (Fig. 5 (right)). The existence of a new level at 1198 keV is further supported by the observation of a 400 keV transition, coincident with both the 2_2^+ to 2_1^+ (631 keV) and 2_2^+ to 0_1^+ (798 keV) transitions.

The 525 keV transition was observed in singles data but is not present in coincidence with any lines in the ^{162}Yb spectrum. Coincidences between the transitions populating the level at 1130 keV and two of the transitions depopulating the level, at 1130 and 963 keV, were observed but no evidence was found for the placement of the 525 keV transition also depopulating this level. Without the two feeding transitions, there is no support for the existence of the 0^+ level at 605 keV. The revised level scheme based on the above arguments is given in Fig. 4 (right).

4 Conclusions

With the identification of a new first exited 0^+ state, the low-lying energy levels of ^{162}Yb are now in good agreement with the predictions of the X(5) critical point model. For both of the energy ratios, $E(4_1^+)/E(2_1^+)$ and $E(0_2^+)/E(2_1^+)$, ^{162}Yb with N=92 is a candidate in the Yb isotopic chain for an X(5) nucleus as shown in Fig. 6. Also in excellent agreement are the energy levels of the yrast sequence (Fig.6 upper right). The one set of observables which exhibit large differences relative

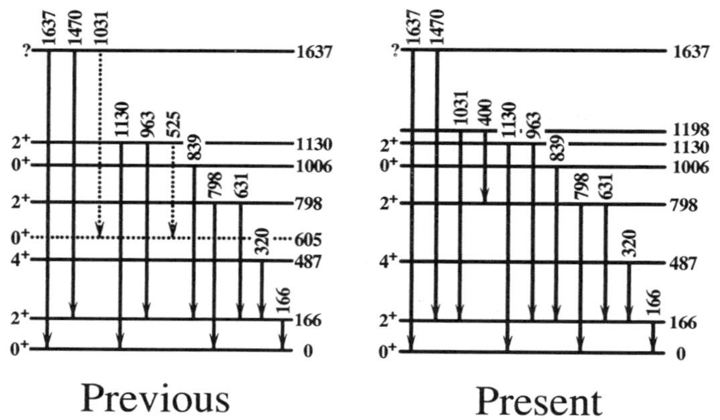

Figure 4. Low-lying levels of ^{162}Yb and their depopulating γ-ray transitions. Left: previously reported level scheme[11]. The 0^+ state at 605 keV and the transitions feeding it were placed in ref. 11 but are showed here as dashed since the present experiment produces the alternate level scheme as shown on the right. Right: modified level scheme constructed from the coincidence data obtained in the present experiment(see text). The energies of the excited states and transitions are given in keV.

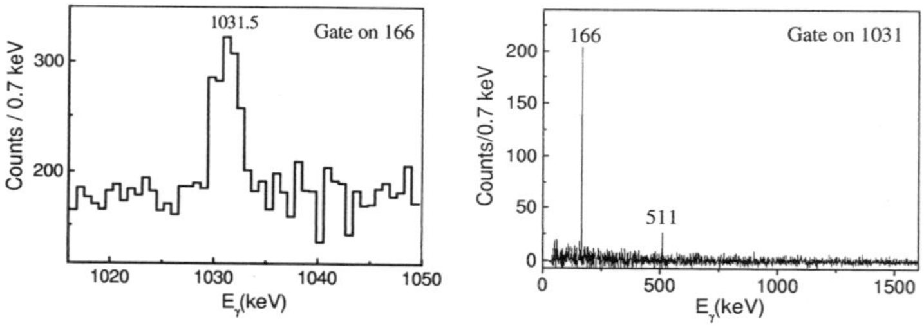

Figure 5. Gated coincidence spectra providing evidence for the placement of the 1031 keV transition as depopulating a new level at 1198 keV. Left: spectrum gated on the 166 keV, 2_1^+ to 0_1^+, transition. Right: spectrum gated on the 1031 keV transition.

to the X(5) predictions are the B(E2) values of the yrast sequence (Fig. 6 bottom right). Above J=4, the experimental values deviate from the X(5) predictions, even dropping below that of the rotor. An experiment is currently in progress at Yale to measure the lifetimes of the excited yrast states in order to obtain new yrast B(E2) values.

Having established ^{162}Yb as a possible candidate for the X(5) symmetry, we are continuing the search for other critical point nuclei. From Fig. 1, the next nucleus along the P$\sim$ 5 curve with a good $R_{4/2}$ ratio ($\sim$2.9-3.0) is ^{166}Hf. Looking to the literature[13] we find a similar problem to that of ^{162}Yb, namely a low-lying first excited 0^+ state which does not agree well with the X(5) predictions, but, further up in energy an excited 0^+ state that does. An experiment was performed again at

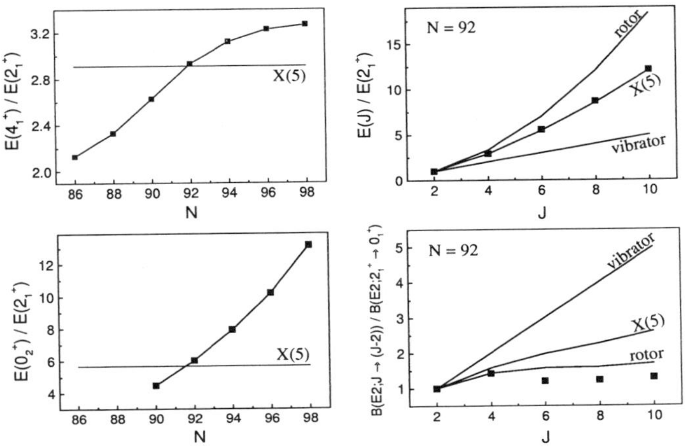

Figure 6. Evolution of basic observables in both the Yb isotopic chain and specifically for N=92.

the Yale Moving Tape Collector to verify the existence of the lowest 0^+ state and analysis is currently underway.

Acknowledgments

We are grateful to F. Iachello for discussions of critical point symmetries and to the other members of the WNSL Nuclear Structure Group for collaboration on these experiments. This work was supported by U.S. DOE Grants No. DE-FG02-91ER-40609 and DE-FG02-88ER40417.

References

1. F. Iachello, *Phys. Rev. Lett.* **87**, 052502 (2001).
2. F. Iachello, *Phys. Rev. Lett.* **85**, 3580 (2000).
3. R.F. Casten and N.V. Zamfir, *Phys. Rev. Lett.* **87**, 052503 (2001).
4. N.V. Zamfir et al., *Phys. Rev.* **C 60**, 054312 (1999).
5. R. Krucken et al., *Phys. Rev. Lett.* **88**, 232501 (2002).
6. A. Dewald et al., *Int. Conf. on Nuclear Structure with Large γ-arrays*, Legnaro, Italy, 2002.
7. M.A. Caprio et al., *Phys. Rev.* **C 66**, 054310, 2002.
8. R.F. Casten, D.S. Brenner, and P.E. Haustein, *Phys. Rev. Lett.* **58**, 658 (1987).
9. R.F. Casten, *Phys. Rev. Lett.* **54**, 1991 (1985).
10. R.G. Helmer and C.W. Reich, *Nucl. Data Sheets*, **87** 317 (1999).
11. H. Behrens, Thesis, Tech. Univ. Munich (1980).
12. N.V. Zamfir and R.F. Casten, *J. Res. Natl. Stand. Technol.* **105**, 147 (2000).
13. E.N. Shurshikov and N.V. Timofeeva, *Nucl. Data. Sheets*, **67** 138 (1992).

TEST OF THE CRITICAL POINT SYMMETRY X(5) IN N=90 NUCLEI AND A≈180 OS ISOTOPES

A. DEWALD, O. MÖLLER, D. TONEV, A. FITZLER, B. SAHA, K. JESSEN, S. HEINZE, A. LINNEMANN, J. JOLIE, K.O. ZELL, P. VON BRENTANO
Institut für Kernphysik der Universität zu Köln, Köln, Germany

P. PETKOV
Institute for Nuclear Research and Nuclear Energy Sofia, Sofia, Bulgaria

R.F. CASTEN, M. CAPRIO, V. ZAMFIR
W.N.S.L., Yale University, New Haven, Connecticut 06520, USA

R. KRÜCKEN
Physik-Department E12, TU München, Garching, Germany

D. BAZZACCO, S. LUNARDI, C. ROSSI-ALVAREZ, F. BRANDOLINI, C. UR
INFN Sezione Padova, Padova, Italy

G. DE ANGELIS, D.R. NAPOLI, E. FARNEA, N. MARGINEAN, T. MARTINEZ, M. AXIOTIS
INFN, Laboratori Nazionali di Legnaro, Legnaro, Italy

Reliable and precise lifetimes of excited states in ^{154}Gd and ^{156}Dy were measured using the recoil distance Doppler shift (RDDS) technique. Excited states of ^{154}Gd were populated via Coulomb excitation with a ^{32}S beam at 110 MeV delivered by the FN tandem accelerator at the University of Cologne. For ^{156}Dy a coincidence plunger experiment was performed at the Laboratori Nazionali di Legnaro with the GASP spectrometer and the Cologne coincidence plunger apparatus using the reaction ^{124}Sn(^{36}S,4n)^{156}Dy at a beam energy of 155 MeV. The measured transition probabilities in ^{156}Dy and ^{154}Gd as well as the corresponding energy spectra are compared with the predictions of the recently proposed X(5) model and in the case of ^{156}Dy also with an IBA fit. In addition, criteria for finding new X(5) regions are given and the Os nuclei around A=180 were found to be very good X(5) candidates.

1 Introduction

The investigation of nuclear phase transition phenomena is one of the new and very challenging topics in nuclear structure physics. A lot of theoretical and experimental work has been devoted to this topic, especially to the question how phase transitions manifest in nuclei. [1,2,3,4,5,6,7,8,9,10,11]

Important contributions were made by F. Iachello who introduced the new dynamical symmetries E(5) and X(5) at the critical points of phase transitions between U(5) and O(6) and between U(5) and SU(3), respectively. The manifestation of these new symmetries allows the definition of specific experimental observables, needed to relate the theoretical picture with existing nuclei, and thus makes a test of the theoretical concepts possible. For the first time the new X(5) symmetry had experimentally been established in ^{152}Sm[7] and ^{150}Nd[13]. The neighboring N=90 nuclei ^{154}Gd and ^{156}Dy were regarded as promising X(5) candidates, too. The aim of the present work was to supply further experimental data, especially electromagnetic transition probabilities, to allow stringent tests of the X(5) predictions.

2 Experimental Details

In this paper we present the results of two lifetime experiments. One was performed at the FN tandem accelerator of the University of Cologne using the recoil distance Doppler shift (RDDS) technique after Coulomb excitation of states in ^{154}Gd with a ^{32}S beam of 110 MeV. In order to fix the kinematics of the reaction, backscattered beam particles were detected by 6 Si detectors mounted down-stream close to the target foil. The experimental set-up was very similar to that described in [14,15]. Particle gated singles spectra were analyzed and lifetimes of six excited states were determined using the differential decay curve method (DDCM)[16]. The obtained results are in good agreement with previous ones [17]. The lifetime of the 4_2^+ state, which is important for the test of the X(5) predictions, was measured for the first time.
For ^{156}Dy an RDDS measurement was performed at the Laboratori Nazionali di Legnaro with the GASP spectrometer and the Cologne coincidence plunger. In addition, a Doppler shift attenuation measurement (DSAM) was carried out. The reaction ^{124}Sn(^{36}S,4n)^{156}Dy at a beam energy of 145 MeV was used for both experiments. The lifetimes in ^{156}Dy were obtained, in the RDDS case, using gated spectra with gates from above the levels of interest in order to eliminate all problems related to discrete and unobserved feeders. From the RDDS experiment, 13 lifetimes were determined with experimental errors between 2-5%. From the DSAM analysis 5 lifetimes were obtained for the 22^+, 24^+,..., 30^+ states of the s-band and 3 for the 20^+, 22^+, and 24^+ states of the ground state band. Because of a lack of statistics only gates from below the level of interest were analyzed. For further details of the procedure employed and the used stopping power we refer to [18].

3 Critical Point Symmetry X(5)

In the case of the X(5) symmetry the nuclear potential is approximated by a square-well and a harmonic oscillator potential with respect to the β and γ degrees of freedom, respectively. In addition, the β and γ degrees of freedom are considered to be decoupled. The corresponding Hamiltonian can be solved analytically, and specific predictions can be made for both the energy spectrum and the transition probabilities [6]. The X(5) energies of the lowest band (S1) are located between the vibrator and symmetric rotor values. As it is shown in [19] the gsb energies normalized to the 2^+ band member of the N=90 isotones ^{150}Nd, ^{152}Sm, ^{154}Gd, and ^{156}Dy follow the X(5) predictions very well. Crucial observables are also the energy ratios $E(0_2^+)/E(2_1^+)$, $E(4_1^+)/E(2_1^+)$, and $R(4/2)_{S2}=(E(4_2^+)-E(0_2^+))/(E(2_2^+)-E(0_2^+))$. As shown in table 1, for the considered N=90 isotones these quantities also agree very well with the X(5) values. Therefore it is of special interest to check for ^{154}Gd and ^{156}Dy, to which extent the electromagnetic transition probabilities determined in this work can be reproduced by the X(5) predictions.

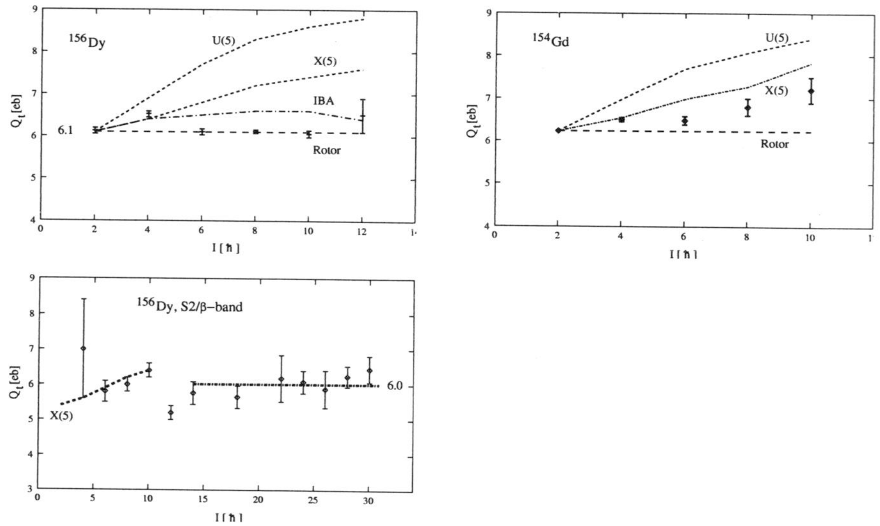

Figure 1. Upper panels: Q_t values of the gsb of ^{156}Dy (left) and ^{154}Gd (right) together with the theoretical values of the X(5) symmetry, the symmetric rotor and the IBA U(5)limit. In the case of ^{156}Dy, the Q_t values of the IBA fit normalized to the experimental $Q_t(2^+ \rightarrow 0^+)$ value are shown, too. Lower panel: Q_t values of the β band (S2 band) and those of an s-band above spin $12\hbar$. The dashed line depicts the X(5) values of the S2 band.

Table 1. Comparison of energy ratios of several N=90 and Os nuclei with the X(5) values.

energy ratios	^{150}Nd	^{152}Sm	^{154}Gd	^{156}Dy	X(5)	^{176}Os	^{178}Os	^{180}Os
$E(4_1^+)/E(2_1^+)$	2.93	3.01	3.03	2.93	2.92	2.93	3.02	3.09
$R(4/2)_{S2}$	2.63	2.69	2.71	2.67	2.80	3.01	3.09	–
$E(0_2^+)/E(2_1^+)$	5.19	5.62	5.53	4.88	5.67	4.45	4.94	–

Fig. 1 shows the Q_t values of the gsb of ^{154}Gd together with the theoretical values of the X(5) symmetry, the symmetric rotor, and the IBA U(5) limit [20] normalized to the experimental $Q_t(2^+ \rightarrow 0^+)$ value. The $Q_t(4^+ \rightarrow 2^+)$ value agrees with the X(5) one. For the states of higher spins the experimental values are just between the X(5) and the rotor values.

Of special importance are also the transition probabilities of the first excited band (S2) and those of the inter-band transitions between S1 and S2. In fig. 2 these quantities as well as the energy spectra of the X(5) S1 and S2 bands are compared with the corresponding experimental values of ^{154}Gd. The overall agreement is found to be very good for both the energies and the transition probabilities. Note that no fit parameter is used. Only two normalization factors were used, one for the energies and one for the transition probabilities, in order to normalize the energy

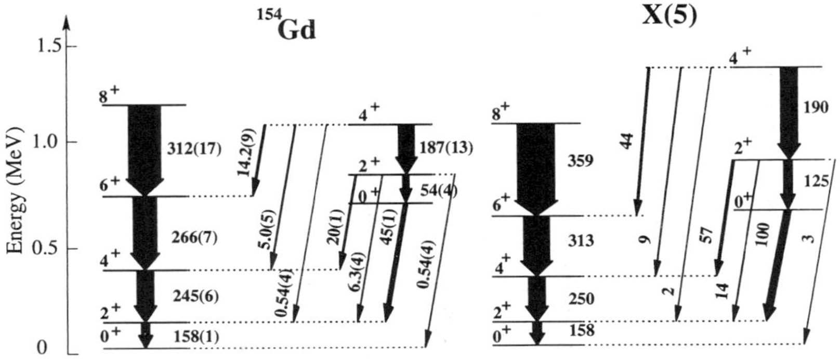

Figure 2. Comparison of experimental data of ^{154}Gd with the X(5) predictions. The E(2_1^+) value (in keV) and the B(E2;$2_1^+ \rightarrow 0_1^+$) value (in W.u.) are normalized to the experimental data. The B(E2) values are given next to the arrows.

spectrum to the 2_1^+ energy and the transition probabilities to the B(E2;$2^+ \rightarrow 0^+$) value. The overall agreement of the experimental values of ^{154}Gd with the X(5) predictions is as good as for ^{150}Nd which was found to be the best example of a X(5) nucleus so far.

Regarding ^{156}Dy the situation looks a bit different. As can be seen in fig. 1, the transition probabilities in the gsb follow those of a symmetric rotor with high precision except the $Q_t(4^+ \rightarrow 2^+)$ value which, as in the case of ^{154}Gd, agrees well with the theoretical X(5) value. The experimental values of the β band, i.e. the S2 band, using the X(5) labeling, also agree very well with the X(5) predictions up the first band-crossing at spin 12 $\hbar$.

In fig. 3, the experimental energies and transition probabilities of ^{156}Dy are compared with the X(5) predictions and with the results of an IBA [20] fit ($\chi = -0.8\sqrt{7/4}$ and η=0.8) using the Hamiltonian

$$H= C[\eta n_d - (1-\eta)/N \cdot Q(\chi)Q(\chi)]$$

where $n_d = d^\dagger \cdot \tilde{d}$ and $Q(\chi) = (s^\dagger\tilde{d} + d^\dagger s)^{(2)} + \chi \cdot (d^\dagger\tilde{d})^{(2)}$.

For the determination of the transition probabilities, intensities from [19] were used. It is obvious that the energy spectrum of the S2 band of X(5) is expanded compared to the one of ^{156}Dy whereas the intra-band transition probabilities agree quite nicely with the X(5) values. The order of magnitude of the inter-band transition probabilities is also in agreement, but one specific difference is observed. In the X(5) symmetry the inter-band transition strengths for the $I_{S2} \rightarrow (I+2)_{S1}$ transitions are larger than for the $I_{S2} \rightarrow I_{S1}$ ones and the $I_{S2} \rightarrow (I-2)_{S1}$ transition strengths are one order of magnitude smaller than the latter ones. This specific behavior of the inter-band transitions can not be observed in ^{156}Dy. Here the transition probabilities

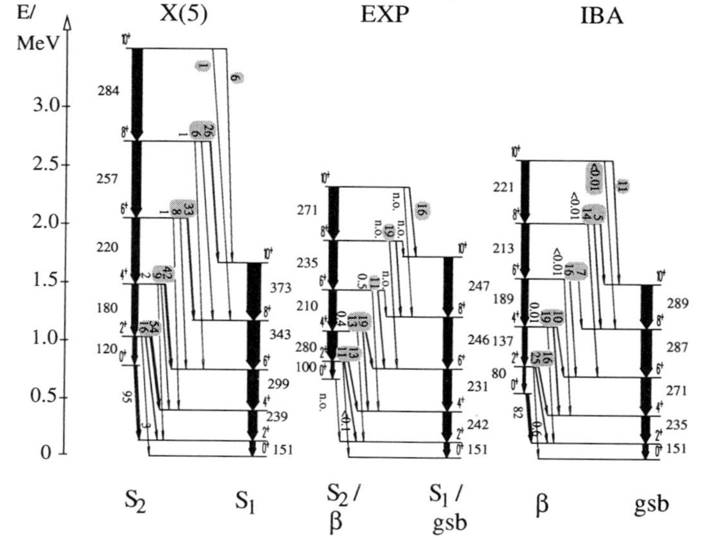

Figure 3. Comparison of the experimental data of ^{156}Dy with the X(5) predictions and the results of an IBA fit described in the text. The calculated values were normalized as described in the caption of fig. 2.

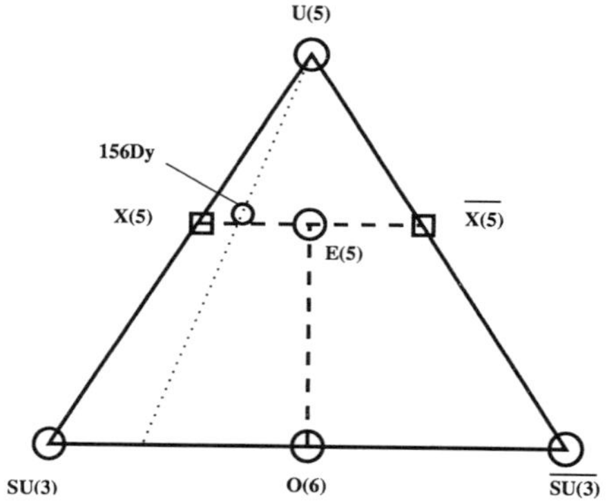

Figure 4. Extended Casten triangle (see text). The dashed lines represent the lines of first-order phase transitions. The dotted line depicts the $\chi = -0.8\sqrt{(7/4)}$ line of the IBA parameter space.

of the $I_{S2} \rightarrow I_{S1}$ transitions are strongest whereas the $I_{S2} \rightarrow (I+2)_{S1}$ ones are a bit weaker or have not been observed. In summary, many features of the X(5) symmetry were observed in ^{156}Dy but some deviations, too.
The comparison with the IBA fit gave an overall good description for the energies

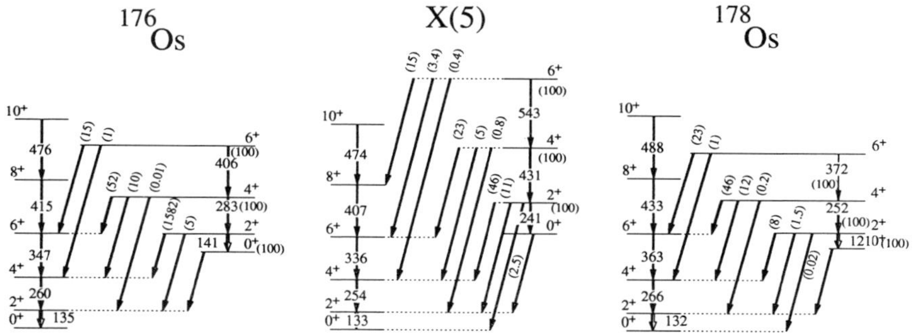

Figure 5. Comparison of the experimental data of ^{176}Os and ^{178}Os with the X(5) predictions. The calculated energies are normalized to the mean value of the energies of the 2_1^+ states of ^{176}Os and ^{178}Os. Relative B(E2) values for transitions depopulating a common level of the β band (S2 band) are given in parenthesis. The B(E2) values of the $I_2 \rightarrow (I\text{-}2)_2$ transitions are set to 100.

and the transition strengths. Also the relative strengths of the $I_{s2} \rightarrow (I+2)_{s1}$ to $I_{s2} \rightarrow I_{s1}$ transitions are reproduced.

J. Jolie et al. related phase transitions and critical points in finite quantal systems to the Landau theory of continuous phase transitions [12]. This work led to an extension of the so called Casten triangle and explicitly includes oblate deformed nuclei. This extended Casten triangle can be considered as a phase diagram for nuclei with the IBM dynamical symmetries U(5), SU(3) and $\overline{\text{SU(3)}}$ located at the corners of the triangle. Three different phases (nuclear deformations) are contained in this triangle which are separated by the border-lines, representing first order phase transitions between spherical and deformed phases and between oblate and prolate deformations. The dynamical symmetry O(6) is located between SU(3) and $\overline{\text{SU(3)}}$ and is considered to indicate a phase transition also. The new dynamical symmetry E(5) appears to be a triple point, because it is located where the two border-lines meet.

It is interesting to compare the IBA parameters used in the fit for ^{156}Dy with those reproducing best the X(5) symmetry ($\chi = -1.0\sqrt{7/4}$ and η=0.75). This allows to locate ^{156}Dy in the phase diagram of the extended Casten triangle close to the critical point of the X(5) symmetry. Its location is a bit shifted from X(5) towards O(6) (see fig. 4) indicating the onset of γ-softness. This is consistent with the fact that the γ band of ^{156}Dy is found to be lower in energy as compared to the N=90 isotones ^{150}Nd, ^{152}Sm, and ^{154}Gd, all of which are closer to the critical point of the X(5) symmetry.

4 New X(5) regions

Concerning the X(5) symmetry, only nuclei with N=90 and A≈150 were found to show the characteristic X(5) features so far. In addition to these cases Bizzeti et al. pointed out, that ^{104}Mo is also a good X(5) candidate due to its energy spectrum

[21].

In order to find other regions where X(5) like nuclei exist, it appears to be reasonable to look for those regions where rather fast transitions from β deformed or SU(3)-like nuclei to spherical ones occur. It has been shown [22,23] that the nuclear deformation is related to the number of valence protons N_π and valence neutrons N_ν by the formula:

$$\beta^2 = \alpha N_\pi N_\nu,$$

where alpha is a constant. Applying this simple relation, one expects a fast transition from spherical to axially symmetric deformation for large values of $N_\pi(N_\nu)$ [proton (neutron) numbers at mid-shell] and small or moderate numbers of $N_\nu(N_\pi)(\approx 3$-$4)$. It will be interesting to test experimentally whether these criteria indeed define the occurrence of the X(5) symmetry.

We note, that for the A=150 region these criteria are fulfilled with $N_\nu=4$ and numbers of protons $n_\pi \approx 64$ (proton mid-shell).

Following the above defined criteria, ^{180}Os and ^{178}Os with neutron numbers of $n_\nu=104$, and 102, respectively (mid-shell), and valence proton numbers of $N_\pi=3$ are expected to be X(5) candidates. It turned out that the experimental energy spectra of the nuclei ^{176,178}Os (fig. 5) and ^{180}Os show the expected X(5) features. In table 1 some crucial energy ratios of ^{176,178,180}Os are compared to the corresponding values of the X(5) predictions and to those of X(5) like nuclei in the A=150 mass region.

So far no experimental information on absolute transition probabilities is available for these nuclei. Therefore aside from the energies only the experimental branching ratios can be compared to the X(5) predictions. Very good agreement is found for ^{176,178}Os which is shown in fig. 5. There is less experimental data known in the case of ^{180}Os. We also would like to mention that in the three nuclei ^{176,178,180}Os the 2^+ state of the β band is lower in energy than the 2^+ state of the γ band. Thus it can be expected that γ-softness plays a minor role in these nuclei. One can conclude that the experimental data of ^{176,178,180}Os, known so far, support very much an X(5) structure. Further tests including absolute B(E2) values are suited to perform more stringent tests of the X(5) predictions.

5 Summary

Precise and reliable lifetimes of excited states in ^{154}Gd were measured using the RDDS technique after Coulomb excitation. Lifetimes in the gsb and β, i.e. S2, bands of ^{156}Dy were determined from a coincidence RDDS and a DSAM experiment. Constant transition quadrupole moments were found in the gsb (Q_t=6.1 eb) as well as in the s-band (Q_t=6.0 eb). A very good agreement is found between experiment and the X(5) predictions in the case of ^{154}Gd. The new experimental data indicate ^{156}Dy to be more γ-soft than the other recently established X(5) nuclei, ^{152}Sm and ^{150}Nd. Nevertheless many features of a typical X(5) nucleus are observed in ^{156}Dy. In addition, criteria for X(5) regions are given, and ^{176,178,180}Os were found to be new and very promising X(5) canditates.

Acknowledgments

This work was supported by BMBF (Germany) under the contract no. 06OK958 and under the European Union TMR Programme, contract HPRI-CT-1999-00083.

References

1. H. Emling et al., Phys. Lett. **B217**, (1989), 33.
2. J.D. Morrison et al., J. Phys.G: Part. Phys.**15**, (1989), 1871.
3. H. R. Andrews et al., Nucl. Phys.**A219**, (1973), 141.
4. F. Iachello, Phys. Rev. Lett. **85**, (2000), 3580.
5. R.F. Casten, N.V. Zamfir, Phys. Rev. Lett. **85**, (2000), 3584.
6. F. Iachello, Phys. Rev. Lett. **87**, (2001), 052502
7. R.F. Casten, N.V. Zamfir, Phys. Rev. Lett. **87**, (2001), 052503.
8. J. Jolie, P. Cejnar, J. Dobes, Phys. Rev. **C 60**, (1999), 0613003.
9. R.F. Casten, D. Kusnezov, N.V. Zamfir, Phys. Rev. Lett. **82**, (1999), 5000.
10. V. Werner et al. Phys. Lett. **B 527**,(2002), 55.
11. J. Jolie et al., Phys. Rev. Lett. **87**, (2001), 162501.
12. J. Jolie et al., Phys. Rev. Lett., **89**,(2002), 182502.
13. R. Krücken et al., Phys. Rev. Lett. **88**, (2002), 232501.
14. T. Klug et al., Phys. Lett. **B 495**,(2000), 55.
15. T. Klug et al., Phys. Lett. **B 524**,(2001), 252.
16. A. Dewald, S. Harissopoulos and P. von Brentano, Z. Phys. **A334**, (1989) 163.
17. NNDC data base, http//www.nndc.bnl.gov
18. P. Petkov et al., Nucl. Phys. **A674**, (2000), 357.
19. M.A. Caprio et al., Phys. Rev. **C66**, (2002), 054310.
20. F. Iachello and A. Arima, *The Interacting Boson Model* (Cambridge University Press, Cambridge, 1987).
21. P. Bizzeti and A.M. Bizzeti-Sona, submitted to Phys.Rev. Lett.
22. R.F. Casten et al., Nucl. Phys. **A443**, 1985,1
23. A. Dewald et al., proceedings of the XIV Intern. School on Nucl.Phys., Neutron Phys. and Nucl.Energy, Varna, 2001

PHASE TRANSITIONS IN THE OCTUPOLE DEGREE OF FREEDOM

P.G. BIZZETI
Dipartimento di Fisica, Universitá di Firenze
I.N.F.N., Sezione di Firenze
Via G. Sansone 1, 50019 Sesto Fiorentino (Firenze), Italy

1 Introduction

1.1 Phase transitions and new symmetries in nuclei

Phase transitions in the nuclear shape can be observed, at the boundary between a spherical and a deformed region, when a proper *order parameter* – *e.g.*, the ratio $E(4^+)/E(2^+)$ – is reported as a function of a proper driving parameter – *e.g.*, the number of neutrons along a chain of isotopes or the number of protons along a chain of isotones. It has been shown by Iachello [1,2] that new dynamic symmetries can be expected at the critical point: the E(5) symmetry for the phase transition between spherical shape and γ–unstable deformation, and the X(5) symmetry for the transition between spherical and deformed axially symmetric shape. Here, we are more interested in the latter case. First examples of X(5) symmetry in transitional nuclei have been found [3,4] in ^{152}Sm and ^{150}Nd. In a quite different region, we have found [5] that also the level scheme of ^{104}Mo is very close to the X(5) predictions, not only in the ground–state band, but also in the two γ–excited bands with $n_\gamma = 1$, $K = 2$ and with $n_\gamma = 2$, $K = 4$.

1.2 Phase transitions in the octupole degree of freedom

After observation of the X(5) symmetry – the signature of the phase transition between spherical and axially deformed shape – in ^{104}Mo ($Z = 42$, $N = 62$) and in ^{152}Sm ($Z = 62$, $N = 90$), it appears reasonable to explore whether evidence of phase transition exists also for $Z = 90$, *i.e.* in the chain of Thorium isotopes. Here, however, the presence of low lying negative–parity states (Fig. 1) shows that octupole (and/or higher order) terms play an important role in the collective nuclear excitations, in addition to that of the collective quadrupole.

The relative importance of the octupole with respect to the quadrupole collectivity is indicated by the inverse of the ratio $E(1^-)/E(2^+)$ (Fig. 1*d*) and reaches its maximum in the region where the critical point of the phase transition in the quadrupole mode could be located on the basis of Fig. 1*b*. Quite a variety of nuclear shapes can be obtained with the combination of the quadrupole and octupole modes. Here, we limit our discussion to axially symmetric shapes, or shapes very close to the axial symmetry. Also in this limited field, we can meet a number of different situations, ranging from quadrupole + octupole vibrations around a spherical equilibrium to a permanent deformation in one or both of the two degrees of freedom, over which quadrupole and octupole vibrations can again take place. The result of a Strutinsky–model calculations of the $V(\beta_2, \beta_3)$ potential for the Th

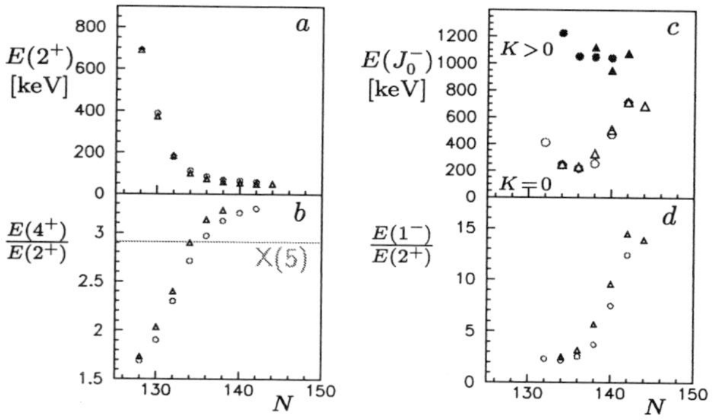

Figure 1. Indicators of the quadrupole collectivity (left) and octupole collectivity (right), as a function of the neutron number N, in the isotope chains of Ra (circles) and Th (triangles): a – Excitation energy of the first 2^+ level; b – Energy ratio $E(4^+)/E(2^+)$; c – Excitation energy of the first level of the $K^\pi = 0^-$ band, $J_0^\pi = 1^-$ (open symbols) and of the lowest known level of other negative–parity bands, $J_0^\pi = 2^-$ or 1_2^- (full symbols); d – Energy ratio $E(1^-)/E(2^+)$.The horizontal line in the part b shows the value expected for the X(5) symmetry.

isotopes [6] show that well developed minima at $\beta_2 > 0$ and either $\beta_3 = 0$ or $\beta_3 \neq 0$ could exist in this region, as well as potentials which are rather soft with respect to quadrupole/octupole vibrations. Moreover, one of the reported cases shows a potential minimum well localized at a finite value of β_2, but extending with a rather flat behaviour over a wide region of β_3 values: just what one should expect for the critical point of the phase transition between reflection–symmetric and reflection–asymmetric shapes, in the presence of a permanent quadrupole deformation.

In the following, this particular situation will be discussed in detail and the predictions of a simple model will be compared with experimental data in the Th isotope chain. This model can be considered as an extension of the one valid for small–amplitude oscillations, as discussed *e.g.* by Eisenberg and Greiner[7]. In order to extend the analysis to situations different from the limit of stable quadrupole deformation, and also to provide a more consistent theoretical frame for the proposed model, we have developed a new parameterization scheme for the combined quadrupole–octupole deformation in the reference frame of the principal axis of inertia, valid in situations close to the axial symmetry. This scheme and some of its implications will be presented in the last part of this talk.

2 Octupole excitations in nuclei with stable quadrupole deformation

2.1 Previous investigations of the Octupole + Quadrupole case

The algebraic approach to the study of nuclear excitations involving reflection–asymmetric shapes, proposed in 1985 by Engel and Iachello [8], has been recently discussed in a paper by Zamfir and Kusnezov [9].

In the frame of the geometrical approach, a number of theoretical investigations of the octupole vibrations around a stable quadrupole deformation have been reported in the last 50 years [10,11,12,13,14,15,16]. Most of them concern models limited to the $K = 0$ (axially symmetric) mode. A consistent theoretical frame has been provided by the work by Donner and Greiner [10]. In their approach, the "intrinsic" reference frame is referred to the principal axes of the quadrupole mode alone which, in the presence of octupole vibrations, do not coincide – in general – with those of the overall tensor of inertia. This theory is very general but, in order to obtain definite predictions for particular cases – such that of octupole vibrations around a quadrupole deformed shape – a number of approximations are necessary. First, the rotation–vibration coupling is neglected. Then, the vibrations in the different octupole modes are only considered in the limit of small amplitude. In this limit, the restoring forces can be expected to be harmonic, but of different strength for the different modes. In addition, Coriolis interaction can couple together states of different bands and alter, therefore, the rotational energy sequence.

There are cases, however, in which the $K^\pi = 0^-$ band lies much lower than other vibrational bands of either parity. This situation can be expected to occur in a transition region between octupole vibration and stable (axial) quadrupole–octupole deformation. In fact, at the latter limit – never met in nuclei, but observed in binary asymmetric molecules – the $K^\pi = 0^-$ band and the ground 0^+ band should merge into a single rotational band with alternate parity.

In such a case, the effect of Coriolis coupling between the states of the 0^- band and the corresponding ones of the 1^- band can be expected to be less important, owing to the large energy denominator. Moreover, if the relevant part of the Hilbert space can be restricted – at least as a first approximation - to the quadrupole modes and the octupole mode with $K^\pi = 0^-$, the symmetry axis of the octupole-induced tensor of inertia coincides with the axis 3 of the "intrinsic" frame of Donner and Greiner, and the distinction between the principal axis of the overall reference frame and those of the quadrupole mode turns out to be irrelevant in this case.

We can hope, therefore, that such an approach would maintain its validity out of the limit of small amplitude, as long as axially symmetric modes are concerned. We admit that a consistent treatment would require to consider together, at least, all the modes of a given tensor rank and that it is a risky procedure that of ignoring part of the dynamical variables. However, all sort of collective model necessarily neglect a substantial part of the ensemble of dynamical variables, those corresponding to the nucleon relative motion. The adiabatic condition is invoked in this case, due to the fact that the classical frequencies (*i.e.* the quanta of excitation energy) for the degrees of freedom considered in the model are very different from those of the degrees of freedom that the model ignores. We hope a similar approximation can be valid also in our case.

2.2 A simple model for the critical point in the axial octupole mode

We now discuss a simple model, that can be considered as an extension of the Donner an Greiner approach valid for the small–amplitude octupole vibration, to the finite–amplitude vibrations in the axial octupole mode. A general discussion of the various quadrupole–octupole degrees of freedom will be postponed to the following

Section. At the moment, we assume:

- – Permanent quadrupole deformation $\bar{\beta}_2$.
- – Amplitude of β_2 vibrations (around $\bar{\beta}_2$) and of γ vibrations (around zero) negligible in comparison to $\bar{\beta}_2$.
- – Axial octupole vibrations (around zero), in a proper potential well.
- – Non axial octupole vibrations of negligible amplitude (in comparison to $\bar{\beta}_2$).
- – For the rotation – vibration wave function we assume $\Psi \propto \psi(\beta_3)\mathcal{D}^{J*}_{M,0} \propto \psi(\beta_3)Y_{J,M}$, with $\psi(-\beta_3) = (-)^J\psi(\beta_3)$.

Two simple forms for the potential $V(\beta_3)$ have been considered: a harmonic potential and a square well (with impenetrable walls) extending from $-\beta_3^{max}$ to β_3^{max}. The latter one is a sensible approximation for the critical–point potential.

The wave equation for rotation and vibration (in the β_3 variable only) can – in principle – be obtained following the Pauli prescriptions for quantization in curvilinear coordinates. This procedure, however, is not always sufficient to obtain a well–defined result, and further uncertainties derive from the variable which are ignored (or "frozen") in the present model. We are going to adopt the simplest form among the ones that could be assumed, provided that the model takes into account the dependence of the moment of inertia on the octupole amplitude, and reduces to the one by Donner and Greiner in the small–amplitude limit.

From now on, in order to simplify the expressions, we consider the inertia parameter B_λ as included in the definition of the amplitudes $a_\mu^{(\lambda)}$ or β_λ: *e.g.*, the variable β_2 is, in the present notation, equivalent to the expression $\sqrt{B_2}\,\beta_2$ in the notation of Bohr. In the frame of Bohr's hydrodynamical model [17], the moments of inertia along the principal axes, in a situation close to the axial symmetry, are given by

$$\mathcal{J}_1 \approx \mathcal{J}_2 \approx 3(\beta_2^2 + 2\beta_3^2)$$
$$\mathcal{J}_3 \approx 4\beta_2^2\gamma_2^2 + ...$$

(here, the contribution of non–axial octupole terms and higher–order terms in γ_2 have been neglected). It is useful to define the new variables $x = \sqrt{2}\,\beta_3/\,\bar{\beta}_2$ and $\epsilon = \frac{1}{\hbar^2}\,\bar{\beta}_2^2\,E$. For the differential equation in the variable x we adopt

$$\frac{d^2\psi}{dx^2} + \frac{2x}{1+x^2}\,\frac{d\psi}{dx} + \left[\epsilon - \frac{J(J+1)}{6(1+x^2)} - v(x)\right]\psi = 0 \qquad (1)$$

with $\psi(-x) = (-)^J\,\psi(x)$.

The theoretical basis for this assumption will be discussed in the next section. Here we note that the Eq. 1 reduces to that of Donner and Greiner [10,7] when $x \ll 1$. With the square–well potential adopted at the critical point ($v(x) = 0$ for $|x| < b$ and $= +\infty$ outside, so that $\psi(\pm b) = 0$) the Eq. 1 is formally identical to that of the *Radial Oblate Spheroidal wavefunctions* [18] with $m = 0$. The eigenvalues and eigenfunctions can be evaluated numerically. They, obviously, depend on b. The calculated ratios $E(J^\pi)/E(2^+)$ for $s = 0$ (s being the number of zeroes of the eigenfunction ψ in the open interval $0 < x < b$) are depicted *versus* b in the Fig. 2. The eq. 1 differs from the small-amplitude limit of the theory of Donner and Greiner in two points: in the value of the moment of inertia, which in the present model

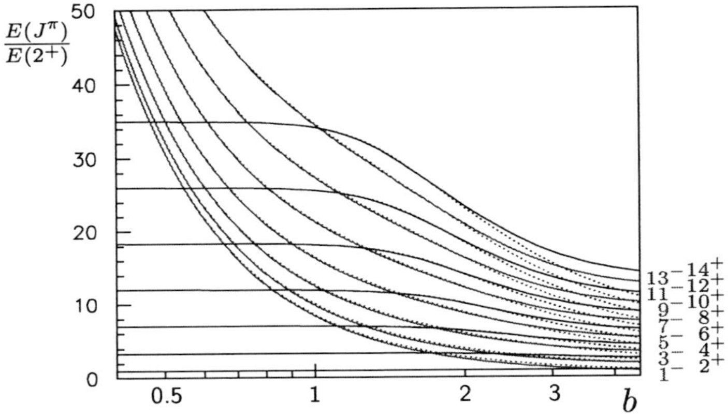

Figure 2. Ratio $E(J^\pi)/E(2^+)$ for $s = 0$, as a function of b, for states of the ground–state band with different J^π. Dotted lines show the results obtained with a differential equation corresponding to that of Eq. 1 *without the first–derivative term.*

depends on β_3; and in the presence of a first–derivative term. While the first point is essential to the model, the second has limited influence on the results as long as *realistic* values of $b < 2$ are considered, as shown, *e.g.*, by the dotted lines of Fig. 2.

2.3 Comparison with the experimental data

Data reported in Fig. 1*a*, *b* show that the heavier isotopes of Th possess a stable quadrupole deformation, while the lighter ones can be interpreted as vibrational or non–collective nuclei. The quadrupole–deformed region extends above the mass 224 (which could correspond to the critical point of phase transition, having $E(4^+)/E(2^+) \approx 2.91$). As for the octupole mode, it certainly shows a vibrational character at $A \geq 230$. For mass 224 to 228, the 1^- band–head of the octupole $K^\pi = 0^-$ band decreases well below all other octupole bands, and is close to the first levels of the positive–parity g.s. band. The mass interval in which our simple model could work is therefore restricted to ^{226}Th and ^{228}Th.

As shown in the fig. 3, ^{226}Th appears to be close to the critical point, when we assume $b = 1.73$ in order to reproduce exactly the position of the first 1^- level. The fit of the high–spin part of the g.s. band is even better for $b = 1.87$, at the expense of a slight disagreement in the energy of the first 1^-. More details can be found in the ref. [19]. Here, we only mention the fact that the observed branching ratios in the γ decay are consistent with those expected at the critical point, within the (admittedly large) experimental uncertainties. As for the $s = 2$ band, only the first two levels (0_2^+, 2_2^+) are known, and their excitation energies are not far from the values predicted for the critical point.

If the critical point of phase transition is close to ^{226}Th, then ^{228}Th should be closer to the vibrational limit. Actually, the right–hand part of fig. 3 shows a reasonable agreement with the results of our model with a harmonic potential of

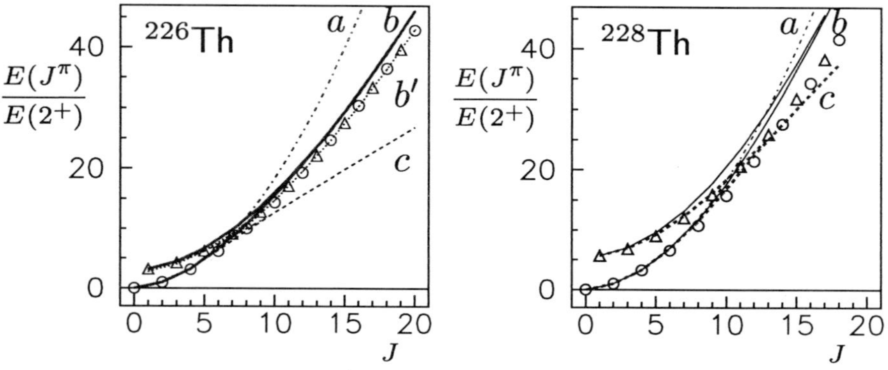

Figure 3. Ratios $E(J^\pi)/E(2^+)$ for $s = 0$, as a function of J, for states of the ground–state bands of ^{226}Th and ^{228}Th, compared with different model calculations: rigid rotor (curve a), critical point in the octupole mode (curves b and b'), present model with harmonic potential in β_3 (curve c). The curves b and c correspond to a fit on the lowest 1^- state, the curve b', on the 20^+ state.

the form $\frac{1}{2}cx^2$, with $c = 12.1$ fitted to reproduce the position of the first 1^-.

2.4 The $A = 224$ puzzle

Lighter isotopes of Th are already out of the region of permanent quadrupole deformation. As one can see from Fig. 1c, for ^{224}Th – and also for ^{224}Ra – the energy ratio $E(4^+)/E(2^+)$ is close to the critical value 2.91 corresponding to the X(5) symmetry for the phase transitions in the quadrupole deformation. Actually, Fig. 4 shows that the X(5) behaviour is not limited to the 4^+ but extends to all positive–parity levels up to $J = 18$. The negative part of the ground-state band merges with the positive–parity one already around $J \approx 6$. Energy ratios for ^{224}Ra are very close to those of ^{224}Th, both for the positive and the negative parity levels, while the absolute energy scales differ by more than 15%. The agreement with X(5) prediction does not extend, however, to the $s = 2$ band: in ^{224}Ra a second 0^+ and a second 2^+ level are known, but their energies are in disagreement with those of X(5) by almost a factor of 2.

There is, therefore, some indication of a critical–point behaviour similar to the X(5) symmetry, but the presence of the octupole mode cannot be neglected. One could suspect that quadrupole and octupole vibration be correlated with one another, with the absolute value of the ratio β_3/β_2 approximately constant during the motion. To explore this possibility, however, it is necessary to build a consistent theoretical frame to describe the simultaneous motion in the quadrupole and octupole degrees of freedom. This work is still underway. Part of it will be summarized in the next section.

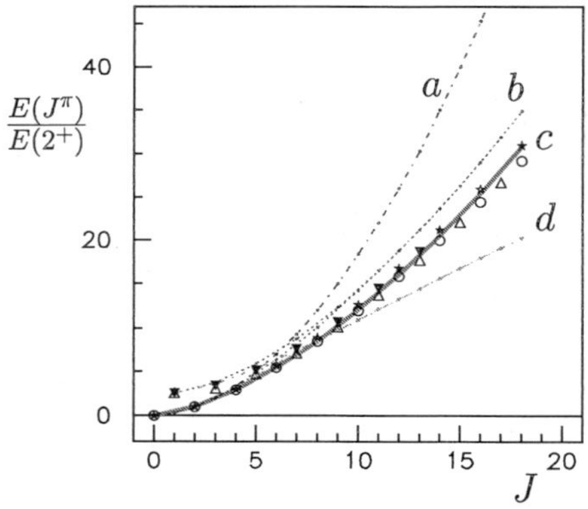

Figure 4. Energy ratios for the ground–state band of ^{224}Ra (full symbols) and of ^{224}Th (open symbols), compared with the predictions of the X(5) model (curve *c*, positive–parity levels only). Other curves correspond to rigid rotations (*a*), and to the model discussed in the section 2: at the critical point (*b*) or with harmonic potential in β_3 (*d*). For curves *b* and *d*, the free parameter has been adjusted to reproduce the position of the 1^- state.

3 A general theoretical frame for simultaneous axial quadrupole and octupole excitations

3.1 The different steps of a consistent calculation

The Bohr collective model of 1956 showed how it is possible to describe the surface vibration (limited, in this case, to the quadrupole mode) and the rotational degrees of freedom of a nucleus, considered as a droplet of condensed matter. We will try to extend this treatment to the quadrupole plus octupole modes, retaining the basic assumptions and approximations of Bohr (*e.g.*, the irrotational flow).

Such a program will include five different steps:

1. – The choice of the intrinsic reference frame (it will be referred to the principal axis of the overall tensor of inertia, at variance with the Donner–Greiner approach).

2. – In this frame, the 5 + 7 amplitudes of the quadrupole plus octupole motion are no longer independent, and are expressed in terms of 9 new parameters ("intrinsic" amplitudes).

3. – The classical kinetic energy is expressed in term of the 9 intrinsic amplitudes and their time derivatives, and of the derivatives of the 3 Euler angles (or, if we prefer, the 3 intrinsic components of the angular velocity along the principal axis of inertia), and takes the form $T = \frac{1}{2} \sum \dot{Q}_\mu \mathcal{G}_{\mu\nu} \dot{Q}_\nu$

4. – The quantum kinetic–energy operator is defined, according to the Pauli prescriptions for quantization in a non–cartesian frame:
$\hat{T} = -\frac{\hbar^2}{2}\, G^{-1/2} \frac{\partial}{\partial Q_\mu} \left[G^{1/2} (\mathcal{G}^{-1})_{\mu\nu} \frac{\partial}{\partial Q_\nu}\right]$ – where $G = \mathrm{Det}\ \mathcal{G}$.

5. – At this point, a particular model can be defined, by a proper choice of the potential energy, and the corresponding set of differential equation must be solved to obtain the model predictions.

Our work is still in progress. Here we are going to discuss in some detail only the first two steps of the calculation, and we will limit the further discussion to a short summary, showing under which assumptions we can justify the model described in the previous Section 2.

3.2 A new parameterization

Here we chose as "intrinsic" reference frame the principal axes of the overall tensor of inertia, resulting from the combined quadrupole and octupole excitation. We shall limit our discussion to situations close to the axial symmetry, and define a parameterization which automatically sets to zero the three products of inertia $\mathcal{J}_{\kappa,\kappa'}$ ($\kappa \neq \kappa'$) *up to the first order in the amplitude of non-axial modes.*

For the quadrupole mode alone, it would be enough to assume $a_2^{(2)} = a_{-2}^{(2)}$ real and $a_{\pm 1}^{(2)} = 0$, or adopt the standard parameterization in terms of β_2 and γ_2. For the octupole mode alone, a parameterization suitable to this purpose has been proposed in 1999 by Wexler and Dussel [20]. We shall adopt here a very similar one. As long as we only consider axial deformations, the intrinsic reference frame of the quadrupole and of the octupole coincide with one another and with the intrinsic axes of the tensor on inertia. It is no longer so if non-axial modes are also considered. However, if their amplitude is small, it will be enough to add a set of *small quantities* $\tilde{a}_\mu^{(\lambda)}$ to the corresponding one given by the axially symmetric parameterization, linearize the three non-linear equations $\mathcal{J}_{\kappa\kappa'} = 0$ and solve them up to the first order in the "small" amplitudes. It is sufficient to consider the correction for those amplitudes which – in the limit of axial symmetry – are either zero or small *of the second order*: the imaginary part of $a_{\pm 2}^{(\lambda)}$ and the real and imaginary parts of $a_{\pm 1}^{(\lambda)}$. These quantities can be expressed in terms of three new parameters ξ_c, η_c, ζ_c in such a way that the conditions $\mathcal{J}_{\kappa\kappa'} = 0$ are automatically fulfilled at least in the first order. We obtain

$$
\begin{aligned}
a_0^{(2)} &= \beta_2 \cos\gamma_2 \approx \beta_2 \left[1 - (1/2)\ \gamma_2^2\right] \\
a_1^{(2)} &= \tilde{a}_1^{(2)} = -\sqrt{2}\ \beta_3\ (\eta_c + i\zeta_c) \\
a_2^{(2)} &= \sqrt{1/2}\ \beta_2\ \sin\gamma_2 + \tilde{a}_2^{(2)} \approx \sqrt{1/2}\ \beta_2\ \gamma_2 + i\sqrt{5}\ \beta_3\ \xi_c \\
a_0^{(3)} &= \beta_3\ \cos\gamma_3 \approx \beta_3 \left[1 - (1/2)\ \gamma_3^2\right] \\
a_1^{(3)} &= -(5/2)\ (X + iY)\ \sin\gamma_3 + \tilde{a}_1^{(3)} \approx -(5/2)\ (X + iY)\ \gamma_3 + \beta_2\ (\eta_c + i\zeta_c) \\
a_2^{(3)} &= \sqrt{1/2}\ \beta_3\ \sin\gamma_3 + \tilde{a}_2^{(3)} \approx \sqrt{1/2}\ \beta_3\ \gamma_3 + i\ \beta_2\xi_c \\
a_3^{(3)} &= X \left[\cos\gamma_3 + (\sqrt{15}/2)\ \sin\gamma_3\right] + iY \left[\cos\gamma_3 - (\sqrt{15}/2)\ \sin\gamma_3\right] \\
&\approx X + iY + (\sqrt{15}/2)\ (X - iY)\ \gamma_3
\end{aligned}
\tag{2}
$$

With this choice, the non-diagonal terms $\mathcal{J}_{\kappa\kappa'}$ are small of the second order in the "small" quantities γ_2, γ_3, X, Y, ξ_c, η_c, ζ_c.

3.3 The classical expression of the kinetic energy

Now it is possible to express the classical kinetic energy (as given by Bohr's hydrodynamic model) in terms of the new variables and of the intrinsic components q_κ

of the angular velocity. The classical expression has the form

$$T = \frac{1}{2}\sum \dot{Q}_\mu \mathcal{G}_{\mu\nu}\dot{Q}_\nu \tag{3}$$

where $\dot{Q} \equiv (\dot{\xi}_1, \dot{\xi}_1, ..., \dot{\xi}_9, q_1, q_2, q_3)$, $\dot{\xi}_\mu$ $(\mu = 1, ..., 9)$ are the time derivative of the nine parameters we have just defined, and q_1, q_2, q_3 are the intrinsic components of the angular velocity of the intrinsic system in a given inertial frame. Also here, we adopt the convention of incorporating the inertia parameter B_λ in the definition of the amplitudes $a_\mu^{(\lambda)}$, so that our $a_\mu^{(\lambda)}$ would correspond to $\sqrt{B_\lambda}\, a_\mu^{(\lambda)}$ in the Bohr notations. The form of the matrix $\mathcal{G}$ obtained in this way is however not satisfactory. In particular, its determinant G – to be used in the Pauli's quantization procedure – does not converge, in the limit $\beta_3 \to 0$, to the corresponding expression of the Bohr's theory of quadrupole excitation [17]. To this purpose, it is convenient to redefine the variables ξ_c, η_c, ζ_c in terms of three new ones, ξ, η, ζ,

$$\xi_c = \xi \Big/ \sqrt{\beta_2^2 + 5\beta_3^2}, \qquad \eta_c = \eta \Big/ \sqrt{\beta_2^2 + 2\beta_3^2}, \qquad \zeta_c = \zeta \Big/ \sqrt{\beta_2^2 + 2\beta_3^2} \tag{4}$$

At this point, the determinant takes the form

$$G = \text{Det } \mathcal{G} \propto \beta_2^2\beta_3^2\left(\beta_2^2 + 2\beta_3^2\right)^2\left(\beta_2^2 + 5\beta_3^2\right)^{-1}\left(\beta_2^2\gamma_2 + \sqrt{5}\,\beta_3^2\gamma_3\right)^2 \xrightarrow[\beta_3 \to 0]{} \beta_2^8\gamma_2^2 \cdot \beta_3^2$$

consistent with that of a pure quadrupole motion. This form of the matrix has some other distinguished advantages. In particular, all non diagonal terms involving the time derivatives of β_2 or β_3 and either the derivative of one of the other intrinsic amplitudes or q_3 turn out to be zero in the present approximation. Other non diagonal elements are small (of the first order) in the "small" amplitudes γ_2, γ_3, X, Y, ξ, η, ζ and have negligible effect on the results, with the only exception of elements of the last line and column. The latter, in fact, are still small of the first order, but must be compared with a diagonal element $\mathcal{J}_3$, which is small of the second order in the "small" non–axial amplitudes. These terms play an important role in the treatment of the intrinsic component of the angular momentum along the approximate axial–symmetry axis. On this subject, we will only mention the fact that γ_2 and γ_3 appear not to be the most convenient variables and it is better to use, in their place, two orthogonal combinations of the form

$$\begin{aligned} \gamma &= \sqrt{5}\gamma_2 - \gamma_3 \\ \gamma_0 &= \left(\beta_2^2\gamma_2 + \sqrt{5}\,\beta_3^2\gamma_3\right)\, f(\beta_2, \beta_3) \end{aligned} \tag{5}$$

It is also possible to identify elementary excitations that contribute one, two or three units of $\hbar$ to the angular–momentum component along the approximate symmetry axis. A complete treatment of this subject will be reported in a future paper [21].

As a consequence of the change of variables, new non-diagonal terms (small of the first order) appear in the matrix $\mathcal{G}$. They have negligible (second–order) effects on the results, or can be eliminated with a slight change in the definition of β_2 and β_3. It is important to stress that all these changes of variables cannot be considered, as it could seem at first sight, as completely arbitrary. In some cases, as here, they are suggested by the structure of the angular momentum operators – or, as we have seen before, by some requirement at a limiting case. Eventually, the choice of the model potential will suggest the best form of the parameters to be used.

3.4 The case of permanent quadrupole deformation

We assume $\beta_2 = \bar{\beta}_2 + \beta_2'$, with $|\beta_2'| \ll \bar{\beta}_2$ confined in a region close to zero by harmonic restoring forces. In this case, also β_2' is one of the "small" quantities which can be neglected and β_2 can be replaced by its average value $\bar{\beta}_2$.

The form of the matrix $\mathcal{G}$ discussed in the previous paragraph is not the most suitable now, if we want that the limit for small β_3 amplitudes reduce to the expression by Donner and Greiner. The choice of the variables, however, is not uniquely determined by this condition. We obtain the differential equation in β_3 used in our simple model (eq. 1) if we choose as independent variables, in addition to β_3 and to the already defined β_2', X, Y, ξ, η, ζ, the quantities u_0 and ξ_1 defined as

$$u_0 = \frac{1}{\beta_2^2 + 5\beta_3^2}\,\gamma_0 \qquad \xi_1 = \frac{\beta_2\beta_3}{2\sqrt{\beta_2^2 + 5\beta_3^2}}\,\gamma \tag{6}$$

and assume that all these quantities remain close to zero, due to harmonic restoring forces. In this way, the model introduced in the previous section 2.2 has been derived from a completely consistent theoretical frame, although with a particular (and, to some extent, questionable) choice of the potential–energy expression.

References

1. F. Iachello, Phys. Rev. Lett. **85**, 3580 (2000).
2. F. Iachello, Phys. Rev. Lett. **87**, 052502 (2001).
3. R.F. Casten and N. Zamfir, Phys. Rev. Lett. **87**, 052503 (2001).
4. R. Krücken et al., Phys. Rev. Lett. **88**, 232501 (2002).
5. P.G. Bizzeti and A.M. Bizzeti–Sona, Phys. Rev. C **66**, 031301 (R) (2002).
6. W. Nazarewicz et al., Nucl. Phys. A **441**, 420 (1985).
7. J.M. Eisenberg and W. Greiner, *Nuclear Theory* (3rd edition, Amsterdam 1987) Vol. I, chapter 10.
8. J.I. Engel and F. Iachello, Phys. Rev. Lett. **54**, 1126 (1985).
9. N. Zamfir and D. Kusnezov, Phys. Rev. C **63**, 054306 (2001).
10. W. Donner and W. Greiner, Z. Phys. **197**, 440 (1966).
11. S. Rohozinski, Rep. Progr. Phys. **51**, 541 (1988), and references therein.
12. P.A. Butler and W. Nazarewicz, Revs. Mod. Phys. **68**, 349 (1996) and references therein.
13. P.O. Lipas and J.P. Davidson, Nucl. Phys. **26**, 80 (1961).
14. V.Yu. Denisov and A.Ya. Dzyublik, Nucl. Phys. A **589**, 17 (1995).
15. R.V. Jolos and P. von Brentano, Phys. Rev. C **60**, 064317 (1999).
16. N. Minkov *et al.*, Phys. Rev. C **63**, 044305 (2000).
17. A. Bohr, Dan. Mat. Phys. Medd. **26**, nr. 14 (1952).
18. M. Abramowitz and I.A. Stegun, *Handbook of mathematical functions* (New York 1970), Section 21.6.
19. P.G. Bizzeti and A.M. Bizzeti–Sona, Proc. Int. Conf. NS2002 (Legnaro 2002), Europ. Phys. J., in the press.
20. C. Wexler and G.G. Dussel, Phys. Rev. C **60**, 014305 (1999).
21. P.G. Bizzeti and A.M. Bizzeti–Sona, to be published.

PROMPT PARTICLE DECAY IN NUCLEI: PRESENT STATUS AND FUTURE PERSPECTIVES

C. FAHLANDER and D. RUDOLPH
Department of Physics, Lund University,
SE-22100 Lund, Sweden
E-mail: claes.fahlander@nuclear.lu.se

During recent years the nuclear decay modes of discrete prompt proton and alpha-particle emission from deformed or superdeformed high-spin states have been discovered in nuclei in the vicinity of doubly-magic ^{56}Ni. The particle decays may be viewed as self-regulated two-dimensional quantum tunnelling processes. Due to the decay the remaining nuclear mean-field potential is rearranged dramatically. Quantum-mechanical tunnelling is a wide-spread phenomenon in the natural sciences. Therefore, a full understanding of this process may be of importance far beyond nuclear physics. Significant experimental progress has been made lately. It illustrates that prompt particle decays are "natural" decays in proton rich nuclei in the mass 60 region. New experiments are being planned, involving new instrumentation for charged particle detection.

1 Introduction

The first case of prompt proton emission was observed in 1998 in the proton decay from an excited superdeformed state in ^{58}Cu [1]. The process is different from direct proton emission, which was discovered already in 1970 in the decay of an isomeric state of ^{53}Co [2], and later also found from ground states of nuclei along the proton drip line [3]. In direct proton radioactivity the proton competes with β decay. It is therefore a slow process, of the order of μs to ms, and the transition takes place from a spherical initial state of the parent nucleus to a spherical final state of the daughter nucleus, or from a deformed to a deformed state. Prompt proton radioactivity, on the other hand, competes with γ rays, which places the time scale of the decay into the fs to ns regime, and allows the study of these decays in prompt coincidence with the preceding γ rays of the parent nucleus and with the subsequent γ rays of the daughter nucleus. Also, the prompt particle decays proceed from highly- or superdeformed initial states to near-spherical daughter states. The decay mode may be viewed as a self-regulated two-dimensional quantum tunnelling process. It is two-dimensional because the initial state is prolate deformed with a long-to-short axis of about 1.5:1. It is self-regulated since there is a dramatic rearrangement of the nuclear mean-field in the course of the decay, from a deformed to a spherical nuclear potential.

2 Present Status of Prompt Particle Decay

The isotope ^{56}Ni is generally accepted to represent a doubly-magic spherical nucleus due to the shell gap at particle number 28, which separates the $1f_{7/2}$ shell from the upper fp shell consisting of the $2p_{3/2}$, $1f_{5/2}$, and $2p_{1/2}$ orbitals. Thus, in ^{58}Cu, which has 29 protons and 29 neutrons, the $1f_{7/2}$ shell is filled with both protons and neutrons, and there are one extra proton and one extra neutron in the upper fp-shell. The configurations of its excited states thus involve particle-hole excitations

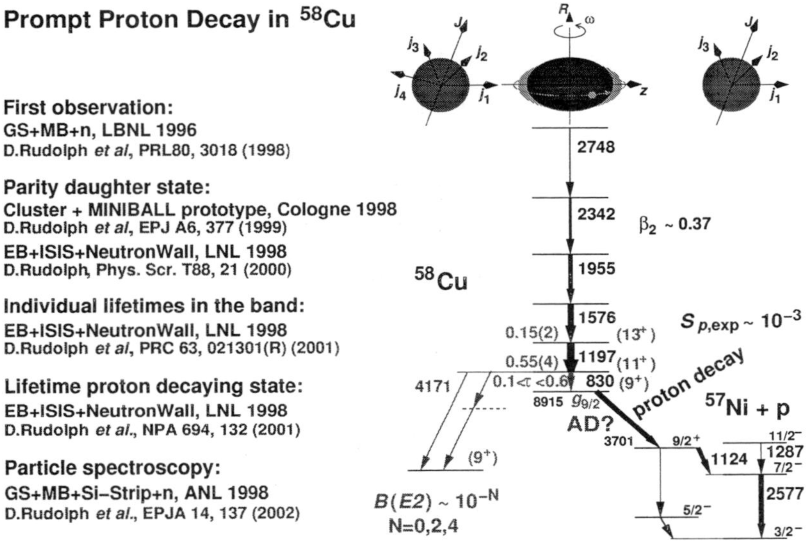

Figure 1: Spectroscopic information on ^{58}Cu obtained from the experiments listed on the left hand side of the figure. They all involved the GAMMASPHERE (GS) or the EUROBALL (EB) γ-ray multi-detector system except the experiment performed in Cologne, which involved cluster and MINIBALL prototype detectors. The experiments also involved the 4π charged-particle detector arrays MICROBALL (MB) [5] at GAMMASPHERE and ISIS [6] at EUROBALL, and in some experiments also neutron detectors at GAMMASPHERE and the NEUTRONWALL [7] at EUROBALL. In the last experiment the three most-forward CsI detector rings of MICROBALL were replaced by four ΔE–E Si-strip telescopes with a total of about 800 active pixels. It was crucial for improving the overall energy resolution for particle detection from about 700 keV to about 300 keV making possible high-resolution particle spectroscopy.

across the 28 shell gap, into the fp shell and into the $1g_{9/2}$ intruder orbital. In the Nilsson scheme the lowest-Ω component of the $1g_{9/2}$ orbital is strongly deformation driving giving rise to shell gaps at large deformation for particle numbers 28 and 30. It plays a very important role for the evolution of the shapes of nuclei in the mass 60 region. The more protons and neutrons that are promoted into the $1g_{9/2}$ orbital, the larger will be the deformation of the state (see e.g. ref. [4]). However, it is also very important for the prompt proton emission from the highly deformed states, because emitting a $1g_{9/2}$ proton naturally helps in reducing the deformation of the system.

Fig. 1 shows the observed rotational band based on the superdeformed state of the second minimum of the nuclear potential of ^{58}Cu. The configuration of the state is a four-particle four hole excitation across the shell gap coupled to one proton and one neutron in the $1g_{9/2}$ orbital. It is unique and commonly used as reference in the mass 60 region [8]. There are two γ-ray linking transitions observed, and the proton decay proceeds from the (9^+) state at the bottom of the band into a spherical state in ^{57}Ni. Applying the Doppler Shift Attenuation Method to levels in ^{58}Cu, lifetimes

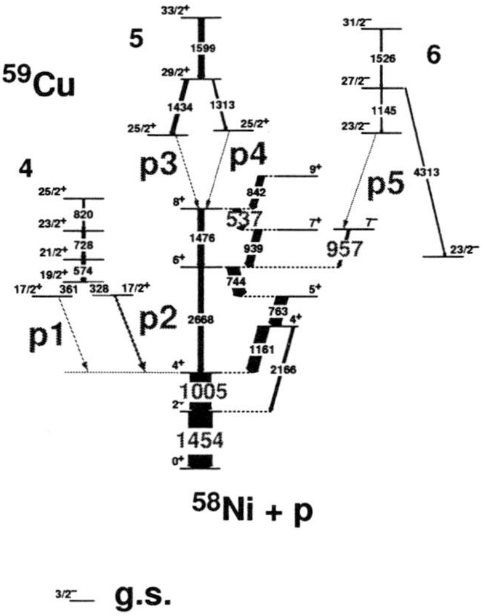

Figure 2: Prompt proton decay scheme of ^{59}Cu.

of individual states could be determined[9]. The 830 keV line, which feeds the proton-decaying state at 8915 keV, reveals both a stopped and a shifted component in its lineshape. If the protons are emitted while the recoils are still moving they have higher energies. This has been utilized to study the energy correlations between the 830 keV γ ray and the 2341 keV proton peak. A detailed DSAM lineshape analysis of the proton line finally yields 0.06 ps $< \tau <$ 0.58 ps for the lifetime of the proton-decaying state [10]. In a separate experiment the spin and parity of the daughter state in ^{57}Ni has been measured to $I^{\pi} = 9/2^{+}$ [11]. Its wave function is thus mainly composed of the $1g_{9/2}$ neutron configuration suggesting that the tunnelling particle indeed is the proton in the lowest-Ω $1g_{9/2}$ orbital, which essentially moves in the equatorial plane of the strongly deformed nucleus. Unfortunately, a model describing both the dynamic shape change associated with the prompt proton decays and the overlap between initial and final wave functions is not at hand. Instead, the present results have been compared to simple, semi-classical WKB estimates of the decay rates [12]. Such predictions depend solely on the Q-value of the decay and the angular momentum of the escaping proton; the larger the Q-value and the smaller the angular momentum, the faster is the calculated decay rate. Next to the observed $1g_{9/2}$ proton decay at 2341 keV there is a possibility for a 2178 keV $1f_{7/2}$ proton decay into the yrast $11/2^{-}$ state of ^{57}Ni. The WKB model predicts an almost exclusive $1f_{7/2}$ decay due to its smaller angular momentum, which obviously is at variance with the observations [12]. The discrepancy between the simple model and the observations clearly hints at nuclear structure and/or deformation effects, which are necessary to counteract the preference of the $1f_{7/2}$ decay strengths.

Significant experimental progress has lately been made based on data from the

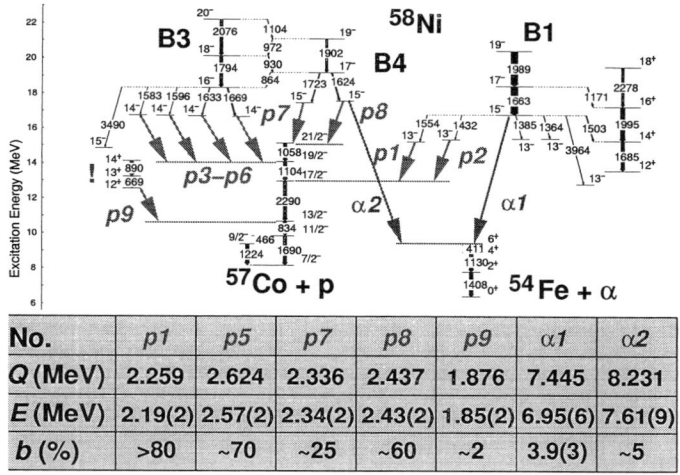

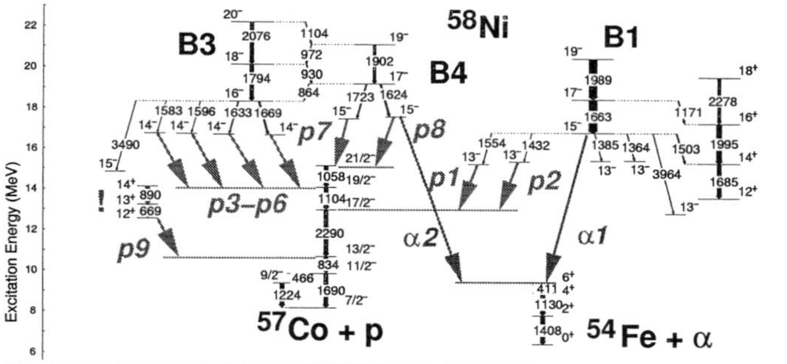

No.	p1	p5	p7	p8	p9	α1	α2
Q (MeV)	2.259	2.624	2.336	2.437	1.876	7.445	8.231
E (MeV)	2.19(2)	2.57(2)	2.34(2)	2.43(2)	1.85(2)	6.95(6)	7.61(9)
b (%)	>80	~70	~25	~60	~2	3.9(3)	~5

Figure 3: Prompt particle decay scheme of ^{58}Ni.

experiments listed in Fig. 1. The last experiment in the list was optimized for the study of the prompt particle decays. It allowed for a revised study of prompt proton particle emission in ^{58}Cu [12], which supports the idea of an exclusive 100% proton branch from the 8915 keV (9^+) state. It also made possible prompt particle spectroscopy in the neighbouring isotope ^{59}Cu, where a total of five prompt proton decays were observed [13]. They imply the first observation of "fine structure" for the new decay mode as shown in Fig. 2. From the spins and parities of the initial and final states it is clear that all five proton decays relate to the emission of $1g_{9/2}$ protons. Prompt proton decay has been observed in two more nuclei in the mass 60 region, namely in ^{56}Ni [14], and in ^{58}Ni [15]. The *preliminary* particle decay scheme of ^{58}Ni is shown in Fig. 3. Nine states, from four different structures, are observed to decay by proton emission. The Q-values of the decays, the measured proton energies, and the relative proton intensities are shown in the figure. All the observed decays, except one, involve a spin difference of 9/2 between the initial and final states, with no parity change, suggesting again the emission of a proton with the expected $1g_{9/2}$ character. In the proton decay from the 12^+ state ($p9$), a near spherical state, to the $13/2^-$ state in ^{57}Co, the spin difference is 11/2, and it involves a change in parity. This strongly hints that this proton decay surprisingly proceeds via a super-intruder $1h_{11/2}$ proton. In ^{58}Ni two α decays were also observed from two 15^- states into the same 6^+ state in ^{54}Fe. One of these α decays ($\alpha 1$) was known from previous studies [16]. Note that one of the 15^- states also decays by proton emission, but the relative intensities of $p8$ and $\alpha 2$ only add up to 65%, suggesting that also γ decay, as yet uncovered, is important. Thus, in ^{58}Ni there is competition between proton, α and γ decay from the same state, an interesting decay, which we eventually would like to understand in more detail.

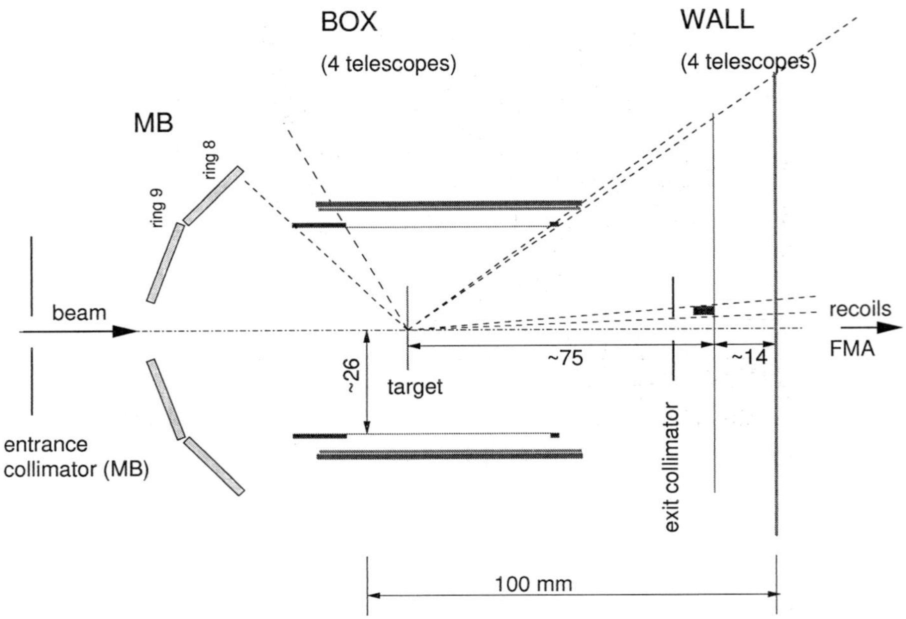

Figure 4: LUSIA, the Lund Silicon Array for prompt particle spectroscopy.

3 Future Perspectives of Prompt Particle Decay

Future studies of prompt particle decay call for much improved particle detection. Therefore, we are presently developing in Lund a new detector system, LUSIA (the Lund Silicon Array), based on ΔE–E Si-strip telescopes (Fig. 4), primarily to be used at GAMMASPHERE. LUSIA will replace essentially all of MICROBALL with a total of eight Si-strip telescopes. Four of them form a wall at forward angles ($5^\circ < \Theta < 40^\circ$), and another four form a box covering the central section around the target position ($40^\circ < \Theta < 120^\circ$). Backward angles are still to be covered with two CsI rings of MICROBALL. LUSIA will have a total of some 2000 pixels. The prompt particle decays are primarily detected in the forward wall. The high granularity in the central section is important for the precise determination of the momenta of the evaporated particles and subsequently the momenta of the recoiling nuclei prior to possible prompt particle decays. This will enable us, on an event-by-event basis, to make a proper determination of the spin direction of the recoiling nucleus, since it is perpendicular to the plane spanned by the beam axis and the recoil vector. This, in turn, will make possible a precise measurement of the angular distribution of the emitted particles with respect to the spin axis of the recoiling (proton emitting) nucleus.

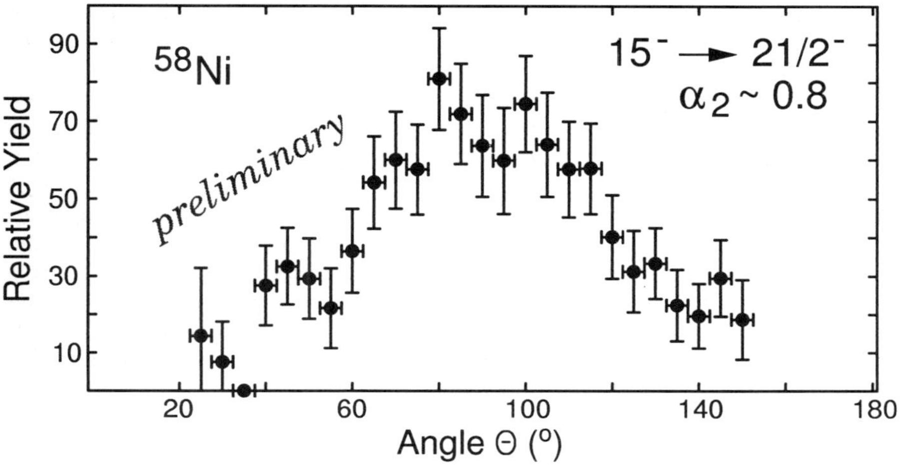

Figure 5: Preliminary experimental proton angular distribution relative to the spin axis of the proton decay $p8$ in ^{58}Ni.

Such a measurement of the proton angular distribution will provide unique tests of contemporary many-body mean-field theories, and it will challenge the theories of the two- or multi-dimensional quantum tunnelling process, which to date only barely is understood. The main complication in describing the new decay process is the fact that it involves such a dramatic rearrangement of the nuclear many-body system and, hence, of the associated mean-field potential. It has been suggested that it can be understood only in terms of a time-dependent approach, and the first theoretical attempts to describe this unprecedented decay mode, based on numerical solutions of the time-dependent two-dimensional Schrödinger equation, indicate that the angular distribution of the emitted particles is extremely sensitive to its initial state, i.e., to its quantum-mechanical wave function prior to the decay [17]. The determination of the wave function of a quantum object in a well-defined quantum state is, of course, the ultimate aim of experimental studies of quantum systems. It has also been shown that the proton angular distribution is very sensitive to the time it takes for the nucleus to change its shape [18]. That time, of course, is related to the tunnelling time, the time it actually takes for the proton to penetrate the potential barrier.

We already have preliminary results of the proton angular distribution from the experiment, which involved the first use of Si-strip detectors. Only a small part of the solid angle was, however, covered with these type of detectors, so it is not a very precise measurement. Nevertheless, the measured angular distribution of the emitted proton ($p8$) relative to the spin axis of the nucleus ^{58}Ni is shown in Fig. 5. It cannot be directly compared to the calculated angular distributions of ref [18], since they were made relative to the symmetry axis of the nucleus, which is more or less perpendicular to the spin axis. However, the overall shape of the measured distribution is consistent with the expectations of a particle with rather high angular momentum. Interestingly, the distribution peaks around 90°, i.e., there are no protons emitted at 0° relative to the spin axis (see also Fig. 1). They are mainly emitted at 90°, but not necessarily, as one may be misled to believe, at the tip of the nucleus (90° relative to the spin axis; 0° relative to the symmetry axis), where the Coulomb barrier is smallest. The time-dependent calculations of Talou et al [18] have shown that the tunnelling direction is dictated by the topology (probability density) of the wave function rather than by the deformation of the nuclear potential. The protons may very well be emitted at some angle relative to the symmetry axis. Interestingly, there is also some structure observed in the measured distribution. However, it remains to be seen whether this structure will be observed in a precise determination, and if so, if it can provide access to the fundamental issue of the nuclear time scale.

4 Summary

Prompt particle decay have been found in four nuclei, namely in ^{58}Cu [1], ^{59}Cu [13], ^{56}Ni [14], and ^{58}Ni [15,16]. They are examples of the two-dimensional quantum tunnelling process. The angular distribution of the tunnelling particle is sensitive to the spatial distribution of the particle wave function inside the nucleus before it was emitted, and it is sensitive to the time it takes for the nucleus to change its shape from strongly deformed to spherical. This time is associated with the tunnelling time, the time that the proton actually stays in the tunnel. Very interesting for the future is therefore a precise measurement of the angular distribution of the tunnelling particle to obtain information on these very fundamental issues. Given the plain number of prompt particle decays identified until now the new decay mode seems to be a common feature at least in proton rich nuclei near ^{56}Ni. The quantitative continuation of the present studies is therefore clearly also the quest for more candidates in the vicinity of ^{56}Ni and in nearby regions.

Acknowledgments

D.G. Sarantites and his co-workers from Washington University deserve a lot of credit for their perfect and persistent work concerning the Si-strip high-resolution experiment. We would also like to thank all friends and collegues in all of the mentioned experiments. This research was supporeted in part by the Swedish Research Council.

References

1. D. Rudolph *et al.*, *Phys. Rev. Lett.* **80**, 3018 (1998).
2. K.P. Jackson *et al.*, *Phys. Lett.* B **33**, 281 (1970).
3. S. Hofman *et al.*, *Z. Phys.* A **305**, 111 (1982).
4. C. Andreoiu *et al.*, Eur. Phys. J. A **14**, 317 (2002).
5. D.G. Sarantites *et al.*, *Nucl. Instrum. Methods* A **381**, 418 (1996).
6. E. Farnea *et al.*, *Nucl. Instrum. Methods* A **400**, 87 (1997).

7. Ö. Skeppstedt *et al.*, *Nucl. Instrum. Methods* A **421**, 531 (1999).
8. A.V. Afanasjev, I. Ragnarsson, and P. Ring, Phys. Rev. C **59**, 3166 (1999).
9. D. Rudolph *et al.*, *Phys. Rev.* C **63**, 021301(R) (2001).
10. D. Rudolph *et al.*, Nucl. Phys. **A694**, 132 (2001).
11. D. Rudolph *et al.*, Eur. Phys. J. A **6**, 377 (1999).
12. D. Rudolph *et al.*, Eur. Phys. J. A **14**, 137 (2002).
13. D. Rudolph *et al.*, Phys. Rev. Lett. **89**, 022501 (2002).
14. D. Rudolph *et al.*, *Phys. Rev. Lett.* **82**, 3763 (1999).
15. D. Rudolph *et al.*, to be published.
16. D. Rudolph *et al.*, Phys. Rev. Lett. **86**, 1450 (2001).
17. N. Carjan, P. Talou, and D. Strottmann, in Proc. *The Nucleus: New Physics for the New Millenium*, Faure, South Africa, January 1999, Eds. F.D. Smit, R. Lindsay, and S.V. Förtsch, Kluwer Academic / Plenum Publishers, New York, 1999, p. 115.
18. P. Talou, in Proc. *International Workshop Pingst 2000 – Selected Topics on* $N = Z$ *Nuclei*, June 2000, Lund, Sweden, eds. D. Rudolph and M. Hellström, (Bloms i Lund AB, 2000), p. 10.

QUADRUPOLE MOMENTS AND DEFORMATIONS OF "SHEARS" STATES IN THE $Z = 82$ PB NUCLEI

D. L. BALABANSKI*, G. NEYENS, K. VYVEY

IKS, University of Leuven, Celestijnenlaan 200 D, B-3001 Leuven, Belgium

We have measured the quadrupole moments of the 11^- isomers in ^{194,196}Pb and have proven that the particle-hole $\pi(3s_{1/2}^{-2}1h_{9/2}1i_{13/2})_{11^-}$ intruder excitation across the $Z = 82$ shell strongly polarizes the core: the quadrupole moments of these states are nearly *one order of magnitude* larger compared to the $\nu(1i_{13/2}^{-n})$ neutron hole states and a *factor of two* larger than the normal $\pi(1h_{9/2}1i_{13/2})_{11^-}$ excitation in the Po nuclei. We have also measured the quadrupole moment of the $29/2^-$ $\nu(1i_{13/2}^{-1}) \otimes \pi(3s_{1/2}^{-2}1h_{9/2}1i_{13/2})_{11^-}$ magnetic-rotational band head in ^{193}Pb, $|Q_s| = 2.84(26)$ eb. The results are compared to different theoretical calculations in an attempt to conclude on the deformation of the "shears" states.

1 Introduction

The microscopic world imposes a severe restriction on rotational motion: a quantal system must be non-spherical in nature. Thus, the rotation of atomic nuclei is associated with the presence of a significant electric quadrupole moment, *i.e.* with the breaking of the spherical symmetry of the charge distribution. As a result, sequences of enhanced electric quadrupole transitions connecting $\Delta I = 2$ levels, known as rotational bands, are observed in experiment. Regions of deformed nuclei where rotational motion is established lie well away from the magic numbers.

It was therefore very surprising when regular sequences of γ rays have been observed in the vicinity of closed shells; see [1,2] for reviews. For the $Z = 82$ Pb nuclei, which are known to be nearly spherical in their ground states, regular bands of strongly enhanced $M1$ γ-ray transitions have been discovered [4], known as magnetic-rotational bands, which were associated with the rotational motion of the system. In the case of the magnetic bands, the magnetic dipole moment has a large component perpendicular to the nuclear spin that specifies an orientation in space. The magnetic dipole rotates around the spin axis, generating the strong $M1$ radiation, which is observed experimentally. Along the bands the angular momentum is increased by a step-by-step alignment of the proton and neutron spins. Since this resembles the closing of a pair of shears, the $M1$ bands are referred to as "shears" bands. The following criteria to characterize these bands have been adopted [2]:

(1) These are $\Delta I = 1$ sequences with strong $M1$ transitions.
(2) The reduced transition probabilities, $B(M1)$, are large ($\approx$ few μ_N^2).
(3) The crossover $E2$ transitions are either absent, or very weak.
(4) The deformation should be small, $\beta_2 \leq 0.15$.
(5) The dynamical moment of inertia is small, ranging from 10 to 25 $\hbar^2$/MeV.

In the lead nuclei alone more than 40 bands have been observed, which fulfill the above criteria. These magnetic rotational bands are built on proton-particle exci-

*PERMANENT ADDRESS: FACULTY OF PHYSICS, ST. KLIMENT OHRIDSKI UNIVERSITY OF SOFIA, BG-1164 SOFIA, BULGARIA// E-MAIL: MITAK@PHYS.UNI-SOFIA.BG tations across the $Z = 82$ shell gap coupled to neutron-holes in the $1i_{13/2}$ sub-shell. Typical configurations are, *e.g.*, the $\{\pi(3s_{1/2}^{-2}1h_{9/2}1i_{13/2})_{11^-} \otimes \nu 1i_{13/2}^{-1}\}_{29/2^-}$ state in ^{191,193}Pb and the $\{\pi(3s_{1/2}^{-2}1h_{9/2}1i_{13/2})_{11^-} \otimes \nu(1i_{13/2}^{-2})_{12^+}\}_{16^-}$ state in ^{194,196,198}Pb [2]. While all other observables in the list above can be deduced from γ-ray spectroscopy, the deformation of the bands is either assumed, or extracted with a large uncertainty from lifetime measurements of the weak $E2$ transitions. Therefore, we set up a project to measure the spectroscopic quadrupole moments of isomeric states, which are either the "blades of the shears", or "shears" states (the $29/2^-$ isomer in ^{193}Pb), in order to obtain information on the deformations of these excitations.

2 Spectroscopic quadrupole moments of the "blades of the shears"

The Pb nuclei are well-known examples for shape coexistence; see [5,6] for reviews. As a result of the shell closure at $Z = 82$, the Pb nuclei are spherical close to their ground states, but at higher excitation energies a subtle interplay between configurations with different shapes - spherical, oblate and prolate - has been predicted [7,8] and observed exprimentally in in-beam and decay studies; see *e.g.* [9,10] and references therein. Potential energy surface calculations [11], as well as Hartree-Fock-Bogoliubov calculations [12] and relativistic Hartree-Bogoliubov calculations [13] yield shape-coexisting minima for the Pb nuclei.

Our experiments aimed at the spectroscopic quadrupole moments of the long-lived states: the proton particle-hole (2p-2h) excitations across the $Z = 82$ shell gap, which appear as 11^- isomers in the Pb nuclei, and the neutron-holes in the $1i_{13/2}$ sub-shell. The latter are either $13/2^+$ isomers (one-hole states in the odd-A Pb nuclei), or 12^+ isomers (two-hole states in the even-mass Pb nuclei). For the $\nu(1i_{13/2}^{-n})$ states in the Pb nuclei small quadrupole moments were measured. These excitations are built in the spherical well and polarize the core towards small prolate deformations, see [14] for a review.

2.1 Deformations of the 11^- isomers in the Pb nuclei

The spectroscopic quadrupole moments of the $I^\pi = 11^-$ isomers in ^{194,196}Pb [15,16] have been measured at the CYCLONE facility at Louvain-la-Neuve by applying the Level Mixing Spectroscopy (LEMS) technique [17]. The experiment yields the ratio of the electric quadrupole interaction frequency, $\nu_Q = eQ_sV_{zz}/h$, to the magnetic moment, μ; V_{zz} is the electric field gradient (EFG) of the investigated nucleus in a non-cubic host lattice, in which the nucleus is implanted. The ^{194,196}Pb isomers were produced in a ^{nat}Re(^{14}N,5n) reaction at a beam energy of 87 MeV.

Fig. 1 shows the LEMS curve obtained for the 337 keV transition mainly fed by the 12^+ isomer (82%). A quadrupole interaction frequency $\nu_Q(12^+) = 38(3)$ MHz could be derived [16]. The quadrupole moment of this state was measured before as $Q_s = 0.65(5)$ eb. The analysis of the 498 keV transition, depopulating the ^{196}Pb(11^-) isomer resulted in $\nu_Q(11^-) = 199(32)$ MHz (see Fig. 1) by using the measured magnetic moment, $\mu = 10.56(88)$ μ_N [18]. As all isomers within the same isotopic chain experience the same EFG, the spectroscopic quadrupole moment of the isomer could be derived from the ratio $\nu_Q(^{196}\text{Pb}; 11^-)/\nu_Q(^{196}\text{Pb}; 12^+)$ as

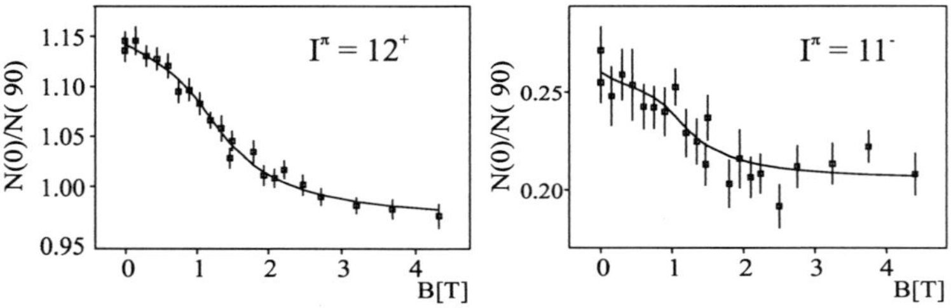

Figure 1. Left side: LEMS curve obtained for the 337 ($E1, 10^+ \rightarrow 9^-$) keV transition. Right side: LEMS curve obtained for the 498 ($E1, 11^- \rightarrow 12^+$) keV transition.

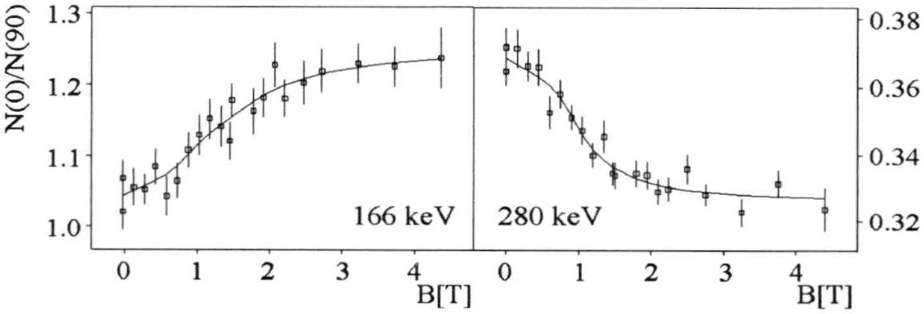

Figure 2. LEMS curves obtained for the 166 keV ($E2; 9^- \rightarrow 7^-$) and the 280 keV ($E1; 5^- \rightarrow 4^+$) transitions in ^{194}Pb.

$Q_s(^{196}\text{Pb}; 11^-) = (-)3.41(66)$ eb [16].

A novel analysis technique has been developed in order to extract the quadrupole moment of the ^{194}Pb 11^- isomer from the same data set [15], as the 352 keV transition, depopulating the ^{194}Pb 11^- isomer, was not intense enough for a sufficiently accurate LEMS analysis. Therefore, we have analyzed the transitions depopulating the 12^+ isomer, which are sensitive to both $\nu_Q(12^+)$ and $\nu_Q(11^-)$ in the ^{194}Pb decay scheme. The quadrupole interaction frequencies, $\nu_Q(^{194}\text{Pb}; 12^+) = 28(2)$ MHz and $\nu_Q(^{194}\text{Pb}; 11^-) = 262(41)$ MHz, are the average values obtained from fitting the obtained LEMS curves. Sample LEMS curves for the 166 keV ($E2$) and the 280 keV ($E1$) transitions are shown in Fig. 2. The quadrupole moment of the 11^- isomer is derived from the ratio $\nu_Q(^{194}\text{Pb}; 11^-)/\nu_Q(^{196}\text{Pb}; 12^+)$ as $\mid Q_s \mid = 4.48(86)$ eb [15].

In order to get a better feeling for the magnitude of the measured quadrupole moments of the 11^- isomers we emphasize that the maximally aligned neutron states $(\nu 1i_{13/2})^2_{12^+}$ in the same nuclei, with a similar value of the angular momentum, have quadrupole moments which are nearly one order of magnitude smaller: $Q_s(12^+;^{196}\text{Pb}) = 0.65(5)$ eb and $Q_s(12^+;^{194}\text{Pb}) = 0.49(3)$ eb. The structure of the 11^- states must therefore be very different from that of neutron two quasiparticle states. The lowest possible proton excitation across the closed shell which gives a spin value 11^- is $\pi(3s^{-2}_{1/2} 1h_{9/2} 1i_{13/2})_{11^-}$ and the measured magnetic moment of the 11^- isomer in ^{196}Pb is very close to that expected for a rather pure excitation in-

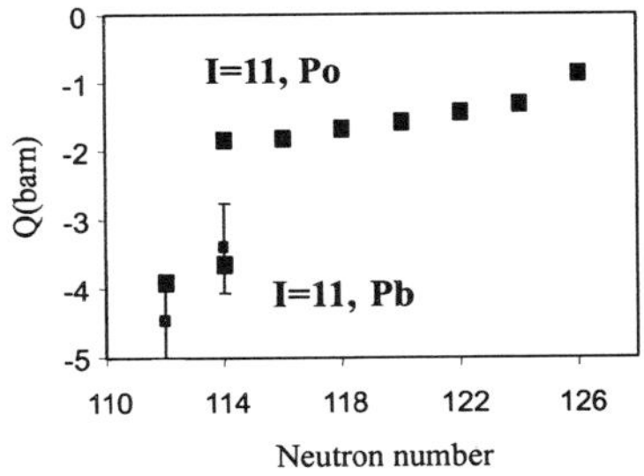

Figure 3. Systematics of the calculated (full squares) and experimental quadrupole moments of the 11_1^- states in the Po and Pb isotopes.

volving these proton orbitals [18]. We conclude that the 2p-2h $\pi(3s_{1/2}^{-2}1h_{9/2}1i_{13/2})_{11^-}$ proton excitations have an oblate deformation $\beta_2 \approx -0.15 \div -0.2$ and are built in the oblate deformed well.

2.2 Intruder vs normal 11^- states in the Pb region

Next we compare the measured quadrupole moments with the quadrupole moment of the pure $\pi(1h_{9/2}1i_{13/2})_{11^-}$ configuration (the "single-particle" estimate); a good approximation for this quantity is the quadrupole moment of the 11_1^- state in the two-proton nucleus ^{210}Po: $Q_s(11_1^-) = (-)0.86(11)$b [19]. The quadrupole moments of the 11^- isomers in Pb are thus *four to five times* larger than the single-particle value. This is in part because the neutron number is close to the neutron mid-shell ($N = 104$). However, the non-magic neutron number by itself is not sufficient to explain a factor of four difference between the single particle estimate and the experimental values. A further important point is that the 11^- isomers in Pb involve an additional pair of proton holes besides the maximally aligned two-proton particle configuration in the corresponding Po nuclei. The shell-model view of intruder states [5,6] explains the large gain in energy of 2p-2h excitations across the closed shell by the additional proton-neutron interaction created with the additional valence particles. The attractive proton-neutron interaction leads to the onset of deformation as soon as the number of valence particles of both types is large enough.

This discussion can be quantified by comparing the measured quadrupole moments of the 11^- isomers in ^{196,194}Pb to the theoretical values of $Q_s(11^-)$ in both the Pb and the Po isotones. This allows us to deduce the influence of the additional two proton holes on the collectivity and deformation of these states. As demonstrated in [20] (see Fig. 3), the quadrupole moment of the intruder state is a *factor of two* larger than the theoretical prediction of the normal state.

3 Spectroscopic quadrupole moments of the "shears states"

We have measured the spectroscopic quadrupole moment of the $29/2^-$ magnetic-rotational band head in ^{193}Pb with the $\nu(1i_{13/2}^{-1}) \otimes \pi(1h_{9/2}1i_{13/2})_{11^-}$ configura-

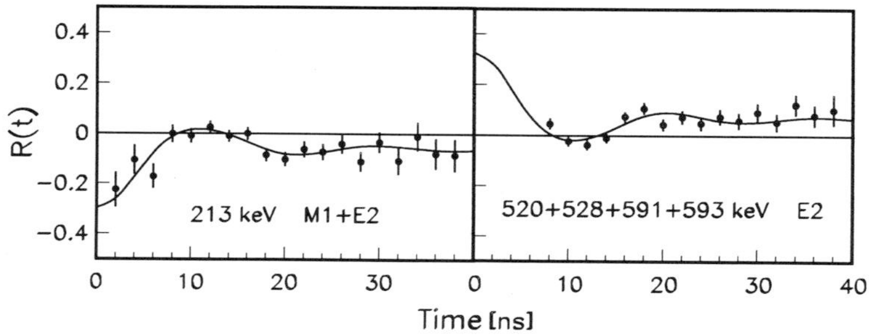

Figure 4. TDPAD spectra for γ lines involved in the decay of the 9.4 ns $29/2^-$ isomer in ^{193}Pb showing the quadrupole interaction in a solid Hg host at T = 170 K.

tion [21], which is a 9.4 ns isomer. The assigned configuration for this state has been confirmed by the g-factor measurement [22]. The quadrupole interaction (QI) of the $29/2^-$ isomeric state has been investigated in the EFG of solid Hg by applying the time-differential perturbed angular distribution (TDPAD) method. The Hg host has been chosen because it provides a large EFG ($V_{zz} = 17.4(9) \cdot 10^{21}$ V/m^2 at T = 170 K) [23], which results in a strong interaction on the short time scale given by the isomer lifetime. The experiment was carried out at the XTU-Tandem of Laboratori Nazionali di Legnaro. The isomers ware populated in the ^{170}Er(^{28}Si,5n) reaction at a beam energy of 143 MeV. The excited ^{193}Pb nuclei recoiled out of the 0.5 mg/cm^2 ^{170}Er foil into a solid 0.2 mm Hg layer mounted on a Cu cold finger at a temperature T = 170(1) K.

The QI frequency ω_0 resulting from the $R(t)$ spectra (see Fig. 4) led to an average value $\nu_Q(29/2^-) = 1.20(9)$ GHz, from which the quadrupole moment of the $29/2^-$ isomer $|Q_s| = 2.84(26)$ eb has been derived [24].

Tilted-axis cranking (TAC) calculations have been performed, using the model as it is described in Ref. [25], except that the electric quadrupole moment is calculated as the expectation value of the proton system. The quadrupole-quadrupole coupling constant, which controls the size of the deformation, was adjusted to reproduce the quadrupole moment of the $\nu(1i_{13/2}^{-2})_{12^+}$ isomers in ^{194,196}Pb. The calculations for ^{196}Pb yield $Q_s^{TAC}(11^-) = -2.86$ eb and $Q_s^{TAC}(16^-) = -1.20$ eb. Note, that the model parameters have been adjusted to the nearly-spherical 12^+ state, and do not reproduce very well the quadrupole moment of the deformed 11^- state ($|Q_s(11^-)| = 3.41(66)$ eb) and may be also of the 16^- state. In the case of ^{193}Pb the calculations yield $Q_s^{TAC}(29/2^-) = -2.85$ eb (in an agreement with the experimental result) and a quadrupole deformation ε_2 ($\approx 0.95\beta_2$) $= -0.13$. The estimated deformation of the "shears" state in ^{193}Pb takes an intermediate value between the experimentally observed oblate ($\beta_2 \approx -0.15 \div -0.2$) and nearly-spherical ($\beta_2 \approx +0.04 \div +0.07$) deformations. This is due to the fact that the shape of the "shears" states is a compromise between the neutron $\nu i_{13/2}^{-1}$ hole state, which drives the shape towards weakly prolate, and the proton $\pi(h_{9/2}i_{13/2})_{11^-}$ particle state, which drives it towards oblate deformation.

In order to estimate the quadrupole moment of the magnetic-rotational band head from the experimental quadrupole moments of the 11^- and the $\nu i_{13/2}^{-n}$ isomers we assume additivity of the E_2^0 quadrupole operator, $E_2^0(tot) = E_2^0(\pi) + E_2^0(\nu)$, similar as it has been used for the magnetic $M1$ operator [22]. Possible admixtures in the wave functions of the 11^- and the 12^+ isomers are automatically taken into account by using the measured quadrupole moments in the calculation. We obtain a value for the quadrupole moment of the "shears" state $Q_s^{add}(16^-) = -0.32(10)$ eb, which is much smaller than the one calculated by the TAC model.

When applying the same approach to ^{193}Pb, by coupling the quadrupole moment of the 11^- isomer in ^{194}Pb, $Q_s = (-)4.48(86)$ eb [15] with that of the $13/2^+$ isomer in ^{193}Pb, $Q_s = +0.195(10)$ eb, we obtain $Q_s^{add}(29/2^-) = -3.2(6)$ eb, which is similar to the experimental value and the TAC calculation. This imposes the question: why is there a difference in the case of ^{196}Pb and not in the case of ^{193}Pb? In ^{194,196}Pb the reduction compared to the much larger value for the 11^- state results partly from the coupling to the 12^+ state with its small and positive quadrupole moment, but it is mainly a geometrical effect. This is best seen in the semi-classical version of the additivity approach. Here, the total spectroscopic quadrupole moment can be written as [25]:

$$Q_s = (1 + 3/2I)^{-1} \left[d_{00}^2(\theta_\pi)Q_{0,\pi} + d_{00}^2(\theta_\nu)Q_{0,\nu}\right] \qquad (1)$$

The Wigner function d_{00}^2 goes through zero near $\theta \approx 55°$, which is close to the angles ($\theta_\pi \approx \theta_\nu \approx 45°$) of the proton and neutron blades with the rotational axis at the band head. In the case of the $29/2^-$ state in ^{193}Pb there is only one neutron hole coupled to the 11^- state. In this way the neutrons affect the deformation of the "shears" state much less, than in the case of ^{196}Pb, and the additivity rule seems to hold.

4 Summary

We have measured the quadrupole moments of the 11^- isomers in ^{194,196}Pb [15,16]. These first measurements of quadrupole moments of 2p-2h excitations across the $Z = 82$ shell gap give unambiguous experimental information on the nature of the 11^- intruder states in the neutron-deficient Pb nuclei. A comparison of the experimental numbers with particle-vibration coupling calculations shows that the coupling of the valence protons with a more collective underlying core is causing a substantial increase of the quadrupole moments of the 11^- isomers in Pb compared to the corresponding quadrupole moments in Po.

Calculations within the framework of the TAC model reproduce the quadrupole moment and the deformation of the 11^- and 12^+ isomers in ^{194,196}Pb. While the 12^+ neutron hole state is nearly spherical $\beta_2^{TAC}(12^+;^{196}\text{Pb}) = 0.044$, the 11^- state is found to be moderately deformed $\beta_2^{TAC}(11^-) = -0.127$. The calculations, using the same model parameters, predict that the deformation of the "shears" states is similar to the deformation of the 11^- intruder states: $\beta_2^{TAC}(16^-;^{196}\text{Pb}) = -0.134$ and $\beta_2^{TAC}(29/2^-;^{193}\text{Pb}) = -0.137$.

We have also measured for the first time the static quadrupole moment of a 'shears' state, the $I^\pi = 29/2^-$ isomer in ^{193}Pb, $|Q_s = 2.84(26)|$ eb. This value supports the conclusion that the deformation of the "shears" state is determined by the polarization effect of the proton excitation. On the other hand, so far magnetic bands have been observed in the Pb nuclei where intruder excitations make the configuration of the band heads. Despite searches, such bands have not been observed for the Po nuclei where the normal proton state is much less deformed. Is this accidental, or larger core polarization is required for this motion in atomic nuclei?

Acknowledgments

This work includes experiments carried out in Louvain-la-Neuve, Belgium by the Leuven-Bonn-Sofia collaboration (experiments PH-152 and PH-174) and at the Legnaro National Laboratory, Italy by the Leuven-Sofia-Bucharest-Bonn-Legnaro/Padova collaboration (experiment 01/50). Support from the Bilateral Fund, contract BIL02/24 and through the EU TMR Programme under Contract No. HPRI-CT-1999-00083 is acknowledged. G.N. and K.V. acknowledge support of the Flemish Science Foundation (FWO-Vlaanderen).

References

1. R.M. Clark and A.O. Macchiavelli, Annu. Rev. Nucl. Part. Sci. **50**, 1 (2000).
2. Amita *et al.*, At. Data Nucl. Data Tables **74**, 283 (2000).
3. S. Frauendorf, Rev. Mod. Phys. **73**, 463 (2001).
4. G. Baldsiefen *et al.*, Phys. Lett. B **275**, 252 (1992).
5. K. Heyde *et al.*, Phys. Rep. **102**, 211 (1983).
6. J.L. Wood *et al.*, Phys. Rep. **215**, 101 (1992).
7. F.R. May *et al.*, Phys. Lett. B **68**, 113 (1977).
8. W. Nazarewicz, Phys. Lett. B **305**, 195 (1993).
9. G.D. Dracoulis *et al.*, Phys. Lett. B **432**, 37 (1998).
10. A.N. Andreyev *et al.*, Nature (London) **405,** 430 (2000).
11. K. Van de Vel *et al.*, Phys. Rev. C **65,** 064301 (2002).
12. N. Smirnova *et al.* Phys. Lett. B (2003) in print.
13. T. Niksic *et al.*, Phys. Rev. C **65,** 054320 (2002).
14. G. Neyens, Prog. Rep. Phys. **66**, 633 (2003).
15. K. Vyvey *et al.*, Phys. Rev. C **65,** 024320 (2002).
16. K. Vyvey *et al.*, Phys. Rev. Lett. **88,** 102502 (2002).
17. F. Hardeman *et al.*, Phys. Rev. C **43**, 130 (1991).
18. J. Penninga, *et al.*, Nucl. Phys. **A471**, 535 (1987).
19. J. A. Becker *et al.*, Nucl. Phys. **A522**, 483 (1991).
20. K. Vyvey *et al.*, Phys. Lett. B **538**, 33 (2002).
21. G. Baldsiefen *et al.*, Phys. Rev. C **54,** 1106 (1996).
22. S. Chmel *et al.*, Phys. Rev. Lett. **79**, 2002 (1997).
23. R. Vianden, Hyp. Int. **35,** 1079 (1987).
24. D.L. Balabanski *et al.*, submitted to Phys. Rev. Lett. (2003).
25. A. Frauendorf, Nucl. Phys. **A677,** 115 (2000).

STRUCTURE OF BANDS IN NEUTRON-RICH EVEN PALLADIUM ISOTOPES

A. GIANNATIEMPO*+, A. NANNINI+ AND P. SONA*+

Dipartimento di Fisica, Università di Firenze and INFN, Sezione di Firenze+*
via G. Sansone 1, Sesto Fiorentino, Florence, Italy
E-mail: giannatiempo@fi.infn.it

The analysis of the new experimental data on neutron-rich palladium isotopes, performed in the framework of the IBA-2 model, supports our previous findings on the importance of mixed-symmetry components for a correct description of the collective positive-parity bands observed in the mass region A $\simeq$ 80 and $\simeq$ 110.

1 Introduction

In the last few years a noticeable effort has been devoted to the experimental study of neutron-rich palladium isotopes produced via spontaneous and heavy-ion-induced fission and investigated via delayed and prompt γ-ray spectroscopy (see, e.g., Refs. [1,2,3,4,5]). In particular, new experimental data on even- and odd-spin bands of positive parity in ^{112,114,116}Pd isotopes and information on ^{118}Pd, a nucleus not studied at all previously, were provided.

We have recently analyzed the positive-parity states in even Pd, Ru and Kr isotopes in the framework of the IBA-2 model [6], focusing on the identification of states of mixed symmetry (MS) character [7,8]. The three chains have in common an U(5) structure for a neutron number close to 50 changing towards an O(6) structure when approaching the middle of the relevant neutron shell. As a result of the analysis, it turned out that the g.s. band in the Ru and Pd chains and the band based on the 3_1^+ state in the Ru and Kr chains have a symmetry character changing in going from low-lying states where fully symmetric (FS) component are large or predominant to high-lying states which have a clear MS character.

The new information on $^{112-118}$Pd is now exploited to test whether it supports our previous findings on the positive-parity bands in these three chains.

2 Results and discussion

In the analysis of $^{100-116}$Pd [7] we used the Hamiltonian

$$H = \varepsilon\,(\hat{n}_{d_\pi} + \hat{n}_{d_\nu}) + \kappa\,\hat{Q}_\pi \cdot \hat{Q}_\nu + w_{\pi\nu}\,\hat{L}_\pi \cdot \hat{L}_\nu + \hat{M}_{\pi\nu}. \tag{1}$$

and the standard expressions of $E2$ and $M1$ transition operators. Six out of the twelve model parameters were kept fixed and two of those which were varied are almost constant.

We have compared the new experimental information on ^{112,114,116}Pd with the predictions of the model keeping the same parameters as in Ref. [7], apart from the Majorana parameters ξ_2 and ξ_3 which were very slightly varied in ^{114,116}Pd.

As to the determination of the parameters for ^{118}Pd, we repeated the same kind of analysis performed for the other isotopes and, for the parameters not kept

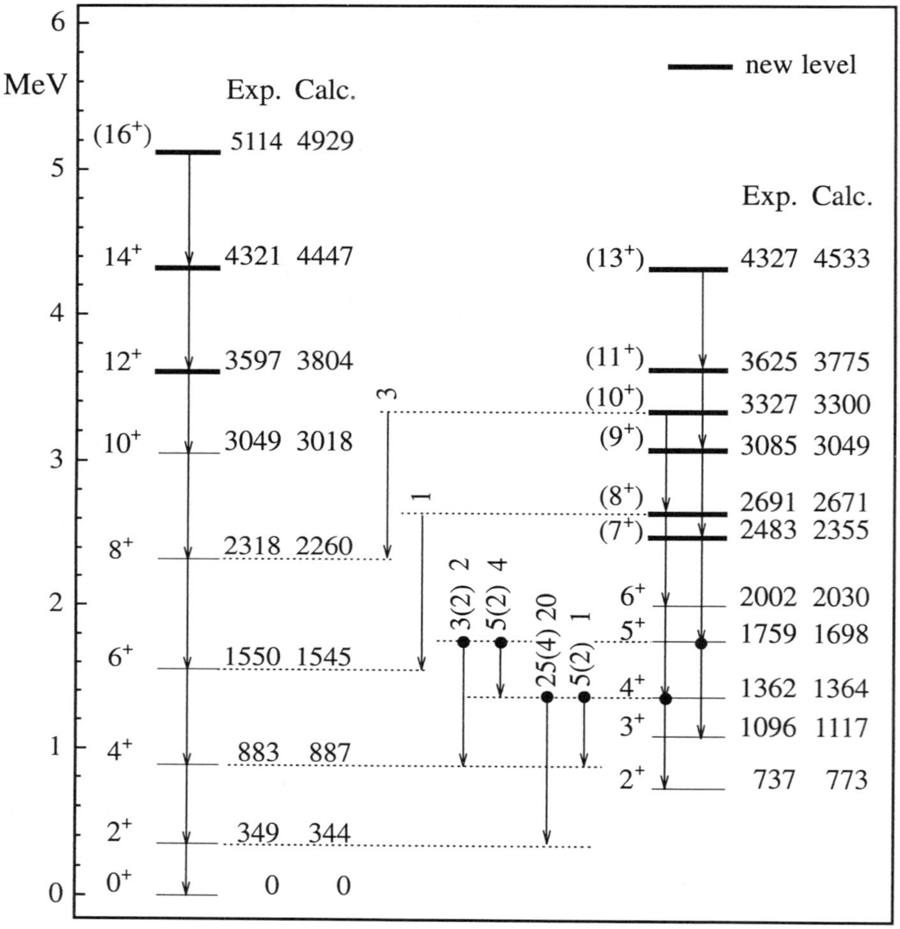

Figure 1: Comparison of experimental and predicted excitation energies and e.m. properties in ^{112}Pd. The intensities of the interband transitions from the 4^+ and 5^+ states in the γ-band are given as percentage of the corresponding intraband transition. The intensity of the interband transitions observed from the states of spin 8 and 10 has not been reported.

fixed, obtained values which are quite close to those of ^{114,116}Pd. Their values makes it possible to recognize the trend past the neutron mid shell at N = 66 of the parameters varying along the isotopic chain so that we can now confidently make predictions for still heavier palladium isotopes whose structure should approach again the U(5) limit.

As an example of the information that the new experimental data provides on the structure of the heavier palladium isotopes we show in Fig. 1 the case of ^{112}Pd. Both the g.s. band and the so called γ-band (built on the 2_2^+ state and made-up by states of alternate even- and odd-spin) have been extended to higher spins. It is seen that the new levels have an energy rather close to that predicted and that also the staggering observed in the γ-band is reasonably reproduced. The calculations predicted a stretched cascade for the g.s. band, for the even-spin and for the odd-spin bands (with some interband branchings from the 4^+ state and 5^+ states) of the γ-band. This is just what has been observed in recent experiments.

The information on the experimental excitation energies of the members of the g.s. band in the heavier Pd isotopes is collected in Fig. 2 and compared to the model predictions. The new data concerns both levels and spin-parity assignments. As expected for states of collective nature, the higher-lying states display a regular trend as a function of the mass number, the excitation energy showing a minimum for A close to the neutron mid-shell. This is also the case for the low-lying states of spin up to 8, which turn out to have FS character. The agreement obtained between the experimental and calculated excitation energies and the stretched character predicted for all the bands, as observed experimentally, makes one confident that a satisfactory description of the states considered has been achieved. We also remark that the agreement on the energies worsen drastically for the states of spin ≥ 10 when assigning to the Majorana parameters values sufficiently high to push MS states at higher energies.

To give an example of the structure of the states belonging to the g.s. band we have reported on the left-hand side of Fig. 3 the F-spin and n_d (d-boson number) components for the relevant wave functions of ^{118}Pd. The low-spin states have a quite pure FS character with one predominant n_d component. A large mixing of F-spin and n_d components is present in the wavefunction of the state of spin 8. The states of spin 10 and 12 have instead a clear MS character: the former has comparable $F_{max} - 1$ and $F_{max} - 2$ components and the $n_d = 6$ component is dominating, while the latter has $F_{max} - 2$, $n_d = 7$ components. To account for the structure of this band we have referred to the U(5) limit of the model and have reported on the right-hand side of the figure the group of states of interest. When the values of the Majorana parameter ξ_2 are positive and those of ξ_3 negative (as we found in the Pd chain) the spacing of the two MS bands reported in the right part of the figure is smaller than that of the FS band (see Ref. [7]). It can therefore happen that the excitation energies of states in the MS bands become very close or even lower than the excitation energies of the states of equal spin of the g.s. band. The calculated structure of the low-lying states in ^{118}Pd closely resembles that of the corresponding states of the FS band in the U(5) limit, on the other hand the structure of the state of spin 10 and 12 is close to that of the states of same spin in the MS bands.

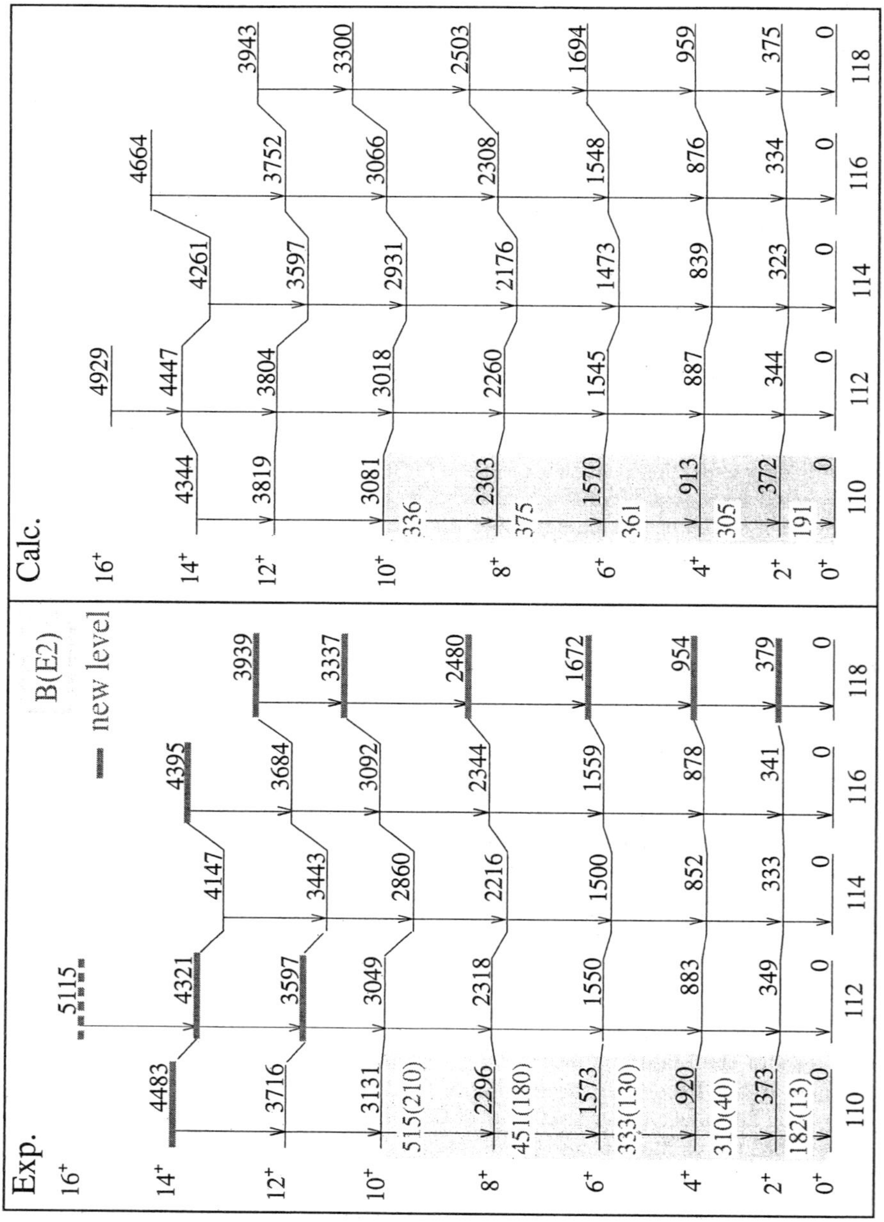

Figure 2: Experimental excitation energies of the g.s. band in the newly studied heavier palladium isotopes are compared to the predicted ones. Data on the g.s. band of ^{110}Pd are also shown since this is the only case in the Pd chain where the comparison on B(E2)'s up to spin 10 is possible. B(E2) values are given in units $10^3 e^2\ b^2$.

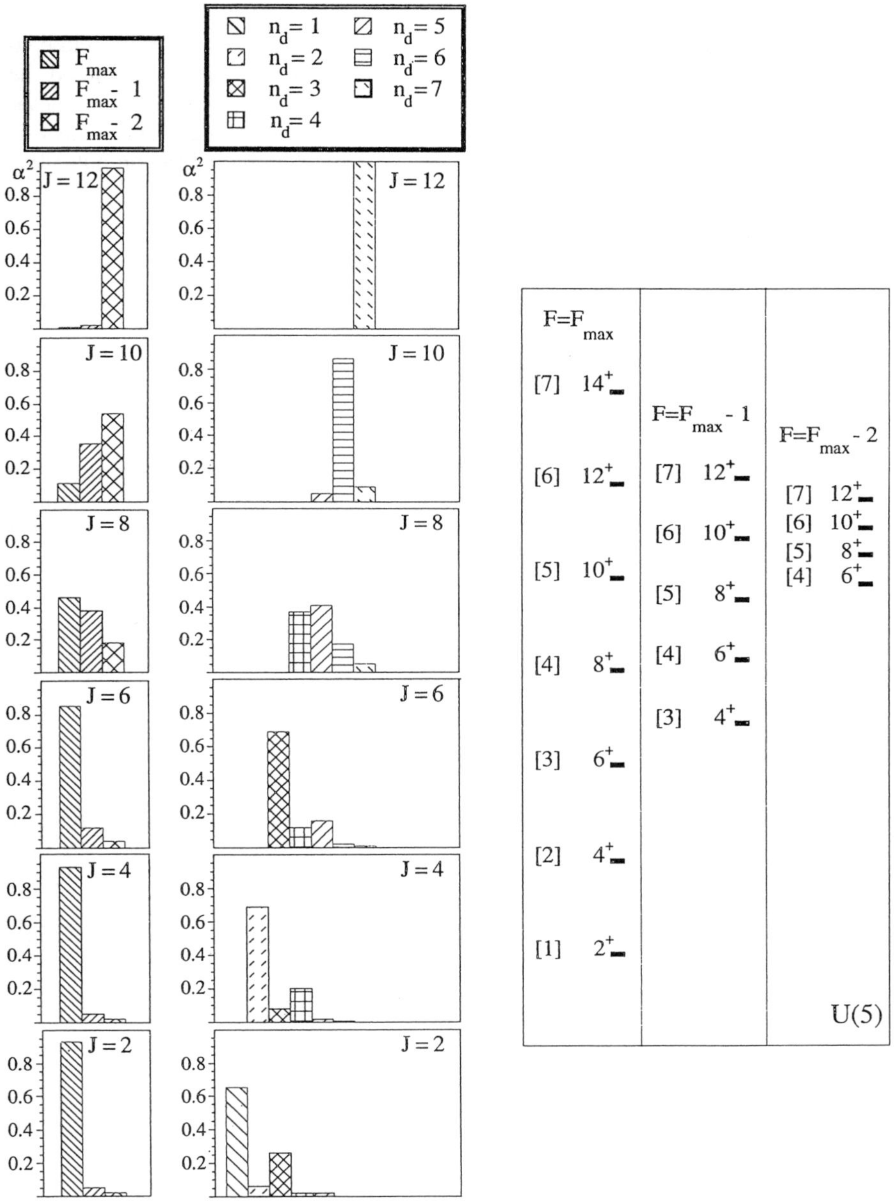

Figure 3: Left: square amplitudes of the F-spin and n_d components of the wave functions of the g.s. band in ^{118}Pd are reported. Right: relevant group of states in the U(5) limit. The d-boson number is reported in square brackets.

The new experimental data on $^{112-116}$Pd clearly establish the presence of two bands built on the 2_2^+ state and on the 3_1^+ state; their energies are quite close to the predicted values. Very small interband transition strengths are predicted, as experimentally observed. It turns out that also in this case the symmetry character of the bands is changing from the low-spin states, where the FS component is dominant or large, to the high-spin states, which have dominant MS components.

3 Conclusions

The new experimental data on

- the g.s. band
- the band based on the 3_1^+ state
- the band based on the 2_2^+ state

in the neutron-rich palladium isotopes have been studied in the framework of the IBA-2 model. The overall agreement between the model predictions and the new experimental spectroscopic data reflects the importance of the MS components for a correct description of the structure of these bands.

References

1. K. Butler-Moore *et al.*, J. Phys. G; Nucl. Part. Phys. **25**, 2253 (1999).
2. M. Houry *et al.*, Eur. Phys. J. A **6**, 43 (1999).
3. A. Jokinen *et al.*, Eur. Phys. J. A **9**, 9 (2000).
4. X.Q. Zhang *et al.*, Phys. Rev. C **63**, 027302 (2001).
5. J.H. Hamilton *et al.*, Eur. Phys. J. A **15**, 175 (2002).
6. F. Iachello and A. Arima, *The interacting boson model* (Cambridge University Press, Cambridge, 1987).
7. A. Giannatiempo, A. Nannini, and P. Sona, Phys. Rev. C **58**, 3316 (1998), *ibidem* **58**, 3335 (1998).
8. A. Giannatiempo, A. Nannini, and P. Sona, Phys. Rev. C **62**, 044302 (2000).

OCTUPOLE TWO PHONON STATES IN DEFORMED NUCLEI

G. GRAW, Y. EISERMANN, R. HERTENBERGER, AND H.-F. WIRTH
Ludwig-Maximilians-Universität München, Am Coulombwall 1, D-85748 Garching, Germany

S. CHRISTEN, O. MÖLLER, D. TONEV, AND J. JOLIE
Institut für Kernphysik, Universität zu Köln, Zülpicher Str. 77, D-50937 Köln, Germany

C. GÜNTHER
ISKP, Universität Bonn, Nußallee 14-16, D-53115 Bonn, Germany

A.I. LEVON
Institute of Nuclear Research, Academy of Science, Kiev

N.V. ZAMFIR
Wright Nuclear Structure Laboratory, Yale University

By means of the (p,t) reaction the excitation spectra of 0^+ states in ^{158}Gd [1], ^{228}Th, ^{230}Th, and ^{232}U have been studied using the Q3D magnetic spectrograph facility at the Munich tandem accelerator. The 0^+ transfer angular distributions have very large cross sections at very small reaction angles, a feature that allows to identify these states in otherwise very complicated and dense spectra. We resolved for each of these nuclei typically 12 excited states with safe 0^+ assignments. The studied excitation energy range is up to 3.1, 2.5, 2.5, and 2.1 MeV, resp. As for ^{158}Gd [2], we compare the data with *spdf*-IBA calculations. The parameters are chosen to reproduce the low lying spectra of these axially symmetric, statically deformed nuclei, especially the bands of negative parity. For the energy ranges considered, the IBA predicts five excited 0^+ states of pure *sd* (quadrupolar) bosonic structure for all these nuclei, but three, six, seven, and four excited 0^+ states resp., which have two bosons in the *pf* boson space. They are related to – or represent – octupole two phonon excitations. The collective model descriptions provide nearly quantitatively the number of the observed excited 0^+ states in these actinide nuclei.

1 Introduction

In the spectra of nuclei the observation of low lying 0^+ states is indicative for the presence of specific modes of nuclear excitations. In a double closed shell nucleus as ^{208}Pb only two excited 0^+ states have been observed up to now: a neutron pairing vibrational mode and an octupole two phonon excitation [3]. Intruder configurations may contribute in addition as observed for ^{40}Ca and in less closed nuclei as ^{90}Zr, ^{96}Zr [4], and ^{112}Cd [5]. In earlier (p,t) studies of the medium weight nuclei ^{146}Nd [6], ^{146}Sm [7], ^{134}Ba, and ^{132}Ba [8], typically nine excited 0^+ states have been observed in the excitation energy range of 2.5 to 4 MeV. Specific particle-hole and particle-particle correlations [6], and particle-core coupling effects [7] had been considered to describe the number of 0^+ states in ^{146}Nd [6] and ^{146}Sm [7].

In deformed nuclei we have to expect additional 0^+ states because of the quantisation with respect to the intrinsic axis: An excitation mode with angular momentum J splits into states distinguished by their K quantum numbers, which range from zero to J. Thus e.g. the quadrupolar (one phonon) vibration with $J = 2^+$ splits into $K = 0^+$ and $K = 2^+$ states, which are known as the heads of the related β and γ vibrational bands. This means that we have to discuss splitting of collective and of non-collective modes of excitation. As it is well established e.g. for the 0^+ in ^{208}Pb and from QRPA-like calculations, comp. Ref. [3], in most cases the lowest state of a given angular momentum J^π is collective. States related to specific two quasiparticle modes follow at much higher excitation energies. Thus, low lying excitations are expected to be described well within models accounting for the respective collective features.

The only case studied up to now was ^{158}Gd [1]. At excitation energies below 3.1 MeV 13 excited 0^+ states had been identified. Zamfir *et al.* [2] compared these data with *spdf*-IBA calculations. Mixing between d and pf bosons was neglected, the f (and p) bosons account for octupole collectivity. The parameters were chosen to reproduce the low lying spectra, especially the bands of negative parity.

The IBA calculation predicts five excited 0^+ states of pure *sd* (quadrupolar) bosonic structure and three excited 0^+ states which have two bosons in the pf boson space. They are related to – or represent – octupole two phonon excitations (OTP). This study showed that the collective model descriptions account for an essential part of the number of the observed excited 0^+ states in the rare earth nucleus ^{158}Gd.

The aim of the present work is to study axially symmetric, statically deformed actinide nuclei and especially to benefit there from the known change in octupole collectivity, which is very large for the lighter of the actinide nuclei and weaker for the heavier ones. We studied so far ^{232}U, ^{230}Th, and ^{228}Th. Compared to $E_x(1^-,{}^{158}\mathrm{Gd}) = 977.1$ keV of ^{158}Gd the excitation energies of their lowest negative parity states are $E_x(1^-,{}^{232}\mathrm{U}) = 563.2$ keV, $E_x(1^-,{}^{230}\mathrm{Th}) = 508.1$ keV, and $E_x(1^-,{}^{228}\mathrm{Th}) = 328.0$ keV. Thus we compare three similar nuclei, which differ only by one neutron pair and/or by one proton pair, but vary an important parameter, the octupole collectivity.

2 Experiments

2.1 Experimental procedure, data taking

The (p,t) experiments have been performed at the Munich tandem accelerator laboratory with a 25 MeV proton beam with an intensity of 1–2 μA on the target. The reaction products have been analyzed with the Q3D magnetic specrograph and detected in its focal plane. We used two different focal plane detectors which are multiwire proportional chambers with readout of a cathode foil structure for position determination and $\Delta E/E_{\mathrm{rest}}$ particle identification. The targets ^{234}U, ^{232}Th, and ^{230}Th had a thickness of 100 μg/cm^2 each, evaporated on 22 μg/cm^2 thick carbon backings. The isotopic purity was about 99 %. The resulting triton spectra have a resolution of 6–7 keV FWHM and are virtually background-free. Angular distributions of the cross sections are extracted from spectra at five different laboratory angles in the case of ^{230}Th(p,t)^{228}Th (7.5°, 12.5°, 16°, 20°, 30°) and at ten angles in the case of ^{232}Th(p,t)^{230}Th and ^{234}U(p,t)^{232}U (5°, 7.5°, 10°, 12.5°, 16°, 20°, 26°, 30°, 35°, 40°). A typical spectrum is shown in Fig. 1.

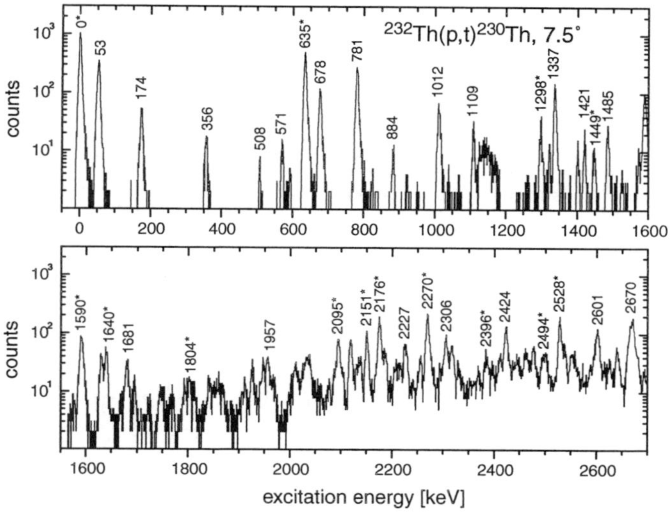

Figure 1. Spectrum for ^{232}Th(p,t)^{230}Th in logarithmic scale. Some levels are labeled with their excitation energy in keV. States assigned as 0^+ are marked with an asterisk.

2.2 Angular distributions, DWBA analysis

Differential cross section angular distributions for those observed transfers to ^{230}Th which are assigned as 0^+ are shown in Fig. 2 together with calculated DWBA angular distributions for the 0^+ transfer. There are a few transitions with very low cross sections below 1 μbarn/sr which show for very low scattering angles a decreasing cross section with increasing angle, but otherwise not the typical pattern. They are omitted in the present state of discussion.

In the range of very low cross sections higher order coupled channel effects may produce angular distributions, which deviate strongly from those for one-step direct excitations. In this way we avoid wrong 0^+ assignments, but may miss to assign 0^+ transfers. We cannot claim to observe all excited 0^+ states. Our analysis is restricted to those, showing up with a spectroscopic strength of about 0.5 percent of the ground state excitation strength or more, defining in this way a class of states.

2.3 Experimentally obtained 0^+ *states*

The excitation energies E_x and the spectroscopic factors of excited 0^+ states we observe for ^{232}U, ^{230}Th and ^{228}Th are displayed in Fig. 3. The spectroscopic factors are normalized to the observed gound state transfer strength. The values given do not refer to a kind of best over all agreement with the observed angular distribution, but refer to the experimental cross sections, measured at $\theta_{lab} = 7°$, and scale these with the calculated excitation energy dependence of DWBA cross sections. The

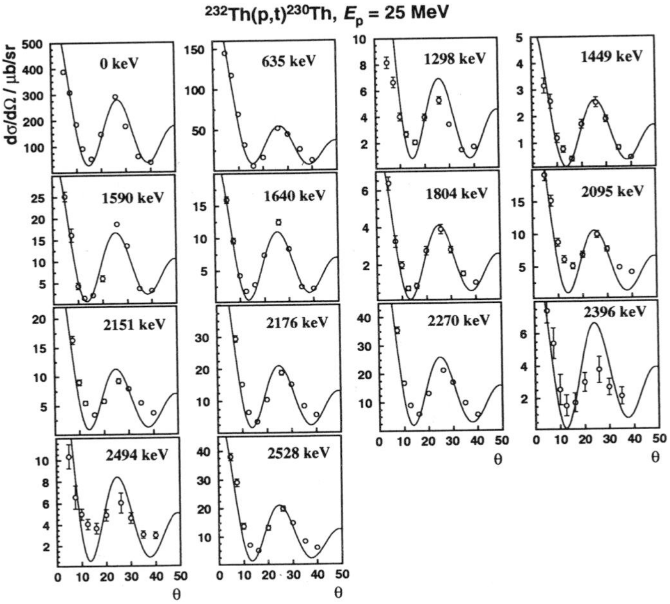

Figure 2. Angular distributions of assigned 0^+ states in ^{230}Th with DWBA calculations.

advantage of the latter way is to provide a definite procedure. This is appropriate since at present we do not yet know about the specific structure of the individual states.

The summed 0^+ transfer strength to excited states adds up to 75 %, 64 %, and 60 % of the ground state transfer strengths for ^{232}U, ^{230}Th, and ^{228}Th, resp. One should note here that ^{232}U and ^{230}Th have the same number of neutrons ($N = 140$). They differ by one proton pair, as ^{230}Th and ^{228}Th do by one neutron pair.

3 Analysis of the data

3.1 Excited 0^+ *states in deformed nuclei*

In statically deformed, axially symmetric nuclei one may expect a number of excited 0^+ states, since any excitation mode with angular momentum J splits energetically into $J+1$ components with projections K relative to the internal symmetry axis, and K ranging in between $K = J, J-1, ..., 1, 0$. In this way any excitation of positive parity causes a $K = 0^+$ state, one of the 0^+ states we are looking for.

The nuclei we study are prolate, as most of the deformed nuclei. For prolate nuclei we have to expect, as for Nillson states, strongest binding for the lowest K state. This enhances the number of 0^+ states at low excitation energies. This effect is especially pronounced for the K-splitting of the octupole vibrational state, as it

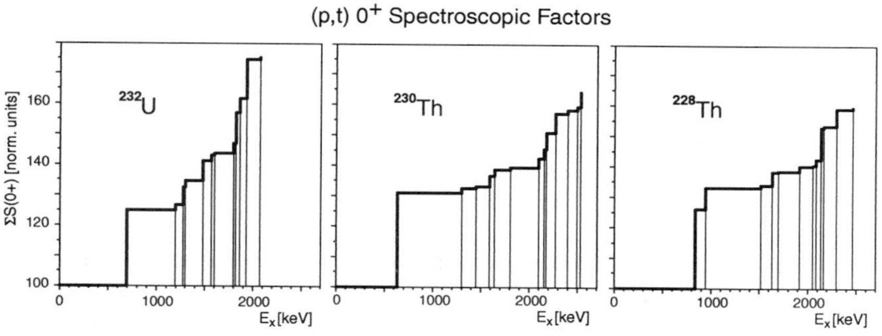

Figure 3. Incremental plot of the transfer strengths to excited 0^+ states in ^{232}U, ^{230}Th, and ^{228}Th.

is discussed by Cottle and Zamfir [9] for rare earth nuclei: The respective $K = 0^-$ and $K = 1^-$ band heads are observed at very low excitation energies, whereas the $K = 3^-$ strength is shifted upwards in energy.

In their interpretation of recent measurements of 0^+ states in ^{158}Gd [1] Zamfir, Zhang and Casten [2] restrict their discussion to multiphonon excitations of quadrupolar and octupolar type only. The excitation energies of 0^+ states resulting from quadrupolar excitations are obtained with nearly identical values both in the geometrical collective model (GCM) and in the conventional interacting boson model (IBA). Qualitatively speaking, one obtains one $K = 0^+$ state from the $J = 2^+$ one-phonon excitation, the β-vibrational state, followed by three $K = 0^+$ states from the three $J = 0^+, 2^+, 4^+$ two-phonon excitations and perhaps further $K = 0^+$ states from still higher phonon excitation.

The octupolar excitations contribute to the number of $K = 0^+$ states via the $K = 0^+$ projections of their two-phonon excitations with $J = 0^+, 2^+, 4^+$ and 6^+, and via the coupling of these octupole-two-phonon excitations with quadrupole-phonon excitations. Formally, this is treated within the IBA. To keep their discusion transparent, they used a Hamiltonian with minimal coupling, especially they excluded coupling of d with f bosons. Then this sdf-IBA calculation gives the $K = 0^+$ excitations of pure quadrupolar type at exactly the same energies as a pure sd-IBA, and all of the additionally calculated $K = 0^+$ states result with pure f^2-boson content, equivalent to octupole-two-phonon excitations. In this way these exploratory calculations accounted for a considerable amount of the number of observed $K = 0^+$ states in the range of excitation energy considered.

One has to expect, however, further $K = 0^+$ states, part of collective nature and others of non-collective nature:

Collective excitations of multipolarity 0^+ and 4^+, the monopole pairing vibrational excitation (MPV) and some hexadecapole vibrational collectivity, has to be expected and shall lead to a respectable number of excited $K = 0^+$ states. The monopole pairing vibration is well established for ^{208}Pb. Of the two known 0^+ states for ^{208}Pb, the lower one is identified as the MPV state, and the higher one as the 0^+ member of the octupole-two-phonon excitation multiplet, compare Ref. [3] and further references there. In the literature one expects at least two kinds of MPV states, one for neutron-pair excitations (n-MPV) and one for proton-pair excitations (p-MPV). The latter one usually is expected at higher excitation energy. Because of its collective (vibrational) nature, in a (p,t) reaction the n-MPV state is expected to be strongly excited. In case of a relatively dense spectrum of 0^+ states the n-MPV state will mix with the nearby states. In our case, compare Fig. 3, mixing is significant and one may discuss as a center of the transfer strength an excitation energy near 1600 keV and equate this with an unperturbed excitation energy of the neutron monopole pairing vibration state (nMPV). The pairing vibrational excitations result from a particle-particle coupling in the residual interaction, which, however, is not included in the usual RPA- or IBA-like structure calculations. Within these frames the pairing vibrational excitation has to be considered as a kind of intruder configuration.

Hexadecapole vibrational collectivity is well established in spherical nuclei, at excitation energies near or weakly above the collective octupole state. In inelastic scattering it is related with large one-step transition strength to excited 4^+ states. The analysis of hexadecapole collectivity is complicated because of the necessity to differentiate against quadrupole-two-phonon 4^+ excitations and related processes. Also the hexadecapole strength may be distributed over a few neighbouring states, in contrast to the octupole strength, which is concentrated in the lowest 3^- state. One may compare with a study of ^{112}Cd by Hertenberger [5] and Garrett *et al.* [10].

As for the octupole vibration, it is reasonable to assume that the $K = 0^+$ member of the $J = 4^+$ collective excitation is pushed down to rather low energies and should be observed within our experimental range. This may be treated formally introducing a g boson and expanding a *sd*-IBA to a *sdg*-IBA, analog to the case of octupole collectivity. In this way additional $K = 0^+$ states, resulting from quadrupole-hexadecapole coupling, will also derive.

As interesting question will remain whether, in addition to these states of collective origin discussed above, we will have to expect also $K = 0^+$ states of non-collective origin from positive parity two-quasiparticle (2QP) excitations. It is the concept of collectivity, that for each given multipolarity J they are expected at considerable higher energy than the respective collective vibrational state. We discussed the relevant multipolarities $J = 0^+, 2^+, 4^+$, $J = 0^+$ and 4^+ are high in energy, noncollective $J = 2^+$ states might contribute at higher excitation energies. Experimentally it will be difficult to differentiate these states against the collective quadrupolar excitations. It will be very interesting to see respective microscopic calculations similar to the QRPA multiphonon calculations, compare Ref. [3], but in a non-spherical basis.

3.2 Deformed nuclei and IBA

The role of the octupole degree of freedom in heavy deformed nuclei, and the related description with f bosons, added to the established IBA in the *sd*-boson space (*sd*-IBA), has been studied systematically for deformed rare earth nuclei by Cottle and Zamfir [9] and for transitional actinides by Zamfir and Kusnezov [11,12].

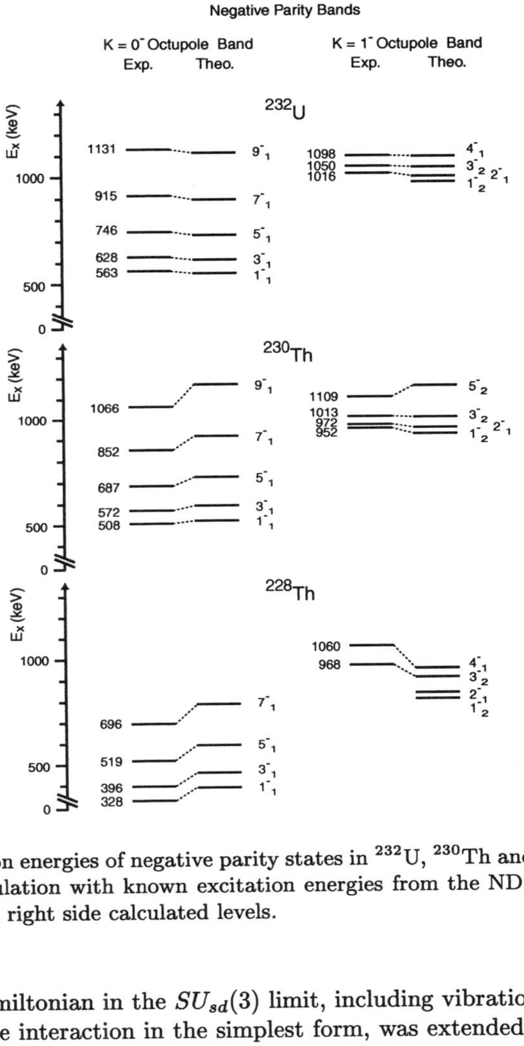

Figure 4. Excitation energies of negative parity states in ^{232}U, ^{230}Th and ^{228}Th comparing an *spdf*-IBA calculation with known excitation energies from the ND compilation. Left side experimental, right side calculated levels.

The IBA Hamiltonian in the $SU_{sd}(3)$ limit, including vibrational contributions and a quadrupole interaction in the simplest form, was extended to the $SU_{spdf}(3)$ limit:

$$H = \epsilon_d \hat{n}_d + \epsilon_p \hat{n}_p + \epsilon_f \hat{n}_f - \kappa \hat{Q}_{spdf} \cdot \hat{Q}_{spdf}, \qquad (1)$$

where ϵ_d , ϵ_p and ϵ_f are the boson energies and $\hat{n}_d$, $\hat{n}_p$ and $\hat{n}_f$ are the boson number operators. Note that the same strength κ of the quadrupole interaction describes the *sd* bosons and the *pf* bosons. The $\hat{Q}_{spdf}$ quadrupole operators

$$\begin{aligned}\hat{Q}_{spdf} &= \hat{Q}_{sd} + \hat{Q}_{pf} = [s^\dagger \tilde{d} + d^\dagger s]^{(2)} - (1/2)\sqrt{7}[d^\dagger \tilde{d}]^{(2)} \\ &+ (3/5)\sqrt{7}[p^\dagger \tilde{f} + f^\dagger \tilde{p}]^{(2)} - (9/10)\sqrt{3}[p^\dagger \tilde{p}]^{(2)} - (3/10)\sqrt{42}[f^\dagger \tilde{f}]^{(2)} \qquad (2)\end{aligned}$$

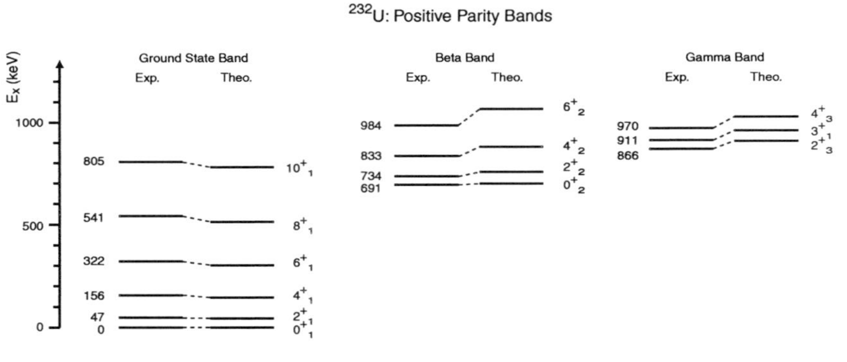

Figure 5. Excitation energies of positive parity states in ^{232}U, comparing an spdf IBA calculation with known excitation energies from ND compilation. Left side experimental, right side calculated levels.

are used as in Ref. [12], the $-\sqrt{7}/2$ factor in front of the $[d^\dagger\tilde{d}]^{(2)}$ may be adjusted introducing an additional parameter χ_{sd}. This Hamiltonian was used in Refs. [2,11,12].

For the rare earth nuclei the IBA in the *sdf*-boson space (*sdf*-IBA) reproduces reasonably well the main features of the observed negative parity states, for the actinide nuclei a better reproduction of the respective data is obtained if one allows in addition to the f boson for a p boson (*spdf*-IBA). The physical nature of the p boson is not clear. It may result as an artefact, or an anharmonicity, of an octupole excitation in a quadrupolar deformed potential. In the present context we treat the f or the combination of a p and an f boson (pf boson) as a way to describe octupole collectivity.

In Fig. 4 we display the reproduction of the excitation energies of negative parity states in ^{228}Th, ^{230}Th, and ^{232}U, comparing an *spdf*-IBA calculation with known excitation energies from the ND compilation. Left side experimental, right side calculated levels. The pf boson parameters are chosen to reproduce the $K = 0^-$ and $K = 1^-$ bandheads, they are determined by the known excitation energies of the $J^\pi = 1^-, K = 0^-$ and $J^\pi = 3^-, K = 1^-$ states. For ^{228}Th the $J^\pi = 1^-, K = 0^-$ excitation energy is 328.0 keV and thus significantly lower than 508.1 keV and

Table 1. Multipole parameters of the *spdf*-boson IBA calculation. The number of negative parity bosons is allowed to range from 0 to 3.

Nucleus Total Number of Bosons	^{232}U 12	^{230}Th 11	^{228}Th 10
ϵ_s	0.0000	0.0000	0.0000
ϵ_p	0.9900	1.0000	1.0500
ϵ_d	0.2500	0.2500	0.2100
ϵ_f	0.9400	0.9000	0.6500
κ	0.0120	0.0140	0.0180
χ_{sd}	1.3228	1.0000	1.3228

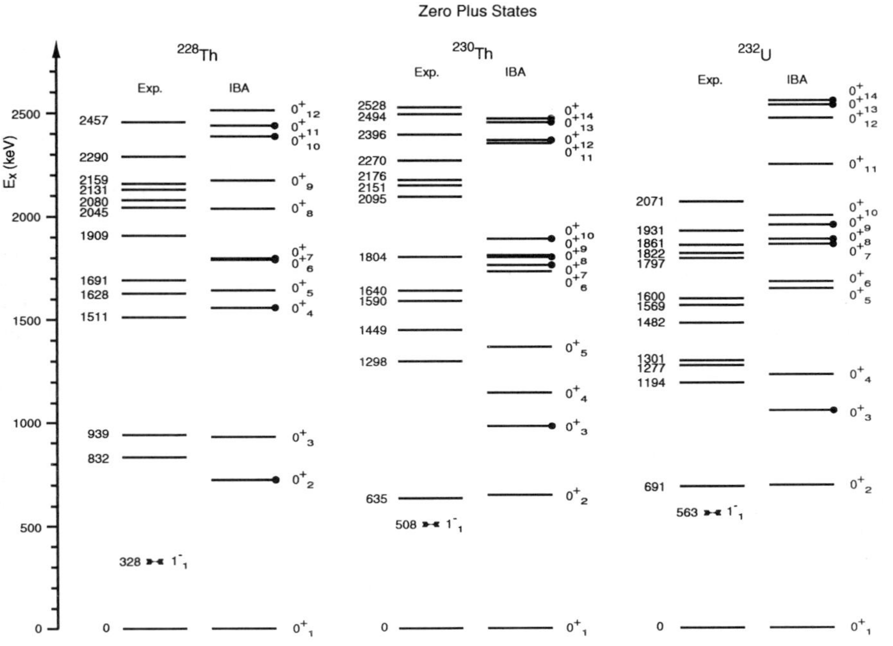

Figure 6. Excitation energies of all known excited 0^+ states in ^{228}Th, ^{230}Th, and ^{232}U, compared with *spdf*-IBA calculations. OTP states are marked by a dot. The excitation energies of the 1^-_1 states are indicated.

563 keV for ^{230}Th and ^{232}U, in contrast to the respective $J^\pi = 3^-, K = 1^-$ state excitation energies of 968.3, 1012.5 and 1050 keV, which are about the same.

The IBA parameters in the *sd* boson space are determined by the low energy spacing of the ground state band and the $J^\pi = 2^+_1, K = 0^+$ and $J^\pi = 2^+_1, K = 2^+$ heads of the β- and γ-bands, respectively. The values, as they are named in the code, are listed in Table 1. The excitation energies of positive parity states in ^{232}U, comparing an *spdf*-IBA calculation with known excitation energies from the ND compilation, are shown in Fig. 5. Left side experimental, right side calculated levels.

4 Comparison of 0^+ states with IBA calculations

The excitation spectra of the 0^+ states obtained for ^{232}U, ^{230}Th, and ^{228}Th and the results of *spdf*-IBA calculations are displayed in Fig. 6.

As for ^{158}Gd [2] in the *spdf*-IBA calculations mixing in between d and pf bosons is neglected and the f (and p) bosons account for octupole collectivity. The parameters are chosen to reproduce the low lying spectra, especially the bands of negative parity, compare Fig. 4. The key quantities for octupole collectivity, the

1^- excitation energies ($E_x(1^-, {}^{228}\mathrm{Th}) = 328.0\,\mathrm{keV}$ $E_x(1^-, {}^{230}\mathrm{Th}) = 508.1\,\mathrm{keV}$, and $E_x(1^-, {}^{232}\mathrm{U}) = 563.2\,\mathrm{keV}$) are also indiated. For the energy ranges considered (2.5, 2.5, and 2.1 MeV, resp.), the IBA predicts five excited 0^+ states of pure *sd* (quadrupolar) bosonic structure for each of these nuclei, but six, seven, and four excited 0^+ states, resp., which have two bosons in the *pf* boson space. They are related to – or represent – octupole two phonon excitations (OTP).

Inspecting the lowest three of the excited states of ^{228}Th, ^{230}Th and ^{232}U we have a good correlation in excitation energy between experiment and calculation. Significant is the much lower excitation energy of the second excited state in ^{228}Th ($E_x(0_3^+, {}^{228}\mathrm{Th}) = 939$ keV) than in ^{230}Th ($E_x(0_3^+, {}^{230}\mathrm{Th}) = 1298$ keV), which relates nicely with twice the excitation energy of the respective 1^- states. Thus, in the energy ranges considered (2.5, 2.5, and 2.1 MeV, resp.) the IBA predicts 11, 13, and 9 excited 0^+ states. Accounting in addition for the presence of a monopole pairing vibrational state, and perhaps one state from hexadecupole collectivity, both not included in the calculation, we have nearly perfect agreement with the number of 12, 13, and 12 observed states.

Further studies have to prove whether this agreement is accidental or physically correct. At present we can conclude that for these actinide nuclei collective model descriptions provide nearly quantitatively the number of the observed excited 0^+ states.

Acknowledgements

Work supported by the DFG (C4-Gr894/2-3), MLL, and US-DOE, contract number DE-FG02-91ER-40609.

References

1. S.A. Lesher *et al.*, *Phys. Rev.* C **66**, 051305 (2002).
2. N.V. Zamfir, J.-Y. Zhang, R.F. Casten, *Phys. Rev.* C **66**, 057303 (2002).
3. B.D. Valnion *et al.*, *Phys. Rev.* C **63**, 024318 (2001).
4. S.A. Fayans *et al.*, *Nucl. Phys.* A **577**, 557 (1994).
5. R. Hertenberger *et al.*, *Nucl. Phys.* A **574**, 414 (1994).
6. V. Yu. Ponomarev *et al.*, *Nucl. Phys.* A **601**, 1 (1996).
7. A.M. Oros *et al.*, *Nucl. Phys.* A **613**, 209 (1997).
8. G. Cata-Danil *et al.*, *Phys. Rev.* C **54**, 2059 (1996).
9. P.D. Cottle and N.V. Zamfir, *Phys. Rev.* C **54**, 176 (1996).
10. P.E. Garrett *et al.*, Conf. Proc. Capture Gamma-Ray Spectr., Prague 2002.
11. N.V. Zamfir and D. Kusnezov *Phys. Rev.* C **63**, 054306 (2001).
12. N.V. Zamfir and D. Kusnezov, *Phys. Rev.* C **67**, 014305 (2003).

IBFM2 STUDY OF ODD-A CS AND XE ISOTOPES AND BETA DECAY IN XE-CS

N. YOSHIDA
Faculty of Informatics, Kansai University,
2-1-1 Ryozenji-cho, Takatsuki-shi, 569-1095 Japan
E-mail: yoshida@res.kutc.kansai-u.ac.jp

L. ZUFFI
Dipartimento de Fisica dell'Universita' di Milano and Istituto Nazionale di Fisica Nucleare, Sezione di Milano, Via Celoria 16, Milano I-20133, Italy
E-mail: zuffi@mi.infn.it

S. BRANT
Department of Physics, Faculty of Science, University of Zagreb, 10000 Zagreb, Croatia
E-mail: brant@sirius.phy.hr

We show the results of calculations of the energy levels and electromagnetic properties of the positive-parity states as well as the beta-decay rates in the odd-mass Xe and Cs isotopes in the proton-neutron interacting boson-fermion model (IBFM2). The properties are well described by the model calculation.

1 Introduction

The Xe region has been extensively studied in the interacting boson model (IBM) and its variations[1,2,3,4,5,6]. The even-even nuclei in this region draw special interest for the IBM because of their O(6) symmetry characters. To describe the odd-mass nuclei, the interacting boson-fermion model (IBFM) is used in which an odd fermion is coupled to the even-even core composed of bosons. So far, however, the calculations in the IBFM, or in the IBFM2 which is the version distinguishing proton bosons from neutron bosons, have mainly concerned the energy levels and the electromagnetic properties.

The application of the IBFM2 to beta-decay was proposed in 1988 [7] where systematic calculations of the nuclei in the region with $52 \leq Z \leq 58$ were presented, and for Ru and Tc nuclei[8,9]. After years, much more experimental data are available.

We recently performed the calculation of the beta-decay in the odd-mass Rh-Pd isotopes using the IBFM2[10,11]. In the present contribution, we extend the calculation to the Xe-Cs region. We will show the results of calculations of the positive-parity states in the odd-mass Xe and Cs isotopes, including the beta-decay from Cs to Xe. The beta-transition rates not only to the ground state but also to some excited states are discussed.

2 Calculation in the Odd-A Xe and Cs Isotopes

The hamiltonian consists of the boson hamiltonian, the hamiltonian of the odd fermion, and the interaction between the bosons and the odd fermion. The IBM2

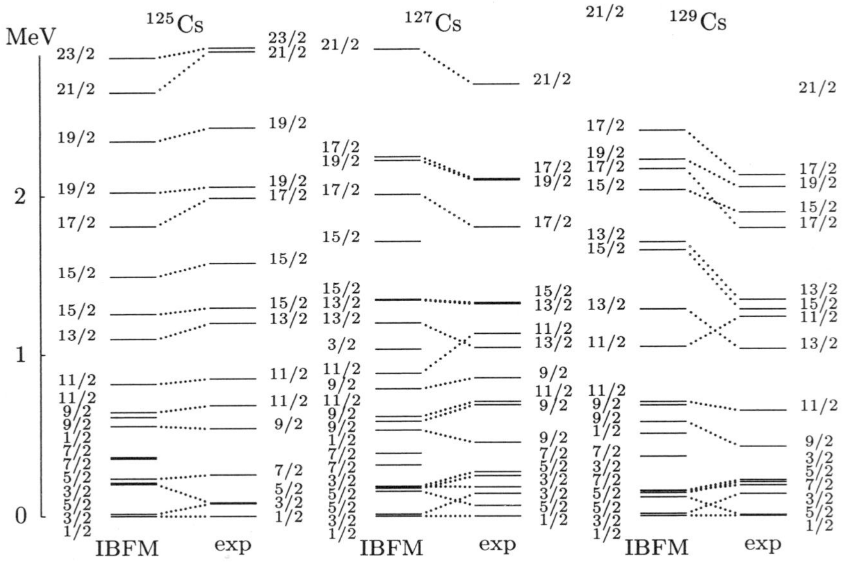

Figure 1. Energy Levels in the Cs Isotopes.

parameters for the even-even Xe isotopes are taken from Ref.[2]. To describe the odd-even isotopes of Cs (Xe), we couple a proton (neutron) to these cores.

2.1 Energy Levels in the Cs Isotopes

The odd-mass Cs isotope are described by coupling an odd proton to the even-even Xe cores. The proton single-particle energies are taken from Ref.[6]. The energy of $d_{5/2}$ has been reduced from the original value of 0.20 MeV to 0.05 MeV later in order to obtain better overall agreement in beta decay. The BCS equations are solved with the orbitals $g_{7/2}$, $d_{5/2}$, $s_{1/2}$, $d_{3/2}$ and $h_{11/2}$, with given $\Delta = 12/\sqrt{A}$ MeV. The first four orbitals are included in the IBFM2 calculation for the positive-parity states. In the boson-fermion interaction, the quadrupole and the monopole interactions are included between the odd proton and the neutron bosons, in addition to the exchange interaction of the quadrupole type. The parameterization is the same as Refs.[3,11]. We allow the interaction strengths vary gradually depending on the mass number. The adopted values are: $\Gamma_0 = 0.9$, 0.76, 0.74; $A = -0.6$, -0.66, -0.8 and $\Lambda_0 = 1.65$, 2.30, 2.92 for the mass numbers 125, 127, 129 respectively, in units of MeV. The results are shown in Fig. 1. The experimental data are taken from Refs.[14,15,16]. Generally reasonable agreement is seen. To see more detail, the calculated $3/2_1^+$ is systematically comes lower than the experimental location, while the calculated $5/2_1^+$ lies higher. The Coriolis effect may explain this difference. The locations of the yrast states with $I \geq 7/2$ are reasonably well reproduced.

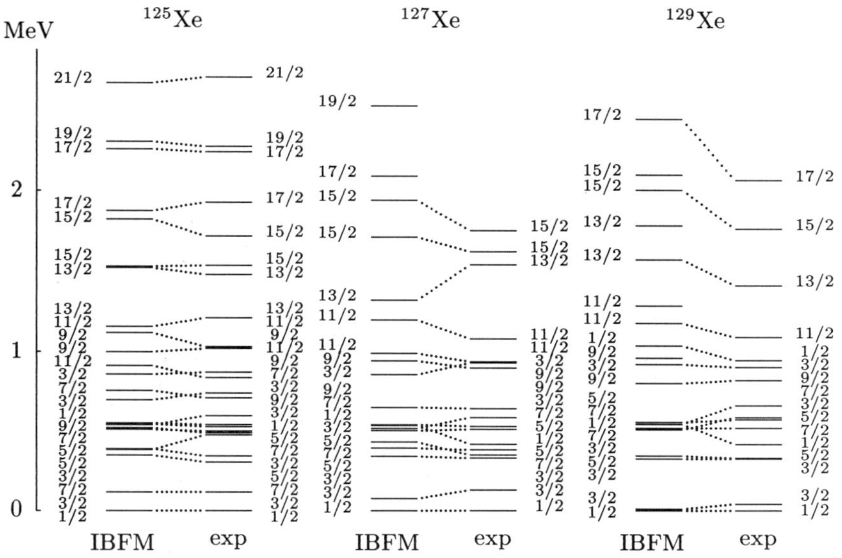

Figure 2. Energy levels in the Xe isotopes.

2.2 Energy Levels in the Xe Isotopes

The odd-mass Xe isotopes are described by coupling an odd neutron hole to the neighboring even-even Xe cores. The same orbitals as in Cs are taken into account. The single-particle energies are taken from Ref.[4], except for $g_{7/2}$ and $h_{11/2}$. The single-particle energy of $g_{7/2}$ has been varied to obtain the correct spins of the ground states. The values 0.3, 0.35, 0.4 MeV are taken for A =125, 127, 129, respectively. The single-particle energy of $h_{11/2}$ is set to 1.30 MeV. In the boson-fermion interaction, the quadrupole and the monopole interactions are included between the odd neutron and the proton bosons, in addition to the exchange interaction of the quadrupole type. The adopted strengtes are $\Gamma_0 = 0.39$, 0.44, 0.50 MeV for the mass numbers 125, 127, 129 respectively. The constant values $A = -0.42$, $\Lambda_0 = 0.40$ (MeV) are taken. The results are shown in Fig. 2. The experimental data are taken from Refs.[17,18,19,20]. Reasonable agreement is seen. In ^{127}Xe, there are two different interpretations about the spin of the 510 keV level. Although Ref.[18] adopts $I = 3/2$, Refs.[4,20] insist $I = 5/2$ because of very weak beta-decay from $I = 1/2$ in ^{127}Cs and the level systematics in neighboring nuclei. We have chosen the latter for better agreement with calculation. However, the spin of 510 keV is still an open problem.

2.3 Electromagnetic Moments and Transitions

We use the boson effective charge $e^{\mathrm{B}} = 0.150$ eb to explain the B(E2) values and the quadrupole moments in these odd-mass nuclei. The value used in our calculations,

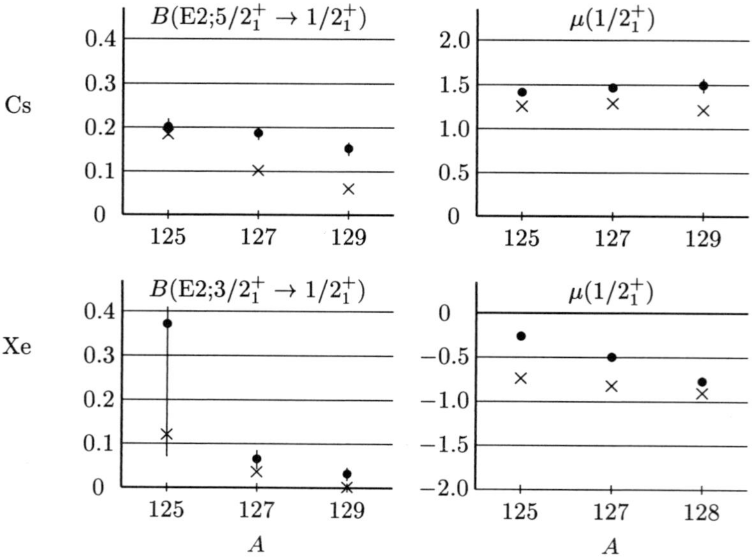

Figure 3. Electromagnetic moments and transitions. data. The symbol • shows experiments while × shows calculation.

is larger, for some isotopes, than the one determined from the corresponding even-even cores (≈0.108 eb). This difference may be due to a polarization effect caused by the odd fermion. For the odd proton in Cs $e_\pi^{\rm F} = 1.5$ e, while for the odd neutron in Xe $e_\nu^{\rm F} = 0.5$ e. For the magnetic dipole operator, the boson g-factors for all the isotopes are: $g_\nu^{\rm B} = 0$, $g_\pi^{\rm B} = 0.8\ \mu_{\rm N}$. For the odd proton in Cs, the spin g-factor is reduced by the factor of 0.85, while for the odd neutron in Xe, the spin g-factor is reduced by the factor of 0.5. Figure 3 shows reasonable agreement.

2.4 Beta Decay

The scheme is essentially the same as Ref.[7]. Namely, the IBFM images of the real particle creation and annihilation operators are constructed[12,13], and then they are coupled to form the Fermi and the Gamow Teller operators. Although some previous works introduced overall normalization factors to account for the absolute values of the beta-transition rates[7], we did not introduce any adjustable parameters in the beta-decay operators. Figure 4 shows beta-decay rates to the yrast and the yrare 1/2+ and $3/2^+$ levels in terms of $\log_{10} ft$ values. The experimental values have been derived by electron capture and β^+ experiments by Refs.[17,18,19]. By properly taking the wave functions so that the decay rates from the ground states $(1/2_1^+)$ to the ground states $(1/2_1^+)$ are reproduced, we obtain reasonable agreement in those to $1/2_2^+$ and to $1/2_2^+$, except for the decay to $3/2_1^+$ in ^{127}Xe.

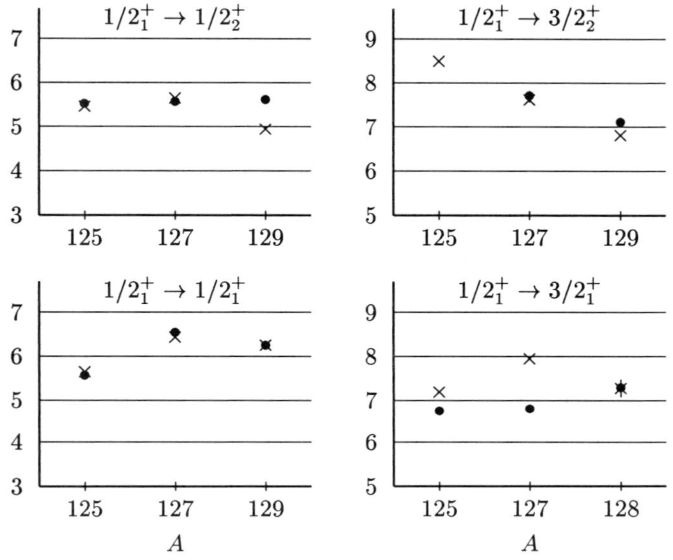

Figure 4. The beta-decay rates from ^{A}Cs to ^{A}Xe shown in terms of $\log_{10} ft$ values. The symbol • shows experiments while × shows calculation.

3 Discussions

Looking at the wave functions, the ground states of ^{125,127,129}Cs are dominated by two orbitals $g_{7/2}$ and $d_{3/2}$ (30~40% each). The orbital $d_{5/2}$ has comparable amount of mixture, too. The mixture of the component is $s_{1/2}$ (10~15%) is small. In the daughter nuclei ^{125,127,129}Xe, dominant component of the ground state is $s_{1/2}$ (80~90%). The main contribution to the Gamow-Teller matrix elements comes from the transition: $\pi d_{3/2} \to [\tilde{d}_\nu \, \nu s_{1/2}]^{(3/2)}$. As for the first excited states with $I = 3/2$, the main component in the wave functions is $d_{3/2}$. The main contributions to the Gamow-Teller matrix elements come from the transitions: $\pi d_{3/2} \to [\tilde{d}_\nu \, \nu d_{3/2}]^{(3/2)}$ and $\pi d_{3/2} \to [\tilde{d}_\nu \, \nu d_{3/2}]^{(5/2)}$. But cancellation occurs in these two contributions. That is one reason why $\log ft$ values to the $3/2_1^+$ states are larger (i.e., the matrix elements are smaller) than $\log ft$ values to the ground states ($1/2_1^+$). It would be interesting if these features could be explained in a simlple way using symmetry.

In the calculation of beta-decay, we have tried to avoid the normalization factors. Allowing some normalizations in a consistent way may improve the agreement in relative transition rates.

4 Conclusions

We have shown the preliminary results of the IBFM2 calculation in Xe and Cs including beta-decay between these isotopes. Our aims are to extend the calculations to odd nuclei of other mass regions, and also to the beta decay between even-even and odd-odd nuclei, including description and predictions.

References

1. F. Iachello and A. Arima, *The interacting boson model* (Cambridge Univ. Press, Cambridge, 1987); F. Iachello and P. Van Isacker, *The interacting boson-fermion model* (Cambridge Univ. Press, Cambridge, 1991).
2. G. Puddu, O. Scholten and T. Otsuka, *Nucl. Phys.* A **348**, 109 (1980).
3. J. M. Arias, C. E. Alonso and R. Bijker, *Nucl. Phys.* A **445**, 333 (1985).
4. Gh. Cata-Danil, D. Bucurescu, A. Gizon and J. Gizon, *J. Phys.* G **20**, 1051 (1994).
5. N. Yoshida, A. Gelberg, T. Otsuka, I. Wiedenhöver, H. Sagawa and P. von Brentano, *Nucl. Phys.* A **619**, 65 (1997).
6. A. Gizon, B. Weiss, P. Paris, C. F. Liang, J. Genevey, J. Gizon, V. Barci, Gh. Cata-Danil, J. S. Dionisio, J. M. Lagrange, M. Pautrat, J. Vanhorenbeeck, Ch. Vieu, L. Zolnai, J. M. Arias, J. Barea and Ch. Droste, *Eur. Phys. J.* A **8**, 41 (2000).
7. F. Dellagiacoma, Ph.D. thesis, Yale Univ. 1988; F. Dellagiacoma and F. Iachello, *Phys. Lett.* B **218**, 299 (1989).
8. G. Maino and L. Zuffi, in *Proc. of the 7th International Conference on Nuclear Reaction Mechanism*, Varenna, 1994, ed. E. Gadioli (University of Milan, Milan, 1994), p. 765.
9. G. Maino, in *Proc. of the International Symposium on Perspective for the Interacting Boson Model*, Padova, 1994, ed. R. F. Casten *et al* (World Scientific, Singapore, 1995), p. 617.
10. N. Yoshida and L. Zuffi, in *Proc. of the Conference: Bologna 2000, Structure of the Nucleus at the Dawn of the Century* eds. G. C. Bonsignori, M. Bruno, A. Ventura and D. Vretenar (World Scientific, Singapore, 2001), p. 233.
11. N. Yoshida, L. Zuffi and S. Brant, *Phys. Rev.* C **66**, 014306 (2002).
12. O. Scholten, Ph. D. thesis, University of Groningen, 1980.
13. R. Bijker, Ph. D. thesis, University of Groningen, 1984.
14. J. Katakura, *Nucl. Data Sheets* **86**, 955 (1999).
15. K. Kitao, M. Oshima, *Nucl. Data Sheets* **77**, 1 (1996).
16. Y. Tendow, *Nucl. Data Sheets* **77**, 631 (1996).
17. J. Katakura, *Nucl. Data Sheets* **86**, 955 (1999).
18. K. Kitao, M. Oshima, Nucl. Data Sheets **77**, 1 (1996).
19. Y. Tendow, *Nucl. Data Sheets* **77**, 631 (1996).
20. P. F. Mantica, Jr., B. E. Zimmerman, W. B. Walters, H. K. Carter, D. Rupnik, E. F. Zganjar, W. L. Croft and Y.-S. Xu, *Phys. Rev.* C **42**, 902 (1990).
21. G. Maino and L. Zuffi, in *Proc. of the 5th International Spring Seminar on Nuclear Physics*, Ravello, 1995, ed. A. Covello (World Scientific, Singapore, 1996), p. 611.
22. N. Yoshida, L. Zuffi and A. Arima, *Czech. J. Phys.* **52**, Suppl. C615 (2002).

ALGEBRAIC DESCRIPTION OF HIGH ANGULAR MOMENTUM STATES IN NUCLEI

D. VRETENAR AND S. BRANT
Physics Department, Faculty of Science, University of Zagreb, Croatia

G. BONSIGNORI
Physics Department and INFN, University of Bologna, Italy

We review the applications of the Interacting boson-fermion plus broken-pairs model in the description of the structure of high angular momentum states in deformed, transitional and spherical nuclei.

1 The Interacting Boson Model plus Broken Pairs

The highly successful algebraic approach to the structure of collective states in nuclei, based on the interacting boson approximation, has been extended to the physics of high angular momentum states [1,2,3,4]. In the formulation of models that describe high-spin states in even-even, odd-even and odd-odd nuclei, one has to go beyond the boson approximation and include selected non-collective fermion degrees of freedom, either those that correspond to the unpaired fermions (in odd-even and odd-odd systems), or/and through successive breaking of correlated S and D pairs. High-spin states are generated not only by the alignment of bosons, but also by the coupling of fermion pairs to the boson core. This approach is especially relevant for nuclei in transitional regions, where single-particle excitations and vibrational collectivity are dominant modes, and the traditional cranking approach to high-spin physics is not adequate. The theoretical framework, based on the interacting boson approximation, and extended with broken pairs, provides a unified and complete microscopic description of both low- and high-spin structures with a single model Hamiltonian.

The treatment of states with $J \approx 10\hbar - 30\hbar$ in terms of interacting bosons and fermions introduces a number of novel features (pair-breaking interactions, residual interactions between unpaired fermions), and also extends the concept of dynamical symmetry to high-spin structures. In this work we review the applications of models based on broken pairs in the analysis of experimental data on excitation energies, moments of inertia, backbending, and electromagnetic transition rates. The model with one and two broken pairs for even-even nuclei has been used in the description of high-spin states in the Hg [3,5], Sr-Zr [6,7,8], Nd-Sm [9,10] and Cd [11,12] regions. The interacting boson-fermion model for odd-even nuclei has been extended with one broken pair, and applied in studies of high-spin bands in Hg isotopes [13], ^{139}Sm [14], ^{137}Nd [15,16], ^{97}Y [17], ^{99}Nb [18], ^{65,67}Ga [19] and ^{101}Ag [20]. The symmetries of high-spin spectra in deformed nuclei have been analyzed in Refs. [1,2], and the onset of chaos in the model space that contains broken-pair states has been investigated in Ref. [21]. Spectral fluctuations have been studied as a function of the quadrupole-quadrupole interaction between the fermion pair and the core and the pair-breaking interaction that mixes states with different number of fermions.

For a description of high-spin states the interacting boson/ interacting boson-fermion (IBM/IBFM) model space has to be extended by including part of the original shell-model fermion space through successive breaking of correlated S and D pairs (s and d bosons). The interacting boson-fermion plus broken-pairs approach is based on the simplest version of the IBM/IBFM models [22]: the boson space consists of s and d bosons, no distinction is made between proton and neutron bosons. High-spin states are generated not only by the alignment of d bosons, but also by coupling fermion pairs to the boson core. A boson can be destroyed, i.e. a correlated fermion pair can be broken, by the Coriolis interaction and the resulting non-collective fermion pair recouples to the core. The structure of high-spin states is described in terms of broken pairs.

The model space for an even-even nucleus with $2N$ valence nucleons reads

$$\begin{aligned} &\mid N\ bosons\ > \oplus \mid (N-1) bosons \otimes 1\ broken\ pair > \\ &\qquad \oplus \mid (N-2) bosons \otimes 2\ broken\ pairs > \oplus\ \ldots \end{aligned}$$

The model Hamiltonian has four terms: the IBM-1 [22] boson Hamiltonian, the fermion Hamiltonian, the boson-fermion interaction, and a pair breaking interaction that mixes states with different number of fermions.

$$H = H_B + H_F + V_{BF} + V_{mix}. \tag{1}$$

The fermion Hamiltonian H_F contains single-fermion energies and fermion-fermion interactions. The interaction between the unpaired fermions and the boson core contains the dynamical, exchange and monopole interactions of the IBFM-1 [22].

The terms H_B, H_F and V_{BF} conserve the number of bosons and the number of fermions separately. In our model only the total number of nucleons is conserved, bosons can be destroyed and fermion pairs created, and vice versa. In the same order of approximation as for V_{BF}, the pair breaking interaction V_{mix} which mixes states with different number of fermions, conserving the total nucleon number only, reads

$$\begin{aligned} V_{mix} = &-U_0 \sum_{j_1 j_2} u_{j_1} u_{j_2} (u_{j_1} v_{j_2} + u_{j_2} v_{j_1}) \langle j_1 \parallel Y_2 \parallel j_2 \rangle^2 \frac{1}{\sqrt{2j_2+1}} \left([a^\dagger_{j_2} \times a^\dagger_{j_2}]^{(0)} \cdot s \right) \\ &-U_2 \sum_{j_1 j_2} (u_{j_1} v_{j_2} + u_{j_2} v_{j_1}) \langle j_1 \parallel Y_2 \parallel j_2 \rangle \left([a^\dagger_{j_1} \times a^\dagger_{j_2}]^{(2)} \cdot \tilde{d} \right) + \ h.c. \end{aligned} \tag{2}$$

If mixed proton-neutron configurations are included in the fermion model space, i.e. there can be both proton and neutron broken pairs, the full model Hamiltonian reads

$$H = H_B + H_{\nu F} + H_{\pi F} + H_{\nu BF} + H_{\pi BF} + H_\nu^{mix} + H_\pi^{mix} + H_{\nu\pi}, \tag{3}$$

where $H_{\nu\pi}$ denotes the proton-neutron interaction term. In the framework of the IBM/IBFM plus broken pairs, calculations are performed in the laboratory frame and they produce, in contrast to the cranking approach, results that can be directly compared with experimental data. All states within the model space and their electromagnetic properties are compared with experiment, rather than just band-head energies. The Interacting boson-fermion plus broken-pairs model represents an important contribution to the theoretical tools that can be employed in the analysis of the structure of high-spin states in nuclei.

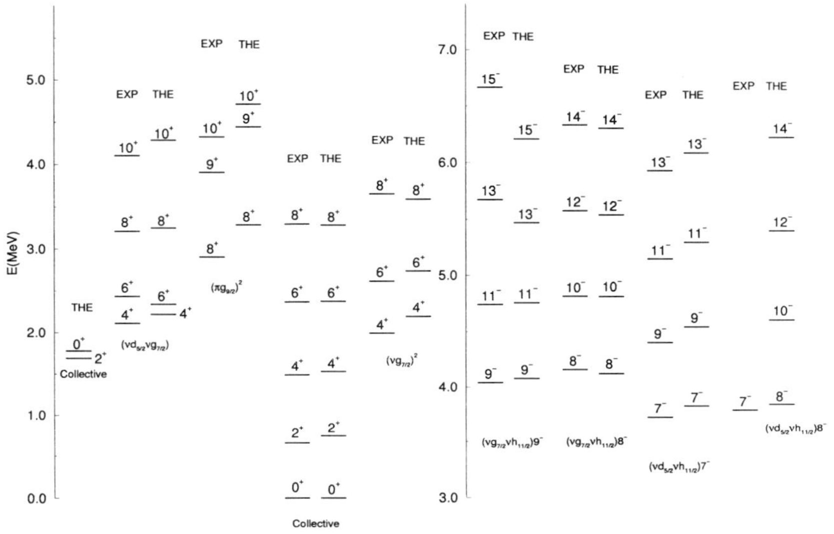

Figure 1. Positive and negative parity states in ^{104}Cd compared with the results of the IBM plus broken-pairs calculation.

2 High-spin structure in ^{104}Cd

The light Cd isotopes, with $Z = 48$ and $A \approx 104$ are, because of the effects of the $Z = 50$ shell closure, almost spherical at low spins. At higher angular momentum weakly deformed structures develop due to the deformation driving effects of the neutron intruder orbital.

In Refs. [11,12] we have employed the interacting boson-fermion model with broken pairs in the analysis of low and high-spin structures of ^{104}Cd. In general, most of the parameters of the Hamiltonian are taken from analyses of the low- and high-spin states in the neighboring even and odd nuclei. For ^{104}Cd the parameters of the boson hamiltonian are: $\epsilon = 0.887$ MeV, $C_0 = 0.44$ MeV, $C_2 = 0.233$ MeV, $C_4 = 0.0333$ MeV, the number of bosons $N = 4$, neutron quasiparticle energies and occupation probabilities $E(\nu d_{5/2}) = 1.211$ MeV, $E(\nu s_{1/2}) = 2.287$ MeV, $E(\nu h_{11/2}) = 2.691$ MeV, $E(\nu g_{7/2}) = 1.500$ MeV, $v^2(\nu d_{5/2}) = 0.57$, $v^2(\nu s_{1/2}) = 0.06$, $v^2(\nu h_{11/2}) = 0.04$, $v^2(\nu g_{7/2}) = 0.23$, the strength of the neutron pair-breaking interaction $U_2^\nu = 0.25$ MeV, the strength of the boson-neutron dynamical interaction for positive parity states $\Gamma_0^\nu = 0.9$ MeV, for negative parity states $\Gamma_0^\nu = 0.2$ MeV, the strength of the boson-neutron exchange interaction $\Lambda_0^\nu = 0.2$ MeV, the strength of the boson-neutron monopole interaction for negative parity states $A_0^\nu = 0.03$ MeV, and $\chi = -0.9$. The parameters of the boson-proton interactions, the quasiparticle energy and the occupation probability of the $\pi g_{9/2}$ orbital, have been adjusted to reproduce the positive parity states in ^{107}In:

$\Gamma_0^\pi = 0.9$ MeV, $\Lambda_0^\pi = 0$ MeV, $A_0^\pi = -0.1$ MeV, $\chi = -0.9$, $E(\pi g_{9/2}) = 1.809$ MeV, $v^2(\pi g_{9/2}) = 0.82$. The strength of the proton pair-breaking interaction is $U_2^\pi = 0.2$ MeV.

In Fig. 1 we display the calculated spectra of positive- and negative-parity states, and compare them with their experimental counterparts. Only few lowest calculated levels of each angular momentum are included. According to the structure of wave functions, states are classified in bands. The agreement between the experimental results and calculations in the framework of the IBM plus broken-pairs is very good. Two-neutron configurations based on the $d_{5/2}$, $g_{7/2}$ and $h_{11/2}$ orbitals have been assigned to the observed structures of negative and positive parity, both two-proton $(g_{9/2})^2$ and two-neutron $(d_{5/2}, g_{7/2})$ configurations have been identified above the collective vibrational sequence. Results of model calculations reproduce in detail the experimental excitation energies, as well as the electromagnetic transition strengths [12].

3 High-spin dipole bands in transitional nuclei

Nuclei in the A = 130-140 mass region are γ-soft and the polarizing effect of the aligned nucleons induces changes in the nuclear shape. Because of the different nature of the excitations (particles for proton, and holes for neutron configurations), the alignment of a pair of $h_{11/2}$ protons induces a prolate shape, whereas the alignment of a neutron pair in the $h_{11/2}$ orbital drives the nucleus towards a collective oblate shape.

In Ref. [10] we have employed the IBM with proton and neutron broken pairs in an analysis the excitation spectrum of ^{136}Nd. In particular, we have investigated the structure of several high-spin dipole bands based on two proton - two neutron configurations. The experimental level scheme of positive-parity states is displayed in Fig. 2. In addition to the ground state band and the quasi γ-band, bands 3, 5, 7 and 8 result from the alignment of two protons or two neutrons in the $h_{11/2}$ orbital, the two four-quasiparticle dipole bands have labels 10 and 11.

There are 6 neutron valence *holes* and 10 proton valence *particles*. The resulting boson number is N=8. The set of parameters for the boson Hamiltonian is: ϵ=0.36, C_0=0.16, C_2=-0.12, C_4=0.19, V_2=0.11 and V_0=-0.3 (all values in MeV). The boson parameters have values similar to those that have been used in the calculation of ^{138}Nd [9], ^{137}Nd [16] and ^{139}Sm [14].

In A $\approx$ 140 nuclei the structure of positive parity high-spin states close to the yrast line is characterized by the alignment of both proton and neutron pairs in the $h_{11/2}$ orbital. For positive-parity states we have only included the proton and neutron $h_{11/2}$ orbitals in the fermion model space. Additional single-nucleon states make the two broken-pairs bases prohibitively large. The single quasiparticle energies and occupation probabilities are obtained from a BCS calculation. Similar to our previous calculations for ^{138}Nd and ^{137}Nd, the resulting quasiparticle energies for the proton and neutron $h_{11/2}$ states had to be slightly renormalized. $E_\nu(h_{11/2}) = 1.75$ MeV, $v_\nu^2(h_{11/2}) = 0.83$, $E_\pi(h_{11/2}) = 1.60$ MeV, $v_\pi^2(h_{11/2}) = 0.07$. In order to further reduce the large size of the space with two-broken pairs, we had prediagonalized the boson Hamiltonian, and the fermion states were then coupled

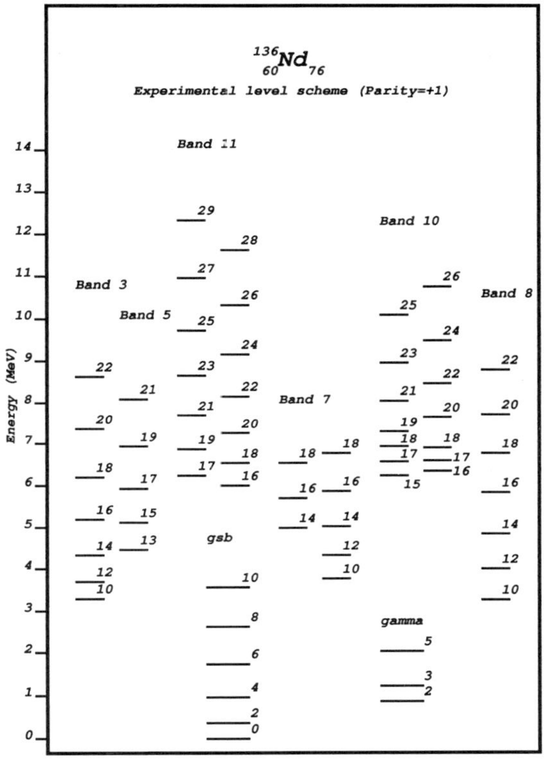

Figure 2. Experimental excitation spectrum of positive-parity states in ^{136}Nd.

to the lowest eigenvectors, i.e. only to the collective ground state band. The parameters of the fermion-boson interactions are determined from IBFM calculations of low-lying negative-parity states in ^{137}Nd and neighboring odd-proton nuclei.

For a quantitative description of the structure of four-fermion dipole bands, it is necessary to modify the quadrupole-quadrupole dynamical interaction

$$V_{dyn} = \Gamma_0 \sum_{j_1 j_2} (u_{j_1} u_{j_2} - v_{j_1} v_{j_2}) \langle j_1 \parallel Y_2 \parallel j_2 \rangle \times \left([a^\dagger_{j_1} \times \tilde{a}_{j_2}]^{(2)} \cdot Q^B \right) , \qquad (4)$$

by an extension of the standard boson quadrupole operator Q^B with a higher order term

$$\chi' \sum_{L_1 L_2} \left[\left[d^\dagger \times \tilde{d} \right]^{(L_1)} \times \left[d^\dagger \times \tilde{d} \right]^{(L_2)} \right]^{(2)} , \qquad (5)$$

The parameters of the neutron dynamical fermion-boson interaction are Γ_0=0.3 MeV, χ=-1 and χ'=-0.2, and for protons: Γ_0= 0.22 MeV, χ=+1 and χ'=+0.2. For the exchange interaction Λ_0^ν = 1.0 and Λ_0^π = 1.5 for neutrons and protons, respectively. The strength parameter of the pair-breaking interaction is U_0 = 0

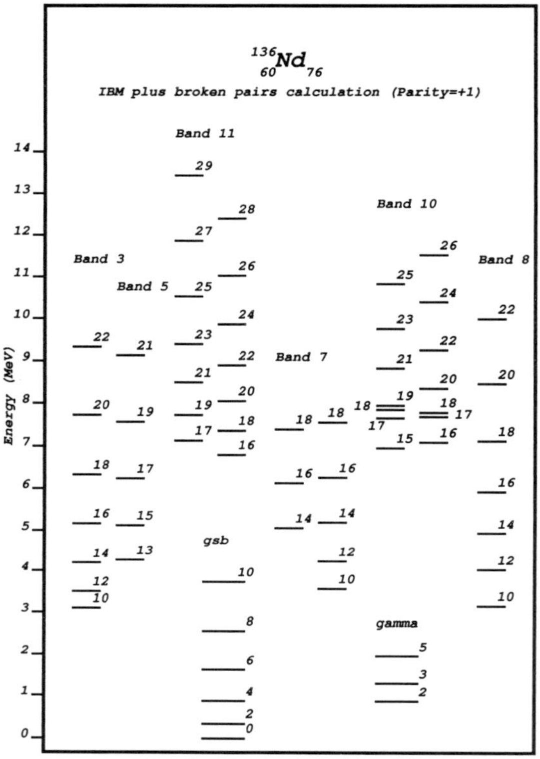

Figure 3. Results of IBM plus broken pairs calculation for positive parity bands in ^{136}Nd.

and $U_2 = 0.2$ MeV, both for protons and neutrons in broken pairs. The residual interaction between unpaired fermions is a surface δ-force with strength parameters: $V_{\nu\nu} = -0.1$, $V_{\pi\pi} = -0.1$ and $V_{\nu\pi} = -0.9$ for neutron-neutron, proton-proton and proton-neutron, respectively.

In Fig. 3 we display the calculated spectrum of positive-parity states. According to the structure of wave functions, states are classified in bands labeled in such a way that a direct comparison can be made with their experimental counterparts. The calculated positive-parity structures 3, 5, 7, 8, 10 and 11, as well as the ground state band an the quasi γ-band, have to be compared with the experimental bands of Fig. 3. The bands 3, 5, and 7 result from the alignment of a pair of protons in the $h_{11/2}$ orbital. The band 8 corresponds to two $h_{11/2}$ neutrons coupled to the boson core. Finally, the two dipole bands 10 and 11 correspond to four-quasiparticle states, two protons and two neutrons in their respective $h_{11/2}$ orbitals, coupled to the ground state band of the core. With the same set of parameters we have also calculated negative parity states based on the neutron orbitals $s_{1/2}$, $d_{3/2}$ and $h_{11/2}$. The resulting bands reproduce the experimental data.

The occurrence of regular dipole bands ($\Delta I = 1$) in nearly spherical and transitional nuclei presents an interesting phenomenon. In the semiclassical picture of the cranked shell model, $\Delta I = 1$ high-spin bands have been described as TAC (Tilted Axis Cranking) solutions. In our model such $\Delta I = 1$ structures are produced by the fermion-boson interactions. However, in order to obtain the correct energy spacings for the bands 10 and 11, it was necessary to include the additional term (5) in the boson quadrupole operator. We have also found that a crucial role in the excitation spectrum of these bands is played by the proton-neutron delta-interaction.

4 High-spin states in ^{101}Ag

The structures of high-spin states in neutron-deficient nuclei below the Sn isotope chain, close to the doubly-magic ^{100}Sn, are dominated by the interplay between proton holes in the $g_{9/2}$ orbit and neutrons distributed over the orbits in the $N = 50 - 82$ shell.

In Ref. [20] we have applied the interacting boson-fermion plus broken-pairs model in the analysis of positive and negative parity structures in ^{101}Ag. The model space for an odd-even nucleus with $2N + 1$ valence nucleons reads

$$|(N) bosons \otimes 1 \; fermion\rangle \; \oplus$$
$$|(N-1) bosons \otimes 1 \; broken \; pair \otimes 1 \; fermion\rangle.$$

The two fermions in the broken pair can be of the same type as the unpaired fermion, resulting in a space with three identical fermions. If the fermions in the broken pair are different from the unpaired one, the fermion basis contains two protons and one neutron or vice versa.

In the analysis of ^{101}Ag we take as the core nucleus: $^{102}_{48}$Cd. This nucleus displays a transitional structure between the pure shell-model spectrum of ^{100}Cd and the vibrational spectrum of ^{104}Cd. The structure of low and high-spin states of both parities in ^{104}Cd has been recently described in the framework of the interacting boson model plus one-broken pair [11,12]. The set of parameters for the boson Hamiltonian is (all values in MeV): $\epsilon = 0.78$, $C_0 = 0.3$, $C_2 = 0.2$, $C_4 = 0$, $V_2 = 0$, and $V_0 = 0$. The calculated excitation spectrum of ^{102}Cd corresponds to the SU(5) dynamical symmetry limit of the IBM, and nicely reproduces the low-lying states observed in the experimental spectrum.

The fermion space of proton single-quasiparticle states contains the orbitals: $f_{5/2}$ ($E = 3.621$ MeV, $v^2 = 0.99$), $p_{3/2}$ ($E = 2.997$ MeV, $v^2 = 0.98$), $p_{1/2}$ ($E = 1.946$ MeV, $v^2 = 0.96$), $g_{9/2}$ ($E = 1.696$ MeV, $v^2 = 0.72$), and $d_{5/2}$ ($E = 7.696$ MeV, $v^2 = 0.003$). The parameters of the proton fermion-boson interactions are: $\Gamma_0 = 0.55$ MeV and $\chi = -0.9$ for the dynamical interaction, $\Lambda_0 = 1.8$ MeV for the exchange interaction, and $A_0 = 0.1$ MeV for the monopole interaction. The value of the parameter χ in the boson quadrupole operator is taken from the calculations of ^{104}Cd [12]. $\chi = -0.9$ in the E2 operator, together with the vibrational charge $e^{vib} = 1.5$, reproduces the B(E2) values for the transitions between the low-lying states of the core nucleus ^{102}Cd, and the calculated quadrupole moment $Q(2_1^+) = -0.172$ *eb* is in agreement with the systematics of this mass region. The strength parameter of the proton pair-breaking interaction is $U_2 = 0.2$ MeV, and the

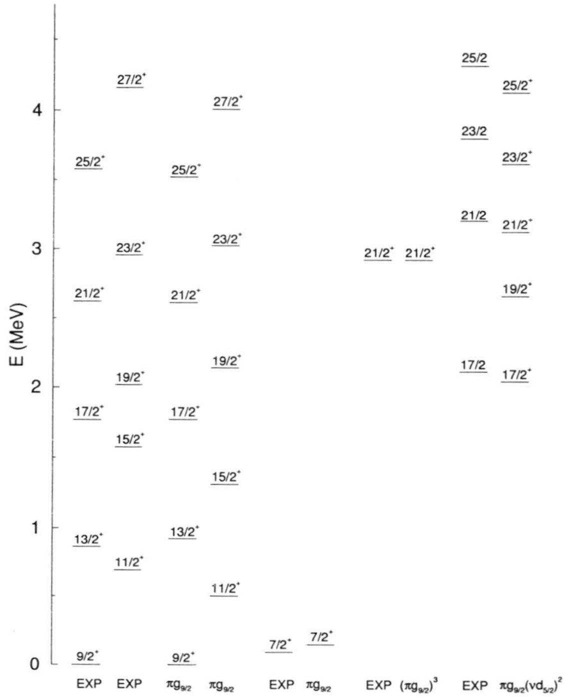

Figure 4. The calculated positive parity levels in ^{101}Ag shown in comparison with the sequence of experimental levels.

residual interaction between unpaired protons is a surface δ-force with the strength $v_0 = -0.15$ MeV.

The calculation includes the neutron orbitals: $d_{5/2}$ ($E = 1.013$ MeV, $v^2 = 0.44$), $g_{7/2}$ ($E = 2.02$ MeV, $v^2 = 0.11$), $h_{11/2}$ ($E = 2.549$ MeV, $v^2 = 0.03$). The parameters of the neutron fermion-boson interactions are: the strength of the dynamical interaction is $\Gamma_0 = 0.5$ MeV for positive parity states, and $\Gamma_0 = 0.2$ MeV in the calculation of states of negative parity, $\chi = -0.9$ in the boson quadrupole operator, $\Lambda_0 = 0.2$ MeV for the exchange interaction, the strength parameter of the monopole interaction is $A_0 = -0.04$ MeV for $\pi = +1$ states and $A_0 = -0.03$ MeV for $\pi = -1$ states. The values of the neutron fermion-boson interaction parameters are very similar to those used in the calculation of high-spin states based on neutron two-quasiparticle states in ^{104}Cd. The strength parameter of the neutron pair-breaking interaction is $U_2 = 0.15$ MeV, and the strength of the δ-interaction between unpaired neutrons is $v_0 = -0.03$ MeV.

In Fig. 4 we compare the experimental spectrum of positive-parity states with results of the present calculation. Only those calculated states are shown which have an experimental counterpart. For the low-spin part, the excitation spectrum displays a weakly coupled structure based on the proton $g_{9/2}$ orbital. The lowest structure of favored states is very well reproduced by the calculated band based

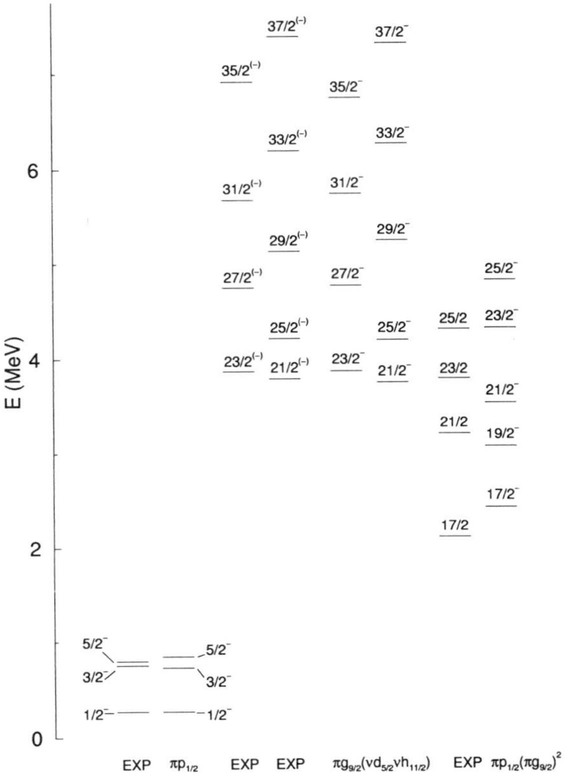

Figure 5. The calculated negative parity states in ^{101}Ag are compared with experimen.

on $9/2_1^+$ state. The band of unfavored states displays an anomaly around 2 MeV (between the states $15/2_1^+$ and $19/2_1^+$) which could not be obtained in the theoretical spectrum. The calculation reproduces the position of the low-lying $7/2_1^+$ state, and predicts the excitation energy of the lowest three-proton state $(\pi g_{9/2})^3$ $21/2^+$. This state, which is the band-head of the lowest $(\pi g_{9/2})^3$ band, has a possible experimental counterpart at 2922 keV, although the experimental level could also belong to the yrare $\pi g_{9/2}$ structure. The parity of the sequence of experimental states (17/2, 21/2, 23/2, 25/2) shown on the right-hand side of Fig. 4, has not been determined. We have compared the experimental structure with the calculated one proton - two neutrons band $\pi g_{9/2}(\nu d_{5/2})^2$. The excitation energies of the experimental sequence are in good agreement with the calculated positive parity band based on the band-head $17/2^+$.

The calculated and experimental states of negative parity are compared in Fig. 5. Again, only those theoretical levels are shown which have a possible experimental counterpart. The calculation reproduces the triplet of low-lying states based on the proton $p_{1/2}$ orbital. Above 4 MeV excitation energy two $\Delta I = 2$ sequences of probably negative parity states are observed, based on the band-heads $21/2^{(-)}$ and $23/2^{(-)}$, respectively. These two sequences are very well reproduced by the two lowest $\Delta I = 2$ bands based on the one proton - two neutrons configuration $\pi g_{9/2}(\nu d_{5/2}, \nu h_{11/2})$. We notice an almost perfect correspondence between the calculated and experimental levels, up to the highest observed angular momenta. In addition to the comparison shown on the right-hand side of Fig. 4, we have also investigated the possibility that the experimental sequence: 17/2, 21/2, 23/2 and 25/2, is of negative parity. In Fig. 5 the experimental levels are compared with the lowest three-proton band: $\pi p_{1/2}(\pi g_{9/2})^2$.

The calculation reproduces the measured B(E2) and B(M1) values for transitions among states based both on single-proton, and three-fermion configurations. In particular, an excellent agreement between the calculated and experimental mean-lives is obtained.

References

1. D. Vretenar, V. Paar, G. Bonsignori, and M. Savoia, *Phys. Rev.* C **42**, 993 (1990).
2. D. Vretenar, V. Paar, M. Savoia, and G. Bonsignori, *Phys. Rev.* C **44**, 223 (1991).
3. F. Iachello and D. Vretenar, *Phys. Rev.* C **43**, R945 (1991).
4. S. Cacciamani, G. Bonsignori, F. Iachello, and D. Vretenar, *Phys. Rev.* C **53**, 1618 (1996).
5. D. Vretenar, G. Bonsignori, and M. Savoia, *Phys. Rev.* C **47**, 2019 (1993).
6. P. Chowdhury *et al*, *Phys. Rev. Lett.* **67**, 2950 (1991).
7. A. A. Chishti *et al*, *Phys. Rev.* C **48**, 2607 (1993).
8. C.J. Lister, P. Chowdury, and D. Vretenar, *Nucl. Phys.* A **557**, 361c (1993).
9. G. de Angelis *et al*, *Phys. Rev.* C **49**, 2990 (1994).
10. D. Vretenar, S. Brant, G. Bonsignori, L. Corradini, and C. M. Petrache, *Phys. Rev.* C **57**, 675 (1998).
11. G. de Angelis *et al*, *Phys. Rev.* C **60**, 014313 (1999).
12. G. A. Müller *et al*, *Phys. Rev.* C **64**, 014305 (2001).
13. D. Vretenar, G. Bonsignori, and M. Savoia, *Z. Phys.* A **351**, 289 (1995).
14. C. Rossi Alvarez *et al*, *Phys. Rev.* C **54**, 57 (1996).
15. C. M. Petrache *et al*, *Z. Phys.* A **352**, 5 (1995).
16. C.M. Petrache *et al*, *Nucl. Phys.* A **617**, 228 (1997).
17. G. Lhersonneau, S. Brant, V. Paar, and D. Vretenar, *Phys. Rev.* C **57**, 681 (1998).
18. G. Lhersonneau *et al*, *Phys. Rev.* C **57**, 2974 (1998).
19. I. Dankó *et al*, *Phys. Rev.* C **59**, 1956 (1999).
20. E. Galindo *et al*, *Phys. Rev.* C **64**, 034304 (2001).
21. Y. Alhassid and D. Vretenar, *Phys. Rev.* C **46**, 1334 (1992).
22. A. Arima and F. Iachello, *Phys. Rev. Lett.* **35**, 10 (1975); F. Iachello and O. Scholten, *Phys. Rev. Lett.* **43**, 679 (1979); F. Iachello and A. Arima, *The Interacting Boson Model* (Cambridge University Press, Cambridge, 1987).

SEARCH FOR SENIORITY ISOMERS: LIFETIME MEASUREMENTS IN ^{93}Tc AND ^{95}Ru

K. P. LIEB, E. GALINDO, M. HAUSMANN AND A. JUNGCLAUS[c],

II. Physikalisches Institut, Universität Göttingen, D-37073 Göttingen, Germany;
E-mail: lieb@physik2.uni-goettingen.de

I. P. JOHNSTONE,

Department of Physics, Queen's University, Ontario, Canada

R. SCHWENGNER

Institut für Kern- und Hadronenphysik, FZ Rossendorf, D-01314 Dresden, Germany

A. DEWALD, A. FITZLER AND O. MÖLLER,

Institut für Kernphysik, Universität zu Köln, D-50937 Köln, Germany

G. DE ANGELIS, A. GADEA, T. MARTINEZ, D. R. NAPOLI AND C. UR

INFN, Laboratori Nazionali di Legnaro, I-35020 Legnaro, Italy

The shell model in N ≈ 50 nuclei with few valence nucleons predicts high-spin states, whose decays are severely inhibited, due to selection rules of the single-particle transitions and/or the need to recouple both types of nucleons. Such seniority isomers play a dominant role in the interpretation of shell model structures, as will be discussed on the basis of lifetime measurements in ^{93}Tc and ^{95}Ru.

1 Introduction

Isomerism in atomic nuclei, made visible by delayed γ-decays usually arises for three reasons:
a) large spin difference between initial and final state, thus requiring high γ-multipolarity L;
b) small energy difference leading to a small phase space factor E_γ^{2L+1} in the transition probability even of low multipolarity L; and
c) large difference in structure, for instance due to very different quadrupole deformations.

In nuclei near closed shells described by the shell model, another class of isomeric states has been found, which one may call seniority isomers [1-4]. The corresponding γ-ray transitions are often severely inhibited and arise because of the number, nature and angular momentum coupling of the participating nucleons.

Let us illustrate this class of isomerism for one of the nuclei considered here, ^{93}Tc. In the sim-plest truncation, the $9/2^+$ ground state of ^{93}Tc has the configuration $[\pi^2 p_{1/2}\, \pi^3 g_{9/2}]$, outside

---

[c] present address: Universidad Autónoma de Madrid and IEM, CSIC, E-28049 Madrid, Spain

the ^{88}Sr (N=50) core. This configuration reaches up to spins 21/1$^+$ and 25/2$^-$ if we allow for the $\pi p_{1/2} \rightarrow \pi g_{9/2}$ recoupling of the proton within the $[\pi^1 p_{1/2}\, \pi^4 g_{9/2}]$ configuration, essentially at no extra energy. To reach higher spins, one may either consider a $\nu g_{9/2} \rightarrow \nu d_{5/2}$ neutron core excitation extending up to spin 35/2$^+$ or 39/2$^-$, or alternatively a $\pi f_{5/2} \rightarrow \pi g_{9/2}$ proton core excitation reaching up 29/2$^-$ or 31/2$^+$. If the high-spin states in ^{93}Tc now have the lowest seniorities possible, the 23/2$^+$ member of the seniority v = 3 $[\pi^3 g_{9/2} \otimes \nu^{-1} g_{9/2}\, \nu d_{5/2}]$ family should be a seniority isomer, since the 23/2$^+$ $\rightarrow$ 21/2$^+$ transition would require a highly forbidden $\nu d_{5/2} \rightarrow \nu g_{9/2}$ M1 neutron transition. Even the neutron-allowed $\nu d_{5/2} \rightarrow \nu g_{9/2}$ E2 transition would be jeopardized by the required recoupling within the $\pi^3 g_{9/2}$ proton part of the wave functions involved (from proton seniority v_π = 1 for 23/2$^+$ to v_π = 3 for 21/2$^+$), which would inhibit this transition, too. This example illustrates that the search and lifetime measurements for seniority isomers near the N = 50 shell closure are of fundamental importance for checking their predicted wave functions. As different substructures overlap within their respective spin ranges, one also may hope that lifetime measurements would help us to disentangle the different modes of excitation.

2 The experiment

Picosecond lifetimes in ^{93}Tc and ^{95}Ru [5,6] have been measured via the recoil distance Doppler shift technique, using the Cologne plunger with a 1.4 μm thin self-supporting ^{64}Zn foil, enriched to 99.8%, and a 7 μm thick Au stopper foil. The ~1 particle nA beam was provided by the LNL XTU accelerator at Legnaro. The residual nuclei have been populated in the reactions ^{64}Zn(^{35}Cl,α2p)^{93}Tc and ^{64}Zn(^{35}Cl, 3pn)^{95}Ru at 135 MeV beam energy. Data have been accumulated for 13 flight distances between 20 μm and 6.17 mm. The γ-radiation has been detected in the GASP array equipped with 40 Compton-suppressed Ge detectors arranged in 7 rings. Whenever possible we used the Differential Decay Curve (DDC) method [7] to arrive at lifetime values free from uncertainties of the feeding conditions. In all the other cases, the feeding scenarios were accounted for by including the discrete and continuum (side) feeding times and intensities.of the states considered [8]. A total of 13 lifetimes and 11 lifetime limits were determined in ^{93}Tc; the corresponding numbers for ^{95}Ru are 11 lifetimes and 15 lifetime limits. This gives access to about 100 E1, M1 and E2 transition strengths in the two nuclei.

3 Transition strengths

The high-spin schemes of both evaporation residues were taken from the literature [9,10]. In the case of ^{95}Ru, considerable modifications above the 29/2$^+$ yrast state were made with respect to the work of Ghugre et al. [9]. We identified a series of stretched M1 and cross-over E2 transitions reaching up to the 9344 keV (41/2$^+$) state, which is separated by a 2.8 MeV energy gap from the presumed 43/2$^+$ level at 12163 keV. The level scheme comprises four families of states separated by energy gaps of 1.3 – 2.8 MeV. Similar substructures evolved for ^{93}Tc and gaps appear above the $21/2^+$ and $25/2^-$ yrast states.

Several shell model calculations labeled SM1 to SM4 were carried out for both nuclei [5,6], in which either neutron core excitation or a proton core excitation (or both) were incorporated. For the model space, single-particle energies (SPE) and two-body-matrix elements (TBME) see [11-19]. The calculated largest partition for the yrast states is typically 50-80%, that for the yrare states 20-60%. For calculating electromagnetic transition strengths, we used effective charges of $e_\pi = 1.72$ e, $e_\nu = 1.44$ e and quenched spin magnetic moments $g^s_{\pi,\nu} = 0.7\, g^{s,free}_{\pi,\nu}$.

3.1 ^{93}Tc and N =50 Isotones

Based on previous lifetime and magnetic moment measurements in the N = 50 isotones ^{94}Ru and ^{95}Rh [1-4] and the arguments detailed before, the $25/2^-$ yrast state in ^{93}Tc and ^{95}Rh are the members of the $[\pi p_{1/2}\, \pi^{4,6} g_{9/2}]$ families with the highest spins. No matter whether neutron or proton core excitation occurs to reach higher spins, the M1 and E2 decays to these states should be a sensitive measure of the seniority and structure of their wave functions. Fig. 1 illustrates the evolution of M1 and E2 strengths in the two isotones and in ^{94}Ru. We find E2 transitions of moderate strengths (100 – 200 e^2fm^4) and rather large M1 strengths (0.1-1 μ_N^2) along all the cascades. A dramatic exception refers to the $27/2^- \rightarrow 25/2^-$ ($13^- \rightarrow 12^-$) M1 and $29/2^- \rightarrow 25/2^-$ ($14^- \rightarrow 12^-$) E2 transitions, which are reduced by 2-3 orders of magnitude! This experimental finding is an excellent indicator that these $25/2^-$ states have indeed the expected $[\pi^1 p_{1/2}\, \pi^{4,6} g_{9/2}]$ structure with the good seniority quantum number v = 5. As to the $27/2^- \rightarrow 25/2^-_1$ M1 strength in ^{93}Tc, let us compare the experimental value, 11(1) $10^{-3}\mu_N^2$, with the predictions of SM1, SM3 and SM4, which range from 5 10^{-4} μ_N^2 up to 24 10^{-3} μ_N^2! In SM2 the $27/2^-$ state has a 50% partition $[\pi^{-1} f_{5/2}\, \pi^4 g_{9/2} \otimes \nu^{-1}$

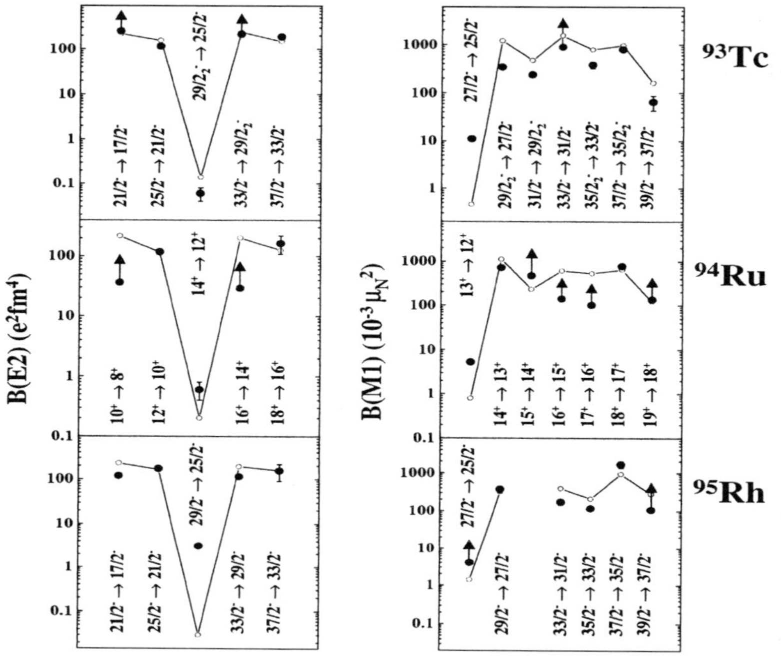

Fig. 1: Comparison of experimental strengths for stretched E2 and M1 transitions in the N = 50 isotones ^{93}Tc, ^{94}Ru and ^{95}Rh with shell model calculations (SM1). Note the strongly reduced strengths of transitions feeding the $25/2^-$ and 12^+ yrast states [1,2,6].

$\nu d_{5/2}$] and a 45% partition [$\pi^{1}p_{1/2}\ \pi^{4}g_{9/2} \otimes \nu^{-1}g_{9/2}\ \nu d_{5/2}$], while in SM3 (SM4) it has an 87% (27%) partition [$\pi^{-1}f_{5/2}\ \pi^{4}g_{9/2}$], without any neutron core excitation. Moreover, the small $29/2^-$ $\rightarrow 25/2^-_1$ E2 strength, B(E2) = 0.06(2) e^2fm^4, has to be compared with the shell model predictions, ranging from 0.003 to 171 e^2fm^4 for the various shell model calculations, if we consider the $29/2^-_2$ yrare state as a seniority isomer. On the other hand, taking the $29/2^-_1$ as the experimental coun- terpart, we arrive at theoretical B(E2) values of 0.14 – 183 e^2fm^4. We conclude that both inhibited decays depend very sensitively on the shell model space and parameters

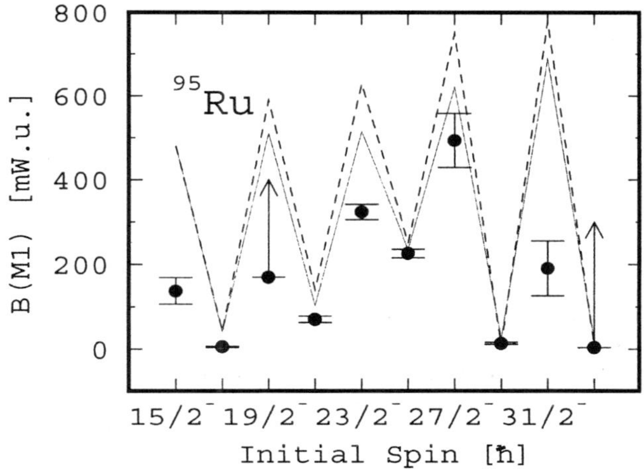

Fig. 2: Staggering of B(M1) values along the negative parity yrast sequence in ^{95}Ru [5] in comparison with shell model calculations (SM1: full line; SM3: dashed line)

3.2 ^{95}Ru

The systematics of the calculated (and if available experimental) M1 strengths presented in Fig. 1 shows another effect in the three N = 50 isotones, namely a staggering with the spin. This alternating increase/decrease of B(M1) by a factor of 2 – 4 is even more pronounced when going to the N = 51 nucleus ^{95}Ru as displayed in Fig. 2. This B(M1) staggering can be traced, in a simple way, to the shell model partitions of the corresponding states. Up to spin 31/2$^-$, all the wave functions are dominated by the $[\pi^1 p_{1/2}\ \pi^5 g_{9/2} \otimes \nu d_{5/2}]$ partition, with the protons coupled to the odd spins J_π = 5, 7, 9, 11 and 13. Large stretched M1 strengths of 0.14 – 0.50 Wu occur between states with equal J_π-value, while much smaller M1 strengths of 0.001 – 0.07 Wu connect states, for which J_π changes by $2\hbar$. The enhancement or reduction of B(M1) thus relies mostly on the recoupling of the six protons within the $[\pi^1 p_{1/2}\ \pi^5 g_{9/2}]$ part. This configuration is exhausted for spin 31/2$^-$ and, indeed, the slow M1 and E2 feedings of this state from the 33/2$^-$ and 35/2$^-$ yrast states appear to indicate two more seniority isomers. Their predicted structures are $[\pi^4 g_{9/2} \otimes \nu h_{11/2}]$ and $[\pi^1 p_{1/2} \pi^5 g_{9/2} \otimes \nu^{-1} g_{9/2} \nu^2 d_{5/2}]$, respectively.

In conclusion, lifetimes of high-spin states in N ≈ 50 nuclei having multi-particle configurations are very sensitive to the angular momentum couplings involved. Families of high-spin states arise, which differentiate between core and valence excitations requiring increasing seniority. This leads to long-lived states, called seniority isomers with strongly forbidden M1 and E2 decays, which represent suitable and very sensitive tests of the shell model calculations. Detailed comparisons of measured and calculated transition strengths are given in [5,6].

The authors gratefully acknowledge the cooperation of the XTU tandem crew and the GASP/EUROBALL measuring team at Legnaro. This work has been funded by BMBF, Bonn.

References

1. A. Jungclaus, et al., *Nucl. Phys.* **A637**, 346 (1998).
2. A. Jungclaus, et al., *Phys. Rev.* **C60**, 014309 (1999).
3. A. Jungclaus, et al., *Eur. Phys. J.* **A6**, 29 (1999).
4. K. P. Lieb, et al., in *Highlights of Modern Nuclear Structure*, S Agata sui due Golfi, A Covello, Ed. (World Scientific, Singapore, 1999) p. 183.
5. E. Galindo, doctoral thesis, Göttingen (2002) unpublished;
   E. Galindo, et al., *Phys. Rev.* **C**, submitted.
6. M. Hausmann, doctoral thesis, Göttingen (2003) unpublished;
   M. Hausmann, et al., *Phys. Rev.* **C**, submitted.
7. G. Böhm, et al., *Nucl. Instr. Meth.* **A329**, 248 (1993); P. Petkov, et al., Nucl. Instr. Meth. **A431**, 208 (1999).
8. K. P. Lieb, in *Experimental Techniques in Nuclear Physics*; D. Poenaru, W. Greiner, Eds., (de Gruyter, Berlin, 1997).
9. S. S. Ghugre, et al., *Phys. Rev.* **C50**, 1346 (1994).
10. H. A. Roth, et al., *Phys. Rev.* **C50**, 1330 (1994)
11. I. P. Johnstone and L. D. Skouras,. *Phys. Rev.* **C55**, 1227 (1997).
12. S. S. Ghugre and S. K. Datta, *Phys. Rev.* **C52**, 1881 (1995).
13. F. J. D. Serduke, R. D. Lawson and D. H. Gloeckner, *Nucl. Phys.* **A256**, 45 (1976); D. H. Gloeckner, *Nucl. Phys.* **A253**, 301 (1975).
14. X. Ji and B. H. Wilenthal, *Phys. Rev.* **C37**, 1256 (1988).
15. J. B. Ball, J. B. McGrory and J. S. Larsen, *Phys. Lett.* **41B**, 581 (1972).
16. G. Winter, et al., Nucl. *Phys. Rev.* **C48**, 1010 (1993).
17. R. Gross and A. Frenkel, *Nucl. Phys.* **A267**, 85 (1976).
18. K. Muto, T. Stimano and H. Horie, *Phys. Lett.* **135B**, 349 (1984).
19. D. Zwarts, *Comput. Phys. Commun.* **38**, 365 (1985).

MAGNETIC MOMENTS FROM THE MEDITERRANEAN TO MT. FUJI

N. BENCZER-KOLLER, M. J. TAYLOR, G. KUMBARTZKI, Y.Y. SHARON, L. ZAMICK

Department of Physics and Astronomy, Rutgers University, New Brunswick, NJ, 08903, USA

E-MAIL:NKOLLER@PHYSICS.RUTGERS.EDU

T. J. MERTZIMEKIS

NSCL, Michigan State University, East Lansing, MI 48824, USA

A. E. STUCHBERY

Department of Nuclear Physics, Australian National University, Canberra, ACT0200, Australia

The question of the integrity of the closed doubly magic cores at $Z, N = 20{,}28$ is examined from the perspective of g factors of 2_1^+ states in ^{44}Ca and other nuclei in the $f_{7/2}$ shell. The g factors were measured by the transient field technique and Coulomb excitation in inverse kinematics. A model which describes the 2_1^+ states as a mixture of approximately equal parts of spherical four valence neutron $(fp)_\nu^4$ configurations and core-excited, deformed, configurations agrees well with the measured value $g(^{44}Ca; 2_1^+)$ = + 0.12(5).

1 Introduction

I first met Franco Iachello at the Weizmann Institute some 28 years ago. Since then our paths have crossed many times, around the Mediterranean, and also in Japan where discussions with Akito Arima on the impact of magnetic moments determinations and on the application of the nascent IBA models had a profound influence on my future work. Franco eventually moved to Yale at a time when I was carrying out magnetic moment measurements at the WNSL accelerator. The laboratory was somewhat rudderless. But Franco moved into the depths of the WNSL mausoleum, joined the experimentalists, encouraged them, hired Rick Casten and reinvigorated the Yale nuclear physics program.

Since our first meeting I have followed his work with great admiration. The great success of IBA in predicting structure and detailed characteristics such as energies, transition probabilities and occasionally electromagnetic moments, provided the driving engine to elucidate nuclear structure at a time when the usual attempts at explanations based on microscopic details of shell model calculations, configuration mixing, pairing effects, meson exchange, etc... came somewhat short. Franco Iachello's inspiration and the constant encouragement of our experimental work supported the search for new techniques, the development of measurements with higher precision and ultimately the application of these techniques toward the exploitation of the new rare isotope facilities.

Magnetic moments have traditionally played a prominent role in the determination of the microscopic components of nuclear wave functions. It was clear from the very first measurements of the anomalous moments of the free proton and neutron that these elementary fermions were quite complex. When embedded in nuclei, their magnetic properties changed even more due to additional orbital magnetic moment components and interactions with other nucleons.

In first order, magnetic moments of single particles can be calculated in the independent particle model and yield the well known *Schmidt* limits. However, the observation of the ground states of many odd nuclei shows that very few magnetic moments actually fall on these limiting lines.

The structure of excited states is more complex, but in general the magnetic moments, μ, of these states can be calculated in the framework of the single particle shell model or from the expression, $g = Z/A$, derived from a collective picture of the nucleus. The g factor is given by $g = \mu/I$, and where I is the nuclear state total angular momentum. Again, few nuclear excited, spherical, single particle or collective states obey these rules.

The interacting boson approximation yields a similar expression,

$$g(2_1^+) \; = \; g_\pi \frac{N_\pi}{N_\pi+N_\nu} + g_\nu \frac{N_\nu}{N_\pi+N_\nu}$$

Here, the g factors of the proton and neutron boson pairs are taken to be $g_\pi \sim 1$ and $g_\nu \sim 0$. N_π and N_ν represent the numbers of proton and neutron boson pairs outside closed shells. In actuality, $g_\pi < 1$ and $g_\nu > 0$, reflecting single-particle characteristics.

The special advantage realized from measurements of magnetic dipole moments arises from the particular distinction between neutrons and protons: the bare magnetic moments of protons for $j > 1/2$ states are positive while for neutrons the moments are positive only for $j = l - 1/2$ states but are negative for $j = l+1/2$ states.

Measurements of the g factors of excited nuclear states can yield precise values for the deviations of the actual magnetic moments from the predicted values and thus supply a wealth of information on the structure of a particular state. In addition, measurements for a sequence of states, as a function of energy or spin, contribute significant information on the evolution of structure with spin.

2 Experimental Procedures

The technique used to obtain the data presented here involves a combination of Coulomb excitation of beams interacting, in *inverse kinematics*, with lighter target nuclei, and the use of the transient magnetic field experienced by the heavier nuclei while they traverse a ferromagnetic material.[1,2,3] The light target ions are scattered forward and are recorded in a detector located at 0° with respect to the beam, while the excited projectile traverses the ferromagnetic layer and stops in the copper (hyperfine-interaction free) backing of the target. The *same* target is used for each isotope studied. The γ–rays are detected in four detectors placed at angles where the angular correlation has optimum slope. The analysis is carried out in a manner similar to that used in the traditional forward-scattering methods. This technique is applicable to nuclear states with mean lifetimes of the order of picoseconds and has been applied to nuclei across the periodic table. The details of the experiment are described in a forthcoming paper.[4]

This technique has the added advantage of providing high statistics, hence higher precision, and of allowing measurements that were unfeasible in previous setups. Furthermore, Coulomb excitation leads to spectral simplicity, as only few states are excited, and backgrounds are generally minimized.

3 General results obtained with the inverse kinematics/Coulomb excitation/transient field technique.

The magnetic moments of low-lying states in nuclei from O to Hg have been measured with varying degrees of precision. Two very general observations can be made.

- In collective systems, the single particle contributions play a major role, especially as a function of increasing energy and spin.
- In systems near shell closure, where single particle excitations dominate, collective excitations are also very important and are manifested via a large degree of core excitation and configuration mixing.

The purpose of carrying out measurements of magnetic moments is precisely to separate with some degree of confidence the different components of the nuclear wave functions and examine the interplay between single particle and collective degrees of freedom.

Recent experiments and recent advances in calculations have aimed to analyze the possible independence of *fp* shell nucleons from the supposedly "closed" ^{40}Ca core which supports them. The magnetic moments of 2_1^+ and 4_1^+ states in ^{46,48,50}Ti, ^{50,52,54}Cr, ^{58,60,62,64}Ni, ^{62,64,66,68,70}Zn and $N = 28$ nuclei have been measured. These data and the related theoretical implications have been summarized and discussed in a recent review article.[5] Theoretical calculations within the framework of the shell model with KB3 and FPD6 interactions have been carried out within a restricted model space as well as using the full *fp* shell. In addition, large scale shell model calculations (LSSM) incorporating some core excitations have been performed[6] in specific cases. While the g factors of the 2_1^+ and 4_1^+ states in ^{50,52}Cr are in good agreement with full *fp* shell model calculations, the calculated values for the Ti isotopes are somewhat larger than the predictions. This deviation was attributed to the fact that particles are excited out of the doubly magic ^{40}Ca core. It has long been suspected that ^{40}Ca has significant core excitation,[9] whilst ^{48}Ca is expected to be a better double magic nucleus. The current experiment was designed to test quantitatively this hypothesis with yet another type of measurement.

The Ca isotopes have provided a fertile ground for calculations for many years. Only recently, however, has it become possible to determine accurately electromagnetic moments of excited states. In this paper, the the g factor of the 2_1^+ state of ^{44}Ca, $g = +\ 0.12(5)$, has been measured and is examined in terms of existing calculations. These results are in disagreement with the earlier data obtained in an experiment using tilted foils[7] which yielded a negative g factor, albeit with a

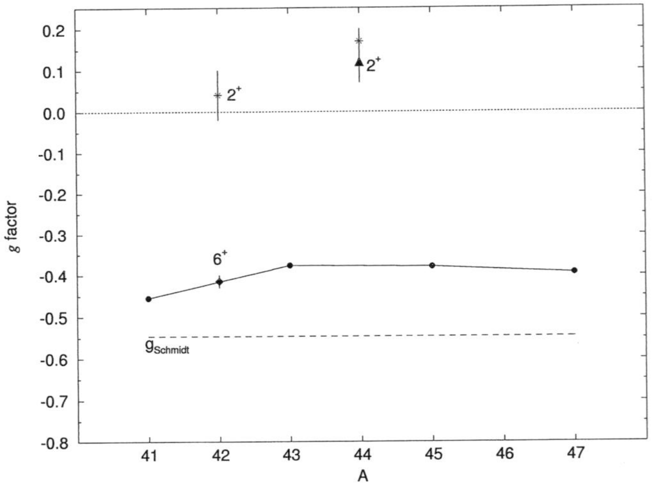

Figure 1: Measured g factors of 2_1^+ states in the even Ca isotopes and of the $(7/2)^-$ ground states in the odd Ca isotopes. For completeness, the g factor of the 6_1^+ state in ^{42}Ca is also indicated, denoted by a diamond. The triangle denotes the current result. The data denoted by a * were measured by the Bonn group (Ref.8).

large error. The cause for this discrepancy is not understood, but tilted foils experiments are intrinsically very difficult. In addition, a measurement of the g factor of the 2_1^+ state of ^{42}Ca, $g = +\ 0.04(6)$ has been recently reported as well as another measurement of $g(^{44}\text{Ca};2_1^+) = +\ 0.17(3)$.[8]

4 ^{44}Ca, midshell between ^{40}Ca and ^{48}Ca

The g factors that have been measured in even and odd Ca isotopes are displayed in Fig. 1. ^{44}Ca has nominally four valence neutrons occupying the $f_{7/2}$ orbit, corresponding to either four particles or four holes in the shell. Sharon and Zamick, using both the KB3 and the FDP6 nucleon-nucleon interactions and either the pure $(f_{7/2})^4_\nu$ or the full fp shell, have calculated the g factor and the quadrupole moment Q for the 2_1^+ state in ^{44}Ca as well as the B(E2;$0_1^+ \to 2_1^+$). In their calculation a neutron effective charge of 0.65 was used as was the case in Ref. 11. Their results are shown in Table 1 together with the results of the LSSM calculation.[6] This latter calculation includes 12p-12h configurations. However, it does not use the "full" space, but excludes smaller components of the wave function from the $f_{5/2}$ and $p_{1/2}$ orbits. That calculation further shows that at least 6p-6h are required to obtain a positive g factor.

If, however, the structure of the ^{40}Ca core is such that a number of particle-hole excitations occur, the 2_1^+ state of ^{44}Ca could be pictured as having a complex wave function consisting of two parts, one corresponding to an excited, core-deformed nucleus and the other involving only four neutrons in the fp shell. It has been known for several decades that indeed ^{40}Ca has such a coexistence structure and that, in fact, the second 0^+ state is mainly a 4p-4h state which supports a rotational band.[9] Without a need to know the details of the deformed core, the following wave function describes the 2_1^+ state in ^{44}Ca:

Table 1: Comparison of experimental data for ^{44}Ca with the results of various theoretical calculations.

	Exp't	KB3 $(f_{7/2})^4_\nu$	KB3 full $(fp)^4_\nu$	FDP6 $(f_{7/2})^4_\nu$	FDP6 full $(fp)^4_\nu$	LSSM [6]
g	+0.12(5)	-0.547	-0.415	-0.547	-0.374	+0.38
Q e^2b	-0.14(7) [11]	0	-0.039	0	-0.067	
B(E2;$0^+_1 \to 2^+_1$) e^2b^2	0.047(2) [13]	0.00636	0.00935	0.00636	0.01175	0.0526

Table 2: Relative amplitudes of the single particle contributions and the core excitations to the 2^+_1 state wave functions in ^{44}Ca, except for the Gerace and Green calculation which applies to ^{42}Ca only (Ref.8).

	C^2	D^2
Gerace and Green	~ 0.50	~ 0.50
Stripping data	0.47	0.53
Q_{meas}	0.43	0.57
g_{meas}	0.39(6)	0.61(6)

$$\langle \Psi(2^+_1) \rangle = \mathrm{C}[(fp)^4_\nu]_{J=2} + \mathrm{D}[\Psi_{def}]_{J=2}$$

where core excitations involve 2p-2h, 4p-4h, etc... The g factor becomes:

$$g(2^+_1)_{meas} = \mathrm{C}^2[g(fp)^4_\nu] + \mathrm{D}^2[g(\Psi_{def})]$$

The coefficients C and D are the amplitudes of the different components of the wave function and $\mathrm{C}^2 + \mathrm{D}^2 = 1$. In this model, $[g(fp)^4_\nu] = -0.395$, an average of the values calculated with the FDP6 and KB3 interactions, and $g(\Psi_{def}) \sim Z/A$.

The coefficients C and D for the 2^+_1 state in ^{44}Ca can now be calculated from the measured g or Q and be compared with the estimates obtained from the original Gerace and Green [9] calculations and stripping data [10] (as interpreted in Ref.11) and are shown in Table 2. Overall, the calculation with the FPD6 interaction agrees better with experimental data than that using the KB3 interaction. The same result was observed in the case of ^{46}Ti. [12]

A similar calculation was carried out by Sharon and Zamick for ^{42}Ca. The results indicate that the amplitude of the single particle contribution to the wave function of the 2^+_1 state is $C^2 = 0.45(6)$ while the deformed amplitude is $D^2 = 0.55(6)$. These results correspond to a slightly less deformed core than obtained in ^{44}Ca, which can be understood since there are only two neutrons available to excite the ^{40}Ca core instead of four. Thus the question of which core is more inert, that of ^{40}Ca or ^{48}Ca may be readily answered by forthcoming measurements of the g factor of the 2^+_1 state of ^{46}Ca. Such an analysis in terms of a two-component wave function involving $(fp)^n$ and deformed parts can be carried out for other nuclei in this region. For example, in the $N = Z$ nucleus ^{44}Ti where the two protons may impede excitation of the core and isospin considerations play a very important role, a pure $(fp)^4$ description works well.[14]

In summary, the measurements of the g factors of 2_1^+ states in nuclei in the $f_{7/2}$ shell yield information concerning the nature of the nucleon-nucleon interaction, and of the relative single particle and core excitation components of the wave functions of the Ca isotopes.

Acknowledgments

The authors thank all their collaborators in many of the experiments described in this paper, in particular C. W. Beausang, M. A. Caprio, C. Hutter, and J. J. Ressler at Yale. The authors acknowledge the support of the U.S. National Science Foundation.

References

1. N. Benczer-Koller, M. Hass and J. Sak, Annu. Rev. Nucl. Part. Sci. **30**, 53 (1980).
2. K.-H. Speidel, N. Benczer-Koller, G. Kumbartzki, C. Barton, A. Gelberg, J. Holden, G. Jakob, N. Matt, R.H. Mayer, M. Satteson, R. Tanczyn, and L. Weissman, Phys. Rev. C**57**, 2181 (1998).
3. N. K. B. Shu, D. Melnik, J. M. Brennan, W. Semmler and N. Benczer-Koller, Phys. Rev. C**21**, 1828 (1980.)
4. M. J. Taylor, N. Benczer-Koller, G. Kumbartzki, T. J. Mertzimekis, S. J. Q. Robinson, Y. Y. Sharon, L. Zamick, A. E. Stuchbery, C. Hutter, C. W. Beausang, J. J. Ressler, M. A. Caprio, submitted to Phys. Lett. B.
5. K.-H.Speidel, O. Kenn and F. Nowacki, Prog. Part. Nucl. Phys. **49**, 91 (2002) and references therein.
6. F. Nowacki and K.-H.Speidel, private communication.
7. Y. Niv *et al.* Phys. Rev. Lett. **43**, 326 (1980).
8. K.-H.Speidel, private communication.
9. W. J. Gerace and A. M. Green, Nuc. Phys. A**93**, 110 (1967).
10. J. H. Bjerrregaard and O. Hansen, Phys. Rev. **155**, 1229 (1967).
11. C. W. Towsley and R. N. Horoshko, Nuc. Phys. A**204**, 574 (1973).
12. Y. Sharon and L. Zamick, private communication.
13. S. Raman, C. W. Nestor Jr. P. Tikkanen, At. Data and Nuc. Data Tables, **78**, 1 (2001).
14. S. Schielke *et al.* submitted to Phys. Lett. B.

PROBING NUCLEAR STRUCTURE BY REAL PHOTONS: SYSTEMATICS OF LOW-LYING DIPOLE MODES IN HEAVY NUCLEI [a]

U. KNEISSL
Institut für Strahlenphysik, Universität Stuttgart, Allmandring 3, D-70569 Stuttgart, Germany
E-mail: kneissl@ifs.physik.uni-stuttgart.de

Recent results are summarized on low-lying electric and magnetic dipole excitations in heavy nuclei studied systematically in Nuclear Resonance Fluorescence (NRF) experiments. The systematics of the $M1$ *Scissors Mode* in deformed even-even and odd-mass nuclei are shown. New results are reported on strong $E1$ excitations in spherical nuclei near the N=82 and Z=50 shell closures. The corresponding $J^\pi = 1^-$ states are interpreted as two-phonon excitations due to the coupling of quadrupole and octupole vibrations ($2^+ \otimes 3^-$). A comprehensive systematics of the lowest $E1$ excitations in the entire mass region $130 \leq A \leq 200$ is discussed in view of various excitation modes. Recent experimental developments and future new applications of low-energy, photon-induced reaction studies are discussed.

1 Motivation

Recent trends in modern nuclear physics mainly focus on the study of nuclei under extreme conditions, at high energies (temperature), at high spins and deformations, and at high isospin (nuclei far from stability). Nevertheless, low-lying dipole excitations in heavy, stable nuclei met with an increased interest in the last two decades. The prediction of the orbital $M1$ *Scissors Mode* in deformed nuclei by Lo Iudice and Palumbo [1] and its subsequent discovery by Richter and coworkers [2] in 1984 stimulated a large number of both experimental and theoretical work (for references see, e.g., [3,4,5,6]). On the other hand, also enhanced electric dipole excitations ($E1$), due to the possible occurrence of reflection asymmetric shapes, were proposed by Iachello already in 1984 as a new class of collective modes in nuclei [7,8]. Furthermore, in spherical nuclei near shell closures strong $E1$ two-phonon excitations were expected and observed due to a coupling of the quadrupole and octupole phonons (see, e.g., ref. [9] and refs. therein).

So far all these enhanced dipole excitations were mainly investigated in even-even nuclei [3,4]. However, in NRF experiments of considerably increased sensitivity nowadays the existence of these dipole modes in the neighboring odd-mass nuclei and their expected fragmentation can be investigated, too.

2 Experimental Techniques

2.1 Nuclear Resonance Fluorescence

The real photon probe offers the principal advantage of the well-known electromagnetic interaction mechanism. Therefore, model-independent information can be extracted from photon-induced reaction studies. Furthermore, the low momentum

[a] Dedicated to Prof. Dr. Francesco Iachello on the occasion of his 60th birthday

transfer of real photons leads to a high spin-selectivity in exciting predominantly dipole modes. Therefore, photon scattering off bound states, nuclear resonance fluorescence (NRF), represents the most sensitive technique to study fundamental low-lying dipole modes in heavy nuclei, both of magnetic and electric character, even at excitation energies of rather high total level densities [3].

Photon scattering experiments provide, as well known, valuable spectroscopic information: Precise excitation energies E_x, integrated cross sections $I_{S,0}$, which are proportional to width ratios Γ_0^2/Γ, spins, parities, decay branching ratios, reduced excitation probabilities $B(\pi L)$, and lifetimes τ of the photoexcited states. The formalism describing photon scattering experiments is summarized in previous reviews [3,10].

The availability of high-flux DC bremsstrahlung photon beams and of Ge-γ-spectrometers and polarimeters of high efficiency and excellent energy resolution enabled a new generation of photon scattering experiments leading to a renaissance of the classical NRF technique. At sophisticated bremsstrahlung facilities like at the Stuttgart DYNAMITRON accelerator, where most of the presented results were achieved, nowadays NRF-experiments of considerably increased sensitivities can be performed. This experimental progress allows to study in detail low-lying dipole modes in heavy nuclei and in particular their fragmentation.

2.2 Experimental Setups at the 4 MV Stuttgart DYNAMITRON

The 4 MV Stuttgart DYNAMITRON delivers high DC electron currents up to 4 mA. However, in practice currents are limited by the thermal capacity of the watercooled bremsstrahlung production target to about 300 – 450 μA in the whole energy range of interest (0.8 – 4.3 MeV). NRF experiments can be run at the Stuttgart photon scattering facility at two different setups simultaneously [3]. At the first NRF-site the scattered photons are detected by three carefully shielded Ge(HP)-γ-spectrometers, with efficiencies ϵ of 100% (relative to a 3"×3" NaI/Tl detector) in each case, placed at scattering angles of 90°, 127°, and 150° with respect to the incident beam. In the most sensitive experiments on ^{163}Dy the detector at 127° was surrounded additionally by a BGO anti-Compton shield [11]. At the second site two sectored single crystal Ge–Compton polarimeters (ϵ=25% and 60%, partially with BGO shields) [12,13], installed at slightly backward angles of ≈95°, measure the linear polarization of the resonantly scattered photons, providing the parity information. An additional Ge-γ-detector (ϵ=38%) allows the measurement of angular distributions at this second site too and hence the simultaneous investigation of a second isotope. For further experimental details see, e.g., refs. [3,13,14].

3 M1 Scissors Mode Excitations

3.1 Scissors Mode Excitations in Deformed Even-Even Nuclei

The *M*1 *Scissors Mode* excitations nowadays are known from numerous photon and electron scattering experiments [3,4] to be a rather general phenomenon and fundamental excitation mode in heavy nuclei. Originally found in the even-even, deformed rare earth nuclei [2], it also has been observed subsequently in the other

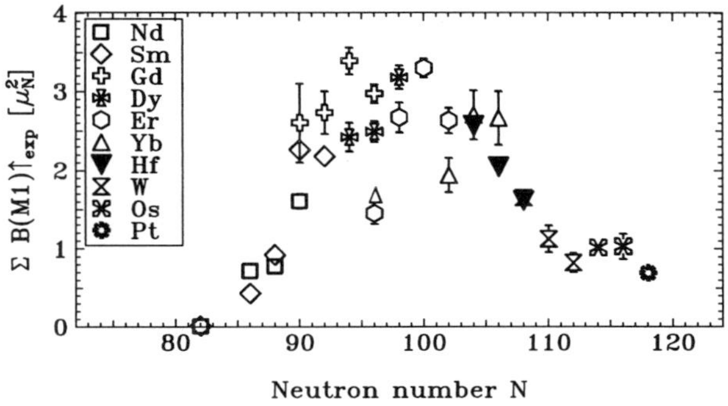

Figure 1: Systematics of the total $B(M1) \uparrow$ strengths observed in even-even nuclei as a function of the neutron number (taken from refs. [3,23] and completed by new data).

island of deformed nuclei, the actinide isotopes [15,16,17], recently in nuclei with a γ-soft triaxiality like ^{194,196}Pt [18,19] and ^{134}Ba [20], and in nuclei in the mass region around $A \approx 100$ [21,22]. The rather collective nature of the *Scissors Mode* has been proven by the smooth A-dependence of integral properties like the total strengths [23,24] and mean excitation energies [25,26]. As can be seen in fig. 1, the summed $B(M1) \uparrow$ values amount to about 3 μ_N^2 in midshell nuclei of the rare earth mass region and drop to lighter and heavier nuclei (see, e.g., [23,24,25]). The detected total strengths exhaust the values given by various sum rule predictions [24,27,28]. These sum rules explicitly contain a proportionality of the total $M1$ strength to the square of the deformation parameter δ, a deformation dependence which has been observed explicitly in NRF-experiments on nuclei in the Sm [29,30] and Nd [31,32] isotopic chains. The socalled "$\delta^2 - law$" corresponds to a clear linear correlation between the total $M1$ strength and the low-lying collective $E2$ strength and can be explained by different theoretical approaches (for references see, e.g., refs. [5,6,30]).

Another basic property of the *Scissors Mode* is its mean excitation energy E_{sc}. It is nicely proportional to $A^{-1/3}$ and hence to the reciprocal nuclear radius (see ref. [25]), a behavior which is reminiscent to the mass dependence of the well-known electric Giant Dipole Resonance (GDR).

The gross features and partially the fragmentation of the *Scissors Mode* in even-even nuclei can be explained quite successful in the framework of microscopic descriptions (see reviews [5,6] and references therein).

3.2 Scissors Mode Excitations in Deformed Odd-Mass Nuclei

The situation in odd-mass nuclei is more complicate. Here the *Scissors Mode* was detected first in the Dy isotopes ^{163}Dy [33] and ^{161}Dy [14] at the Stuttgart bremsstrahlung facility. The observed concentration of dipole strength fitted into the systematics of the *Scissors Mode* in the neighboring even-even isotopes ^{164,162,160}Dy [14,34]. However, the detected total strengths were reduced by a factor of roughly 2–3 as compared to those in the even-even Dy-nuclei [34]. Moreover, subsequent investigations of the odd-mass Gd nuclei ^{157,155}Gd surprisingly revealed an extreme fragmentation of the low-lying dipole strength into about 100 transitions in the energy range 2–4 MeV [14,35]. Further NRF-studies in Darmstadt[36,37] and Stuttgart[35] including odd-proton nuclei established a certain systematics for odd-mass nuclei in the mass range 169 $\leq A \leq 151$.

A common observation was an increasing fragmentation and in particular a strong reduction of the detected strengths with decreasing mass number A. A solution of the problem of missing strength as compared to the neighboring even-even nuclei was offered by the Darmstadt group[38,39], who could show by a statistical fluctuation analysis of the NRF spectra that a considerable portion of the *Scissors Mode* may be hidden in the continuous background of the spectra due to the extreme fragmention.

The aim of a recent joint effort of the Stuttgart and Darmstadt groups[11] was to extend the systematics to the lower mass nuclei ^{151,153}Eu and in particular to resolve and detect a large part of the hidden strength by performing NRF experiments with considerably enhanced sensitivity. In the most sensitive new experiment on ^{163}Dy detection limits could be achieved of about $B(M1)\uparrow \simeq 0.003\,\mu_N^2$ (at excitation energies near 3 MeV). This corresponds to a gain in sensitivity of about a factor 20 compared to the first investigation[33] of ^{163}Dy in 1991.

With this improved setup in total 161 excitations in ^{163}Dy and 138 excitations in ^{165}Ho were detected in the excitation energy range from 2 to 4 MeV. The detected total dipole strength in ^{163}Dy could be doubled as compared to the previous experiment[33]. The new data are included in fig. 2 where all Stuttgart results are depicted. Plotted are as a function of the excitation energy the quantities $g \cdot \Gamma_0^{red} = \frac{(2J+1)}{(2J_0+1)} \cdot (\Gamma_0/E_\gamma^3)$ which are directly proportional to the interesting reduced excitation probabilities $B(M1)\uparrow$ and $B(E1)\uparrow$, respectively (see, e.g. [3,35]). The new data for ^{163}Dy show, besides the previously known strength concentration, a flat and broad distribution of weak excitations which now could be detected and were missed in our previous study[33]. The systematics demonstrates that an extreme fragmentation of the *Scissors Mode* seems to be quite general, whereas strength concentrations like in ^{161,163}Dy represent exceptional cases.

Since in NRF experiments on odd-mass nuclei in general no parity assignments are possible, the values have to be corrected for the comparison with data for even-even nuclei by assuming the same ratio of total electric to magnetic dipole strengths as in the neighboring even-even nuclei. Furthermore, the summation of $M1$ strength has to be limited to the appropriate energy range (2.7–3.7 MeV). The strength ascribed in this manner to the *Scissors Mode* can be compared with the strength predicted by the sum rule for even-even nuclei given in ref. [27]. Only the total dipole strength in ^{163}Dy, as observed in the experiment with the highest sensitivity and ascribed to the *Scissors Mode*, exhausts within the uncertainties the sum rule predictions. From this one may conclude that the problem of the so-called missing *Scissors Mode* strength in odd-mass deformed nuclei is simply due a lack of sensitivity in most of the NRF experiments so far. Sum-rule predictions have also been derived for odd-mass rare-earth nuclei within the Interacting-Boson-Fermion model (IBFM)[40], although not for the nucleus ^{163}Dy discussed here. Nevertheless, these results suggest $B(M1)$ strengths similar or even slightly larger than for even-mass nuclei.

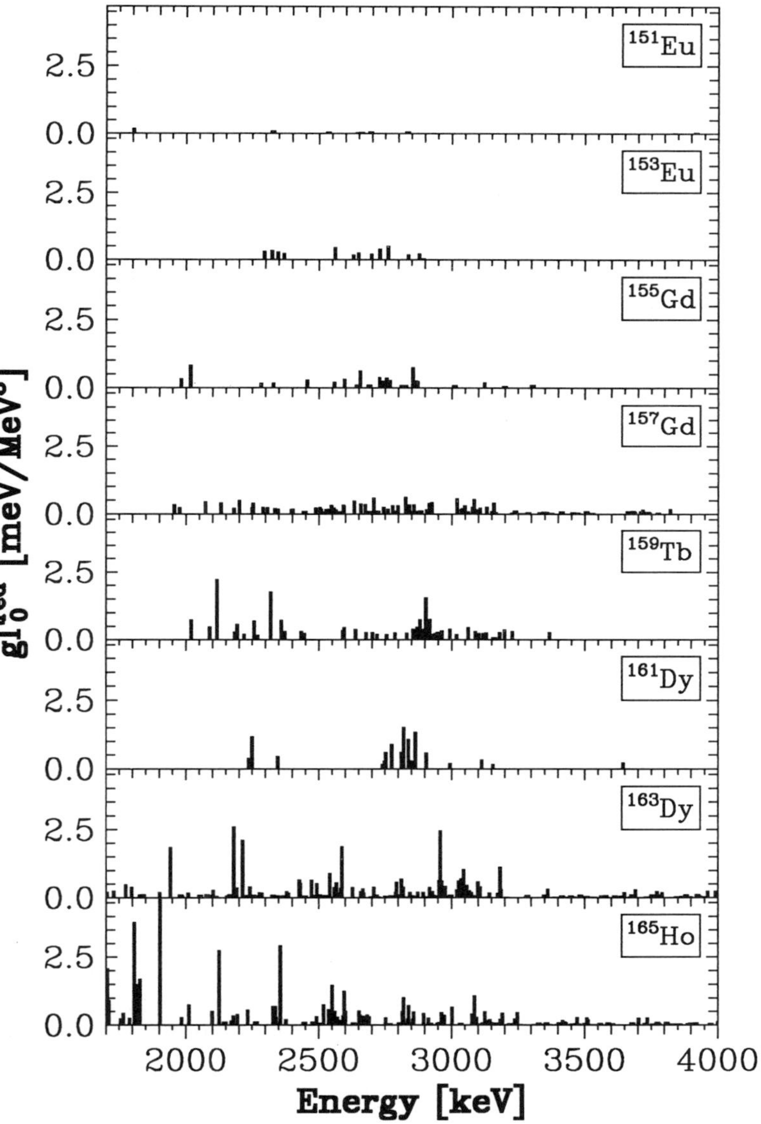

Figure 2: Systematics of the dipole strengths distributions in odd-mass rare-earth nuclei as detected in NRF experiments performed at the Stuttgart facility. Recent results on ^{151,153}Eu, ^{163}Dy and ^{165}Ho [11] are compared with data from previous studies [14,35]. Equal ordinate scales were chosen intentionally to demonstrate the rapidly increasing fragmentation and reduction of detectable strengths for nuclei with decreasing mass number A.

Fluctuation analyses were performed [11] of the NRF spectra observed for ^{163}Dy and ^{165}Ho. The new data [11] on ^{165}Ho provide an important test case for the application of this method to photon scattering spectra. For similar input parameters the analysis of two independent measurements of the $^{165}\text{Ho}(\gamma,\gamma')$ reaction [11,37] with different endpoint energies, energy resolution and sensitivity limits leads to consistent predictions for the total *Scissors Mode* strength within the uncertainties of the method. On the other hand, the new measurement of ^{163}Dy confirms impressively the exceptional character of the *Scissors Mode* strength distribution in this nucleus. The fluctuation analysis produces unphysical background shapes. This may either be due to a breakdown of the underlying statistical assumptions or result from a lack of unresolved strength suggested by the large $B(M1)$ value detected in resolved transitions. An explanation must be sought in the structure properties of the Dy isotopes because a minimum of the fragmentation with respect to the neighboring isotones is also observed for the even-mass cases.

QPM- and sdg-IBFM-calculations [41,42] for odd-mass rare-earth nuclei up to now are not able to provide a realistic description of the experimental strength fragmentation, even not for the most favourable case of ^{163}Dy.

4 E1 Excitations

4.1 E1 Two-Phonon Excitations in Spherical Nuclei Near Shell Closures

The lowest excitations in even-even nuclei near closed shells are usually a $J^\pi=2^+$ quadrupole vibration and a $J^\pi=3^-$ octupole vibration of the nuclear shape. In a simple collective picture one should expect the coupling of the single phonons to a two-phonon excitation multiplet $(2^+ \otimes 3^-)$ consisting of five states with $J^\pi = 1^- \dots 5^-$. In photon scattering experiments the 1^- member of the $2^+ \otimes 3^-$ multiplet can be excited by an $E1$ transition from the ground state. Already in the late seventies Metzger [43] interpreted the enhanced $E1$ excitations observed in his pioneering photon scattering experiments off nuclei around the $N=82$ shell closure as two-phonon excitations of the type $2^+ \otimes 3^-$ (see, e.g., ref. [43]). Moreover, the recent observation of the expected one-phonon decays to the 2^+ quadrupole phonon and the 3^- octupole phonon [44] with transition rates consistent with the phonon coupling picture add additional credence to the interpretation of these 1^- states as quadrupole-octupole coupled excitations. In addition, calculations in the framework of the quasiparticle-phonon-model (QPM), performed for nuclei around the $N=82$ shell closure [45,46] and the Sn-isotopes ($Z=50$) [47] support the interpretation of these $J^- = 1^-$ states as two-phonon excitations.

The new generation of NRF experiments [3] provided a rich amount of data on energies and decay properties of $J^\pi = 1^-$ levels in even-even nuclei around closed shells. The level energy E_{1^-} was observed to be approximately equal to the sum $E_{2^+} + E_{3^-}$ corresponding to a nearly harmonic coupling. In a recent letter [9] all data, available from NRF experiments, on systematic properties of the $E1$ decay of these 1_1^- levels in even-even nuclei around closed shells are summarized and discussed. In fig. 3 the reduced transition probabilities $B(E1, 1^- \rightarrow 0^+)$ are shown as observed in nuclei around the $Z=50$ shell closure. New results for ^{108}Cd [48] and ^{116}Cd [49]

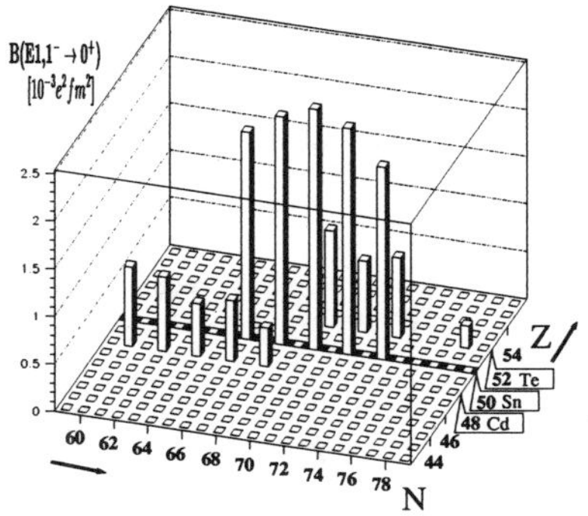

Figure 3: Reduced transition probabilities $B(E1, 1^- \rightarrow 0^+)$ for low-lying $E1$ ground-state transitions in spherical nuclei near the Z=50 shell closure, emphasized by the line in boldtype, from ref. [9], completed by new data for ^{108}Cd [48] and ^{116}Cd [49].

were included. As can be clearly seen the strengths are maximal in magic nuclei at the shell closure and are reduced in nuclei away from the magic number. The same behavior is found generally at all shell closures Z, N=28; Z, N=50 and N=82 [9]. On the other hand, the N- and Z-dependence of the $B(E1, 1_1^- \rightarrow 2_1^+)$ values, the transition probabilities for the decay to the first excited 2^+-states, is entirely opposite. The salient features of these systematics find a natural explanation on the basis of the QPM-calculations of ref. [46] as due to the dipole core polarization associated with the Giant Dipole Resonance (GDR) (see ref. [9]).

4.2 E1 Excitations in Deformed Even-Even Nuclei

In deformed nuclei the octupole vibration can couple to the static quadrupole deformation leading to four octupole vibrational bands in even-even nuclei. These rotational bands are characterized by their K quantum numbers. The 1^- states, representing the band heads of the K=0 and K=1 bands can be easily studied in NRF experiments. All data on spherical, transitional and well-deformed nuclei in the mass region A = 130–200 have been summarized and discussed in a recent publication [50]. The excitation energy in spherical nuclei near shell closures is close to the sum of the quadrupole and octupole phonons. In the transitional region the 1^- energies are lowered mainly due to the lowering of the first 2^+ state. In the region of deformed nuclei the 1^- energies follow the trend of the K=0 octupole bands and increase slightly with the mass number A (see figure 4).

As discussed above, for the two-phonon excitations in the spherical semi-magic N=82 isotones large values for the E1 excitations strength of $B(E1)\uparrow \approx 20 \cdot 10^{-3}$ $e^2 fm^2$ were observed. Nearly the same strengths were found for the well-deformed nuclei in the middle of the N=82–126 neutron major shell. A decrease of the $E1$ strength of about one order of magnitude was observed in the transitional region to γ-soft nuclei. An additional dip of the $E1$ strength appears around N=86 at the transition from spherical to well-deformed nuclei (see figure 4). This transition is also clearly seen in the decay branching of the 1^- state to the first 2^+ state

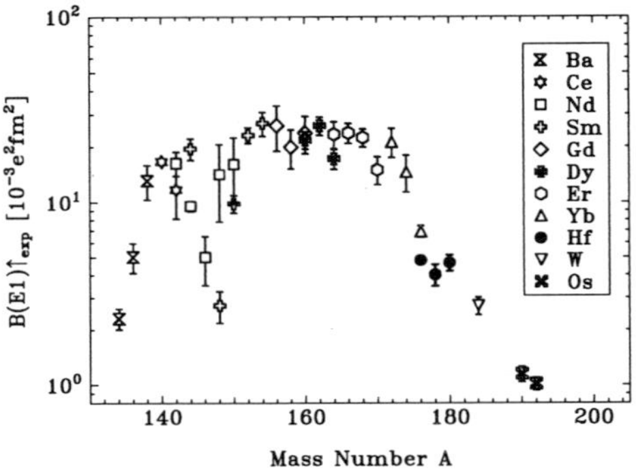

Figure 4: Systematics of the reduced electric dipole excitation strengths $B(E1)\uparrow$ of the lowest (or lowest two) observed $J^{\pi} = 1^-$ states in even-even nuclei of the N=82–126 shell as a function of the mass number A, according to ref. [50].

[32]. Whereas in spherical nuclei this branching is small it reaches for well-deformed nuclei the value of two as expected from the Alaga rules [51].

A special kind of strong E1 excitations near 2.5 MeV in deformed nuclei was systematically observed in NRF-experiments [52,13] exhibiting an uncommon decay branching, hinting to a K mixing [53]. These excitations are candidates for a novel two-phonon excitation caused by the coupling of the octupole and γ-quadrupole vibrations. This conclusion is based on the nearly quantitative agreement of the experimental excitation energies with the sum of the $K = 1$ octupole and $K = 2$ γ-vibration as suggested by the collective model [54] and on the results of sdf-IBA calculations [52], which reproduce the experimental energies and the structure of the states.

5 Concluding Remarks and Outloook

In the past years the NRF-technique has proven to represent an unique tool to investigate low-spin states in heavy nuclei and to extract model-independent spectroscopic information on fundamental dipole excitations. Detailed information exist for both, well deformed and spherical nuclei. In addition, dipole strength distributions were and are studied in isotopic chains of nuclei exhibiting transitions from spherical or deformed to more γ-soft triaxial shapes. Topics of current interest, which can be tackled by NRF eperiments, are, e.g., investigations of mixed-symmetry 2^+ states [21,22], studies of the socalled $E1$ *"Pigmy Resonance"* [55] near the particle emission threshold and of soft dipole modes in nuclei with a high neutron excess [56].

In future, the development of high-flux, quasimonocromatic photon sources, tunable in energy and of nearly 100% linear polarization by Laser-Compton-Back-scattering [57] will considerably extend present applications of the photon scattering technique in nuclear structure physics.

The present availability of intense bremsstrahlung beams and sensitive photon detectors allows also to extend the application of the real photon probe to other problems like photoactivation of long-lived isomers [58]. Combined NRF and photoactivation experiments enable to study the surprisingly high population of these isomers in low-energy photon-induced reactions which are caused by the underlying nuclear structure. The photoactivation of long-lived isomers, furthermore, is of special interest in view of pumping schemes of possible γ-lasers [59] and, in particular, for nuclear astrophysics. Here isomers can play an important role for the nucleosynthesis processes of various isotopes [60,61]. Another new, fascinating application of the photoactivation technique is the measurement of averaged (γ, n)-cross sections and stellar reaction rates, where the photon bath in the environment of the astrophysical so-called γ-process is simulated by an appropriate superposition of bremsstrahlung spectra of different endpoint energies [62].

The addressed interesting physics and the recent experimental developments will open a wide field of applications and will guarantee a promising future for low-energy photo-induced reaction studies.

Acknowledgments

It is a pleasure to thank all colleagues from Cologne, Darmstadt, Dubna, Gent, Karlsruhe, Lexington, Moscow, Rossendorf, Sofia, and Youngstown, for the longstanding, stimulating and fruitful collaborations in the Stuttgart experiments. Special thanks are due to my Stuttgart longterm, present and former coworkers Daniela Belic, H. von Garrel, C. Kohstall, H. Maser, A. Nord, M. Scheck, F. Stedile, and, last not least, to H.H. Pitz for their outstanding engagement. The financial support of the Stuttgart projects by the Deutsche Forschungsgemeinschaft (DFG) under contracts Kn 154/30,31 is gratefully acknowledged.

References

1. N. Lo Iudice *et al*, *Phys. Rev. Lett.* **41**, 1532 (1978).
2. D. Bohle *et al*, *Phys. Lett.* **137B**, 27 (1984).
3. U. Kneissl *et al*, *Prog. Part. Nucl. Phys.* **37**, 349 (1996).
4. A. Richter, *Prog. Part. Nucl. Phys.* **34**, 261 (1995).
5. D. Zawischa, *J. Phys. G: Nucl. Part. Phys.* **24**, 683 (1998).
6. N. Lo Iudice, Riv. Nuovo Cimento **23**, 1 (2000).
7. F. Iachello, *Phys. Rev. Lett.* **53**, 1427 (1984).
8. F. Iachello, *Phys. Lett.* **160B**, 1 (1985).
9. W. Andrejtscheff *et al*, *Phys. Lett.* **B 506**, 239 (2001).
10. U.E.P. Berg and U. Kneissl, *Ann. Rev. Nucl. Part. Sci.* **37**, 33 (1987).
11. A. Nord *et al*, *Phys. Rev.* C **67**, 034307 (2003).
12. B. Schlitt *et al*, *Nucl. Instr. a. Meth. in Phys. Res.* A **337**, 416 (1994).
13. H. Maser *et al*, *Phys. Rev.* C **53**, 2749 (1996).
14. J. Margraf *et al*, *Phys. Rev.* C **52**, 2429 (1995).
15. R.-D. Heil *et al*, *Nucl. Phys.* **A476**, 39 (1988).
16. J. Margraf *et al*, *Phys. Rev.* C **42**, 771 (1990).
17. J. Margraf *et al*, *Phys. Rev.* C **45**, R521 (1992).
18. A. Linnemann *et al*, *Phys. Lett.* **B 554**, 15 (2003).

19. P. von Brentano *et al, Phys. Rev. Lett.* **76**, 2029 (1996).
20. H. Maser *et al, Phys. Rev.* C **54**, R2129 (1996).
21. N. Pietralla *et al, Phys. Rev. Lett.* **83**, 1303 (1999).
22. C. Fransen *et al, Phys. Rev.* C **67**, 024307 (2003).
23. N. Pietralla *et al, Phys. Rev.* C **52**, R2317 (1995).
24. P. von Neumann-Cosel *et al, Phys. Rev. Lett.* **75**, 4178 (1995).
25. N. Pietralla *et al, Phys. Rev.* C **58**, 184 (1998).
26. J. Enders *et al, Phys. Rev.* C **59**, R1851 (1999).
27. N. Lo Iudice *et al, Phys. Lett.* B **304**, 193 (1993).
28. N. Lo Iudice, Phys. Rev. C **57**, 1246 (1998).
29. W. Ziegler *et al, Phys. Rev. Lett.* **65**, 2515 (1990).
30. W. Ziegler *et al, Nucl. Phys.* **A564**, 366 (1993).
31. J. Margraf *et al, Phys. Rev.* C **47**, 1474 (1993).
32. T. Eckert *et al, Phys. Rev.* C **56**, 1256 (1997) and *Phys. Rev.* C **57**, 1007 (1998).
33. I. Bauske *et al, Phys. Rev. Lett.* **71**, 975 (1993).
34. C. Wesselborg *et al, Phys. Lett.* **207 B**, 22 (1988).
35. A. Nord *et al, Phys. Rev.* C **54**, 2287 (1996).
36. C. Schlegel *et al, Phys. Lett.* B **375**, 21 (1996).
37. N. Huxel *et al, Nucl. Phys.* **A645**, 239 (1999).
38. J. Enders *et al, Phys. Rev. Lett.* **79**, 2010 (1997).
39. J. Enders *et al, Phys. Rev.* C **57**, 996 (1998).
40. J. N. Ginocchio *et al, Phys. Rev. Lett.* **79**, 813 (1997).
41. V. Soloviev *et al, Nucl. Phys.* **A613**, 45 (1997).
42. Y.D. Devi *et al, Nucl. Phys..* **A600**, 20 (1996).
43. F.R. Metzger, *Phys. Rev.* C **14**, 543 (1976).
44. M. Wilhelm *et al, Phys. Rev.* C **54**, R449 (1996).
45. M. Grinberg *et al, Nucl. Phys.* **A573**, 231 (1994).
46. V.Yu. Ponomarev *et al, Nucl. Phys.* **A635**, 470 (1998).
47. J. Bryssinck *et al, Phys. Rev.* C **59**, 1930 (1999).
48. A. Gade, *et al, Phys. Rev.* C **67**, 034304 (2003).
49. C. Kohstall, *et al, Physics of Atomic Nuclei* **64**, 1141 (2001) and to be published.
50. C. Fransen *et al, Phys. Rev.* C **57**, 129 (1998).
51. G. Alaga *et al, Dan. Mat. Fys. Medd.* **29** No.9, 1 (1955).
52. U. Kneissl *et al, Phys. Rev. Lett.* **71**, 2180 (1993).
53. A. Zilges *et al, Phys. Rev.* C **42**, 1945 (1990).
54. W. Donner *et al, Z. Phys.* **197**, 440 (1966).
55. A. Zilges *et al, Phys. Lett.* **B 542**, 43 (2002).
56. T. Hartmann *et al, Phys. Rev. Lett.* **85**, 274 (2000).
57. N. Pietralla *et al, Phys. Lett. Rev.* **88**, 012502 (2002).
58. D. Belic *et al, Nucl. Instr. a. Meth. in Phys. Res.* A **463**, 26 (2001).
59. F. Stedile *et al, Phys. Rev .* C **63**, 024320 (2001).
60. D. Belic *et al, Phys. Rev. Lett.* **83**, 5242 (1999).
61. D. Belic *et al, Phys. Rev.* **65**, 035801 (2002).
62. P. Mohr *et al, Phys. Lett.* **B 488**, 127 (2000).

ELECTRIC DIPOLE EXCITATIONS CLOSE TO THE PARTICLE THRESHOLD

A. ZILGES
Institut für Kernphysik, TU Darmstadt, Schlossgartenstr. 9,
D-64285 Darmstadt, Germany
E-mail: zilges@ikp.tu-darmstadt.de

Improved experimental techniques allow to measure the dipole strength distribution in nuclei up to the particle threshold with very high sensitivity and precision. A concentration of E1 strength exhausting up to 1 % of the isovector energy weighted sum rule is found in the 1 $\hbar\omega$ region. Results on N=82 isotones are presented and different interpretations of the strength are discussed.

1 Introduction

The structure of electric dipole excitations contains important information about the symmetry behaviour of protons and neutrons in atomic nuclei [1]. In the most simple microscopic picture E1 excitations consist of particle-hole excitations across one major shell. The residual interaction shifts most of the strength from the 1 $\hbar\omega$ region to the 2 $\hbar\omega$ region, i.e. in heavy nuclei from about 7 MeV to 14 MeV. The resulting resonance structure around 14 MeV is denoted Giant Dipole Resonance (GDR) and its gross features are well studied [2]. In a collective picture the GDR can be described as an out of phase oscillation of the proton versus the neutron fluid. The GDR exhausts nearly 100 % of the isovector Energy Weighted Sum Rule (EWSR).

However, recent experiments have shown that considerable parts of the E1 strength can be found below the GDR region. The lowest E1 excitation in spherical nuclei has a dominantly isoscalar two phonon quadrupole–octupole character, whereas in deformed nuclei the coupling of the octupole degree of freedom to the nuclear deformation leads to the octupole vibrational bands characterized by the K quantum number. A concentration of E1 strength at slightly higher energies around the 1$\hbar\omega$ region is denoted Pygmy Dipole Resonance (PDR) frequently. A detailed understanding of the PDR can help to understand nuclear structure phenomena at higher energies. In addition such a concentration of strength would have important implications on photonuclear reaction rates relevant for the nucleosynthesis of heavy neutron deficient isotopes in explosive stellar events [3,4]. This is because the relevant energy window for photoinduced reactions in the stellar radiation bath is around the threshold energies and not in the GDR region. Another interesting aspect of collective E1 strength is that analogue excitation modes exist in other finite fermion systems, e.g. in metallic clusters [5,6,7] and quantum dots [8,9].

This paper will first describe the experimental method which allows the high precision studies of E1 and M1 excitations. Then it will briefly summarize the knowledge on two phonon quadrupole–octupole states in nuclei. The next part will focus on the results about the E1 strength at higher energies up to the particle threshold and its interpretation. Finally, the paper will close with a short outlook.

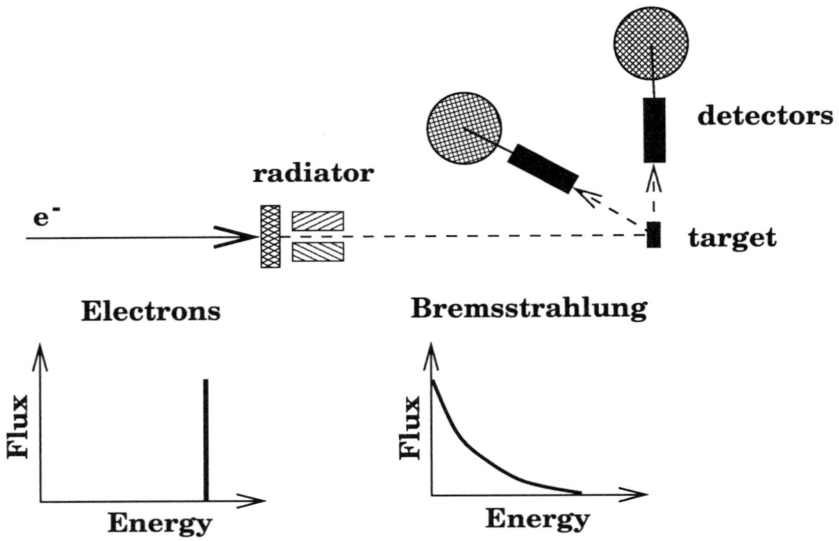

Figure 1: The real photon scattering set up at the S–DALINAC: The electron beam with an energy of up to 10 MeV is converted into a continuous bremsstrahlung photon spectrum. Resonant photon scattering is observed with high resolution Ge(HP) detectors. An extension up to an endpoint energy of 14 MeV will be realized in the near future.

2 Experimental method

The experiments have been performed at the injector of the superconducting Darmstadt linear accelerator S–DALINAC [10]. Electrons with a maximum energy of 10 MeV and currents of up to 40 μA are completely stopped in a massive rotating copper disk and converted into a continuous bremsstrahlung spectrum [11]. An upgrade to a 14 MeV bremsstrahlung set up is planned for the near future. The maximum energy of the photons generated in this way is equal to the electron energy, typical photon fluxes are about 10^6photons/keV·s. A few hundred milligram of the isotope of interest is mounted in the photon beam, and dipole and to less extent quadrupole transitions are induced from the groundstate of the nucleus to higher lying states. The subsequent γ decay of these states back to the ground state or to excited states is observed with two high resolution, high efficiency Ge(HP) semiconductor detectors which are actively shielded by a BGO crystal. The setup is shown in figure 1 schematically.

Figure 2 shows a typical spectrum of the photons scattered of the semi–magic N=82 nucleus ^{144}Sm. Lines stemming from the γ decay of ^{144}Sm and from the decay of a ^{11}B target measured simultaneously for photon flux calibration are sitting on a continuous background. This background is mainly due to nonresonantly scattered bremsstrahlung photons. From such a spectrum one can directly deduce the reduced absolute transition strengths, i.e. the B(E1), B(M1), or B(E2) values.

Real photon scattering (or Nuclear Resonance Fluorescence, NRF) combines

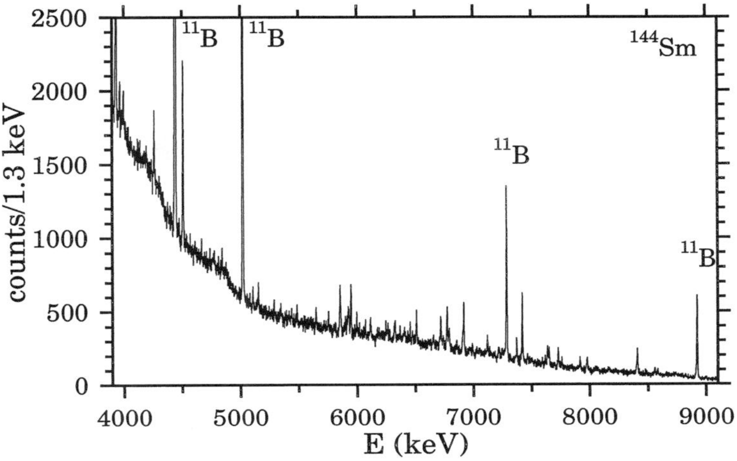

Figure 2: Spectrum of photons scattered of the N=82 nucleus ^{144}Sm. The background increases with decreasing energy mainly due to nonresonantly scattered photons. The labelled lines stem from the photon flux calibration using ^{11}B which is done simultaneously with the measurement on the ^{144}Sm target.

the following advantages[12]:

- A pure electromagnetic excitation mechanism.
- Selectivity with respect to spin and excitation strength.
- A wide energy region is measured in a single run with very high energy resolution.
- A straightforward and model independent data evaluation and absolute strength determination.

Therefore photon scattering is ideally suited for systematic investigations of strength distributions, even in the energy region close to the particle threshold where the level density is already quite high.

3 Two phonon quadrupole–octupole excitations

In all examined nuclei near shell closures one strong electric dipole excitation has been identified very close to the sum energy of the lowest 2^+ state and the lowest 3^- state. It has been shown that this state is the 1^- member of the $2_1^+ \otimes 3_1^-$ two phonon multiplet arising from a nearly harmonic coupling of the two fundamental surface vibrations. The expected decay pattern for such a two phonon state (e.g. a strong E2 decay to the 3_1^- state) has been observed in (p,p'γ) experiments[13]. It has been discussed by Iachello[1], that the dynamic dipole moment induced by such a coupling

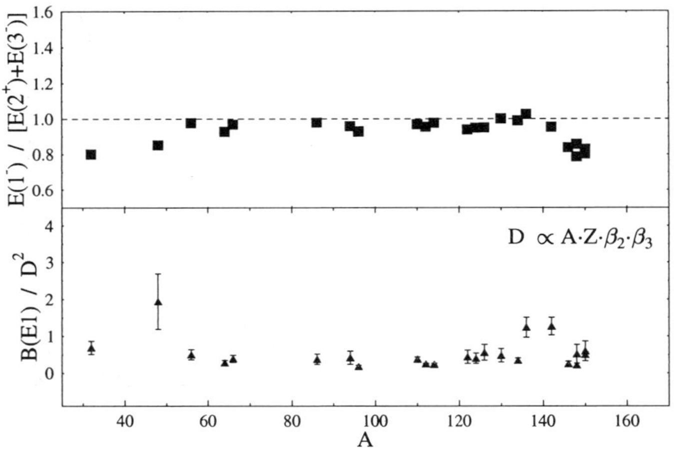

Figure 3: Energy and strength systematics of the lowest 1^- state in nuclei near shell closures. The lower part gives the ratio of the measured B(E1) groundstate transition strength and the dipole moment calculated from the dynamic deformation parameters β_2 and β_3 (figure taken from ref. [15]).

should scale with $\beta_2 \times \beta_3$, i.e. with the product of the quadrupole and octupole deformation parameters. The wealth of experimental data from (γ, γ') experiments collected during the last decade opens now the possibility for a systematic study of this strength behaviour over a wide range of nuclei. The upper part of figure 3 shows the ratio of the energy of the lowest 1^- state to the sum energy of the 2_1^+ and 3_1^- states for all investigated non magic nuclei near shell closures. One can see that this ratio is always very close to unity indicating a high degree of harmonicity of the two phonon coupling. This is a consequence of the reduced Pauli blocking because two phonons with very different underlying single particle structures are involved. The lower part shows the measured B(E1) strength divided by the dynamic dipole moment calculated from the quadrupole and octupole deformation parameters [15]. One can see that the ratio is nearly constant over a wide range of nuclei including very different β_2 and β_3. This supports the simple picture of generating a dynamic dipole moment by the coupling of the two fundamental excitations. We note that quadrupole–octupole excitations can be found in nuclei away from closed shells as well. A detailed discussion is given in refs. [16] and [17].

4 E1 strength close to the particle threshold

The improvement of the experimental technique of photon scattering allows it today to extend the high precision strength measurements from the low energy region (see

Ulrich Kneissl's contribution to these proceedings) up to the particle threshold. We have performed detailed studies on the N=82 isotones ^{138}Ba, ^{140}Ce, and ^{144}Sm up to an energy of about 10 MeV recently [18]. Previous experiments on ^{138}Ba and ^{140}Ce limited to a maximum photon energy of 6.5 MeV have shown first evidence for E1 strength in the $1\hbar\omega$ region in these nuclei [19,20]. A similar concentration has been observed in the closed proton shell Sn nuclei by the Gent group [21]. Please keep in mind that the level density in the examined energy region is very high even for semi–magic isotopes and that one relies on the selectivity of the photon probe in combination with high resolution detectors to get useful information.

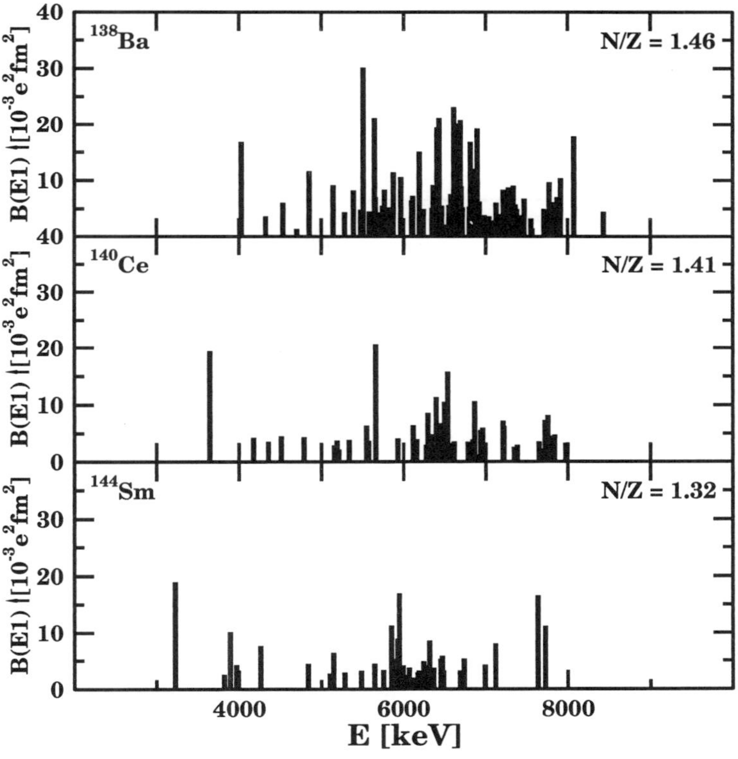

Figure 4: B(E1) strength in the three investigated N=82 isotones. The neutron to proton ratio N/Z is given for each isotope (figure taken from ref. [18]).

The results deduced from our new experiments are shown in figure 4 which gives the B(E1)↑ strength distributions. A typical excitation pattern can be observed in the three investigated nuclei: First an isolated E1 transition below about 4 MeV which corresponds to the two phonon quadrupole–octupole excitation discussed in the previous chapter. And second a concentration of strength with a resonance like structure between 5.5 and 8 MeV. Only negligible strength could be found above about 8 MeV up to the neutron separation energy which shows that the observed states are not just statistical E1 excitations sitting on the tail of the GDR, but a collective excitation mode with a different structure. The summed strength of the transitions amounts to $0.58(0.11)e^2fm^2$, $0.25(0.04)$ e^2fm^2, and $0.20(0.04)$ e^2fm^2 for ^{138}Ba, ^{140}Ce, and ^{144}Sm, respectively. This amounts to about 0.78(15)%, 0.33(5)%, and 0.24% of the isovector EWSR. The negative parity of all states below 6.5 MeV in ^{138}Ba has been proven by Norbert Pietralla et al. in experiments performed at the HIGS facility at Duke University using a monoenergetic photon beam [22].

What is the origin of the observed E1 strength? Franco Iachello suggested to fold our discrete strength distributions with a Lorentzian of a fixed width [23]. Figure 5 shows the result for a folding width of Γ=250 keV. This plot suggests that probably several different modes are responsible for the observed E1 strength: Besides the low lying $2^+\otimes3^-$ state, structures may be identified around 5.5 MeV (with decreasing strength when going from ^{138}Ba to ^{144}Sm), 6.5 MeV, and 7.5 MeV.

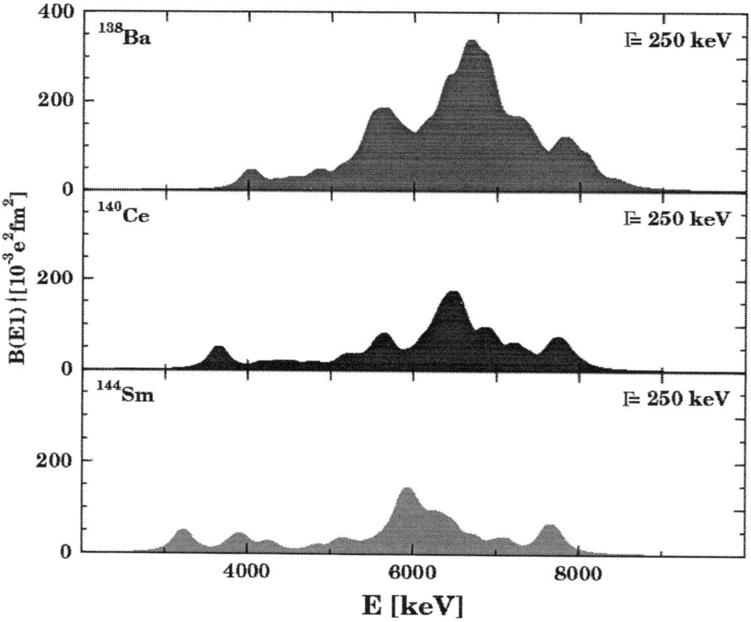

Figure 5: Strength distribution from fig. 4 folded with a Lorentzian of 250 keV width.

One possible source of E1 strength is the breaking of the symmetry of the proton and neutron distributions in the nucleus due to an enrichment of one type of nucleons in the nuclear surface. Such a "skin" could lead to an isovector E1 mode [24,25]. Systematic experiments on stable nuclei using antiprotons at the LEAR facility at CERN have indeed shown an enrichment of neutrons in the periphery for nuclei with a neutron separation energy $S_n < 10$ MeV and a proton enrichment for certain relative proton rich nuclei with a closed neutron shell, i.e. ^{144}Sm [26]. Similar results are yielded by self consistent HFB calculations [27]. Our experimental studies so far include nuclei ranging from the relatively neutron rich nucleus ^{138}Ba (N/Z=1.46, S_n=8.6 MeV) to the relatively proton rich ^{144}Sm (N/Z=1.32, S_n=10.5 MeV). In a simplified picture one could therefore expect a minimum for the E1 strength in the nucleus ^{142}Nd where the skin effects should be rather small. An excitation mode which may be related to this macroscopic picture are the F–vector E1 transitions in the sdf IBA–2 introduced by Norbert Pietralla [28].

A local breaking of the isospin symmetry due to clustering has been proposed by Franco Iachello as a possible other source of E1 strength [1]. Here a cluster with a different neutron to proton ratio than the remaining core can lead to a dipole moment. The energy of such a mode is expected in the vicinity of about 6 MeV in the A=140 mass region [23]. In this mass region an enhanced probability for α–clustering can be deduced from the α instability of some neighbouring isotopes.

Microscopic calculations in the RPA show a concentration of isoscalar E1 strength at slightly higher energies [29,30,31]. The strength originates from the $3\hbar\omega$ isoscalar giant dipole resonance partly shifted down to lower energies by residual interactions.

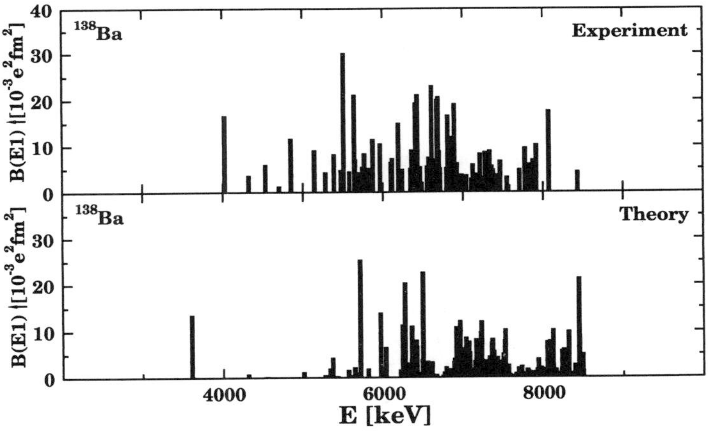

Figure 6: Comparison of experimental B(E1) strength distribution (upper part) and a calculation in the quasiparticle phonon model QPM (lower part) for ^{138}Ba.

In Darmstadt a calculation has been performed in the framework of the quasiparticle phonon model QPM with an extended configuration space [32]. The energies of single particle states near the Fermi level have been varied to obtain an improved description of the strongest excitations. The results for ^{138}Ba are shown in figure 6 and are in good agreement with the experiment. However, the systematic trend of the summed strength in the N=82 nuclei could not be reproduced. An inspection of the averaged transition charge densities for neutrons and protons shows that surface neutrons are important to explain the excitations around 7 MeV, thus pointing to an excitation dominated by neutrons in the nuclear periphery. However, the situation is much less pronounced than in the case of ^{208}Pb [33].

5 Summary and Outlook

In high resolution photon scattering experiments at the S–DALINAC at Darmstadt University of Technology a concentration of dipole strength exhausting up to 1% of the EWSR has been observed in three semi–magic N=82 nuclei between 5.5 and 8 MeV. Various model predictions and other experimental observations seem to show that at least part of the strength is due to a structure dominated by neutron excitations in the periphery of the nucleus. More systematic measurements to establish this excitation mode as a fundamental mode in nuclei are necessary. A photon scattering experiment on ^{142}Nd is presently under way at Darmstadt, first very preliminary results point indeed to a lower summed E1 strength in comparison with the neighbouring isotopes ^{140}Ce and ^{144}Sm as discussed above in connection with nucleon skin excitations. Similar experiments on ^{136}Xe and on non–magic nuclei are scheduled for the immediate future.

In addition it is planned to learn more about the isospin character of the mode by means of $(\alpha,\alpha'\gamma)$ and (p,p'γ) coincidence experiments at medium energies using the AGOR cyclotron at KVI. In earlier experiments at the KVI strong isoscalar E1 excitations have been observed in ^{40}Ca, ^{58}Ni, ^{90}Zr, and ^{208}Pb in the energy region around 6 MeV [34]. Form factor measurements can be done in electron scattering experiments at the S-DALINAC.

In summary it can be shown that a study of E1 excitations close to the threshold with a combination of various experimental probes is one approach to learn more about the symmetry character of protons and neutrons in nuclei.

Acknowledgments

Much of the research work has been inspired by numerous discussions with Franco Iachello. This article is dedicated to him on the occasion of his 60th birthday. I want to take the opportunity to thank him and wish him the best for the future!

I thank the members of my research group, especially S. Volz, M. Babilon, T. Hartmann, P. Mohr, and K. Vogt for their engagement and enthusiasm which enabled the realization of the experiments. I thank U. Kneissl, N. Pietralla, V.Yu. Ponomarev, A. Richter, P. von Brentano, and J. Wambach for valuable contributions. This work was supported by the Deutsche Forschungsgemeinschaft (contracts Zi 510/2–1 and FOR 272/2–1).

References

1. F. Iachello, *Phys. Lett.* B **160**, 1 (1985).
2. S. S. Dietrich and B. L. Berman, *At. Dat. Nucl. Dat. Tab.* **38**, 199 (1988).
3. S. Goriely, *Phys. Lett.* B **436**, 10 (1998).

4. P. Mohr, K. Vogt, M. Babilon, J. Enders, T. Hartmann, C. Hutter, T. Rauscher, S. Volz, and A. Zilges, *Phys. Lett.* B **488**, 127 (2000).
5. W. A. de Heer, K. Selby, V. Kresin, J. Masui, M. Vollmer, A. Chatelain, and W. D. Knight, *Phys. Rev. Lett.* **59**, 1805 (1987).
6. F. Iachello, E. Lipparini, and A. Ventura, *Phys. Rev.* B **45**, 4431 (1992).
7. H. Haberland, *Nucl. Phys.* A **649**, 415c (1999).
8. A. Delgado, L. Lavin, R. Capote, and A. Gonzalez, *Physica* E **8**, 342 (2000).
9. R. Capote, A. Delgado, and A. Gonzalez, *Mod. Phys. Lett.* B **15**, 81 (2001).
10. A. Richter, in: S. Myers et al. (Ed.), Proc. 5th European Particle Accelerator Conf. Barcelona, Spain, 1996, IOP Publishing, Bristol.
11. P. Mohr, J. Enders, T. Hartmann, H. Kaiser, D. Schiesser, S. Schmitt, S. Volz, F. Wissel, and A. Zilges, *Nucl. Instrum. Methods* A **423**, 480 (1999).
12. U. Kneissl, H. H. Pitz, and A. Zilges, *Prog. Part. Nucl. Phys.* **37**, 349 (1996).
13. M. Wilhelm, S. Kasemann, G. Pascovici, E. Radermacher, P. von Brentano, and A. Zilges, *Phys. Rev.* C **57**, 577 (1998).
14. A. Bohr and B. R. Mottelson, *Nucl. Phys.* **4**, 529 (1957).
15. M. Babilon, T. Hartmann, P. Mohr, K. Vogt, S. Volz, and A. Zilges, *Phys. Rev.* C **65**, 037303 (2002).
16. C. Fransen, O. Beck, P. von Brentano, T. Eckert, R.–D. Herzberg, U. Kneissl, H. Maser, A. Nord, N. Pietralla, H. H. Pitz, and A. Zilges, *Phys. Rev.* C **57**, 129 (1998).
17. W. Andrejtscheff, C. Kohstall, P. von Brentano, C. Fransen, U. Kneissl, N. Pietralla, H. H. Pitz, *Phys. Lett.* B **506**, 239 (2001).
18. A. Zilges, S. Volz, M. Babilon, T. Hartmann, P. Mohr, and K. Vogt, *Phys. Lett.* B **542**, 43 (2002).
19. R.-D. Herzberg, P. von Brentano, J. Eberth, J. Enders, R. Fischer, N. Huxel, T. Klemme, P. von Neumann–Cosel, N. Nicolay, N. Pietralla, V. Yu. Ponomarev, J. Reif, A. Richter, C. Schlegel, R. Schwengner, S. Skoda, H. G. Thomas, I. Wiedenhöver, G. Winter, and A. Zilges, *Phys. Lett.* B **390**, 49 (1997).
20. R.-D. Herzberg, C. Fransen, P. von Brentano, J. Eberth, J. Enders, A. Fitzler, L. Käubler, H. Kaiser, P. von Neumann–Cosel, N. Pietralla, V. Yu. Ponomarev, H. Prade, A. Richter, H. Schnare, R. Schwengner, S. Skoda, H. G. Thomas, H. Tiesler, D. Weisshaar, and I. Wiedenhöver, *Phys. Rev.* C **60**, 051307 (1999).
21. K. Govaert, F. Bauwens, J. Bryssinck, D. De Frenne, E. Jacobs, W. Mondelaers, L. Govor, and V. Yu. Ponomarev, *Phys. Rev.* C **57**, 2229 (1998).
22. N. Pietralla, Z. Berant, V. N. Litvinenko, S. Hartman, F. F. Mikhailov, I. V. Pinayev, G. Swift, M. W. Ahmed, J. H. Kelley, S. O. Nelson, R. Prior, K. Sabourov, A. P. Tonchev, and H. R. Weller, *Phys. Rev. Lett.* **88**, 012502 (2002).
23. F. Iachello, private communication (2002).
24. P. van Isacker, M. A. Nagarajan, and D. D. Warner, *Phys. Rev.* C **45**, R13 (1992).
25. J. Chambers, E. Zaremba, and J. P. Adams, *Phys. Rev.* C **50**, R2671 (1994).

26. P. Lubinski, J. Jastrzebski, A. Trzcinska, W. Kurcewicz, F. J. Hartmann, W. Schmid, T. von Egidy, R. Smolanczuk, and S. Wycech, *Phys. Rev.* C **57**, 2962 (1998).
27. J. Dobaczewski, W. Nazarewicz, and T. R. Werner, *Z. Phys.* A **354**, 27 (1996).
28. N. Pietralla, see contribution to these proceedings.
29. D. Vretenar, A. Wandelt, and P. Ring, *Phys. Lett.* B **487**, 334 (2000).
30. D. Vretenar, N. Paar, P. Ring, and T. Niksic, *Phys. Rev.* C **65**, 021301 (2002).
31. G. Colò, N. Van Giai, P. F. Bortignon, and M. R. Quaglia, *Phys. Lett.* B **485**, 362 (2000).
32. V. Yu Ponomarev, private communication (2003)
33. N. Ryezayeva, T. Hartmann, Y. Kalmykov, H. Lenske, P. von Neumann-Cosel, V. Yu. Ponomarev, A. Richter, A. Shevchenko, S. Volz, and J. Wambach, *Phys. Rev. Lett.* **89**, 272501 (2002).
34. T. D. Poelhekken, S. K. B. Hesmondhalgh, H. J. Hofmann, A. van der Woude, and M. N. Harakeh, *Phys. Lett.* B **278**, 423 (1992).

MIXED-SYMMETRY MULTIPHONON STRUCTURES AND FIRST EVIDENCE FOR F-VECTOR $E1$ TRANSITIONS

N. PIETRALLA, C. FRANSEN, P. VON BRENTANO
Institut für Kernphysik, Universität zu Köln, Zülpicher Str. 77, 50937 Köln, Germany
E-mail: pietrall@ikp.uni-koeln.de

Recent γ-ray spectroscopy of off-yrast low-spin states of ^{94}Mo yielded evidence for one-phonon and two-phonon states with mixed proton-neutron symmetry, a phenomenon anticipated by Franco Iachello and collaborators in the framework of the interacting boson model (IBM-2). The mixed-symmetry assignments are based on the measurement of absolute $M1$ matrix elements, $\approx 1\ \mu_N$ in size. Corresponding structures were lateron found in other soft nuclei. We give a brief overview over the recent investigations of mixed-symmetry multiphonon structures and we discuss for the first time the observation of F-vector $E1$ transitions involving mixed-symmetry states.

1 Introduction

The interacting boson model (IBM) which has been developed by Franco Iachello and collaborators [1] provides a richly-structured, yet, simple and elegant framework for the understanding of collective phenomena in many-body quantum systems, in general, and of atomic nuclei, in particular. Despite its parametric flexibility, the IBM has a strong predictive power suggesting structures and phenomena that can be tested by experiment. A beautiful example is the prediction [2,3] of a fundamental class of collective nuclear excitation modes, called *mixed-symmetry states* [4,5], and their specific properties. The mixed-symmetry states (MSSs) correspond to collective isovector-type excitations of the valence shell of heavy nuclei.

The proton-neutron symmetry of IBM-2 wave functions is characterized by the representations of the underlying symmetry group and can be quantified by the F-spin (*Franco's spin*) quantum number [6], which is for "elementary bosons" the analog to the nucleonic isospin, T, for "elementary nucleons" introduced by Heisenberg. The analogy between F-spin and T-spin has been elucidated also by the observation of F-spin multiplets [7,8,9] with rather constant energies corresponding to the well known T multiplets. IBM-1 states correspond to the subset of states with $F = F_{\max} = (N_\pi + N_\nu)/2$ in the IBM-2, where $N_{\pi(\nu)}$ denotes the number of proton (neutron) bosons with F-spin projection $F_z = +1/2(-1/2)$. IBM-2 states with $F = F_{\max}$ have wave functions that are symmetric with respect to any pairwise exchange of proton and neutron boson labels.

Mixed symmetry states (MSSs) do not have this symmetry. They are characterized by F-spin quantum numbers $F < F_{\max}$. We will restrict ourselves to MSSs with $F = F_{\max} - 1$. MSSs form a whole class of collective states with similar wave functions [3]. They are connected among themselves by strong electric quadrupole ($E2$) transitions (in the absence of further selection rules) and can decay by weakly-collective $E2$ and strong magnetic dipole ($M1$) transitions to symmetric states with $F = F_{\max}$. This feature enables us to uniquely identify MSSs out of surrounding $F_{\max}$ states because $M1$ transitions between symmetric states are forbidden. The last statement follows from the structure of the $M1$ transition operator

$$T(M1) = \sqrt{\frac{3}{4\pi}}\,(g_\pi L_\pi + g_\nu L_\nu)$$
$$= \sqrt{\frac{3}{4\pi}}\left[\frac{g_\pi N_\pi + g_\nu N_\nu}{N} L^{\text{tot}} + (g_\pi - g_\nu)\frac{N_\pi N_\nu}{N}\left(\frac{L_\pi}{N_\pi} - \frac{L_\nu}{N_\nu}\right)\right] \quad (1)$$

and from the fact [10] that the matrix element of every one-body operator $\hat{O}_\rho = b_\rho^+ \tilde{b}_\rho$ for $\rho \in \{\pi, \nu\}$ between any two symmetric states is proportional to the number of bosons, b, with isospin label ρ:

$$\langle F_{\text{max}}, F_z, \alpha | \hat{O}_\rho | F_{\text{max}}, F_z, \beta \rangle = c^N_{\alpha,\beta}\; N_\rho \;. \quad (2)$$

The angular momentum operators $L_\rho = \sqrt{10}[d_\rho^+ \times \tilde{d}_\rho]^{(1)}$ are one-body operators and, therefore, the second term in Eq. (1) has vanishing matrix elements between symmetric states. On the other hand, $L^{\text{tot}} = L_\pi + L_\nu$ represents the total angular momentum operator in the IBM-2 which cannot induce transitions between different states, either, because it is diagonal. Thus, the search for MSSs focuses on the observation of strong $M1$ transitions with large matrix elements of the order of $1\,\mu_N$ since the relevant difference of the boson g-factors, $g_\pi - g_\nu$, is roughly of that size, if one uses bare orbital values as a starting point.

In the early 1980s Richter and coworkers discovered a strong $M1$ excitation mode in heavy deformed even-even nuclei in electron scattering experiments performed in Darmstadt [11,12]. This $M1$ mode is called *scissors mode* due to its geometrical picture in rotors. The scissors mode was subsequently investigated in great detail using the photon scattering technique, mostly in experiments by Kneißl and collaborators in Stuttgart [13]. The scissors mode is known to be usually fragmented over several $J^\pi = 1^+$ states at energies around 3 MeV [14] and its total $M1$ strength to the ground state reaches at mid-shell a maximum value of $\sum B(M1; 1^+ \to 0_1^+) \approx 1\,\mu_N^2$ and correlates to the collective $B(E2; 2_1^+ \to 0_1^+)$ value [15]. Increasing sensitivity in the photon scattering experiments in the 1990s made the discovery of the scissors mode even in odd-mass nuclei [16] and in O(6) nuclei [17,18] possible.

Vibrational behavior represents another benchmark of nuclear structure and the phonon concept is a simple and useful scheme for the understanding of vibrational excitations. In soft nuclei, MSSs arrange into a multiphonon structure with the one-phonon $2^+_{1,\text{ms}}$ state representing the building block of such structures. Unique signatures for MS multiphonon states are the presence of strong $M1$ transitions to symmetric states with the same number of phonons (or to states with the same d-parity quantum number [19] in nuclei with O(5) symmetry in the absence of further selection rules).

Experimental hints at $2^+_{1,\text{ms}}$ states from the observation of small $E2/M1$ multipole mixing ratios of $2^+ \to 2_1^+$ transitions have first been discussed by Hamilton *et al.* [20] for vibrational $N = 84$ isotones and for Barium nuclei by Molnár and collaborators [21]. Later on, Giannatiempo *et al.* [22,23] suggested an extensive set of MSSs in nuclei of the $A \approx 100$ mass region on the basis of small $E2/M1$ multipole mixing ratios and extensive IBM-2 calculations with many low-energy eigenstates having considerable amounts of mixed-symmetry components. However, small mixing ratios do not unambiguously identify large $M1$ matrix elements (the $E2$ strength can be very small, instead) and small δ's are, thus, not a sufficient signature for MSSs. In fact, the IBM-2 descriptions of low-energy levels by Giannatiempo *et al.* was achieved by invoking a considerable quenching of the IBM-2 $M1$ transition operator by a factor of about 20, which is inconsistent with systematics of the scissors mode. The suggested MSSs in the Cd nuclei were subsequently interpreted [24] differently, namely, as a typical intruder structure outside of the ordinary IBM-2 model space. Indeed, a recent study [25] of the $2^+ \to 2_1^+$ $M1$ strength distribution in the nucleus ^{114}Cd obtained with the Doppler Shift Attenuation Method in Inelastic Neutron Scattering (DSAM-INS) technique showed that the 2_6^+ state at 2219 keV carries the main part of the $2^+ \to 2_1^+$ $M1$ strength and the earlier conclusions about a one-phonon $2^+_{1,\mathrm{ms}}$ state below 1.5 MeV, at least in this nucleus, are, thus, in disagreement with the new data.

Unambiguous identifications of $2^+_{1,\mathrm{ms}}$ states were possible in a handful of nuclides from lifetime measurements using Coulomb excitation [26] or the Doppler shift attenuation method (DSAM), *e.g.*, [27,28,29]. A step forward in the investigation of $2^+_{1,\mathrm{ms}}$ states was done by the demonstration that this fundamental state can be well investigated also in high-resolution photon scattering as was first done for ^{136}Ba at Stuttgart University [30].

The subsequent combination of photon scattering data on the $N = 52$ nucleus ^{94}Mo with very clean and high-statistics off-beam γ-ray spectroscopy following β-decay performed at the University of Cologne led to an unprecedented richness of information on the lowest 1^+_{ms} and $2^+_{1,\mathrm{ms}}$ states [31]. In particular, a strong γ transition between these MSSs could be observed for the first time representing first direct evidence that these states belong to a class of states with similar wave functions.

Significant progress resulted from the addition of in-beam γ-ray spectroscopy on ^{94}Mo studied in the ^{91}Zr$(\alpha, n\gamma)$ reaction at the University of Cologne. Besides a considerable enlargement of the known level scheme of ^{94}Mo the combination of in-beam and off-beam $\gamma\gamma$-coincidence studies resulted in the observation of many new branching ratios and of unambiguous assignments of multipole mixing ratios. In addition the crucial lifetime information could be deduced from the analysis of Doppler shifts. This information enabled us to discover further MSSs with spin and parity quantum numbers 3^+_{ms} [32] and $2^+_{2,\mathrm{ms}}$ [33]. Subsequent investigations of ^{94}Mo at the University of Kentucky using inelastic neutron scattering confirm [34] in general the previous findings, in many cases with considerably smaller uncertainties. The level lifetimes obtained in that work with the DSAM-INS technique extend particularly the dipole and quadrupole strength distributions between low-spin states. In combination, the MSSs can be identified on the basis of comprehensive $M1$ strength distributions between off-yrast low-spin states until about 4 MeV. Figure 1 shows the data relevant for the identification of the $2^+_{1,\mathrm{ms}}$ one-phonon state. Similar data were used for the identification of the 1^+_{ms}, $2^+_{2,\mathrm{ms}}$, and 3^+_{ms} states of ^{94}Mo [35].

Other, recent γ-ray spectroscopy work on ^{96}Ru [36,37] and ^{92}Zr [38] led to the observation of corresponding structures in these $N = 52$ isotones of ^{94}Mo, too. Of particular experimental interest is the successful identification of the one-phonon $2^+_{1,\mathrm{ms}}$ state of ^{96}Ru done in inverse-kinematics Coulomb excitation of a ^{96}Ru-beam

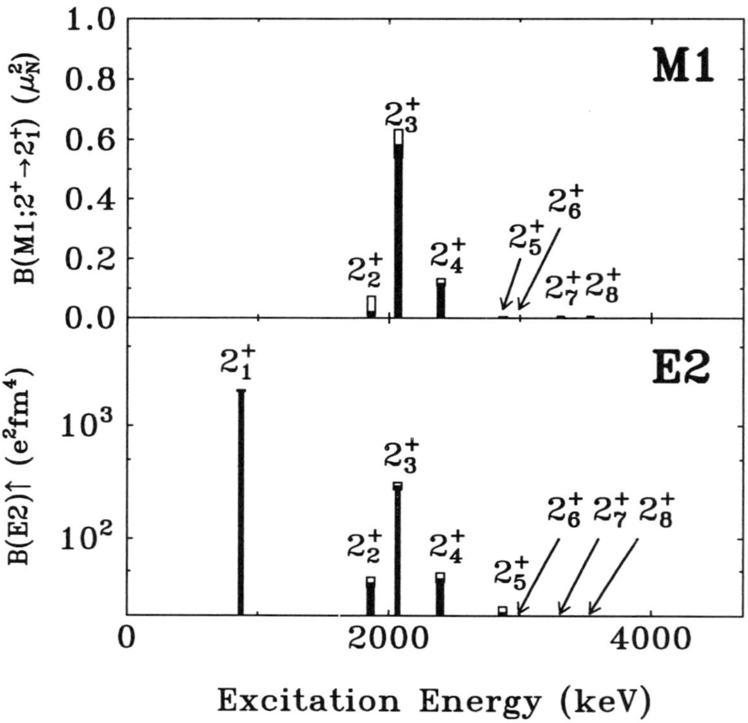

Figure 1. Identification of the one-phonon $2^+_{1,ms}$ state of ^{94}Mo. Complete low-spin spectroscopy using photon scattering and γ-ray spectroscopy following β-decay, (α,$n\gamma$) fusion evaporation, and inelastic neutron scattering reactions yields the $2^+ \rightarrow 2^+_1$ $M1$ strength distribution until about 4 MeV at the top. The 2^+_3 state at 2067.4 keV is the main fragment of the $2^+_{1,ms}$ state of ^{94}Mo with only little fragmentation. The measured $E2$ excitation strength distribution at the bottom (log-scale) confirms this one-phonon assignment.

at Yale [36]. The one-phonon MSS was identified as the 2^+_3 state of ^{96}Ru at 2284 keV from the large matrix element $|\langle 2^+_1 \parallel M1 \parallel 2^+_3 \rangle| = 2.0(3)\ \mu^2_N$ obtained from the measurement of the Coulomb excitation cross section, and the decay pattern. The technique of Coulomb excitation in inverse kinematics develops into a major spectroscopic method for the investigation of exotic nuclei at present and future high-intensity radioactive-ion-beam (RIB) facilities. The ^{96}Ru-experiment has demonstrated that this technique enables us to study MSSs. Corresponding experiments on unstable neutron-rich RIBs of $N = 52$ isotones to search for MSSs are approved parts of the planned RISING campaign at GSI. A one-phonon $2^+_{1,ms}$ and a two-phonon 1^+_{ms} state have also been identified [39] in the $Z = 30$ nucleus ^{66}Zn similar as in ^{94}Mo.

The new data initiated detailed theoretical studies on the formation of mixed-symmetry collectivity using microscopic models such as the nuclear shell model

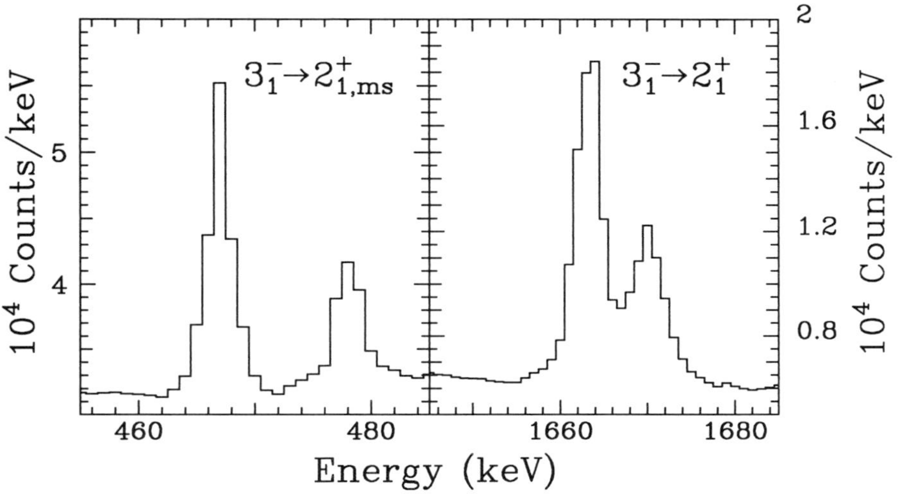

Figure 2. Spectrum of γ-rays from ^{94}Mo taken with the Cologne CUBE spectrometer in the ^{91}Zr$(\alpha, n\gamma)^{94}$Mo reaction at 15 MeV in coincidence with the $2_1^+ \to 0_1^+$ transition. The coincidence relations establish the 466.4-keV and and 1662.6-keV γ-rays as dipole transitions between the 3_1^- state at 2533.8 keV and the MS and symmetric one-phonon 2^+ states of ^{94}Mo. From Ref. [44].

[40] and the quasiparticle phonon model [41] and yielded new developments in the framework of the IBM with respect to the formulation of new $M1$ sum rules for any excited symmetric state [42]. Comparison of these sum rules to the measured $M1$ strength distributions to the 0_1^+, 2_1^+, and 2_2^+ states in ^{94}Mo could be used to conclude on the transitional character of this nucleus in a largely parameter-free fashion [42].

The observed MSSs of ^{94}Mo indicate a multiphonon structure. The available data enable us to first measure the energy anharmonicity, $\epsilon = \langle E_x(J)\rangle/[E_x(2_1^+) + E_x(2_{1,\text{ms}}^+)] - 1$, and the energy splitting, $\Delta_J = \langle E_x(J) - E_x(J_0)\rangle$, of the two-phonon multiplet with mixed symmetry. The energy splitting of the largest fragments of the MS two-phonon multiplet is small [35]. It amounts to only 10% of the excitation energy. The energy anharmonicity is even smaller making the MS two-phonon multiplet a beautiful example of *inhomogeneous phonon coupling*, *i.e.*, the coupling of two different kinds of phonons, besides that of the well known quadrupole-octupole coupling (cf. contributions of U. Kneissl *et al.* and A. Zilges *et al.*).

In addition to the observation of small energy anharmonicities, there are other analogies between quadrupole-octupole coupled and mixed-symmetric quadrupole vibrational structures. Both decay by comparably strong electromagnetic dipole transitions to symmetric quadrupole collective structures at lower energies, where the overall scale of the dipole strengths is set by the $3_1^- \to 2_1^+$ and $2_{1,\text{ms}}^+ \to 2_1^+$ transitions between the one-phonon states [43]. It is an intriguing question whether or not also a coupling of the octupole degree of freedom to the MS structures can

Table 1. Data on $3_1^- \to 2_{1,ms}^+$ F-vector $E1$ transitions in all even-even nuclei for which the $2_{1,ms}^+$ state has been identified on the basis of large $M1$ transition strength and where a transition between these two states has been observed. The data are from Refs. [45,34,37,46], respectively. Except for ^{142}Ce, the $B(E1)$ ratios stem from measured dipole decay intensity ratios of the 3_1^- state. The $B(E1)$ values are in milli Weisskopf units [$1\text{mW.u.}(E1) = 0.065\,A^{2/3}\,10^{-3}\,e^2\text{fm}^2$]. The nuclei ^{100,132}Sn were considered as the appropriate doubly closed shell cores to determine the boson numbers. The quantities α_π and α_ν represent the effective $E1$ boson charges for the one-body terms of the quadrupole-octupole coupled $E1$ operator in the U(5) limit of the *sdf*-IBM-2.

Nuclide		^{92}Zr	^{94}Mo	^{96}Ru	^{142}Ce
$E(2_1^+)$	(keV)	934.5	871.1	832.6	641.2
$E(2_{1,ms}^+)$	(keV)	1847.3	2067.4	2284.2	2004.9
$E(3_1^-)$	(keV)	2339.6	2533.7	2650.0	1652.9
$\frac{B(E1,3_1^- \to 2_{1,ms}^+)}{B(E1,3_1^- \to 2_1^+)}$		2.7(2)	26.0(7)	6.7(10)	> 13.7
$B(E1, 3_1^- \to 2_{1,ms}^+)$	(mW.u.)	0.80(10)	1.31(21)	–	1.86(22)
N_ν/N_π		1/5	1/4	1/3	1/4
α_ν/α_π		-1.55(6)	-2.59(2)	-1.40(7)	[-4, -2.25]
α_π	(efm)	0.049	0.045	–	[0.044, 0.067]
α_ν	(efm)	-0.075	-0.115	–	[-0.174, -0.151]

be observed. The decay strength of such structures might again be correlated to the decay strength between the corresponding one-phonon states, *i.e.*, here to the $3_1^- \to 2_{1,ms}^+$ $E1$ transition strength with, $\Delta F = 1$, F-vector character. Due to the isovector nature of the MS states, quadrupole-octupole coupled $E1$ transitions involving MSSs can help to study the proton-neutron structure of the effective $E1$ transition operator between low-energy states. The observation of such transitions has not yet been discussed in the literature to our knowledge.

In the $N = 52$ and $N = 84$ isotones the energy difference of about 300 – 500 keV between the 3_1^- state and the $2_{1,ms}^+$ states is somewhat larger than in other nuclei. This fact facilitates the search for $3_1^- \to 2_{1,ms}^+$ $E1$ transitions in the $N = 52$ and $N = 84$ isotones. Indeed, γ-ray spectroscopy following inelastic neutron scattering reactions performed at the University of Kentucky and $\gamma\gamma$-coincidence studies performed with the Cologne OSIRIS-CUBE spectrometer have permitted the observation of $E1$ γ-ray transitions between the octupole vibrational state and the $2_{1,ms}^+$ state. Figure 2 shows the relevant parts of the γ-ray spectrum from the ^{91}Zr(α,$n\gamma$)^{94}Mo reaction that enabled us to measure the $E1$ branching ratio from the 3_1^- state to the $2_{1,ms}^+$ and 2_1^+ states of ^{94}Mo. At γ-ray energies of 466.4 and 1662.6 keV, one clearly observes [34] the peaks corresponding to the $3_1^- \to 2^+$ transitions of interest. Their pure dipole character has been established from the measurement of $\gamma\gamma$-angular correlations and also from γ-ray angular distributions [34]. Corresponding $3_1^- \to 2_{1,ms}^+$ transitions have also been found in the level schemes of ^{92}Zr [45], ^{96}Ru [37], and ^{142}Ce [46]. The relevant data are summarized in Table 1.

The striking observation is that, in all cases, the branching ratio $B(E1; 3_1^- \to$

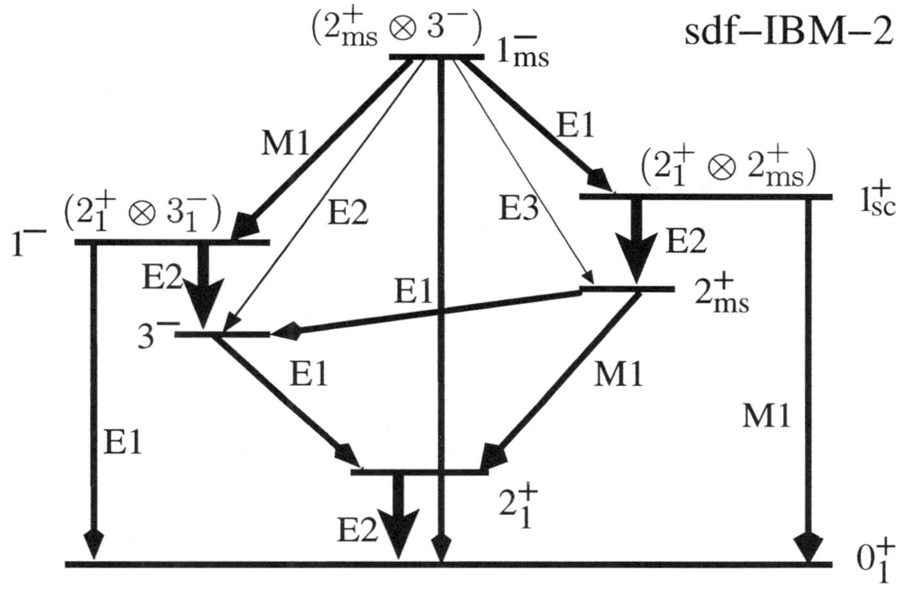

Figure 3. Idea of the *sdf*-IBM-2 [47], which considers monopole (s), quadrupole (d), and octupole (f) bosons and the proton-neutron degree of freedom. This version of the model provides the framework for a simultaneous description of MS structures, quadrupole-octupole structures, and transitions between them. It further predicts, yet unobserved, multiphonon structures with mixed-symmetry and negative parity.

$2^+_{1,ms})/B(E1, 3^-_1 \to 2^+_1)$ is larger than one. Except for ^{92}Zr, where the 2^+_1 state has predominantly neutron character [38] due to the $Z = 40$ subshell closure, the enhancement of the F-vector $3^-_1 \to 2^+_{1,ms}$ $E1$ transition over the F-scalar $3^-_1 \to 2^+_1$ $E1$ transition is about an order of magnitude. This is interesting because the stronger transitions systematically populate particular non-yrast levels, namely the $2^+_{1,ms}$ states. These $E1$ transitions are not used for making conclusions about the structure of the wave functions involved. Rather, their structures were established previously.

Recently, the question of F-scalar and F-vector electromagnetic transitions between octupole phonon states and quadrupole-collective MS states in vibrational-type nuclei has been addressed theoretically by Smirnova *et al.* [47] by considering the proton-neutron degree of freedom for systems with monopole (s) bosons, quadrupole (d) bosons, and octupole (f) bosons in the framework of the *sdf*-interacting boson model-2. The idea of the model is sketched in Fig. 3. Weak coupling of the octupole degree of freedom to the quadrupole degree of freedom was studied. Analytical expressions for excitation energies and electromagnetic transitions were derived [47] in the U(5) and SO(6) dynamical symmetry limits relevant for the discussion of soft nuclei. In that paper, it was argued that a proper description of $E1$ strengths within the *sdf*-IBM-2 framework, particularly those involving two-phonon states generated by quadrupole-octupole coupling, requires a boson $E1$ operator with two-body terms, namely,

$$T(E1) = \alpha_\pi D_\pi + \alpha_\nu D_\nu + \frac{\beta}{2N}[T_{E2}(\eta), T_{E3}(\chi)]_+ \qquad (3)$$

with the one-body terms

$$D_\rho = [d_\rho^+ \times \tilde{f}_\rho + f_\rho^+ \times \tilde{d}_\rho]^{(1)} \qquad (4)$$

where $\tilde{b}$ and b^+ denote boson annihilation and creation operators, respectively, and $\rho \in \{\pi, \nu\}$. The full form of the two-body part, $[T_{E2}, T_{E3}]_+$, can be found in Ref. [47]. For the discussion of the observed $E1$ transitions between the one-phonon states and for an estimate of the effective one-body $E1$ charges, we restrict ourselves here to the one-body part of the transition operator, thus, keeping $\beta = 0$.

In that case the ratio, α_ν/α_π, of the effective $E1$ boson charges for the one-body term can be unambigously obtained from the measured $E1$ branching ratio, if one makes the reasonable assumption, $N_\pi \alpha_\pi \geq |N_\nu \alpha_\nu| \geq 0$, *i.e.*, of a positive effective $E1$ charge for proton bosons and a not too large effective $E1$ charge for neutron bosons. The charge ratios determined from the measured F-vector/F-scalar $E1$ branching ratio are tabulated in Table 1. They lie at values of about -2 indicating the predominantly isovector character of the one-body part of the effective $E1$ transition operator in the *sdf*-IBM-2. This observation [44] is very intriguing because it seems to resemble the isovector character of the $E1$ operator in miscroscopic models.

Absolute values for the effective $E1$ charges α_ρ were then determined from the $B(E1; 3^- \to 2^+_{1,\mathrm{ms}})$ values and are included in Table 1. It is noted that the values for α_π are very similar, ≈ 0.05 *e*fm, for ^{92}Zr, ^{94}Mo, and ^{142}Ce in spite of the quite varying values for the observables they have been derived from. The neutron boson $E1$ charges α_ν are all found to be negative, ≈ -0.1 *e*fm within about a factor of two, with some dependence on the nuclear mass number.

In addition to the transitions discussed so far there exist other F-scalar quadrupole-octupole coupled $E1$ transitions that certainly require two-body terms in the effective $E1$ operator in the U(5) limit, *e.g.*, the $1^-_1 \to 0^+_1$ transition. Inclusion of two-body parts in the boson $E1$ operator results in some numerical changes of the one-body charges derived above. A comprehensive set of $B(E1)$ values in the nuclide ^{142}Ce can be well reproduced by considering the full $E1$ operator from Eq. (3) with a pure F-vector one-body part. It can be expected that the detailed investigation of $E1$ properties of MSSs yields further insights into the proton-neutron structure of the effective $E1$ operator in the valence shell of heavy nuclei in the future.

Acknowledgments

We are grateful to all those who have contributed to this research. We thank Franco Iachello for years of stimulating discussions and inspiration. The contributions of the collaborators on the experiments, in particular, C.J. Barton, R.F. Casten,

A. Gade, J. Jolie, H. Klein, U. Kneissl, H.H. Pitz, V. Werner, and S.W. Yates are gratefully acknowledged. We further thank N.A. Smirnova, A. Gelberg, J. Ginocchio, R.V. Jolos, A. Leviatan, A.F. Lisetskiy, T. Mizusaki, T. Otsuka, P. Van Isacker, and N.V. Zamfir for collaboration and discussions. This work has been supported by the *Deutsche Forschungsgemeinschaft* under support Nos. Pi 393/1-1/1-2.

References

1. F. Iachello and A. Arima in *The interacting boson model* (Cambridge University Press, Cambridge, 1987).
2. F. Iachello, Nucl. Phys. **A358**, 89c (1981).
3. F. Iachello, Phys. Rev. Lett. **53**, 1427, (1984).
4. A. Arima, T. Otsuka, F. Iachello, and I. Talmi, Phys. Lett. **B66**, 205 (1977).
5. T. Otsuka, A. Arima, F. Iachello, and I. Talmi, Phys. Lett. **B 76**, 139 (1978).
6. T. Otsuka, A. Arima, and F. Iachello, Nucl. Phys. **A309**, 1 (1978).
7. P. von Brentano, A. Gelberg, H. Harter, P. Sala, J. Phys. G: Nucl. Phys. **11**, L85 (1985).
8. J. Jolie, P. Van Isacker, K. Heyde, and A. Frank, Phys. Rev. Lett. **55**, 1457 (1985).
9. P.O. Lipas, P. von Brentano, and A. Gelberg, Rep. Prog. Phys. **53**, 1355 (1990).
10. P. Van Isacker, K. Heyde, J. Jolie, and A. Sevrin, Ann. Phys. (NY) **171**, 253 (1986).
11. D. Bohle, A. Richter, W. Steffen, A.E.L. Dieperinck, N. LoIudice, F. Palumbo, and O. Scholten, Phys. Lett. **B137**, 27 (1984).
12. A. Richter, Prog. Part. Nucl. Phys. **34**, 261 (1995).
13. U. Kneissl, H.H. Pitz, and A. Zilges, Prog. Part. Nucl. Phys. **37**, 349 (1996).
14. N. Pietralla, P. von Brentano, R.-D. Herzberg, U. Kneissl, N. LoIudice, H. Maser, H.H. Pitz, and A. Zilges, Phys. Rev. C **58**, 184 (1998).
15. C.Rangacharyulu, A.Richter, H.J.Wörtche, W.Ziegler, R.F.Casten, Phys. Rev. C **43**, R949 (1991).
16. I. Bauske *et al.*, Phys. Rev. Lett. **71**, 975 (1993).
17. P. von Brentano *et al.*, Phys. Rev. Lett. **76**, 2029 (1996).
18. H. Maser, N. Pietralla, P. von Brentano, R.-D. Herzberg, U. Kneissl, J. Margraf, H.H. Pitz, and A. Zilges, Phys. Rev. C **54**, R2129 (1996).
19. N. Pietralla, P. von Brentano, A. Gelberg, T. Otsuka, A. Richter, N.A. Smirnova, and I. Wiedenhöver, Phys. Rev. C **58**, 191 (1998).
20. W.D. Hamilton, A. Irbäck, and J.P. Elliott, Phys. Rev. Lett. **53**, 2469 (1984).
21. G. Molnár, R.A. Gatenby, and S.W. Yates, Phys. Rev. C **37**, 898 (1988).
22. A. Giannatiempo, A. Nannini, A. Perego, P. Sona, and G. Maino, Phys. Rev. C **44**, 1508 (1991).
23. A. Giannatiempo, A. Nannini, and P. Sona, Phys. Rev. C **58**, 3316 (1998); **58**, 3335 (1998).
24. H. Lehmann and J. Jolie, Nucl. Phys. **A 588**, 623 (1995).
25. D. Bandyopadhyay, C.C. Reynolds, C. Fransen, N. Boukharouba, M.T. McEl-

listrem, and S.W. Yates, Phys. Rev. C **67**, 034319 (2003).
26. W.J. Vermeer, C.S. Lim, R.H. Spear, Phys. Rev. C **38**, 2982 (1988).
27. K.P. Lieb, H.G. Börner, M.S. Dewey, J. Jolie, S.J. Robinson, S. Ulbig, and Ch. Winter, Phys. Lett. **B215**, 50 (1988).
28. B. Fazekas, T. Belgya, G. Molnár, A. Veres, R.A. Gatenby, S.W. Yates, T. Otsuka, Nucl. Phys. **A548**, 249 (1992).
29. I. Wiedenhöver, A. Gelberg, T. Otsuka, N. Pietralla, J. Gableske, A. Dewald, and P. von Brentano, Phys. Rev. C **56**, R2354 (1997).
30. N. Pietralla *et al.*, Phys. Rev. C **58**, 796 (1998).
31. N. Pietralla, C. Fransen *et al.*, Phys. Rev. Lett. **83**, 1303 (1999).
32. N. Pietralla, C. Fransen, P. von Brentano, A. Dewald, A. Fitzler, C. Frießner, and J. Gableske, Phys. Rev. Lett. **84**, 3775 (2000).
33. C. Fransen, N. Pietralla, P. von Brentano, A. Dewald, J. Gableske, A. Gade, A. Lisetskiy, and V. Werner, Phys. Lett. **B 508**, 219 (2001).
34. C. Fransen *et al.* Phys. Rev. C **67**, 024307 (2003).
35. N. Pietralla, in *Nuclear Structure Physics*, edts. R.F. Casten, J. Jolie, U. Kneissl, K.P. Lieb (World Scientific, Singapore, 2001), p.243.
36. N. Pietralla, C.J. Barton III., R. Krücken, C.W. Beausang, M.A. Caprio, R.F. Casten, J.R. Cooper, A.A. Hecht, H. Newman, J.R. Novak, and N.V. Zamfir, Phys. Rev. C **64**, 031301(R) (2001).
37. H. Klein, A.F. Lisetskiy, N. Pietralla, C. Fransen, A. Gade, and P. von Brentano, Phys. Rev. C **65**, 044315 (2002).
38. V. Werner *et al.* Phys. Lett. **B 550**, 140 (2002).
39. A. Gade, H. Klein, N. Pietralla, and P. von Brentano, Phys. Rev. C **65**, 054311 (2002).
40. A.F. Lisetskiy, N. Pietralla, C. Fransen, R.V. Jolos, P. von Brentano, Nucl. Phys. **A 677**, 100 (2000).
41. N. LoIudice and Ch. Stoyanov, Phys. Rev. C **62**, 047302 (2000); Phys. Rev. C **65**, 064304 (2002).
42. N.A. Smirnova, N. Pietralla, A. Leviatan, J.N. Ginocchio, C. Fransen, Phys. Rev. C **65**, 024319 (2002).
43. N. Pietralla, Phys. Rev. C **59**, 2941 (1999).
44. N. Pietralla, C. Fransen, A. Gade, N.A. Smirnova, P. von Brentano, V. Werner, and S.W. Yates, submitted for publication.
45. C. Fransen *et al.*, in preparation.
46. J.R. Vanhoy *et al.*, Phys. Rev. C **52**, 2387 (1995).
47. Nadya A. Smirnova, Norbert Pietralla, Takahiro Mizusaki, and Piet Van Isacker, Nucl. Phys. **A 678**, 235 (2000).

SPIN-ISOSPIN EXCITATIONS, PAIRING AND SHAPE COEXISTENCE

E. MOYA DE GUERRA, P. SARRIGUREN, R. ALVAREZ-RODRIGUEZ, A. ESCUDEROS

Instituto de Estructura de la Materia, C.S.I.C., Serrano 123, 28006 Madrid, Spain

We study nuclear responses to the spin-isospin dependent Gamow-Teller operator. We focus on proton rich nuclei of mass around 70-80. We perform QRPA calculations on a selfconsistent deformed single particle basis. Pairing correlations in T=1 channel are included in the BCS approach, while J=1 pairing force is included as a residual particle-particle force in the QRPA calculation, along with the particle-hole force. The sensitivity of the results to pairing and deformation is discussed. Oblate and prolate shape isomers are found in several isotopes. We argue that present data on half-lives and beta-decay spectra can be used to conclude on the strength of pairing and on the amount of shape coexistence in those nuclei.

1 Introduction

Knowledge of single as well as double $\beta-$decay properties of nuclei is essential for progress in many science branches. This is a vast subject where much work has been done and a variety of models, from shell models to statistical models, have been applied. So far most of the theoretical work has focused on spherical nuclei. We discuss here a selfconsistent approach for deformed nuclei that has conceptually been around for some time but whose implementation to realistic calculations has only been done recently. In this approach one starts from an effective, density dependent, two-body interaction to construct the deformation dependent selfconsistent mean field (with its quasiparticle wave functions and energies) and the residual interaction acting among the selfconsistent quasiparticle basis.

The need for such an approach stems from the fact that near the proton drip line extrapolations of phenomenological parametrizations of mean fields and residual interactions (valid for nuclei in the stability valley) are not reliable for this region. In addition, deformed shapes and shape coexistence may appear and have in fact been observed in the proton rich mass region around A=72. One needs to predict both the ground state and the $\beta-$decay properties far from stability in a reliable way. This can be achieved using the selfconsistent deformed mean field approach to predict the shapes, and using residual interactions that are consistent with the mean field to predict the lifetimes.

2 Theoretical approach

In this Section we summarize briefly the theory involved in the microscopic calculations presented in the next Sections. Our method [1,2,3] consists in a selfconsistent formalism based on a deformed Hartree-Fock mean field obtained with a Skyrme interaction including pairing correlations in the BCS approximation. We consider in this paper the force SG2 of Van Giai and Sagawa, that has been successfully tested against spin and isospin excitations in spherical and deformed nuclei. Comparison to calculations obtained with other Skyrme forces have been made, showing that the results do not differ in a significant way. The single particle energies, wave functions, and occupations probabilities are generated from this mean field.

Time reversal and axial symmetry are assumed. The single-particle wave functions are expanded in terms of the eigenstates of an axially symmetric harmonic oscillator in cylindrical coordinates. We use eleven major shells. The method also includes pairing between like nucleons in the BCS approximation with fixed gap parameters for protons Δ_π, and neutrons Δ_ν, which are determined phenomenologically from the odd-even mass differences through a symmetric five term formula involving the experimental binding energies.

In a previous work [2] we analyzed the energy surfaces as a function of deformation for all the isotopes under study here. For that purpose, we performed constrained HF calculations with a quadrupole constraint and we minimized the HF energy under the constraint of keeping fixed the nuclear deformation. Calculations in this paper are performed for the equilibrium shapes of each nucleus, that is, for the solutions, in general deformed, for which we obtained minima in the energy surfaces. Most of these nuclei present oblate and prolate equilibrium shapes [2] that are very close in energy.

We add to the mean field a spin-isospin residual interaction, which is expected to be the most important residual interaction to describe GT transitions. This interaction contains two parts. The particle-hole part is responsible for the position and structure of the GT resonance and is derived selfconsistently from the same energy density functional (and Skyrme interaction) as the HF equation, in terms of the second derivatives of the energy density functional with respect to the one-body densities. The residual interaction is finally written in a separable form by averaging the Landau-Migdal resulting force over the nuclear volume,

$$V_{GT}^{ph} = 2\chi_{GT}^{ph} \sum_{K=0,\pm 1} (-1)^K \beta_K^+ \beta_{-K}^- , \tag{1}$$

where

$$\beta_K^+ = \sum_{\pi\nu} \langle \nu | \sigma_K | \pi \rangle \, a_\nu^+ a_\pi \, . \tag{2}$$

The coupling strength is given by [1,2]

$$\chi_{GT}^{ph} = -\frac{3}{8\pi R^3} \left\{ t_0 + \frac{1}{2} k_F^2 \left(t_1 - t_2 \right) + \frac{1}{6} t_3 \rho^\alpha \right\} . \tag{3}$$

The particle-particle part is a neutron-proton pairing force in the $J^\pi = 1^+$ coupling channel,

$$V_{GT}^{pp} = -2\kappa_{GT}^{pp} \sum_K (-1)^K P_K^+ P_{-K} \, , \tag{4}$$

where

$$P_K^+ = \sum_{\pi\nu} \left\langle \pi \left| (\sigma_K)^+ \right| \nu \right\rangle a_\nu^+ a_{\bar\pi}^+ \, . \tag{5}$$

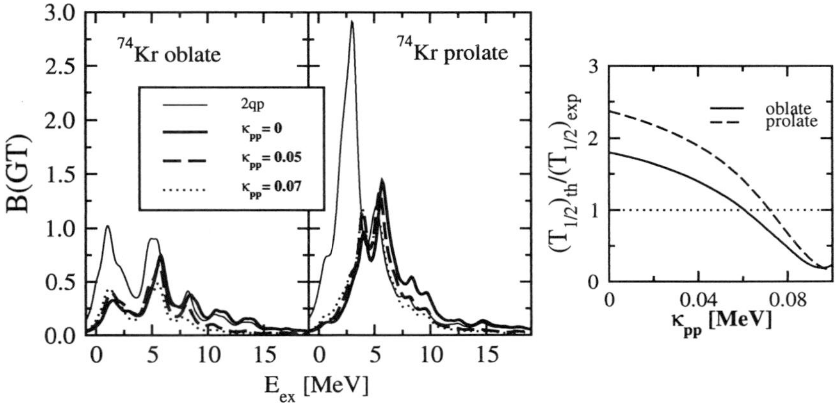

Figure 1. Left: Gamow-Teller strength distributions in ^{74}Kr calculated in RPA with the force SG2 for various values of the strength of the *pp* force. Right: Ratios of calculated to experimental half-lives as a function of κ_{pp}.

A careful search of the optimal strength can certainly be done for each particular case. Instead, we have chosen the same coupling constant for all nuclei considered here. This value has been obtained under the requirements of improving in general the agreement with the experimental half-lives while being still far from the values leading to the collapse.

The Ikeda sum rule is always fulfilled in our calculations

3 The role of J=1 pairing and residual interactions

Figure 1 illustrates the effect of the residual interactions. The calculations are done for the oblate and prolate shapes of ^{74}Kr. The coupling strength of the *ph* residual interaction χ^{ph}_{GT} is obtained from Eq. (3), and its value for $A = 74$ and Skyrme force SG2 is $\chi^{ph}_{GT} = 0.37$ MeV. The coupling strength of the *pp* residual interaction is varied from $\kappa^{pp}_{GT} = 0$ to $\kappa^{pp}_{GT} = 0.07$ MeV.

If we compare first the calculation with only the *ph* residual interaction (solid line) to the uncorrelated one (thin solid line), we can see that the repulsive *ph* force introduces two types of effects. One is a shift of the GT strength to higher excitation energies with the corresponding displacement of the position of the GT resonance. The other is a reduction of the total GT strength. If we now introduce the J=1 *pp* residual interaction (dashed and dotted lines), we can see that its effect, being an attractive force, is to shift the strength to lower excitation energies, reducing the total GT strength as well. The GT strength is also pushed a bit by the *pp* interaction to lower energies in the low energy excitation region. This effect, although small, is of great relevance in the calculation of the β^+/EC half-lives, which are only sensitive to the distribution of the strength contained in the energy region below the Q_{EC} window. The half-life decreases with increasing values of κ^{pp}_{GT}.

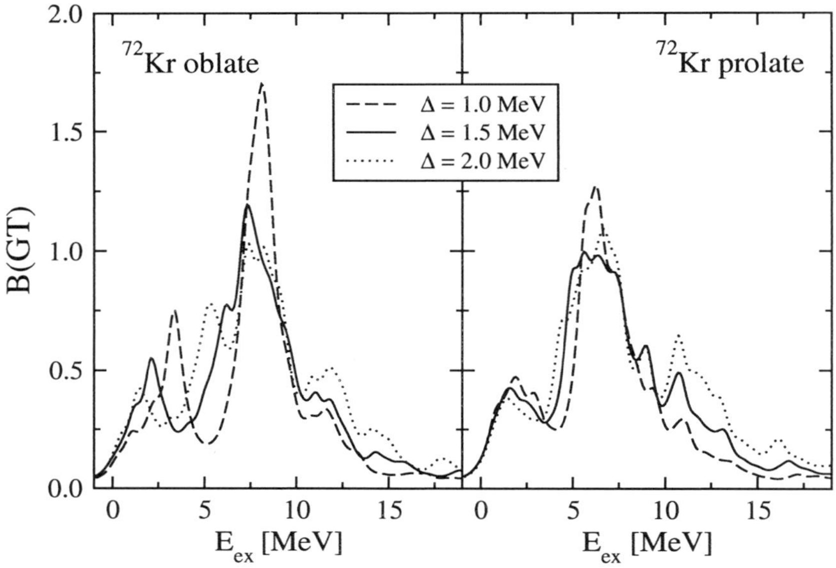

Figure 2. Pairing effect in the QRPA Gamow-Teller strength distribution in the N=Z isotope ^{72}Kr. Solid lines correspond to calculations with empirical pairing gaps.

This is clear because as κ_{GT}^{pp} increases the strength becomes more concentrated at low excitation energies below Q_{EC} and therefore the half-lives are smaller. This is true up to values around $\kappa_{GT}^{pp} = 0.1$ MeV, where the collapse of the QRPA takes place. In the case of ^{74}Kr, we get an optimum value of $\kappa_{GT}^{pp} = 0.06$ MeV and $\kappa_{GT}^{pp} = 0.07$ MeV for the oblate and prolate shapes, respectively. This value will of course depend, among other factors, on the nucleus, shape, and Skyrme interaction and a case by case fitting procedure could be carried out. Nevertheless, we have done calculations for other cases and found that, in general, values around $\kappa_{GT}^{pp} = 0.07$ MeV improve the agreement with experiment in most cases and since this is a valid value far from collapse, we have chosen $\kappa_{GT}^{pp} = 0.07$ MeV as the value of the coupling constant of the *pp* residual interaction.

4 The role of T=1 pairing and BCS correlations

Our quasiparticle basis only includes neutron-neutron and proton-proton pairing correlations in the BCS approach. In principle, one could extend the BCS treatment to include also neutron-proton pairing correlations in the mean field. This may be important particularly for N=Z nuclei.

We have studied [1,2,3] bulk properties (binding energies, charge radii, quadrupole moments, moments of inertia,...) of the nuclei considered here and found that the agreement between theory and experiment is as good for the $N = Z$ as for the $N = Z + 2, Z + 4, Z + 6$ isotopes. We concluded that the effect of neutron-proton

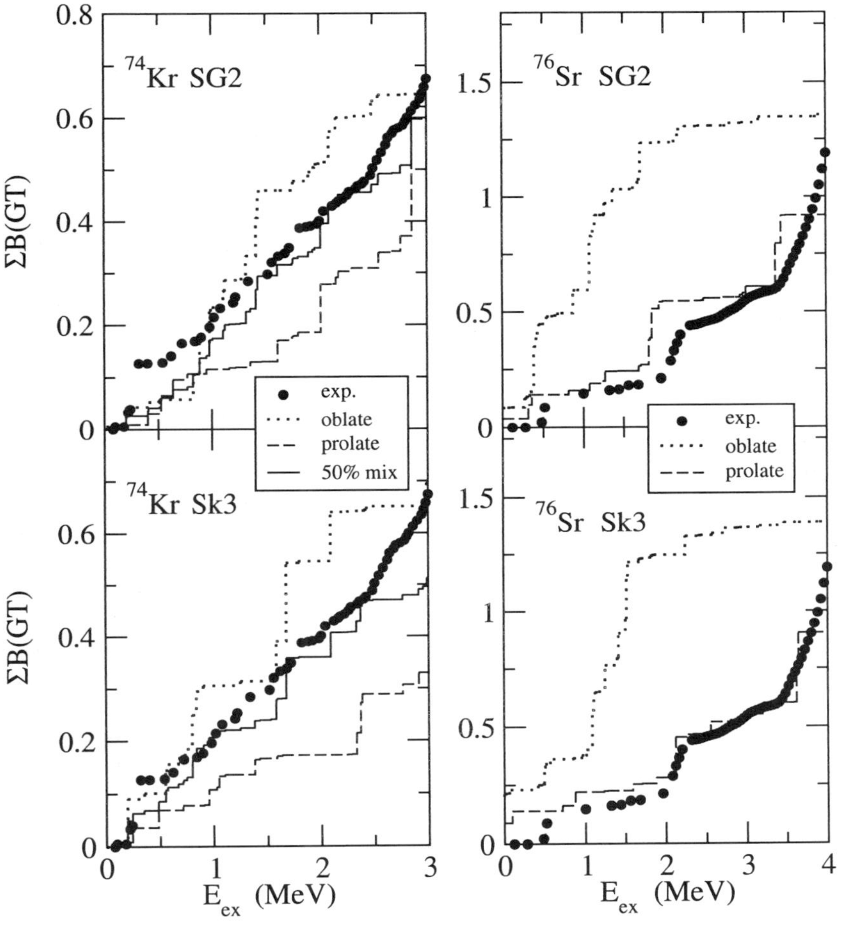

Figure 3. Experimental data on summed β^+ strengths are shown as a function of energy and compared to our theoretical results corresponding to oblate and prolate shapes. Also shown for ^{74}Kr is the result of maximal shape coexistence (50% oblate and prolate).

pairing correlations on these bulk properties is roughly taken into account by the use of the phenomenological gap parameters Δ_π, Δ_ν.

The main effect of taking into account neutron-proton pairing in the HF+BCS calculation would be to increase the diffuseness of the Fermi surface. This diffuseness is proportional to the gap parameters. It is therefore interesting to study the sensitivity of the GT strength to the gap parameters in the $N = Z$ nuclei. To this end we compare in Fig. 2 for ^{72}Kr the QRPA results obtained with the SG2 force in those nuclei for various values of the gap parameters differing by ± 0.5 MeV from the values extracted from the phenomenology. To make the discussion easier we did not include the *pp* residual interaction in those figures.

The main effect of the BCS correlations is to create new transitions that were forbidden in the absence of such correlations. Since the occupation probabilities are now different from 0 or 1, the already existing peaks at $\Delta = 0$ will decrease when $\Delta > 0$, while new strength will appear at other energies and will increase with increasing gap parameters. This new strength is in general located at high excitation energy, while the strength already present at $\Delta = 0$ is mainly concentrated at lower energy. As a consequence, the main effect of increasing the Fermi diffuseness is to smooth out the profile of the GT strength distribution, increasing the strength at high energies and decreasing the strength at low energies.

5 The role of deformation and shape coexistence

The role of deformation on the GT strength distributions can be summarized, in general, in two types of effects. First, deformation breaks down the degeneracy of the spherical shells and this implies that the GT strength distributions corresponding to deformed nuclear shapes will be much more fragmented than the corresponding spherical ones. Second, the energy levels of deformed orbitals coming from different spherical shells cross each other in a way that depends on the magnitude of the quadrupole deformation as well as on the oblate or prolate character. This level crossing may lead in some instances to similar profiles in the GT strength distributions of the various coexisting nuclear shapes but in other cases it may lead to sizable differences between the GT strength distributions corresponding to different shapes of the same parent nucleus. This fact can be exploited to gain information on what is the nuclear shape of a nucleus by just looking at the structure of its β-decay. In particular, in Fig. 3 we compare our results for summed strengths with available experimental data [4]. The figure clearly shows that data for ^{74}Kr agree better with the assumption of oblate and prolate shape coexistence in the ground state. On the contrary, for ^{76}Sr, the data favor a prolate ground state.

Acknowledgments

This work was supported by Ministerio de Ciencia y Tecnología (Spain) under contract number BFM2002-03562.

References

1. P. Sarriguren, E. Moya de Guerra, A. Escuderos, and A.C. Carrizo, Nucl. Phys. **A635** (1998) 55.
2. P. Sarriguren, E. Moya de Guerra, and A. Escuderos, Nucl. Phys. **A658** (1999) 13.
3. P. Sarriguren, E. Moya de Guerra, and A. Escuderos, Nucl. Phys. **A691** (2001) 631; Phys. Rev. C **64** (2001) 064306.
4. E. Poirier, Ph. D. thesis, IRES Strasbourg (2002).

Dipole Symmetry Near Threshold [a]

Moshe Gai

Laboratory for Nuclear Science, Dept. of Physics, University of Connecticut, 2152 Hillside Rd, U3046, Storrs, CT 06269-3046, USA
gai@uconnvm.uconn.edu - http://www.phys.uconn.edu

In celebrating Iachello's 60th birthday we underline many seminal contributions for the study of the degrees of freddom relevant for the structure of nuclei and other hadrons. A dipole degree of freedom, well described by the spectrum generating algebra U(4) and the Vibron Model, is a most natural concept in molecular physics. It has been suggested by Iachello with much debate, to be most important for understanding the low lying structure of nuclei and other hadrons. After its first observation in ^{18}O it was also shown to be relevant for the structure of heavy nuclei (e.g. ^{218}Ra). Much like the Ar-benzene molecule, it is shown that molecular configurations are important near threshold as exhibited by states with a large halo and strong electric dipole transitions. The cluster-molecular Sum Rule derived by Alhassid, Gai and Bertsch (AGB) is shown to be a very useful model independent tool for examining such dipole molecular structure near thereshold. Accordingly, the dipole strength observed in the halo nuclei such as ^{6}He, ^{11}Li, ^{11}Be,^{17}O, as well as the N=82 isotones is concentrated around threshold and it exhausts a large fraction (close to 100%) of the AGB sum rule, but a small fraction (a few percent) of the TRK sum rule. This is suggested as an evidence for a new soft dipole Vibron like oscillations in nuclei.

1 Molecular Dipole Symmetry

A molecular degree of freedom is characterized by excitations that involves the relative motion of two tightly bound constituents and not the excitation of the objects themselve. Hence it is associated with a polarization vector known as the separation vector. Such a vector can be classicaly described in a geometrical model in three dimensions or by using the corresponding group U(4) [1] and the very succesful Vibron model of molecular Physics [2]. This model has two symmetry limits that correspond to the geometrical description of Rigid Molecules, the O(4) limit, or Soft Molecules, the U(3) limit.

A most comprehensive discussion of such molecular structure and the Vibron model can be found in Iachello-Levine book [2] on "Algebraic Thoery of Molecules". In Fig. 1 taken from that book we show the characteristic dimensions of the Ar-benzen molecule. The argon atom is losely bound to the (tightly bound) benzen molecule by a van der Waalls polarization and thus this molecular state lies close to the dissociation limit. We note that the relative dimension and indeed the very polarization phenomena are reminscent of a halo structure where the argon atom creates a "halo" around the benzen molecule.

[a] Work Supported by USDOE Grant No. DE-FG02-94ER40870.

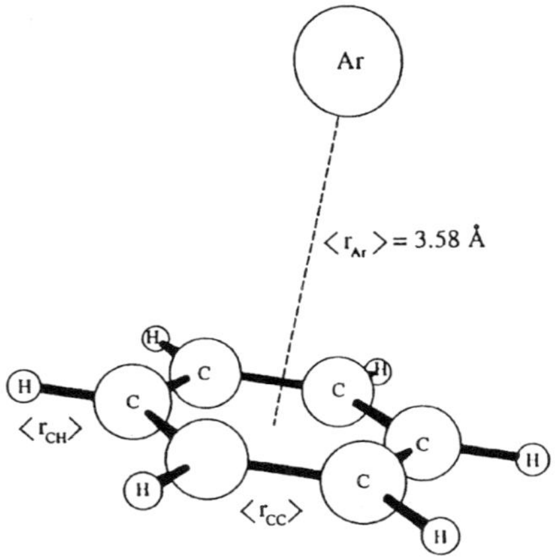

Fig. 1: Characterstics dimensions of the Ar-benzen molecule, adopted from Iachello and Levine [2].

2 The AGB Cluster Sum Rule

The polarization phenomena associated with a molecular state implies that it should be associated with dipole excitations of the separation vector. In this case expectation values of the dipole operator do not vanish as the center of mass and center of charge of the polarized molecular state do not coincide [3,4]. Hence molecular states give rise to low lying dipole excitations. While the high lying Giant Dipole Resonace (GDR) is associated with a Goldhaber-Teller [5] excitation of the entire neutron distribution against the proton distribution, a molecular excitation involves a smaller fraction of the nucleus at the surface and is expected to occur at lower excitation than the GDR; i.e. a soft dipole mode [6,7].

The GDR exhausts the Thomas-Reiche-Kuhn (TRK) [8] Energy Weighted Dipole Sum Rule as applied to nuclei:

$$\begin{aligned} S_1(E1;A) &= \Sigma_i\ B(E1:0^+ \rightarrow 1_i^-) \times E^*(1_i^-) \\ &= \tfrac{9}{4\pi}\ \tfrac{NZ}{A} \times \tfrac{e^2\hbar^2}{2m} \end{aligned} \qquad \text{(equ. 1)}$$

And for a molecular state Alhassid, Gai and Bertsch [9] derived sum rules by subtracting the individual sum rules of the contituents from the total sum rule:

$$\begin{aligned} S_1(E1;A_1 + A_2) &= S_1(A) - S_1(A_1) - S_1(A_2) \\ &= \tfrac{9}{4\pi}\ \tfrac{(Z_1A_2 - Z_2A_1)^2}{AA_1A_2} \times \tfrac{e^2\hbar^2}{2m} \end{aligned} \qquad \text{(equ. 2)}$$

$$S_1(E1;\alpha + A_2) = \tfrac{9}{4\pi}\ \tfrac{(N-Z)^2}{A(A-4)} \times \tfrac{e^2\hbar^2}{2m} \qquad \text{(equ. 3)}$$

$$S_1(E1;n + A_2) = \tfrac{9}{4\pi}\ \tfrac{Z^2}{A(A-1)} \times \tfrac{e^2\hbar^2}{2m} \qquad \text{(equ. 4)}$$

$$S_1(E1;2n + A_2) = \tfrac{9}{4\pi}\ \tfrac{2Z^2}{A(A-2)} \times \tfrac{e^2\hbar^2}{2m} \qquad \text{(equ. 5)}$$

The molecular sum rule, equ (2), was shown to be useful in elucidating molecular (cluster) states in ^{18}O where the measured B(E1)'s and B(E2)'s exhaust 13% and 23%, respectively, of the molecular sum rule[10]. Similarily, these molecular states in ^{18}O have alpha widths that exhaust 20% of the Wigner sum rule. The branching ratios for electromagnetic decays in ^{18}O were also shown to be consistent with predictions of the Vibron model in the U(3) limit[11]. Indeed the manifestation of a molecular structure in ^{18}O has altered our undertsanding of the coexistence of degrees of freedoms in ^{18}O[12]. Similar observations were also made in the heavy nucleus ^{218}Ra[13].

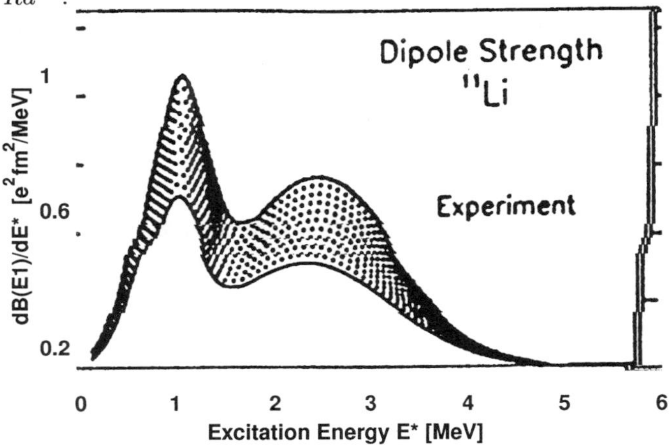

Fig. 2: Dipole strength measured in ^{11}Li[14].

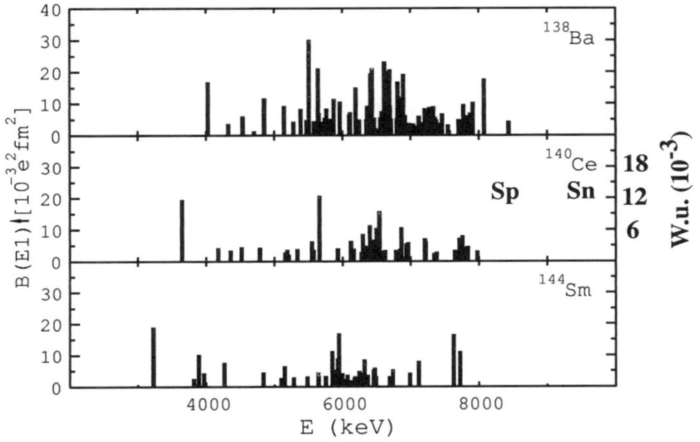

Fig. 3: Dipole strength measured in N=82 isotones[20].

The dipole strength at approximately 1.2 MeV in ^{11}Li[14], shown in Fig. 2, exhausts approximately 20% of the molecular sum rule, and the total strength integrated up to 5 MeV exhausts approximately 100% of the cluster sum rule[15,16], but it only exhausts approximately 8% of the TRK sum rule, see table 1. We emphasize that the experimental efficiency at for example 6 MeV is very large (30%), but no strength is found at higher energies beyond 100% of the molecular sum rule. These two facts strongly suggest the existence of a low lying soft dipole mode in ^{11}Li. Similar observation are reported in ^{11}Be[17], oxygen isotopes[18] and ^{6}He[19], believed to exhibit a halo structure. The N=82 isotones also show a diople strength near threshol as shown in Fig 3[20]. These results are summarized in Table 1.

The ratio of the TRK/AGB sum rules is given by:

$$\begin{aligned} \text{TRK/AGB} &= NZA_1A_2/(Z_1A_2 - Z_2A_1)^2 && \text{(equ. 6)} \\ (\alpha) &= (N-Z)^2/NZ(A-4) \\ \text{(1n)} &= \text{N(A-1)/Z} \\ \text{(2n)} &= \text{N(A-2)/2Z} \end{aligned}$$

Table 1: Measured E1 strength in nuclei.

Nucleus	$< E^* >$	TRK	TRK/AGB	AGB
^{11}Li[14,15]	1.2 MeV	$8.0 \pm 2.0\%$	(2n) 12	$96 \pm 24\%$
^{11}Be[17]	1.0 MeV	5.0%	(1n) 18	90%
^{17}O[18]	< 15 MeV	4%	(1n) 18	72%
^{138}Ba[20]	6.5 MeV	$0.78 \pm 0.15\%$	(1n) 200	$156 \pm 30\%$

3 Conclusions

In conclusions we demonstrate that molecular configurations play a major role in the structure of light and heavy nuclei. Unlike the Giant Dipole Resonance that involves oscillation of the entire neutron-proton distributions, these Vibron states involve only oscillations of the surface of the nucleus, and hence they lie at lower energies than the GDR. Similarly, while the GDR exhausts the TRK sum rule, the Vibron states exhausts the ABG cluster sum rule.

1. F. Iachello, and A.D. Jackson; Phys. Lett. **108B**(1982)151.
2. F. Iachello and R.D. Levine, Algebraic Theory of Molecules; Oxford University Press, 1995.
3. L.A. Radicati; Phys. Rev. **87**(1952)521.
4. M. Gell-Mann and V.L. Telegdi; Phys. Rev. **91**(1953)169.
5. M. Goldhaber and E. Teller; Phys. Rev. **74**(1948)1046.
6. K. Ikeda Nucl. Phys. **A538**(1992)355c.
7. P.G. Hansen; Nucl. Phys. **A588**(1995)1c.
   P.G. Hansen and A.S. Jensen; Annu. Rev. Nucl. Part. Sci. **45**(1995)591.
8. W. Kuhn; Zeit. f. Phys. **33**(1925)408. F. Reiche, W. Thomas; Zeit. f. Phys. **34**(1925)510.
9. Y. Alhassid, M. Gai, and G.F. Bertsch ; Phys. Rev. Lett. **49**(1982)1482.

10. M. Gai, M. Ruscev, A.C. Hayes, J.F. Ennis, R. Keddy, E.C. Schloemer, S.M. Sterbenz and D.A. Bromley; Phys. Rev. Lett. **50**(1983)239.
11. M. Gai *et al.*; Phys. Rev. **C43**(1991)2127.
12. M. Gai *et al.*; Phys. Rev. Lett. **62**(1989)874.
13. M. Gai *et al.*; Phys. Rev. Lett. **51**(1983)646.
14. M. Zinser *et al.*; Nucl. Phys. **A619**(1997)151.
15. G.F. Bertsch and J. Foxwell; Phys. Rev. **C41**(1990)1300.
16. M. Gai; Rev. Mex. Fis. Supp. **45**(1999)106.
17. T. Nakamura *et al.*; Phys. Lett. **B331**(1994)296. N. Gan *et al.* http://www.phy.ornl.gov/progress/ribphys/reaction/rib023.pdf.
18. T. Aumann *et al.*; Nucl. Phys. **A649**(1999)297c. A. Leistenscheneider *et al.*; Acta. Phys. Pol. **B32**(2001)1095.
19. S. Nakayama *et al.*; Phys. Rev. Lett. **85**(2000)262.
20. A. Zilges *et al.* Phys. Lett. **B542**(2002)43.

MEASUREMENT OF THE SPIN ENTANGLEMENT OF TWO-PROTON SYSTEM

H. SAKAI, T. SAITO AND A. TAMII

Department of Physics, University of Tokyo, Hongo 7-3-1, Bunkyo, Tokyo 113-0033, Japan

E-mail:sakai@phys.s.u-tokyo.ac.jp

T. KAWABATA

Center for Nuclear Study, University of Tokyo, Wako Branch at RIKEN, Hirosawa 2-1, Wako, Saitama 351-0198, Japan

E-mail:kawabata@cns.s.u-tokyo.ac.jp

Y. SATOU

Department of Physics, Tokyo Institute of Technology, Ookayama 2-12-1, Meguro, Tokyo 152-8551, Japan

E-mail:ysatou@rarfaxp.riken.go.jp

We are planning to measure the spin entanglement of two protons (p–p) in the spin-singlet state [$^1S_0$]. Results will be compared with predictions by the quantum mechanics as well as by the Bell's inequality. The [$^1S_0$] state will be produced by the $(d, {}^2\mathrm{He})$ reaction and their spin-correlation will be measured by the newly constructed polarimeter EPOL.

1 Introduction

In 1935 A. Einstein, B. Podolsky and N. Rosen (EPR) published a paper[1] entitled "Can Quantum-Mechanical Description of Physical Reality be Considered Complete?" in which they gave a thought experiment and questioned the completeness of quantum mechanics. This so-called "EPR paradox" was immediately responded by N. Bohr by exactly the same title in Ref. 2. This lead to a lot of discussions among physicists. However without experiments subsequent discussions looked like philosophical arguments until 1964. In 1964, John S. Bell proposed a theorem based on Einstein's locality principle, now called Bell's Inequality,[3] which allows us to prove or disprove the EPR paradox on the scientific bases by experiments.

In a standard quantum mechanics (QM) text book[4], the EPR paradox is explained by using spin-correlation experiments as shown in Fig. 1, first illustrated by D. Bohm.[5] Here a two-spin $\frac{1}{2}$ particle system in a spin-singlet state

$$|{}^1S_0\rangle = \frac{1}{\sqrt{2}}(|\uparrow\rangle_1|\downarrow\rangle_2 - |\downarrow\rangle_1|\uparrow\rangle_2) \tag{1}$$

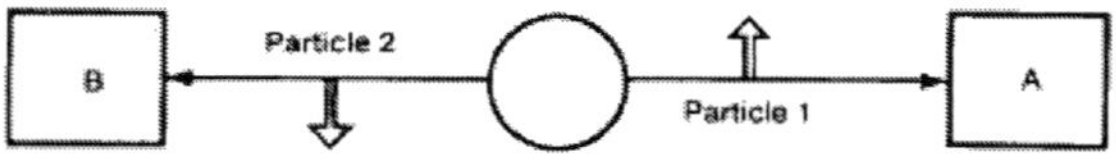

Figure 1. Spin correlation in a spin-singlet state. Figure is taken from Ref. 4.

decays into a back-to-back direction. Subsequently, a spin component of the particle 1(2) is measured by a spin polarimeter A(B), which yields a spin-correlation function.

Quantum mechanics predicts this correlation function as

$$P_{\rm QM}(\phi) = \langle \vec{\sigma}_1 \cdot \hat{a}\ \vec{\sigma}_2 \cdot \hat{b}\ \rangle = -\hat{a}\cdot\hat{b} = -\cos\phi, \tag{2}$$

where the unit vectors $\hat{a}$ and $\hat{b}$ denote the directions of the reference axes of the two polarimeters.

A classical correlation in terms of the Bell's theorem can be estimated by using a so-called color-ball model[6] as

$$P_{\rm BI}(\phi) = -1 + 2\phi/\pi. \tag{3}$$

Figure 2 shows the $P_{\rm QM}(\phi)$ and $P_{\rm BI}(\phi)$ by solid and dashed lines, respectively. Thus one can discriminate each other, if the spin correlation measurements were performed with enough accuracy.

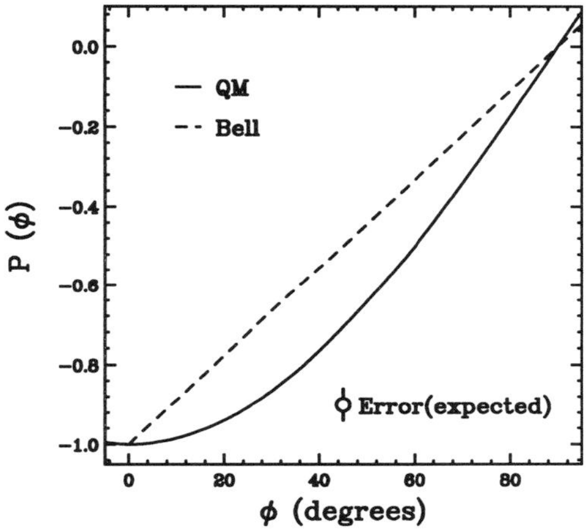

Figure 2. Spin correlations expected for the quantum mechanics (solid line) and classical (dashed line) as a function of ϕ.

2 Experiment by Lamehi-Rachti and Mittig

Many experiments to test Bell's inequality were performed until now. They are all spin correlation measurements of two-photon system, not two-spin $\frac{1}{2}$ particle system as illustrated in the QM text book. Almost all results were consistent with standard QM predictions.[7]

There exists only one experiment by Lamehi-Rachti and Mittig (LRM)[8] who measured polarization correlations of two-proton system (see Fig 3) as is illustrated in the QM text book. The entangled proton system in the spin-singlet state was produced by elastic proton-proton (p–p) scattering at 13.5 MeV.

The spin polarizations of both scattered and recoiled protons were measured in coincidence by two independent polarimeters. The polarimeter consisted of a carbon scatterer and catcher detectors. The degree of polarization was deduced from a left-right asymmetry of scattering. The protons scattered by the carbon foils were detected with four solid-state detectors at positions labeled L_1 or R_1 for the left side polarimeter, and L_2 or R_2 for the right side polarimeter as shown in Fig. 3. The right side polarimeter was kept in the reaction plane, while the left side one was rotated by an angle ϕ with respect to the p–p scattering plane.

LRM defined a spin-correlation function as

$$P_{\rm exp}(\phi) = \frac{N_{LL} + N_{RR} - N_{LR} - N_{RL}}{N_{LL} + N_{RR} + N_{LR} + N_{RL}}, \tag{4}$$

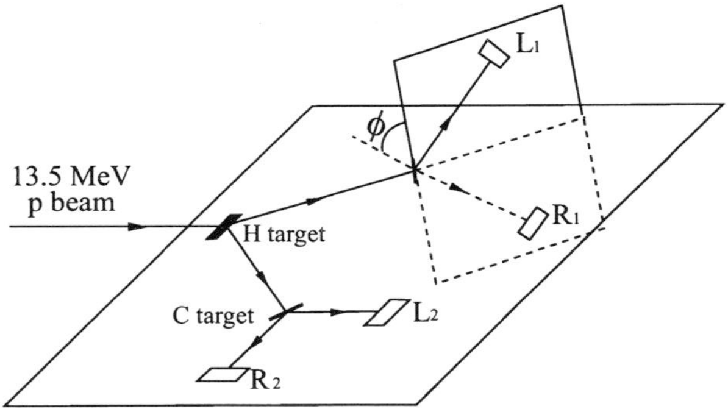

Figure 3. Schematic experimental setup of the LRM's experiment for the measurement of the spin-correlation in p-p scattering.

where N_{LR} is the number of coincidence events between L_1 and R_2 detectors and so on. This pioneering experiment has a number of deficiencies. First, the incident energy of 13.5 MeV, which corresponds to a relative energy of 6.25 MeV between the two protons, is so high that triplet final states $|\,^3P_{0,1,2}\,\rangle$ contribute to the scattering, although LRM estimated their contribution to be only 2%. Second, LRM measured the polarizations of the two protons with two separate polarimeters. It is very difficult for this arrangement to cancel any asymmetry produced by geometrical misalignment. Third, since the polarimeters with small geometrical acceptance were fixed at particular angles with respect to the proton beam axis, the system was able to cover only a small part of the available phase space.

3 Our Experiment

The purpose of the present experiment is to solve some of these problems and to provide a stringent test of the Bell's inequality in a two-proton system.

3.1 Use of the $(d, ^2\mathrm{He})$ Reaction

When we try to produce a two-proton system in a $^1S_0$ state with high purity by p-p elastic scattering, we must reduce the incident proton energy. However, this reduction of the incident proton energy makes polarization measurement very difficult. Thus the requirement to keep the purity of the $^1S_0$ state high is incompatible with efficient measurement of proton spin polarization. One solution to this dilemma is to boost the total scattering system towards higher energies. In this way, the two protons could have sufficiently high kinetic energies in the laboratory frame, while their relative energy can be kept small.

This method can be realized by employing the $\mathrm{A}(d, ^2\mathrm{He})\mathrm{B}$ reaction. This notation means that a deuteron bombards a target A and after the charge-exchange reaction a ^{2}He particle is emitted leaving a residue B. The ^{2}He particle is simply two protons that are coupled into the $^1S_0$ state by the strong final state interactions. Since the ^{2}He is a *quasi*-stable particle, it decays into two protons after $\sim 10^{-20}$ seconds.

The use of the $(d, {}^2\text{He})$ reaction has the following advantages:

- In the LRM experiment, the relative energy of the two protons was fixed by the incident proton energy, while the $(d, {}^2\text{He})$ reaction enables us to select the relative energy of two protons arbitrarily by choosing appropriate kinematic conditions. Thus we can prepare the entangled proton pair (${}^1S_0$-state) with high purity.

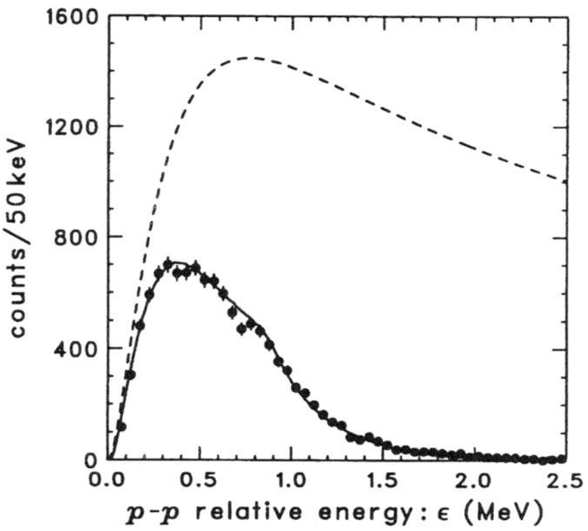

Figure 4. Distribution of p-p relative energy (ϵ). See text for detail.

- Since the kinetic energy of the protons in the laboratory frame is high, the proton polarizations can be measured efficiently.
- Since a pair of protons enter the polarimeter after the momentum analysis by the magnetic spectrometer, the polarizations of the two protons can be measured simultaneously in the same polarimeter. This arrangement can largely eliminate the systematic uncertainties due to the geometrical asymmetry of the detectors.

We have been using the $(d, {}^2\text{He})$ reaction as a spectroscopic tool to study the spin-isospin excitations in nuclei.[9] Figure 4 shows a typical p-p relative energy (ϵ) distribution obtained for the ${}^{90}\text{Zr}(d,{}^2\text{He})$ reaction at $E_d = 270$ MeV and $\theta = 0°$ by Okamura *et al.*[10] The dashed curve represents the prediction of the ϵ distribution based on the Migdal-Watson formalism which utilizes the p-p scattering ${}^1S_0$ phase shift values. The experimental spectrum (closed circles) is well reproduced by the simulation (solid curve) which takes the ^{2}He detection efficiency into account. As can be seen from the figure, we can select the p-p relative energy by setting a gate on ϵ. In other words we can prepare the ${}^1S_0$ state with a purity greater than 99% by selecting $\epsilon \leq 0.5$ MeV.

3.2 Experimental Setup

The present experiment[11] will be carried out at the RIKEN Accelerator Research Facility. A deuteron beam with an energy of 270 MeV from the Ring Cyclotron bombards a liquid hydrogen target. The ^{2}He particle emitted from the target,

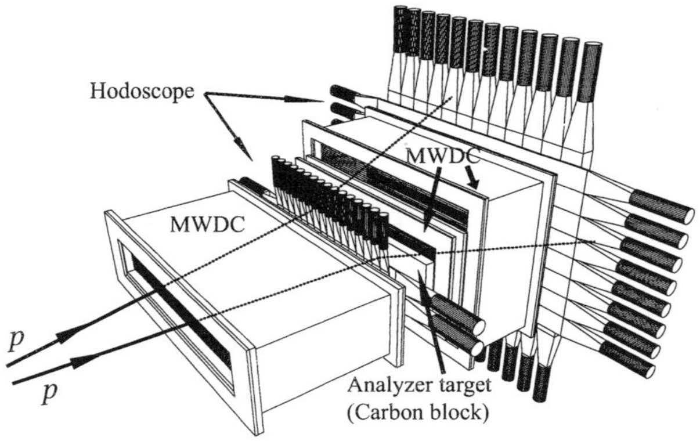

Figure 5. Schematic view of the proton spin-correlation polarimeter EPOL.

which subsequently decays into a pair of protons, is momentum analyzed by the magnetic spectrometer SMART[12] and detected at the focal plane by the proton polarimeter EPOL, which allows us to measure the spin-correlation function of the proton pair.

Since a pair of protons reaches the focal plane symmetrically with each other with respect to the central axis of the spectrometer, the positions of the incident protons are strongly correlated with the relative energy ϵ of the proton pair. Hence we can select proton pairs with, for example, $\epsilon < 1$ MeV by confining the distance between the positions of the proton pair at the focal plane within 60 cm.

A schematic view of EPOL is shown in Fig. 5. EPOL consists of three multi-wire drift chambers (MWDCs), two sets of hodoscopes (arrays of plastic scintillators), and an analyzer target (carbon block). The hodoscopes are used for particle identification and generation of event triggers for data acquisition. The incident protons are scattered from the analyzer target. Three MWDCs measure the trajectories of the two protons before and after scattering.

The spin-correlation function $P_{\text{exp}}(\phi)$ of a pair of protons can be derived according to Eq. (4). Here the angle ϕ is defined as the angle between the reference axes $\hat{a}$ and $\hat{b}$ associated with the pair of protons. In principle one can, after the experiment, choose these reference axes freely during the off-line analysis. This is one of the most characteristic feature of the present unbiased polarimeter system. In practice, however, usually either $\hat{a}$ or $\hat{b}$ is chosen to be a laboratory reference axis such as the direction perpendicular to the horizontal plane.

4 Remarks and Future Prospects

As mentioned above, the $(d, {}^2\text{He})$ reaction can significantly reduce the systematic uncertainties due to the mixture of triplet states and the geometric asymmetry of the detectors compared to LRM's experiment. Using this advantage, we want to discriminate between the Bell's inequality and quantum mechanics by five standard deviations. Expected size of uncertainty is indicated in Fig. 2 by an error bar. The experiment is expected in this fall.

As is usual with this kind of Bell's inequality test, our experiment can not be a complete one. Some loopholes unfortunately remain open, such as 'lightcone' and/or 'detection' loopholes. To close loopholes, we must make some plausible assumptions in the interpretation of experimental result. Further theoretical consideration is needed.

Acknowledgements

The experiment will be carried out under the collaboration of University of Tokyo, RIKEN, CNS, Saitama University, TIT, Aizu University, University of Saskatchewan and KVI.

This work is supported in part by the Mitsubishi Foundation.

References

1. A. Einstein, B. Podolsky, and N. Rosen, Phys. Rev. **47**, 777 (1935).
2. N. Bohr, Phys. Rev., **48**, 696 (1935).
3. J. S. Bell, Physics **1**,195 (1964).
4. For example, J.J. Sakurai, Modern Quantum Machanics, 1985, Benjamin Pub. Comp.
5. D. Bohm: Phys. Rev. **85**, 166 (1957).
6. A. Peres, Quantum Theory: Concepts and Method, 1995, Kluwer Academic Pub.
7. F. Laloë, Am. J. Phys. **69**, 655 (2001) and references therein.
8. M. Lamehi-Rachti and W. Mittig, Phys. Rev. D **14**, 2543 (1976).
9. For example, H. Sakai *et al.*, Nucl. Phys. **A599**, 197c (1994) and H. Okamura *et al.*, Phys. Lett. B **345**, 1 (1995).
10. H. Okamura *et al.*, Nucl. Instr. and Meth. A**406**, 78 (1998).
11. H. Sakai *et al.*, proposal to RIKEN PAC (R350n), 2001.
12. T. Ichihara *et al.*, Nucl. Phys. **A569**, 287c (1994).

SU(6)-BREAKING SYMMETRY AND THE RATIO OF PROTON MOMENTUM DISTRIBUTIONS

M.M.Giannini, E. Santopinto, A. Vassallo
Dipartimento di Fisica dell'Università di Genova, I.N.F.N. Sezione di Genova, Italy

M. Vanderhaeghen
Institut für Kernphysik, Johannes Gutenberg Universität, D-55099 Mainz, Germany

The ratio between the anomalous magnetic moments of proton and neutron has recently been suggested to be connected to the ratio of proton momentum fractions carried by valence quarks. This relation has been obtained within a parametrization of the Generalized Parton Distributions (GPD) [1], but it is completely independent of such a parametrization.
It will be shown that using different CQMs this relation holds within a few percent accuracy. This agreement is based on what all the CQMs have in common: the effective degrees of freedom of the three constituent quarks and the underlying SU(6) symmetry. On the other hand, the experimental value of the ratio is not reproduced by CQMs. This means that the SU(6)-breaking mechanism contained in the phenomenological partonic distributions does not correspond to the SU(6) breaking mechanism implemented in the CQMs we have analyzed [2].
We will also show how this relation can be used in order to understand in which way to implement an $SU(6)$-breaking mechanism and to test models.

1 Introduction

The static properties of baryons are an important testing ground for QCD based calculations in the confinement region. However, different CQMs[3,4,5,6] are able to obtain a comparable good description of the low energy data, so that it is difficult to discriminate among them. A fundamental aspect of the theoretical description is the introduction of terms in the quark Hamiltonian which violate the underlying $SU(6)-$symmetry. It is therefore important to find out observables which are sensitive to the various SU(6)-breaking mechanisms.

In this respect, the relation proposed recently by Goeke, Polyakov and Vanderhaeghen[1] between the anomalous magnetic moments of the proton and the neutron and the proton momentum fractions carried by valence quarks, $M_2^{q_{val}}$, might be a good candidate for testing SU(6)-breaking effects.

Quark models are able to reproduce in a extraordinary way the static low energy properties of baryons with very few parameters and this gives us confidence that they are a good effective representation of the low energy strong interaction dynamics. The QCD based parton model reproduces in a beautiful way the Q^2 dependence of the high energy properties even with naive input. However the perturbative approach to QCD does not provide absolute values of the observables; one can only relate data at different momentum scales. The description based on the Operator Product Expansion (OPE) and the QCD evolution require the input of non-perturbative matrix elements which have to be predetermined[7] and therefore the parton distributions are usually obtained in a phenomenological way from fits to deep inelastic lepton nucleon scattering and Drell-Yan processes. The basic steps are to find a parametrization[8] which is appropriate at a sufficiently large momentum

$Q_0{}^2$, where it is expected that perturbation theory is applicable, and then QCD evolution techniques are used in order to obtain the parton distribution at higher Q^2. Using these parametrizations a large body of data is reasonably described, even if at the origin this parametrization is purely phenomenological.

Gluck, Reya and Vogt [9] started from a parametrized distribution of partons at a very low scale $\mu^2{}_0$, which resembles that of a naive Quark Model of hadron structure, in the sense that the contribution of the valence quarks to the structure function is dominant. As suggested by Parisi and Petronzio [10], the hadronic μ_0^2 scale is defined such that the fraction of the total momentum carried by the valence quarks is unity. This procedure opens the possibility of using Constituent Quark Models as input in order to calculate the nonperturbative (twist-two) nucleon matrix elements, as proposed by Jaffe and Ross [11].

The scheme developed by Traini et al.[12] takes into account all these aspects: it uses as input the quark model results in order to determine the non perturbative matrix elements at the hadronic scale [10], then an upwards NLO evolution procedure at high momentum transfer ($Q^2 = 10$ GeV^2) is performed[12].

Starting from three different Constituent Quark Models [3,6,4], we have calculated the parton distributions at the hadronic scale and we have evaluated the ratio of the proton momentum fractions carried by valence quarks. A NLO evolution has been performed up to $Q^2 = 10$ GeV^2.

All models give a good description of the spectrum and have been used also to describe various observables (elastic and inelastic form factors, strong decays). In particular, the different results for the electromagnetic transition form factors indicate that the models have a quite different Q^2-behaviour. However, the ratio of the proton momentum fractions carried by valence quarks is independent of the scale Q^2, therefore we expect that the study of this relation will give important information on general aspect of CQM.

2 Ratio of proton momentum fractions carried by valence quarks

In Ref. [1], a relation has been proposed between the ratio of the proton and neutron anomalous magnetic moments and the momentum fractions carried by valence u- and d-quark distributions, as follows :

$$\frac{\kappa^p}{\kappa^n} = -\frac{1}{2}\,\frac{4\,M_2^{d_{val}} + M_2^{u_{val}}}{M_2^{d_{val}} + M_2^{u_{val}}}\,, \tag{1}$$

with the proton momentum fraction carried by the valence quarks defined as

$$M_2^{q_{val}} = \int_0^1 dx\, x\, q_{val}(x)\,. \tag{2}$$

In Fig. 1, we show the scale dependence of the *rhs* of Eq. (1), which we shall henceforth denote with R, for various recent parametrizations of next-to-leading order (NLO) and next-to-next-to-leading order (NNLO) parton distributions. Fig. 1 shows that the scale dependence drops out of the *rhs* of Eq. (1), although the numerator and denominator separately clearly have a scale dependence. Furthermore, it is seen from Fig. 1, for all NLO and one NNLO parametrizations of parton distributions, that the relation of Eq. (1) is numerically verified to an accuracy at the one percent level! In particular, the most recent MRST01 NLO, the MRST01 NNLO, and the CTEQ6M NLO parton distributions (which appeared after the writing of Ref. [1]), nicely confirm the finding of Ref. [1]. Although the relation Eq. (1) was originally derived within a parametrization of generalized parton distributions, it is in fact completely independent of such a parametrization, as the *rhs* of Eq. (1) is expressed in terms of moments of forward valence quark distributions alone.

The above observations from phenomenology suggest that Eq. (1) holds and

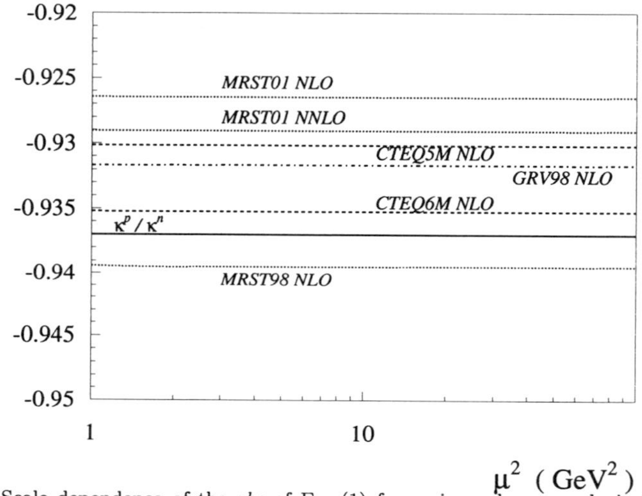

Figure 1: Scale dependence of the *rhs* of Eq. (1) for various phenomenological forward parton distributions as indicated on the curves. Dotted curves : MRST parton distributions (MRST98 NLO, MRST01 NLO, MRST01 NNLO). Dashed curves : CTEQ parton distributions (CTEQ5M NLO, CTEQ6M NLO). Dashed-dotted curve : GRV98 NLO($\overline{\mathrm{MS}}$)[2]. Also shown is the *lhs* of Eq. (1), i.e. the experimental value for κ^p/κ^n (constant solid curve).

that the unpolarized valence $u-$ and d-quark forward distributions contain a nontrivial information about the anomalous magnetic moments of the proton and neutron. It is the aim of the present work to investigate the relation of Eq. (1) in different quark models.

Let us firstly consider the simplest quark model, with exact $SU(6)$ symmetry. In this limit, $M_2^{u_{val}} = 2\, M_2^{d_{val}}$, and $\kappa^p =$ - $\kappa^n = 2$, so that one immediately verifies that Eq. (1) holds.

In reality, the ratio of anomalous magnetic moments deviates from the $SU(6)$ limit by about 6.5 %. The smallness of this deviation is the main reason why constituent quark models are quite successful in predicting nucleon (and more generally baryon octet) magnetic moments. In quark model language, the relation of Eq. (1) implies that the small breaking of the $SU(6)$ symmetry follows some rule which is encoded in the valence quark distributions. In particular, it is interesting to investigate a possible correlation between the ratio of valence $d-$ and u-quark distributions, and the ratio of proton to neutron anomalous magnetic moments in different models. To this end, we turn in the next section to the calculation of parton distributions in quark models with different $SU(6)$ breaking mechanisms.

3 Parton distributions from quark models

The approach, recently developed by M. Traini et al. for the unpolarized distributions [12], connects the model wave functions and the parton distributions at the input hadronic scale through the quark momentum density distribution. In the unpolarized case one can write the parton distributions [12]:

$$q_V(x,\mu_0^2) = \frac{1}{(1-x)^2} \int d^3k \; n_q(|\mathbf{k}|) \; \delta(\frac{x}{1-x} - \frac{k_+}{M}) \tag{3}$$

where k_+ is the light-cone momentum of the struck parton, and $n_q(|\mathbf{k}|)$ represents the density momentum distribution of the valence quark of q-flavour:

$$n_{u/d}(|\mathbf{k}|) = \langle N, J_z = +1/2| \sum_{i=1}^{3} \frac{1 \pm \tau_i^z}{2} \; \delta(\mathbf{k} - \mathbf{k}_i)|N, J_z = +1/2\rangle \tag{4}$$

τ_i^z is the third component of the isospin Pauli matrices, k_i is the momentum of the ith constituent quark in the CM frame of the nucleon, $|N, J_z = +1/2\rangle$ is the nucleon wave function (in momentum space) with $J_z = +1/2$ component.
Using $k_+ = k_0 + k_z$, one can integrate eq. 3 over the angular variables and get:

$$q_V(x,\mu_0^2) = \frac{2\pi M}{(1-x)^2} \int_{k_m(x)}^{\infty} d|\mathbf{k}||\mathbf{k}| \; n_q(|\mathbf{k}|), \tag{5}$$

where

$$k_m(x) = \frac{M}{2} \left| \frac{x}{1-x} - \left(\frac{m_q}{M}\right)^2 \frac{1-x}{x} \right|,$$

M and m_q are the nucleon and (constituent) quark masses respectively.
Eq. (5) can be applied to a large class of quark models and satisfies some important requirements: it vanishes outside the support region $0 \leq x \leq 1$ and it has the correct integral property in order to preserve the number normalization.

In Ref. [2] we have shown that the ratio of the moments of the proton momentum fractions is Q^2 independent (up to NLO evolution) since the Q^2 dependent part of the parton distributions can be factorized.

We discuss the results obtained using different models for the valence quark contributions, namely the Isgur-Karl (IK) model [3], which has been largely used in the past to study the low-energy properties of hadrons and also deep inelastic polarized and unpolarized scattering[12], a hypercentral Coulomb-like plus linear confinement potential model [4] inspired by lattice QCD [13] and an algebraic model [6]; the wave functions of the last two models give a rather good description of the electromagnetic elastic and transition form factors [14 6 16].

The validity of Eq. (1) for the hCQM is analyzed in Table 1 The two members are equal within 0.2 %, although the κ-ratio differs by about 7 % from the experimental value (~ -0.937).

Similar results, reported in Table 1, hold for the other models, with the exception of the U(7) model, where the κ-value is correctly reproduced by construction, while the equation is violated up to a few percent.

	I.K.	HCQM + OGE	HCQM + Isospin	U7
Model prediction for $\frac{\kappa_p}{\kappa_n}$	-1.0	-1.0	-1.0	-0.9372
R-ratio at $Q^2 = 0.5$ GeV^2	-1.0098	-1.0030	-0.9983	-0.9881
R-ratio at $Q^2 = 5.0$ GeV^2	-1.0098	-1.0030	-0.9983	-0.9881
R-ratio at $Q^2 = 10.0$ GeV^2	-1.0098	-1.0030	-0.9983	-0.9881

Table 1: Different CQM predictions for the R-ratio and for the κ-ratio κ^p/κ^n

In order to test if this feature depends on the choice of the CQMs or is a general characteristic, we have used the analytic expression supplied by the Isgur-Karl model and tried to reproduce the experimental value of the two ratios by leaving the amplitudes a'_S,a_M and a_D free. One can also vary the h.o. constant α, with α^{-1} being a measure of the confinement radius. The Q^2-behaviour of the I.K. model is unrealistic because of the gauss-factors, however also in this case the ratio is quite scale independent. The procedure of fitting the amplitudes corresponds to introduce implicitly quite different hamiltonians. The anomalous magnetic moments have the following expressions:

$$\kappa_p = 2(1 - a_M^2) - 4a_D^2 \qquad \kappa_n = -2(1 - a_M^2) + 3/2\ a_D^2\ . \tag{6}$$

If one adopts a model where the only SU(6) breaking comes from the a_M, it is immediately seen from equation (6) that the κ-ratio is exactely equal to -1, like in the SU(6) limit. The crucial quantity seems then to be the a_D amplitude. Assuming that the D-wave amplitude is the only SU(6)-breaking term (*D-model*), we have that: $\frac{2a_S^2-2a_D^2}{-2a_S^2-1/2\ a_D^2} = -0.937$ if $a_S = 0.975$ and $a_D = 0.255$. Calculating the *rhs* of Eq. (1), which we refer as R in the following, with these two values of the parameter and varying α in a quite large interval, the best value obtainable is $R = 0.9988$, with $\alpha = 2.1\ fm^{-1}$, differing by about 7% from the κ-ratio. Finally, leaving completely free the amplitudes a'_S, a_M and a_D in order to fit the κ-ratio and R separately, the resulting amplitudes turn out to be complex.

Therefore, the proposed Equation (1) seems to be valid (up to few percent) for all Constituent Quark Models provided that the SU(6)-violation is not too strong, but both values are quite far from the experimental value of the κ-ratio of -0.937. If one tries to force the SU(6)-violation to reproduce the experimental value, one is apparently faced with too strong constraints coming from the CQM itself. This is a possible indication that the degrees of freedom introduced in the current CQM may be inadequate since one has to take into account pion cloud effects.

The relation Eq. (1) between the ratio of the proton and neutron anomalous magnetic moments and the momentum fractions carried by valence quarks, $M_2^{q_{val}}$, is exactly verified in the SU(6)-invariant limit, where both are equal to -1.

In the currently used Constituent Quark Models, SU(6) violations are introduced in different ways (One-Gluon-Exchange interaction, spin and/or isospin dependent terms, Gürsey-Radicati mass formula, One-Boson-Exchange ...). Such

SU(6) violation is necessary in order to bring the anomalous proton and neutron magnetic moments closer to the experimental values or to reproduce important features of the spectrum, such as the N-Δ mass difference.
In all the models we have considered in this paper (see Table 1) the equality of Eq.(1) holds within a few percent accuracy. This agreement is based on what all the CQMs have in common: the effective degrees of freedom of the three constituent quarks and the underlying SU(6) symmetry.
On the other hand, the experimental value of the ratio is not reproduced by CQMs, at variance with the calculations based on phenomenological parton distributions reported in Fig. 1. This means that the SU(6)-breaking mechanism contained in the phenomenological partonic distributions does not correspond to the SU(6) breaking mechanism implemented in the CQMs we have analyzed.
To conclude, it seems that all CQMs are too strongly constrained by the presence of the standard degrees of freedom corresponding to three constituent quarks. Therefore additional degrees of freedom should be introduced, in particular quark antiquark pairs and/or gluons and the discussed equation of Ref. [1], being sensitive to the SU(6)-breaking mechanism, will provide a useful tool for testing the new models.

1. K. Goeke, V. Polyakov and M. Vanderhaeghen, Prog. Part. Nucl. Phys. **47** (2001) 401.
2. M. M. Giannini, E. Santopinto, A. Vassallo and M. Vanderhaeghen, Phys. Lett. B **552** (2003) 149, and references quoted therein.
3. N. Isgur and G. Karl, Phys. Rev. **D18**, 4187 (1978); S. Capstick and N. Isgur, Phys. Rev. **D 34**,2809 (1986).
4. M. Ferraris, M. M. Giannini, M. Pizzo, E. Santopinto and L. Tiator, Phys. Lett. B **364**, 231 (1995); E. Santopinto, F. Iachello and M. M. Giannini, Eur. Phys. J. **A 1**,307 (1998); Nucl. Phys. A **623**, 101c (1997); M. M. Giannini, E. Santopinto and A. Vassallo, Nucl. Phys. A **699**, 308 (2002).
5. L. Ya. Glozman and D.O. Riska, Phys. Rep. **C268**, 263 (1996); L. Ya. Glozman, *et al.*, Phys. Rev. **C57**, 3406 (1998).
6. R. Bijker, F. Iachello and A. Leviatan, Ann. Phys. (N.Y.) **236**, 69 (1994).
7. A.J. Buras, Rev. Mod. Phys. 50 (1980) 199.
8. A.D. Martin, W.J. Stirling, R.G. Roberts, Ral Report 94-055.
9. M. Glueck and E. Reya, Phys. Rev. D 14 (1976) 3024; M. Glueck, E. Reya and A. Vogt, Z. Phys. C 67 (1995) 433.
10. G. Parisi and R. Petronzio, Phys. Lett. B 62 (1976) 331.
11. R.L. Jaffe and G.C. Ross, Phys. Lett. B 93 (1980) 313.
12. M. Traini, V. Vento, A. Mair and A. Zambarda, Nucl. Phys. A 614 (1997) 472; A. Mair, M. Traini, Nucl. Phys. A **628**, 296 (1998).
13. G. S. Bali, Phys. Rept. **343**, 1 (2001).
14. M. D. Sanctis, M. M. Giannini, L. Repetto and E. Santopinto, Phys. Rev. C **62**, 025208 (2000).
15. M. M. Giannini, E. Santopinto and A. Vassallo, Eur. Phys. J. A **12**, 447 (2001).
16. R. Bijker, F. Iachello, A. Leviatan, Phys.Rev. C **54** 1935 (1996).

THE HYPERCENTRAL CONSTITUENT QUARK MODEL AND ITS SYMMETRY

M.M. Giannini and E. Santopinto
Dipartimento di Fisica dell'Università di Genova, I.N.F.N. Sezione di Genova, Italy

The hypercentral CQM, which is inspired by Lattice QCD calculations for quark-antiquark potentials, is presented, stressing its underlying symmetry. Its results for the spectrum, the helicity amplitudes and the elastic form factors are briefly reported. In the latter case the model has allowed to show, for the first time in the framework of a quark model, that relativistic effects are responsible for a deviation from the usually accepted dipole behaviour, in agreement with recent data taken at the Jefferson Lab.

1 Constituent Quark Models

Looking at the baryon spectrum, one notices that the best known resonances (4* and 3*, according to the PDG classification) can be neatly arranged in $SU(6)$-multiplets, containing states only partially degenerate. This means that any CQM must provide a good description of the average values of the energies, by means of a $SU(6)$-independent interaction, that is spin-flavour independent, to which a $SU(6)$-breaking term is added, in order to reproduce the splittings within the various multiplets.

The various CQMs differ in the treatment of both the $SU(6)$-invariant and the $SU(6)$-breaking interaction.

In order to construct the $SU(6)$-configurations it is sufficient to determine the space part of the wave function, since the spin-flavour parts are standard. The relative motion of the three quarks is described by the Jacobi coordinates

$$\vec{\rho} = \frac{1}{\sqrt{2}}(\vec{r}_1 - \vec{r}_2) , \quad \vec{\lambda} = \frac{1}{\sqrt{6}}(\vec{r}_1 + \vec{r}_2 - 2\vec{r}_3) . \tag{1}$$

and therefore here are six space degrees of freedom. This is the starting point of the algebraic approach [1], which introduces $u(7)$ as the spectrum generating algebra and the totally symmetric representation of $u(7)$ as the corresponding space of the three-quark states. Moreover, the $u(7)$-algebra admits at least the following two subalgebra chains:

$$\begin{array}{ccccc} & & U(6) & & & (I) \\ & \nearrow & & \searrow & & \\ U(7) & & & & SO(6) \, , & \\ & \searrow & & \nearrow & & \\ & & SO(7) & & & (II) \end{array} \tag{2}$$

The first chain corresponds to a spherical oscillator (h.o.) in six dimensions, while the second chain to a $SO(7)$ dynamical symmetry, as for the hyperCoulomb potential (hC).

In the $u(7)$ model developed by Iachello, Bijker and Leviatan [1], the baryon mass operator is written in terms of the normal vibrations of a Y-shaped symmetric top, to which a rotation band is superimposed:

$$M^2_{vibr} = N[k_1 n_u + k_2(n_v + n_w)]N + \alpha L + M_0^2 \tag{3}$$

where n_u, n_v and n_w are the vibration quantum numbers. To the operator of Eq.(3) a Gürsey-Radicati term is added:

$$M^2_{sf} = a[C_2(SU_{sf}(6)) - 45] + b[C_2(SU_f(3)) - 9] + c[C_2(SU_s(2)) - \frac{3}{4}] \tag{4}$$

where the quantities C_2 are the quadratic Casimir operators of the groups indicated within parentheses. The term in Eq.4 is $SU(6)$ violating, since it introduces a dependence of the energy on the spin and isospin, but it does not mix the $SU(6)$-configurations. The $u(7)$ model leads to a good description of the spectrum and of other baryon properties.

In order to follow one of the two chains of Eq. (2), one has to introduce an explicit quark interaction. To this end it is convenient to substitute the Jacobi coordinates with the hyperspherical ones, which keep the four angles Ω_ρ and Ω_λ, but replace ρ and λ with the hyperradius x and the hyperangle ξ

$$x = \sqrt{\vec{\rho}^2 + \vec{\lambda}^2} \ , \qquad \xi = arctg(\frac{\rho}{\lambda}). \tag{5}$$

Using these coordinates, the kinetic term in the three-body Schrödinger equation can be rewritten as [2]

$$-\frac{1}{2m}(\Delta_\rho + \Delta_\lambda) = -\frac{1}{2m}(\frac{\partial^2}{\partial x^2} + \frac{5}{x}\frac{\partial}{\partial x} - \frac{L^2(\Omega_\rho, \Omega_\lambda, \xi)}{x^2}) \ . \tag{6}$$

where $L^2(\Omega_\rho, \Omega_\lambda, \xi)$ is the quadratic Casimir operator of $O(6)$; its eigenfunctions are the well known hyperspherical harmonics [2] $Y_{[\gamma]l_\rho l_\lambda}(\Omega_\rho, \Omega_\lambda, \xi)$ having eigenvalues $\gamma(\gamma+4)$, with $\gamma = 2n + l_\rho + l_\lambda$ (n is a non negative integer).

In the hypercentral Constituent Quark Model (hCQM) [3], the quark interaction is assumed to depend on the hyperradius x only $V = V_{3q}(x)$. The hyperradius x depend on the coordinates of all the three quarks and then $V_{3q}(x)$ is a three-body interaction. Actually, three-body mechanisms are generated by the fundamental multi-gluon vertices predicted by QCD. On the other hand, flux tube models lead to Y-shaped three-quark interactions. Furthermore, a two body potential, treated in the hypercentral approximation [4], leads to a x-dependent potential, since averaging $\sum_{i<j}(r_{ij})^n$ over angles and hyperangle one gets something proportional to x^n.

For a hypercentral potential the hyperradial wave function, $\psi_\gamma(x)$ is factored out and is a solution of the hypercentral equation

$$[\frac{d^2}{dx^2} + \frac{5}{x}\frac{d}{dx} - \frac{\gamma(\gamma+4)}{x^2}] \ \psi_\gamma(x) \quad = \quad -2m\,[E - V_{3q}(x)] \ \psi_\gamma(x) \ . \tag{7}$$

The Eq. (7) can be solved analytically in two cases. The first is the six-dimensional harmonic oscillator (h.o.)

$$\sum_{i<j} \frac{1}{2}\, k\, (\vec{r_i} - \vec{r_j})^2 \ = \ \frac{3}{2}\, k\, x^2 \ = \ V_{h.o}(x) \tag{8}$$

and the second one is the hyperCoulomb (hC) potential

$$V_{hyc}(x) = -\frac{\tau}{x}. \tag{9}$$

It is interesting to observe that the energy levels of the potential (9) can be obtained generalizing the procedure used in the case of the three-dimensional Coulomb problem [5]. In n dimensions the symmetry group is $O(n)$, with the degeneracy group $O(n+1)$, implying the existence of a conserved n-dimensional Runge-Lenz vector. The energy levels can then be written in terms of the eigenvalues of the quadratic Casimir operator

$$E_{\nu\gamma} = -\frac{m\tau^2}{2(C_2(O(n+1)) + (\frac{n-1}{2})^2)} \tag{10}$$

Comparing the first radially excited state (with $\nu = 1$ and $\gamma = 0$) and the first negative states (with $\nu = 1$ and $\gamma = 1$) in the two potentials, one sees that they are perfectly degenerate in the hC potential while in the h.o. case the negative state is much lower. Since the observed first radially excited nucleon state (the Roper) is slightly lower than the negative parity resonances, the hC potential seems to be a better starting point for studying the baryon spectrum. Furthermore, the h.o. levels are too degenerate in comparison with the experimental spectrum.

We shall consider in particular three CQMs:

A) The Isgur-Karl model [6]:

$$V_{3q} = V_{h.o}(x) + U + H_{hyp}, \tag{11}$$

where the strong degeneracy of the h.o. levels is modified by means of the shifting two-body potential U and the splitting within the mujltiplets is produced by the spin dependent hyperfine interaction H_{hyp}.

B) The analytical hypercentral model [7,8]

$$V_{3q} = -\frac{\tau}{x} + \beta x + Ae^{-\alpha x}\sum_{i<j}\vec{\sigma}_i\vec{\sigma}_j + tensorint, \tag{12}$$

the linear confinement is treated as a perturbation and therefore the problem can be solved analytically. The perturbation treatment is justified for the lower levels.

C) The hypercentral Constituent Quark Model [3]

$$V_{3q} = -\frac{\tau}{x} + \alpha x + H_{hyp} \tag{13}$$

because of the confinement term the hyperradial equation (7) has to be solved numerically. The form of the hCQM Eq. (13) is supported by recent Lattice QCD calculations [9], which results in a quark-antiquark potential containing a coulomb-like term plus a linear confinement. Model C) contains only three free parameters which can be fitted in order to reproduce the experimental spectrum[3]. An improved version of the model includes also isospin dependent terms [10], leading to a very good agreement with the experimental levels, including the correct order and position of the Roper and the negative parity resonances.

2 The electromagnetic excitation of baryon resonances

The spectrum is well reproduced by many CQMs, therefore in order to distinguish among them it is necessary to consider other physical quantities of interest, such as the photocouplings, the helicity amplitudes, the strong decays and the nucleon elastic form factors.

The helicity amplitudes are defined as

$$A_M(Q^2) \;=\; \langle B, J', J'_z = M | H^t_{em} | N, J \;=\; \frac{1}{2}, J_z = M-1 \rangle \qquad M \;=\; \frac{1}{2}, \frac{3}{2} \qquad (14)$$

The transverse transition operator is assumed to be

$$H^t_{em} \;=\; -\sum_{i=1}^{3} \left[\frac{e_j}{2m_j} \left(\vec{p_j} \cdot \vec{A_j} \;+\; \vec{A_j} \cdot \vec{p_j}\right) \;+\; 2\mu_j \; \vec{s_j} \cdot (\vec{\nabla} \times \vec{A_j}) \right] \;, \qquad (15)$$

where spin-orbit and higher order corrections are neglected [11,12,13]. In Eq. (15) m_j, e_j, $\vec{s_j}$, $\vec{p_j}$ and $\mu_j \;=\; \frac{ge_j}{2m_j}$ denote the mass, the electric charge, the spin, the momentum and the magnetic moment of the j-th quark, respectively, and $\vec{A_j} \;=\; \vec{A_j}(\vec{r_j})$ is the photon field.

In the case of models B) and C), the parameters fitted to the spectrum are used and therefore the helicity amplitudes are given by parameter-free calculations, while in model A) the h.o. constant $\frac{1}{\alpha} \simeq 0.5 fm$ is adjusted in order to reproduce the amplitude $A_{3/2}$ for the D_{13} resonance at the photon point [11,12]. The photocouplings calculated with these models (and with other models as well) follow qualitatively the behaviour of the experimental data [13]: there is however generally an underestimate of the observed strength. The similarity of the results is due to the fact that the models have basically the same underlying spin-isospin symmetry. As for the transition form factors, the results for to negative parity resonances [14] show that the helicity amplitudes calculated with the h.o. potential have a Q^2 behaviour completely different from data. The models B) and C) reproduce the experimental data for medium-high Q^2, showing that the hypercoulomb interactions apparently leads to more realistic three quark wavefunctions. At low Q^2 there is often a lack of strength, in agreement to what happens at the photon point, specially in the case of the $A_{3/2}$ amplitudes. This discrepancy indicates that some mechanism, important at low Q^2, is missing, such as the quark-antiquark pair production [3,15]. In some cases, the missing strength is less evident, as for the S_{11} resonances. In Fig. 1 we show the helicity amplitude calculated with the hypercentral model [14] for the $S_{11}(1535)$ state in comparison with the model of ref. [16] and the data. It should be reminded that the curve has been published three years earlier than the recent Jlab data [18]. Also the amplitude for the $S_{11}(1650)$ state is well reproduced [14] and this is a sensible test of $SU(6)$ violation, since in absence of any configuration mixing the amplitude should be exactly zero. All these results are only slightly modified by the introduction of relativistic corrections [19].

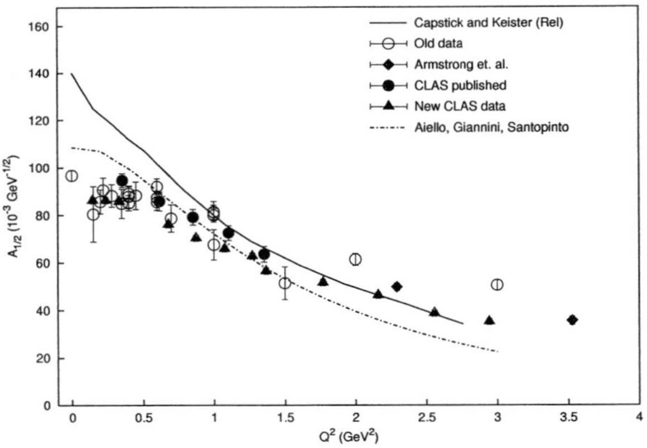

Figure 1: The helicity amplitude for the $S_{11}(1535)$ state, calculated with the hypercentral CQM [14], in comparison with the model of ref. [16] and with data, including the recent Jlab ones. (Data from a compilation by [17].)

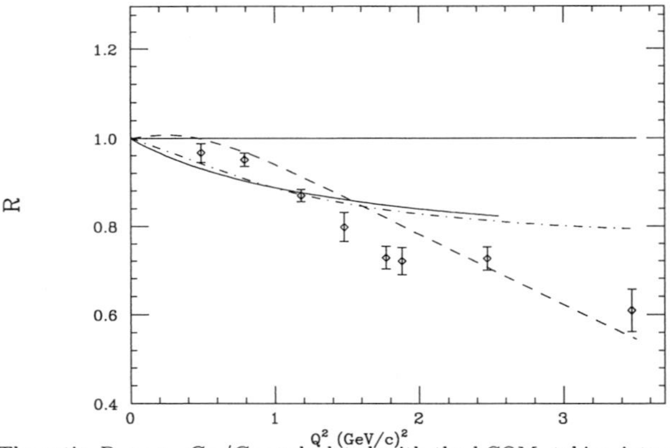

Figure 2: The ratio $R = \mu_p \, G_E/G_M$ calculated with the hCQM, taking into account the relativistic kinematical corrections (full line, ref. [21]). The horizontal full line represents the ratio for CQM without relativistic corrections. The dashed curve is the fit of ref. [24], the dot-dashed curve is the dispersion relation fit of ref. [25]. The points are the data of the recent Jlab experiment Ref. [23]

3 The nucleon elastic form factors

The hypercentral CQM has been used also for the calculation of the elastic nucleon form factors. To this end some relativistic corrections have been introduced, boosting the initial and final state to a common Breit frame and expanding the quark current at the lowest order in the quark momentum [20]. In this way it has been possible to show, for the first time in the framework of quark models [21], that the ratio R between the electric and magnetic form factors of the proton deviates from the standard dipole value ($\simeq 1$) because of relativistic effects (see Fig. 2). This behaviour is in agreement with the recent Jlab data [22].

The relativistic corrections to the hCQM have been further improved by introducing the correct relativistic kinetic energy and using a fully relativistic quark current. Introducing quark small form factors corresponding, the calculated nucleon elastic form factors describe the data very well up to $4GeV^2$ and the ratio R reaches a value of about 0.6, in better agreement with the new Jlab data [23].

1. R. Bijker, F.Iachello and A. Leviatan, Ann. Phys. (N.Y.) **236**, 69 (1994).
2. J. Ballot and M. Fabre de la Ripelle, Ann. of Phys. (N.Y.) **127**, 62 (1980).
3. M. Ferraris, M.M. Giannini, M. Pizzo, E. Santopinto and L. Tiator, Phys. Lett. **B364**, 231 (1995).
4. M. Fabre de la Ripelle and J. Navarro, Ann. Phys. (N.Y.) **123**, 185 (1979); J.-M. Richard, Phys. Rep. **C 212**, 1 (1992).
5. E. Santopinto, M.M. Giannini and F. Iachello, in "Symmetries in Science VII", ed. B. Gruber, Plenum Press, New York, 445 (1995); F. Iachello, in "Symmetries in Science VII", ed. B. Gruber, Plenum Press, New York, 213 (1995).
6. N. Isgur and G. Karl, *Phys. Rev.* **D18**, 4187 (1978); *Phys. Rev.* **D19**, 2653 (1979); S. Godfrey and N. Isgur, Phys. Rev. **D32**, 189 (1985); S. Capstick and N. Isgur, Phys. Rev. **D 34**,2809 (1986).
7. E. Santopinto, F. Iachello and M.M. Giannini, Nucl. Phys. **A623**, 100c (1997).
8. E. Santopinto, F. Iachello and M.M. Giannini, Eur. Phys. J. **A1**, 307 (1998).
9. Gunnar S. Bali et al., Phys. Rev. **D62**, 054503 (2000); Gunnar S. Bali, Phys. Rep. **343**, 1 (2001).
10. M.M. Giannini, E. Santopinto and A, Vassallo, Eur. Phys. J. **A12**, 447 (2001).
11. L. A. Copley, G. Karl and E. Obryk, Phys. Lett. **29**, 117 (1969).
12. R. Koniuk and N. Isgur, Phys. Rev. **D21**, 1868 (1980).
13. M. Aiello, M. Ferraris, M.M. Giannini, M. Pizzo and E. Santopinto, Phys. Lett. **B387**, 215 (1996).
14. M. Aiello, M. M. Giannini, E. Santopinto, J. Phys. G: Nucl. Part. Phys. **24**, 753 (1998).
15. F. Iachello, Proceedings of the 1st International Conference on oerspectives in Hadronic Physics, ICTP, Trieste 12-16 May 1997 (S. Boffi, C. Ciofi degli Atti and M.M. Giannini eds.), World Scientific, Singapore 1998,363.
16. S. Capstick and B.D. Keister, Phys. Rev. **D51**, 3598 (1995).
17. V. Burkert, arXiv hep-ph/0207149
18. R.A. Thompson et al., Phys. Rev. Lett. 86, 1702 (2001).
19. M. De Sanctis, E. Santopinto, M.M. Giannini, Eur. Phys. J. **A1**, 187 (1998).
20. M. De Sanctis, E. Santopinto, M.M. Giannini, Eur. Phys. J. **A2**, 403 (1998).
21. M. De Sanctis, M.M. Giannini, L. Repetto, E. Santopinto, Phys. Rev. **C62**,025208 (2000).
22. M.K. Jones et al., Phys. Rev. Lett. **B84**,1398 (2000).
23. O. Gayou et al., Phys. Rev. Lett. **88**, 092301 (2002).
24. F. Iachello, A. D. Jackson and A. Lande, Phys. Lett. **B43**, 191 (1973).
25. H.-W. Hammer, U.-G. Meissner and D. Drechsel, Phys. Lett. **B385**, 343 (1996); P. Mergell et al., Nucl. Phys. **A596**, 367 (1996).

MANIFESTATION OF SYMMETRY PROPERTIES OF NUCLEON STRUCTURE IN STRONG AND ELECTROMAGNETIC PROCESSES

EGLE TOMASI-GUSTAFSSON AND MICHAIL P. REKALO *
DAPNIA/SPhN, CEA/Saclay, 91191 Gif-sur-Yvette Cedex, France
E-mail: etomasi@cea.fr

In this contribution we present a specific application of a result obtained by Franco Iachello (in collaboration with R. Bijker and A. Leviatan), which concerns the inelastic electromagnetic form factors on the nucleons. In particular we show examples where symmetries inherent to the structure of the nucleon resonances can manifest in complicated processes of the strong interaction.

1 Introduction

Although nucleon resonances are preferentially investigated with electromagnetic probes (due to the fact that the reaction mechanism, the one-photon exchange, is well know), there are also advantages in using hadronic probes, such as large cross sections, absence of radiative corrections, development of good polarized beams, targets, polarimeters, and, in case of α and deuterons, isoscalar selectivity.

We will show through the example of two reactions $\vec{d}+p \to d+X$ and $p+\alpha \to p+\alpha+\pi^0$, how polarization phenomena, together with adequate kinematical choices, can sort out unambiguous information on nucleon resonances.

For an incident energy of a few GeV, over the threshold for the excitation of the Roper resonance, the cross section, in these reactions, shows two structures, one related to the coherent excitation of the target, with pion production (Deck effect[1]), and a second one, of interest here, related to the N^*-excitation. These mechanisms can be described by t-channel exchange of isovector mesons, in case of Deck mechanism, and of isoscalar mesons for nucleon resonances.

The most probable mechanism to describe the N^*-excitation is, in our opinion, the ω-exchange: the $\omega NN-$ coupling is large; the exchange of spin one particles allows to obtain large polarization phenomena and an energy independent cross section. The ω-meson can be considered as as 'isoscalar photon', therefore the cross section and the polarization observables can be related through the vector dominance model (VDM) to the electromagnetic properties of the hadrons.

2 Symmetry and reaction mechanism: $\vec{d}+p \to d+X$

The inclusive production of deuterons in $\vec{d}+p \to d+X$ is dominated by one-nucleon exchange in backward scattering, and by t-exchanges at forward angles. For Roper excitation in framework of ω-exchange, the tensor analyzing power in $d+p \to d+X$,

*PERMANENT ADDRESS: NATIONAL SCIENCE CENTER KFTI, 310108 KHARKOV, UKRAINE

T_{20}, can be written in terms of the electromagnetic form factors (FFs) as[2]:

$$T_{20} = -\sqrt{2}\frac{V_1^2 + (2V_0V_2 + V_2^2)r(t)}{4V_1^2 + (3V_0^2 + V_2^2 + 2V_0V_2)r(t)}, \qquad (1)$$

where $V_0(t)$, $V_1(t)$ and $V_2(t)$ are related to the standard electromagnetic deuteron FFs, G_c (electric), G_m (magnetic) and G_q (quadrupole) by:

$$V_0 = \sqrt{1+\eta}\left(G_c - \frac{2}{3}\eta G_q\right),\; V_1 = \sqrt{\eta}G_m, V_2 = \frac{\eta}{\sqrt{1+\eta}}\left[-G_c + 2\left(1 - \frac{1}{3}\eta\right)G_q\right],$$

and $\eta = -t/4M^2$. The ratio $r(t)$ characterizes the relative role of longitudinal and transversal isoscalar excitations in the transition $\omega + N \to X$. For any nucleon resonance N^* it can be written as follows:

$$r(t) = \frac{|A_S^p + A_S^n|^2}{|A_{1/2}^p + A_{1/2}^n|^2 + |A_{3/2}^p + A_{3/2}^n|^2} \equiv \sigma_L(t)/\sigma_T(t), \qquad (2)$$

where $A_{1/2}$ and $A_{3/2}$ are the two possible transversal FFs, corresponding to total $\gamma^* + N$-helicity equal to 1/2 and 3/2.

From Eq. (1) it appears that: - all information about the ωNN^*-vertex is contained in the function $r(t)$ only; - T_{20} is especially sensitive to the small value of $r(t$ in the interval $0.0 \leq r(t) \leq 0.5$; - a zero value of $r(t)$ results in a $t-$independent value $T_{20} = -1/2\sqrt{2}$, for any value of the deuteron electromagnetic FFs; - the position of points where $T_{20} = 0$ is determined by the model for the deuteron FF. In Fig. 1 we report the theoretical predictions, using Eq. (1), together with the existing experimental data. In such approximation T_{20} is a universal function of t only, without any dependence on the initial deuteron momentum. The experimental values of T_{20} for $p(\vec{d}, d)X$ [3,4], for different momenta of the incident beam are shown as open symbols. These data show a scaling as a function of t, with a small dependence on the incident momentum, in the interval 3.7-9 GeV/c. On the same plot the data for the elastic scattering process $e^- + d \to e^- + d$ [5] are shown (filled stars). All these data show a very similar behavior: negative values, with a minimum in the region $|t| \simeq 0.35\ GeV^2$, increasing toward zero at larger $|t|$.

The full line is the result of the ω-exchange model for the $d + p \to d + X$ process, taking into account the resonances: $S_{11}(1535)$, $D_{13}(1520)$ and $S_{11}(1650)$ which overlap in this energy region. When $r \gg 0$ or if the contribution of the deuteron magnetic FF $V_1(t)$ is neglected, then T_{20} does not depend on the ratio r, and coincides with t_{20} for the elastic *ed*-scattering (with the same approximation).

The deuteron FFs have been taken from [6]. The values of r, are predicted by the algebraic string model of baryons [7] and give a very good description of the data, when taking into account the contribution of all considered resonances (Fig. 2).

One can see that, of the four resonances, only the Roper resonance has a nonzero isoscalar longitudinal FF. The isoscalar longitudinal amplitudes of $S_{11}(1535)$ and $D_{13}(1520)$ vanish because of spin-flavor symmetry, while both isoscalar and isovector longitudinal couplings of $S_{11}(1650), D_{15}(1675)$ and $D_{13}(1700)$ vanish identically. This behavior of the isoscalar FFs is essential for the correct description of the existing experimental data on the $t-$ dependence of T_{20} for the process $d+p \to d+X$.

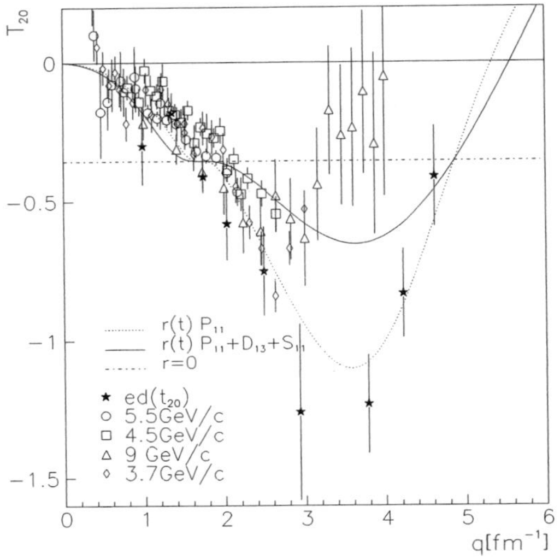

Figure 1. Experimental data on T_{20} for $d+p \to d+X$ at incident momenta of 3.75 GeV/c (open diamond) [3] 5.5 GeV/c (open circles), 4.5 GeV/c (open squares), and 9 GeV/c (open triangles) [4]. The prediction of the $\omega-$exchange model for $r=0$ is represented by the dashed-dotted line. The calculation with r from [7] is represented by the dotted line for the Roper excitation and by the solid line for the excitation of the resonances quoted in the text. The t_{20} data from *ed* elastic scattering (filled stars) are from [5].

From Fig. 1 it appears that the $t-$behavior of T_{20} is very sensitive to the value of r especially at relatively small r, $r \leq 0.5$. These data, in any case, exclude a very small value of r, $r \ll 0.1$ as well as very large values of r. Such sensitivity of T_{20} to the ratio of the corresponding isoscalar FFs of the N^*-excitation gives an evident indication of the excitation of the Roper resonance in this process.

This description depends on how the resonances are excited and not on their decays. An exclusive experiment, last from LNS, shows that T_{20} barely change selectioning one or two pions final state [8].

A temptative was done to select by the effective mass, the region where one of the resonances is predominantly excited [9]. T_{20} for different mass regions, for different incident beam momenta, was measured in the reaction $d+Be \to d+X$, at small forwaed angle. The results are nicely in agreement with our predictions, in the region of the Roper resonance, as well as outside, where the longitudinal FF vanishes and from Eq. (1), we find $T_{20} \simeq 0.35$.

2.1 Exclusive processes: polarization phenomena in $\vec{p}+d \to \vec{p}+M^0+d$, $M^0=\sigma$ *or* π

The polarization properties of the produced protons in the processes $\vec{p}+d \to \vec{p}+M^0+d$, $M^0=\sigma$, η or π^0 depend essentially on the kind of the produced meson and on the quantum numbers ($\mathcal{J}^P$) of the nucleonic resonance in the intermediate

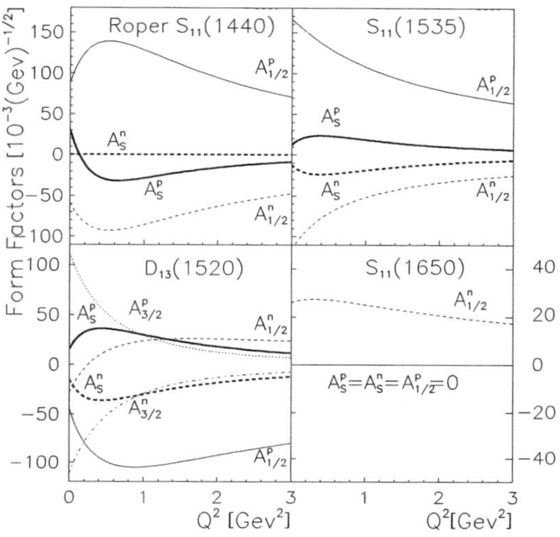

Figure 2. Longitudinal and transversal electromgnetic FFs in units $[10^{-3}\mathrm{GeV}^{-1/2}]$, as a function of Q^2 $[\mathrm{GeV}^2]$ for the four considered resonances. The thick solid (dashed) line is the longitudal FF for proton(neutron). Corresponding thin lines are the transversal FFs, for helicity 1/2. The dotted (dash-dotted) lines are transversal FF for helicity 3/2, for proton and neutron respectively.

state: $p + d \to N^*(\mathcal{J}^P) + d \to p + M^0 + d$. Let us consider the case of Roper excitation, with $\mathcal{J}^P = \frac{1}{2}^+$.

In framework of ω-exchange, the polarization transfer coefficients which do not vanish in case of σ-production: $\vec{p} + d \to \vec{p} + \sigma + d$ are :

$$K_y^{y'} = K_x^{x'} = \frac{\mathcal{R}}{4+\mathcal{R}},\ K_z^{z'} = -\frac{-4+\mathcal{R}}{4+\mathcal{R}},\ \mathcal{R}^{-1} = \frac{V_1^2}{(3V_0^2 + V_2^2 + 2V_0V_2)r(t)}, \tag{3}$$

i.e. these coefficients depend only on the momentum transfer t.

On the other hand a dependence from the M^0-production angle is contained in the decays $N^*(1/2^+) \to p + \pi^0$ (or $p + \eta$), with P-wave production of the pseudoscalar meson. In this case we all $K_a^{a'}$ coefficients allowed by the P-invariance of strong interaction are nonzero:

$$K_y^{y'} = -\frac{\mathcal{R}}{4+\mathcal{R}},\ K_x^{x'} = (1 - 2cos^2\theta_\pi)\frac{\mathcal{R}}{4+\mathcal{R}},\ K_z^{z'} = (1 - 2cos^2\theta_\pi)\frac{4-\mathcal{R}}{4+\mathcal{R}},$$

$$K_x^{z'} = (sin\ 2\theta_\pi)\frac{-4+\mathcal{R}}{4+\mathcal{R}},\ K_z^{x'} = (sin\ 2\theta_\pi)\frac{\mathcal{R}}{4+\mathcal{R}},$$

where we used a coordinate system with the $z-$axis along the momentum transfer $\vec{k}$, $y||\vec{n}$, with $\hat{\vec{n}} = \vec{k} \times \vec{q}/|\vec{k} \times \vec{q}|$, ($\vec{q}$ is the meson 3-momentum) and $x||\vec{n} \times \vec{k}$.

An important property of the $\omega-$exchange model is the universal dependence of all the $K_a^{a'}$ from t, through the ratio $\mathcal{R}(t)$. An experimental check of the θ_π-

dependence of the polarization transfer coefficient would be a signature of the validity of this model.

3 Symmetry and kinematics: the reaction $p + \alpha \to p + \alpha + \pi^0$

We give here a general formalism for the study of polarization phenomena in three body reactions and derive general properties. The application to $p + \alpha \to p + \pi^0 + \alpha$ and in particular to polarization phenomena for different meson exchanges in Roper excitation will allow to conclude that out-of-plane measurements can sort out the reaction mechanism [11].

The main feature of a process with three particles in final state: 1+2 $\to$ 3+4+5 is the non-coplanarity of the kinematics. This can be expressed, for the case of $p+\alpha \to p + \pi^0 + \alpha$, introducing the following combination of 3-momenta: $a = \dfrac{\vec{q} \cdot \vec{p}_1 \times \vec{p}_2}{E_1 E_2 E_\pi}$, where $\vec{p}_1$ and $\vec{p}_2$ are the three momenta of the initial and final proton, $\vec{q}$ is the 3-momentum of the produced pion and E_1, E_2, E_π are the corresponding energies. This expression enters in the definition of five independent kinematical variables which are necessary for the complete description of a process $1 + 2 \to 3 + 4 + 5$.

The variable a is connected with the azymuthal angle ϕ, between the two reaction planes which characterize the process $p + \alpha \to p + \pi^0 + \alpha$: the first is the scattering plane of the proton (i.e. the plane defined by the 3-momenta $\vec{p}_1$ and $\vec{p}_2$) and the second is the plane defined by the pion three-momentum $\vec{q}$ and the transferred momentum $\vec{p} = \vec{p}_1 - \vec{p}_2$. The angle ϕ can be identified with the Treiman-Yang angle [12], which is currently used in the description of the properties of one-meson exchange in high energy collisions. This angle is also convenient for the description of the possible mechanisms for the Roper excitation, in $p + \alpha \to p + \pi^0 + \alpha$. The parameter a is not only a pseudoscalar quantity, but it is a T-odd variable, as it is the product of three 3-momenta.

The non-coplanarity of the general kinematics for $1 + 2 \to 3 + 4 + 5$ results in specific properties of the polarization phenomena, different from the binary collisions. For example, in $p + \alpha \to p + \pi^0 + \alpha$, the vector of polarization of the final proton can have, in the general case, all non-zero components, whereas, for any binary process, the proton polarization (for a P-invariant interaction) has only one non-zero component along the normal to the scattering plane, due to the presence of only one reaction plane.

The presence of the non-coplanarity ($a \neq 0$) has to be taken into account in establishing the spin structure of the matrix element for $p + \alpha \to p + \pi^0 + \alpha$. If the P-invariance of the strong interaction holds, the matrix element is described by the following general parametrization (in the CMS of the considered reaction):

$$\mathcal{M} = \chi_2^\dagger \left[\vec{\sigma} \cdot \vec{m} f_1 + \vec{\sigma} \cdot \vec{k} f_2 + a \left(i \tilde{f}_1 + \vec{\sigma} \cdot \vec{n} \tilde{f}_2 \right) \right] \chi_1, \tag{4}$$

where χ_1 and χ_2 are the 2-component spinors of the protons in the initial and final states; f_1, f_2, $\tilde{f}_1$ and $\tilde{f}_2$ are the scalar independent amplitudes for $p+\alpha \to p+\pi^0+\alpha$, which are functions of the 5 kinematical variables; the unit vectors $\vec{m}$, $\vec{n}$ and $\vec{k}$ are defined as: $\vec{n} = \vec{p}_1 \times \vec{p}_2 / |\vec{p}_1 \times \vec{p}_2|$, $\vec{k} = \vec{p}_1 / |\vec{p}_1|$, $\vec{m} = \vec{n} \times \vec{k}$, The P-invariance of the strong interaction requires that all the amplitudes are even functions of the variable a.

From Eq. 4 we can calculate any polarization observable in terms of the scalar amplitudes and of the parameter a, analyzing powers, polarization transfer etc.. For example, the dependence of the differential cross section on the polarization $\vec{P}$ of the proton beam, in the general case of non-coplanar kinematics, is characterized by three independent analyzing powers, i.e.:

$$\frac{d\sigma}{d\omega}(\vec{p}\alpha \to p\pi^0\alpha) = \left(\frac{d\sigma}{d\omega}\right)_0 [1 + P_n A_n + a(P_m A_m + P_k A_k)], \tag{5}$$

where $\left(\frac{d\sigma}{d\omega}\right)_0$ is the differential cross section (with unpolarized proton beam), $d\omega$ is the element of the phase space for the 3-particle final state and P_n, P_m and P_k (A_n, A_m and A_k) are the three independent and non-zero components of the initial proton polarization vector (analyzing powers). The components P_m and P_k appear multiplied by the parameter a, therefore non contributing in the case of coplanar kinematics.

The dependence of the components of the final proton polarization $\vec{P}_f$ on the initial polarization $\vec{P}$ can be parametrized in the following way:

$$\begin{aligned} \vec{m}\cdot\vec{P}_f &= aD_{mm}\vec{m}\cdot\vec{P} + D_{mk}\vec{k}\cdot\vec{P} + D_{mn}\vec{n}\cdot\vec{P}, \\ \vec{n}\cdot\vec{P}_f &= aD_{nm}\vec{m}\cdot\vec{P} + aD_{nk}\vec{k}\cdot\vec{P} + D_{nn}\vec{n}\cdot\vec{P}, \\ \vec{k}\cdot\vec{P}_f &= D_{km}\vec{m}\cdot\vec{P} + D_{kk}\vec{k}\cdot\vec{P} + aD_{kn}\vec{n}\cdot\vec{P}, \end{aligned} \tag{6}$$

where D_{ij}, $i,j = m,n,k$, are the coefficients of polarization transfer from the initial to the final proton.

Let us give here only the expressions for the coefficient D_{nn}, in terms of the scalar amplitudes f_i, $\tilde{f}_i$ and the parameter a:

$$D_{nn}\left(\frac{d\sigma}{d\omega}\right)_0 = -|f_1|^2 - |f_2|^2 + a^2\left(|\tilde{f}_1|^2 + |\tilde{f}_2|^2\right),$$

One can see that $D_{nn} = -1$, for coplanar kinematics ($a = 0$). This is a known result, which follows from the P-invariance of strong interaction and this result is valid for any amplitudes f_1 and f_2 and for any model of the considered process and for any kinematical conditions, provided $a = 0$. In non-coplanar kinematics, in general, the presence of non-coplanar amplitudes gives $D_{nn} \geq -1$, therefore the quantity $1 + D_{nn}$ characterizes the relative role of non-coplanar amplitudes $\tilde{f}_1$ and $\tilde{f}_2$.

From the general properties of $\pi\alpha$-scattering (for the Deck mechanism) and of the process $\sigma + N \to N + \pi$ (for the Roper excitation), one can show that the non-coplanar amplitudes $\tilde{f}_1$ and $\tilde{f}_2$ are zero for both mechanisms, independently on their parametrizations. The numerical values of the amplitudes f_1 and f_2, and their dependence on the kinematical variables are different for σ and π-exchanges, but for any amplitudes f_1 and f_2 the polarization phenomena have some general properties: - $D_{nn} = 1$, in the whole region of kinematical variables (for coplanar and non-coplanar kinematics); - the polarization of the final proton has only one non-zero component, in the $\vec{n}$-direction, i.e. along the normal to the proton scattering plane; - the sign and absolute value of this component depend on the relative role of the considered mechanisms, and this dependence is very sensitive to the details of the corresponding amplitudes.

This 'coplanar-like' behavior of σ -and π- exchanges in $p + \alpha \to p + \pi^0 + \alpha$ is related to the fact that these mediators are spinless particles. Such mechanisms can not connect different reaction planes. This conclusion does not depend on details, approximations, values of the constants or shape of FFs which are typically taken in the numerical applications, because it is based only on the value of the spin of the exchanged particles.

The most important difference of ω-exchange with respect to σ-exchange for the Roper excitation is due to the spin and has evident implications for the polarization phenomena: a vector particle exchange induces all four amplitudes different from zero, in the general case.

Let us consider the spin structure of $\omega-$exchange, taking into account, for simplicity, in the ωNN^*-vertex only the transverse, i.e. M1 form factor. In this case the matrix element for $p + \alpha \to p + \pi^0 + \alpha$ can be written in the following form:

$$\mathcal{M}_\omega = \chi_2^\dagger \vec{\sigma} \cdot \vec{q} \vec{\sigma} \cdot \vec{k}_1 \times \vec{k}_2 \chi_1 F_\alpha(t) F_{NN^*}(t) \frac{1}{t - m_\omega^2} \frac{f_{N^*}}{w_1 - m^* + i\frac{\Gamma}{2}}, \qquad (7)$$

where $\vec{k}_1$ and $\vec{k}_2$ are the 3-momenta of the initial and final α-particles in CMS of the considered reaction, $F_\alpha(t)$ and $F_{NN^*}(t)$ are the FFs of the $\omega\alpha\alpha$ and ωNN^*-vertexes, f_{N^*} is the constant for the decay $N^* \to N + \pi$, m^* and Γ are the mass and the width of the Roper resonance N^*. For ω-exchange all four scalar amplitudes, coplanar and non-coplanar, are present. This is due to the exchange by vector particles, which connects strongly the different planes of the considered reaction and it depends only on the spin 1 nature of the exchanged particles. In general, ω-exchange, induces acoplanarity and, therefore, deviations from the relation $D_{nn} + 1 = 0$.

4 Conclusions

High energy $d + p \to d + X$-reactions, which are driven by the strong interaction and electromagnetic processes as elastic $e + d$ scattering and electroexcitation of nucleonic resonances, $e + N \to e + N^*$, can be treated in a unified dynamical description. We show that the inelastic electromagnetic FFs of protons and neutrons, calculated in the algebraic model, can be successfully applied to the field of the hadronic interaction for quantitative predictions of different polarization observables. As an example we predict polarization phenomena for forward deuteron emission, in the GeV range, in $d + p \to d + X$, where X contains the possible nucleonic resonances, starting from the Roper (the Δ-resonance being forbidden by the conservation of the isotopic spin). We showed that the tensor analyzing power in this process, is especially sensitive to the Roper resonance excitation. This is due to the fact that among the resonances which can be excited in this kinematical region, only the Roper resonance has a non zero longitudinal isoscalar FF, due to the symmetry properties of the quark structure of these resonances, which are implicitely contained in the algebraic model. We generalized this model for exclusive channels as $d + p \to p + \pi + d$, $d + p \to p + \sigma + d$ and $\alpha + p \to \alpha + p + \pi^0$. We can consider the existing data about T_{20} not only as an evidence for the Roper resonance excitation, but also a specific indication of the properties of the isoscalar form factors for the excitation of N^* resonances, complementary to the inelastic electron-nucleon scattering, $e^- + N \to e^- + N^*$.

The possibility to unify in a common picture such different processes, as $e^- + d \to e^- + d$ and $e^- + N \to e^- + N^*$, from one side, and a hadronic process as $d + p \to d + X$, from another side, suggests a new perspective to study nucleon structure through electromagnetic and hadron excitation of nucleonic resonances.

In the process $p + \alpha \to p + \pi^0 + \alpha$, we showed that the matrix element for σ−exchange, often advocated to describe the Roper excitation and for the π-exchange (Deck-mechanism), has an evident 'coplanar-like' form, with vanishing non-coplanar amplitudes $\tilde{f}_1$ and $\tilde{f}_2$. But the ω-exchange (which seems the most probable physical candidate for the Roper excitation) induces a very rich spin structure of the corresponding contribution to the matrix element (with all four non-zero amplitudes), and specific polarization phenomena, which differ essentially from the case of σ−exchange. Only ω-exchange can induce non-coplanar polarization phenomena.

Future experimental data on polarization observables for $p + \alpha \to p + \pi^0 + \alpha$, which require a detection system in non-coplanar kinematics, will constitute a crucial test in order to disentangle the mechanisms involved.

Experiments, which confirm these predictions, were done in Saturne and Dubna. Further measurements are planned at COSY.

References

1. R.T. Deck, Phys. Rev. Lett. **13**, 169 (1964).
2. M. P. Rekalo and E. Tomasi-Gustafsson, Phys. Rev. C54, 3125 (1996).
3. Exp. LNS250 (unpublished).
4. L.S. Azghirey et al., Phys. Lett. B **361**, 21 (1995); Phys. Lett. B **387**, 37 (1996); JINR Rapid Communications No. 2[88]-98.
5. D. Abbott *et al.* Phys. Rev. Lett. **84**, 5053 (2000) and refs. herein.
6. P.L. Chung et al., Phys. Rev. C **37**, 2000 (1988).
7. R. Bijker, F. Iachello, A. Leviatan, Annals Phys. 236, 69 (1994).
8. L.V. Malinina et al. (The SPES4-pi Collaboration), Phys. Rev. C64, 064001 (2001).
9. V.P. Ladygin et al , Eur. Phys. J. A8, 409 (2000).
10. E. Tomasi-Gustafsson, M.P. Rekalo, R. Bijker, A. Leviatan, F. Iachello, Phys. Rev. C59, 1526 (1999).
11. M. P. Rekalo and E. Tomasi-Gustafsson, Phys. Rev. C63, 054001 (2001).
12. S. B. Treiman and C. N. Yang, Phys. Rev. Lett. **8**, 140 (1962).

REGULARITY AND CHAOS IN LOW–LYING 2^+ STATES OF EVEN–EVEN NUCLEI

A. Y. ABUL–MAGD
Faculty of Science, Zagazig University, Zagazig, Egypt

H. L. HARNEY
Max–Planck–Institut für Kernphysik, Heidelberg, Germany

M. H. SIMBEL
Faculty of Science, Zagazig University, Zagazig, Egypt

H. A. WEIDENMÜLLER
Max–Planck–Institut für Kernphysik, Heidelberg, Germany

Using all the available empirical information, we analyse the spacing distributions of low-lying 2^+ levels in even–even nuclei by comparing them with a theoretical distribution characterized by a single parameter (the chaoticity parameter f). We use the method of Bayesian inference. We show that the necessary unfolding procedure generally leads to an overestimate of f. We find that f varies strongly with the ratio $R_{4/2}$ of the excitation energies of the first 4^+ and 2^+ levels and assumes particularly small values in nuclei that have one of the dynamical symmetries of the Interacting Boson Model.

1 Introduction

The interplay between regular and chaotic motion in nuclei has been a long–standing problem in Nuclear Physics. There is, on the one hand, overwhelming evidence in favour of simple dynamical models especially in the ground–state domain. The evidence derives from the agreement between calculated and measured spectral properties. There is, on the other hand, equally strong evidence for the validity of a random–matrix description, especially from the spectral statistics of slow neutron resonances [1,2]. This success of random–matrix theory negates a dynamical description in terms of simple and (nearly) integrable models and has raised the question: Where in the spectrum of a nucleus with mass number A does the chaotic region start? The statistical analysis of spectra needed to answer this question requires complete (few or no missing levels) and pure (few or no unknown spin–parities) level schemes. Some 15 years ago, complete and pure level schemes were available for only a limited number of nuclei (see, e.g., Refs. [3,4]). The work of Ref. [5] then suggested that the nearest–neighbour spacing (NNS) distribution of low–lying nuclear levels lies between the Wigner and the Poisson distributions which are characteristic, respectively, of fully regular and fully chaotic motion. Through the work of Refs. [6,7,8,9,10,11,12], the evidence presented in Ref. [5] has since become an established fact.

The wealth of spectroscopic data now available in the Nuclear Data tables [13] has motivated us to investigate once again the nuclear ground–state domain. We are able to make more definitive and precise statements about regularity versus chaos in this domain than has been possible so far. As in Ref. [5], we focus attention on 2^+

states of select even–even nuclei. These nuclei are grouped into classes. The classes are defined in terms of the ratio $R_{4/2}$, i.e., the ratio of the excitation energies of the first 4^+ and the first 2^+ level in each nucleus. We argue below that the classes define a grouping of nuclei that have common collective behaviour. The sequences of 2^+ states are unfolded and analysed with the help of Bayesian inference. The chaoticity parameter f defined below is determined for each class. The present paper summarizes two research papers [14,15] where further details may be found.

2 Data Set and Classification of Nuclei

The data on low–lying 2^+ levels of even–even nuclei are taken from the compilation by Tilley *et al.* [16] for mass numbers $16 \leq A \leq 20$, from that of Endt [17] for $20 \leq A \leq 44$, and from the Nuclear Data Sheets [13] for heavier nuclei. We considered nuclei for which the spin–parity J^π assignments of at least five consecutive 2^+-levels are unambiguous. In cases where the spin-parity assignments were uncertain and where the most probable value appeared in brackets, we accepted this value. We terminated the sequence when we arrived at a level with unassigned J^π, or when an ambiguous assignment involved a 2^+ spin–parity among several possibilities, as e.g. $J^\pi = (2^+, 4^+)$. We made an exception when only one such level occurred and was followed by several unambiguously assigned levels containing at least two 2^+ levels, provided that the ambiguous 2^+ level is found in a similar position in the spectrum of a neighboring nucleus. However, this situation occurred for less than 5% of the levels considered. In this way, we obtained 1306 levels of spin–parity 2^+ belonging to 169 nuclei. The composition of this ensemble is as follows: 5 levels from each of 47 nuclei, 6 levels from each of 32 nuclei, 7 levels from each of 22 nuclei, 8 levels from each of 22 nuclei, 9 levels from each of 16 nuclei, 10 levels from each of 14 nuclei, 11 levels from each of 5 nuclei, 12 levels from each of 2 nuclei, and sequences of 13, 14, 15, 17, 20, 21, 24, 30, and 32 levels, each belonging to a single nucleus.

A class of nuclei is defined by choosing an interval within which the ratio

$$R_{4/2} = E(4_1^+)/E(2_1^+) \tag{1}$$

of excitation energies of the first 4^+ and the first 2^+ excited states, must lie. The width of the intervals was taken to be 0.1 when the total number of spacings falling into the corresponding class was about 100 or more. Otherwise, the width of the interval was increased. The use of the parameter (1) as an indicator of collective dynamics is justified both empirically and by theoretical arguments. We recall the reasons in turn.

(i) Casten *et al.* [18] plotted $E(4_1^+)$ versus $E(2_1^+)$ for all nuclei with $38 \leq Z \leq 82$ and with $2.05 \leq R_{4/2} \leq 3.15$. The authors found that the data fall on a straight line. This suggests that nuclei in this wide range of Z–values behave like anharmonic vibrators with nearly constant anharmonicity. As the ratio $R_{4/2}$ approaches the rotor limit $R_{4/2} = 3.33$, the slope of the curve showing $E(4_1^+)$ versus $E(2_1^+)$ decreases within a narrow range of $E(2_1^+)$–values, asymptotically merging the rotor line of slope 3.33. In a subsequent paper [19] it was found that a linear relation between $E(4_1^+)$ and $E(2_1^+)$ holds for pre–collective nuclei with $R_{4/2} < 2$.

Thus, from an empirical perspective, the dynamical structure of medium–weight and heavy nuclei can be quantified in terms of $R_{4/2}$.

(ii) Theoretical calculations based on the Interacting Boson Model (the IBM–1 model [20]) support the conclusion that $R_{4/2}$ is an appropriate measure for collectivity in nuclei. The model has three dynamical symmetries, obtained by constructing the chains of subgroups of the $U(6)$ group that end with the angular momentum group $SO(3)$. The symmetries are labeled by the first subgroup appearing in the chain which are $U(5)$, $SU(3)$, and $O(6)$ corresponding, respectively, to vibrational, rotational and γ-unstable nuclei. Extensive numerical calculations for the classical as well as the quantum-mechanical IBM Hamiltonian by Alhassid *et al.* [21] indeed showed a considerable reduction of the standard measures of chaoticity when the parameters of the IBM model approach one of the three cases just mentioned. The IBM calculation of energy levels yields values of $R_{4/2} = 2.00$, 3.33, and 2.50 for the dynamical symmetries $U(5)$, $SU(3)$, and $O(6)$, respectively. Thus, we may expect increased regularity of nuclei having one of these values of $R_{4/2}$.

One might expect that the chaoticity parameter f defined in Eq. (3) below also assumes small values for nuclei near magic numbers. For mass numbers in this domain, our data set is unfortunately too small to allow us to draw definitive conclusions.

3 Statistical Analysis

3.1 Chaoticity Parameter f

To analyze the data, we need a guess for the form of the NNS distribution $p(s, f)$. Here, s is the spacing of neighboring levels in units of the mean level spacing. The distribution $p(s, f)$ depends on one or more parameters f which describe the transition from Poissonian to Wigner–Dyson form. Several proposals have been advocated for $p(s, f)$. Here we are guided by the following considerations.

We consider a spectrum S containing levels which have the same spin and parity but may differ in other conserved quantum numbers which are either unknown or ignored. The K–quantum number serves as an example. The spectrum S can then be broken down into m subspectra S_j of independent sequences of levels. Let $f_j, j = 1 \ldots m$ with $0 < f_j \leq 1$ and $\sum_{j=1}^{m} f_j = 1$ denote the fractional level number, let $p_j(s)$, $j = 1 \ldots m$ denote the NNS distribution for the subspectrum S_j and $p(s)$ the NNS distribution of S. Both $p(s)$ and $p_j(s_j)$ are defined for spectra with unit mean spacing. We assume that each of the distributions $p_j(s)$ is determined by the Gaussian orthogonal ensemble (GOE). To an excellent approximation, the p_j's are then given by Wigner's surmise [23]

$$p_{\mathrm{W}}(s) = \frac{\pi}{2} s \exp\left(-\frac{\pi}{4} s^2\right) . \tag{2}$$

The construction of $p(s, f)$ for the superposition is due to Rosenzweig and Porter [24]. It depends on the $(m-1)$ unknown parameters f_j, $j = 1, ..., (m-1)$. This fact poses a difficulty because in practice, we do not know the composition of the spectrum. We are not even sure of how many quantum numbers other than spin and parity are conserved. To overcome the difficulty, we use an approximate scheme first proposed in Ref. [25]. Effectively, we replace the $(m-1)$ parameters f_j by a single one, the mean fractional level number $f = \sum_j f_j^2$. This leads to an approximate NNS distribution for S,

$$p(s,f) = \left[1 - f + f\,(0.7+0.3f)\,\frac{\pi s}{2}\right] \exp\left\{-\,(1-f)\,s - f\,(0.7+0.3f)\,\frac{\pi s^2}{4}\right\} . \tag{3}$$

We use f as a fit parameter.

For a large number m of subspectra, f is of the order of $1/m$. In this limit, $p(s,f)$ approaches the Poisson distribution as it should. On the other hand, when $f \to 1$ the spectrum approaches the GOE behaviour as it must. This is why we refer to f as to the chaoticity parameter. If the spectrum S is not pure but rather a superposition of subsequences corresponding to different values of an ignored or unknown quantum number then the mean value f of the fractional density of the superimposed sequences is smaller than unity, and the composite sequence looks rather like a sequence of levels with mixed dynamics.

3.2 Unfolding

Prior to the actual statistical analysis, every sequence of levels has to be unfolded [22] to obtain a new sequence with unit mean level spacing. In the case of a single long spectrum, unfolding is a standard procedure. It consists in fitting a slowly varying function $\epsilon(E,\alpha)$ to the experimental staircase function $N(E)$ of the integrated level density. The fit is obtained by optimizing a set of parameters α. The function ϵ depends monotonically on the energy E. Therefore, we can transform E to ϵ. With respect to the new energy variable ϵ, the level density is uniform and equal to unity.

If the available ensemble of spacings consists of many short sequences of levels (we call this a "composite ensemble"), unfolding is not standard nor is it altogether irrelevant. To test the standard unfolding procedure, we have generated short sequences of levels from three artificial ensembles containing 50, 100, and 200 spacings. Construction of the latter involves an artificially chosen chaoticity parameter f_0 and is described in the following paragraph. These are referred to as the "initial" ensembles. Each short sequence is then artificially folded with a monotonically increasing function of energy. An unfolding procedure is subsequently applied to each sequence. The unfolding procedure does not trivially reproduce the initial ensembles and yields the "final" ensembles. The chaoticity parameter f is then determined for the final ensembles using a χ^2 fit and the Bayesian method described below.

The ensembles of spacings are constructed with the help of a random–number generator. We choose average spacing unity and $f_0 = 0.6$ for the chaoticity parameter. This value is close to what has been obtained in the previous analysis [9] of low–lying nuclear levels. We generate a set of spacings that obeys the probability distribution (3) with $f = f_0$. In this way, we generate three "initial" artificial ensembles of 50, 100, and 200 spacings. Our procedure is open to the criticism that our construction does not pay attention to the stiffness of GOE spectra. We are in the process of rectifying this shortcoming.

The test of the unfolding procedure leads to the following conclusions. (i) Using several unfolding functions leads to nearly the same values for f. This confirms the insensitivity of the final ensemble of spacings to the form of the unfolding function. (ii) The unfolding procedure introduces a bias towards the GOE, i.e. the best-fit value of f is larger than f_0. This is borne out by both, the Bayesian inference and the χ^2-analysis of the spacing histograms for the final distributions. The trend increases as the lengths of the short sequences is decreased. This is simply understood: The unfolding of sequences of just two levels each would give a delta–function peaked at the value of unity (the mean level spacing) and, thus, show strong preference for the GOE. The trend becomes weaker as the sequences become longer but disappears only in the limit of very long sequences. As a consequence, the analysis of the nuclear data set will reliably yield only relative values of f.

The actual unfolding of the data was done by fitting a theoretical expression to the number $N(E)$ of levels below excitation energy E. The expression used here is the constant–temperature formula [3],

$$N(E) = N_0 + \exp\left(\frac{E - E_0}{T}\right) . \tag{4}$$

The three parameters N_0, E_0 and T obtained for each nucleus vary considerably with mass number. Nevertheless, all three show a clear tendency to decrease with increasing mass number. For the effective temperature, for example, we find, assuming a power–law dependence, the result $T = (15 \pm 4)A^{-(0.62\pm0.05)}$ MeV. This value is consistent with an analysis of the level density of nuclei in the same range of excitation energy carried out by von Egidy *et al.* [4]. These authors find $T = (19 \pm 2)A^{-(0.68\pm0.02)}$ MeV.

3.3 Bayesian Analysis

Given Eq. (3) for the proposed distribution, we apply Bayesian analysis to the data. Let $\mathbf{s} = (s_1, s_2, ..., s_N)$ denote a set of spacings s_j. We take the experimental spacings s_j to be statistically independent. This assumption does not apply in general. Indeed, the GOE produces significant correlations between subsequent spacings. However, we recall that we are interested only in the NNS distribution. This distribution is only weakly affected by correlations. We calculate the posterior distribution for f given the events $\mathbf{s}$. We first determine the conditional probability distribution $p(\mathbf{s}\,|f)$ of the set of spacings $\mathbf{s} = (s_1, s_2, ..., s_N)$ for a fixed f. We accordingly write

$$p(\mathbf{s}\,|f) = \prod_{i=1}^{N} p(s_i, f) \ , \tag{5}$$

with $p(s_i, f)$ given by Eq. (3). Bayes' theorem then provides the posterior distribution

$$P(f|\mathbf{s}) = \frac{p(\mathbf{s}|f)\mu(f)}{M(\mathbf{s})} \tag{6}$$

of the parameter f given the events $\mathbf{s}$. Here, $\mu(f)$ is the prior distribution and

$$M(\mathbf{s}) = \int_0^1 p(\mathbf{s}\,|f)\,\mu(f)\;\mathrm{d}f \tag{7}$$

is the normalization. We use Jeffreys' rule [26]

$$\mu(f) \propto \left| \int p(\mathbf{s}\,|f)\ \left[\partial \ln p(\mathbf{s}\,|f)\,/\partial f\right]^2\ \mathrm{ds}\ \right|^{1/2} \tag{8}$$

to find the prior distribution. The latter can be interpreted as the distribution ascribed to f in the absence of any observed s. It is approximated by

$$\mu(f) = 1.975 - 10.07f + 48.96f^2 - 135.6f^3 + 205.6f^4 - 158.6f^5 + 48.63f^6\ . \tag{9}$$

Even for only moderately large N, it is useful to write $p(\mathbf{s}|f)$ in the form

$$p(\mathbf{s}|f) = e^{-N\phi(f)}\ , \tag{10}$$

where

$$\phi(f) = (1-f)\langle s\rangle + \frac{\pi}{4}f(0.7+0.3f)\langle s^2\rangle - \langle \ln[1-f+\frac{\pi}{2}f(0.7+0.3f)s]\rangle\ . \tag{11}$$

Here the notation $\langle x\rangle = (1/N)\sum_{i=1}^{N} x_i$ has been used. By calculating the mean values $\langle\cdots\rangle$ in Eq. (11) for various spectra, one finds that the function $\phi(f)$ has a deep minimum, say at $f = f_1$. One can therefore represent the numerical results in analytical form by parametrizing ϕ as

$$\phi(f) = A + B(f-f_1)^2 + C(f-f_1)^3\ . \tag{12}$$

We then obtain

$$P(f|\mathbf{s}) = c\mu(f)\exp(-N[B(f-f_1)^2 + C(f-f_1)^3])\ , \tag{13}$$

where $c = e^{-NA}/M(\mathbf{s})$ is a normalization constant.

The last step of the Bayesian analysis consists in determining the best–fit value of the chaoticity parameter f and its error for each NNS distribution. When $P(f\,|\mathbf{s})$ is not Gaussian, the best–fit value of f cannot be taken as the most probable value. Rather we take the best–fit value to be the mean value $\overline{f}$ and measure the error by the standard deviation $\sigma$of the posterior distribution (6), i.e.

$$\overline{f} = \int_0^1 fP(f\mid \mathrm{s})\,\mathrm{d}f \quad \text{and} \quad \sigma^2 = \int_0^1 \left(f-\overline{f}\right)^2 P(f\mid \mathrm{s})\,\mathrm{d}f\ . \tag{14}$$

This is not optimal but provides a useful approximation.

4 Results and Discussion

The results obtained for $\overline{f}$ and σ are given in Figure 1 of Ref. [14]. Figure 2 of that reference shows a comparison of the spacing distributions conditioned by $\overline{f}$ and the histograms for each class of nuclei. In view of the small number of spacings within each class, the agreement seems satisfactory.

We recall that the analysis of many short sequences of levels tends to overestimate $\overline{f}$. Therefore, we focus attention not on the absolute values of $\overline{f}$ but on the way $\overline{f}$ changes with $R_{4/2}$. The graph of $\overline{f}$ against $R_{4/2}$ in the Ref. [14] has deep minima at $R_{4/2} = 2.0, 2.5$, and 3.3. These values of $R_{4/2}$ are associated with the dynamical symmetries of the Interacting Boson Model mentioned above. Another minimum of statistical significance occurs for $2.25 \le R_{4/2} \le 2.35$. This minimum may indicate that nuclei which lie between the limiting cases of the $U(5)$ and $O(6)$ dynamical symmetries, are relatively regular. One may associate this region with the critical point of the $U(5)$–$O(6)$ shape transition in nuclei. Iachello [27] has recently shown that this transition is approximately governed by the "critical" $E(5)$ dynamical symmetry. Nuclei with $E(5)$ dynamical symmetry have $R_{4/2} = 2.2$. Experimental examples of this critical symmetry have been found by Casten and Zamfir [28].

In summary, we have determined the chaoticity parameter f for 2^+ levels of even–even nuclei at low excitation energy with the help of a systematic analysis of the NNS distributions. While in a single nucleus the number of states with reliable spin–parity assignments is not sufficient for a meaningful statistical analysis, a combination of sequences of levels taken from similar nuclei provides a sufficiently large ensemble. As the measure of similarity we have taken the ratio $R_{4/2}$ of the excitation energies of the lowest 4^+ and 2^+ levels in each nucleus. The mean chaoticity parameter $\overline{f}$ is found to be indeed dependent on $R_{4/2}$. It has deep minima at $R_{4/2} = 2.0$, 2.5, and 3.3. These minima correspond, respectively, to the $U(5)$, $SO(6)$, and $SU(3)$ dynamical symmetries of the IBM. A further minimum may relate to the critical $E(5)$ symmetry.

Acknowledgments

The authors thank Professor J. Hüfner for useful discussions. A. Y. A.–M. and M. H. S. acknowledge the financial support granted by Internationales Büro, Forschungszentrum Jülich which permitted their stay at the Max–Planck–Institut für Kernphysik, Heidelberg.

References

1. R. U. Haq, A. Pandey, and O. Bohigas, Phys. Rev. Lett. **48**, 1086 (1982).
2. E.P. Watson III, E. G. Bilpuch, and G. E. Mitchell, Z. Phys. **A 300**, 89 (1981).
3. T. von Egidy, A. N. Behkami, and H. H. Schmidt, Nucl. Phys. **A 454**, 109 (1986).
4. T. von Egidy, H. H. Schmidt, and A. N. Behkami, Nucl. Phys. **A 481**, 189 (1988).
5. A. Y. Abul-Magd and H. A. Weidenmüller, Phys. Lett. **B 162**, 223 (1985).
6. G. E. Mitchell, E. G. Bilpuch, P. M. Endt, and J. F. Shriner, Jr., Phys. Rev. Lett. **61**, 1473 (1988); Z. Phys. **A 335**, 393 (1990).
7. S. Raman, T. A. Walkiewicz, S. Kahane, E. T. Jurney, J. Sa, Z. Gàcsi, J. Wei, K. Allaart, G. Bonsignori, and J. F. Shriner, Jr., Phys. Rev. **C 43**, 521 (1991).
8. J. F. Shriner, Jr., G. E. Mitchell, and T. von Egidy, Z. Phys. **A 338**, 309 (1991).
9. A. Y. Abul-Magd and M. H. Simbel, J. Phys. **G 22** , 1043 (1996); **24**, 576 (1998).
10. J. D. Garrett, J. Q. Robinson, A. J. Foglia, and H.-Q. Jin, Phys. Lett. **B 392**, 24 (1997).

11. J. Enders, T. Guhr, N. Huxel, P. von Neumann-Cosel, C. Rangacharyulu, and A. Richter, Phys. Lett. **B 486**, 273 (2000).
12. J. F. Shriner, C. A. Grossmann, and G. E. Mitchell, Phys. Rev. **C 62**, 054305 (2000).
13. Nuclear Data Sheets, until July 2002.
14. A. Y. Abul–Magd, H. L. Harney, M. L. Simbel and H. A. Weidenmüller, nucl-th/0212057 and Phys. Lett. B (submitted).
15. A. Y. Abul–Magd, H. L. Harney, M. L. Simbel and H. A. Weidenmüller, physics/0212049 and Phys. Rev. E (submitted).
16. D. Tilley, H. R. Weller, and C. M. Cheves, Nucl. Phys. **A 564**, 1 (1993); D. Tilley, H. R. Weller, C. M. Cheves, and R. M. Chaster, Nucl. Phys. **A 595**, 1 (1995); D. Tilley, C. M. Cheves, J. H. Kelley, S. Raman, and H. R. Weller, Nucl. Phys. **A 636**, 424 (1998).
17. P. M. Endt, Nucl. Phys. **A 633**, 1 (1998).
18. R. F. Casten, N. V. Zamfir, and D. S. Brenner, Phys. Rev. Lett. **71**, 227 (1993).
19. N. V. Zamfir, R. F. Casten, and D. S. Brenner, Phys. Rev. Lett. **72**, 3480 (1994).
20. F. Iachello and A. Arima, The Interacting Boson Model (Cambridge University Press, Cambridge 1987).
21. Y. Alhassid, A. Novoselsky, and N. Whelam, Phys. Rev. Lett. **65**, 2971 (1990); Y. Alhassid and N. Whelam, ibid **65**, 2971 (1991); Phys. Rev. **C 43**, 2971 (1991); Y. Alhassid and A. Novoselsky, ibid **45**, 1677 (1992); N. Whelam and Y. Alhassid, Nucl. Phys. **A 556**, 42 (1993).
22. O. Bohigas and M.-J. Giannoni, Lecture Notes in Physics **209**, 1 (1984).
23. E. P. Wigner, Oak Ridge National Laboratory Report No. ORNL-2309, 1957.
24. N. Rosenzweig and C. E. Porter, Phys. Rev. **120**, 1698 (1960).
25. A. Y. Abul-Magd and M. H. Simbel, Phys. Rev. **E 54**, 3292 (1996); Phys. Rev. **C 54**, 1675 (1996).
26. H. Jeffreys, Proc. of the Roy. Soc. **A 186**, 453 (1946); H. Jeffreys Theory of Probability, 3rd Edition, Oxford University Press, Oxford 1961.
27. F. Iachello, Phys. Rev. Lett. **85**, 3580 (2000); ibid **87**, 052502 (2001).
28. R. F. Casten and N. V. Zamfir, Phys. Rev. Lett. **85**, 3584 (2000).

SHAPE-PHASE AND ORDER-TO-CHAOS TRANSITIONS IN NUCLEI

GIUSEPPE MAINO

ENEA Physics Division, via Fiammelli 2, 40129 Bologna, Italy, and University of Bologna, via Mariani 5, 48100 Ravenna, Italy

E-mail: giuseppe.maino@bologna.enea.it

Nuclear dynamics is investigated in shape-transitional regions of the Interacting Boson Model - version 2 (IBM-2). Strong suppression of chaotic behavior is found even far from the dynamical symmetry limits, thus confirming the role played by partial dynamical symmetries and suggesting that also possible symmetries at the critical point may contribute to the persistence of regular motion in transitional nuclei.

1 Introduction

In mass region around A=100, nuclei such as ruthenium isotopes belong to shape-transitional chains intermediate between vibrational and gamma-unstable patterns[1]. Moreover, features typical of triaxiality are found and seem to point out this possibility[2]. Strong evidence of peculiar states like mixed-symmetry levels and relevant bands has been observed due to excitation energy patterns and electromagnetic transition properties among them[3]. On the other hand, a detailed investigation of statistical properties of level spectra has shown that in this transitional region, even far from the limiting dynamical symmetries of nearly exact vibrator or gamma-unstable nucleus (corresponding to the U(5) and O(6) algebras of the Interacting Boson Model, IBM), the quantum chaos does not occur and only a fraction of symmetry breaking is observed[4]. This unexpected characteristic has been possibly explained by resorting to partial dynamical symmetries, whose presence in this mass region is well known.

Another possibility lies in the recently proposed examples of new symmetries at shape-transition critical point corresponding to phase transitions[5]. In any case, nuclei in this mass region as well as in the A=190-200 range present very interesting and important features whose analysis can provide useful information and a deeper insight into collective properties of quantum many-body systems. A detailed study based on an extended comparison between experimental and theoretical data about spectroscopic properties for a long isotope chain represents the needed approach to clarify these problems. In this work, the shape-transitional chains between the dynamical symmetries of the IBM-2 are studied in order to identify regions of regular motion even when large breaking of the limiting symmetries occurs, thus pointing out the possible existence of hidden symmetries preventing the onset of quantum chaos.

2 Theoretical calculations and data analysis

A quantum investigation is presented of the regular and chaotic dynamics of medium- and heavy-mass ($A > 40$) even-even nuclei within the framework of the interacting boson model version 2 (IBM-2) [6], which is a realistic model of nuclear structure where the isospin degree of freedom is explicitly introduced. This analysis has been performed by studying the fluctuation properties (departures from spectral uniformity) of nuclear levels and electromagnetic transition intensities by means of the Random Matrix Theory (RMT) [7]. In particular, use has been made of the Gaussian Orthogonal Ensemble (GOE) since chaotic many-body systems with time reversal symmetry (like nuclei) are associated with it.

Nuclei belonging to shape-transitional chains are described by means of the IBM-2 Hamiltonian in the usual Talmi form:

$$H_{TALMI} = E_0 + \varepsilon_\pi \widehat{n}_{d_\pi} + \varepsilon_\nu \widehat{n}_{d_\nu} + k\, \widehat{Q}^\chi_\pi \cdot \widehat{Q}^\chi_\nu + \lambda' \widehat{M}_{\pi\nu} + \widehat{V}_{\pi\pi} + \widehat{V}_{\nu\nu}, \qquad (1)$$

where, quite often, the assumption $\varepsilon_\pi = \varepsilon_\nu = \varepsilon$ is made. In eq.(1), E_0 is a function which does not contribute to the excitation energy, while $\varepsilon_\pi \widehat{n}_{d_\pi} + \varepsilon_\nu \widehat{n}_{d_\nu}$ is the one-body term originating from the pairing between identical nucleons. The quadrupole operator, $\widehat{Q}^\chi_\rho$ $(\rho = \pi, \nu)$, enters in the quadrupole-quadrupole interaction between different nucleons, while χ_π (χ_ν) represents the proton (neutron) quadrupole deformation parameter. The form assumed by the nucleus essentially depends on the value taken by χ_π and χ_ν, since a strong connection exists between the parameter χ_ρ $(\rho = \pi, \nu)$ and the IBM-2 dynamical symmetries, $U_{\pi+\nu}(5)$, $SU_{\pi+\nu}(3)$, $O_{\pi+\nu}(6)$, and $SU_{\pi+\nu}(3)^*$. In particular, the $U_{\pi+\nu}(5)$ dynamical symmetry (vibrational nucleus) arises when $\chi_\rho = 0$ $(\rho = \pi, \nu)$. In the $SU_{\pi+\nu}(3)$ limit (rotational nucleus), one has $\chi_\rho = \pm\frac{\sqrt{7}}{2}$ $(\rho = \pi, \nu)$, while the $SU_{\pi+\nu}(3)^*$ case (triaxial nucleus) corresponds with $\chi_\pi = -\frac{\sqrt{7}}{2}$ and $\chi_\nu = +\frac{\sqrt{7}}{2}$, or vice-versa. Finally, in the $O_{\pi+\nu}(6)$ dynamical symmetry ($\gamma-$unstable nucleus), the χ_π and χ_ν parameters have opposite sign ($-\frac{\sqrt{7}}{2} < \chi_\pi < 0$ and $0 < \chi_\nu < +\frac{\sqrt{7}}{2}$). Moreover, in eq.(1) $\widehat{M}_{\pi\nu}$ represents the so-called Majorana operator and $\widehat{V}_{\rho\rho}$ $(\rho = \pi, \nu)$ term describes the residual interaction between like bosons. The IBM-2 exhibits four dynamical symmetries[8] associated with two different types of group chains, the F-spin[a] symmetry limit (I_1, II_1, III_1, and IV) and the coupled chains (I_2, II_2, III_2, and IV) [6]. The IV chain is a distinctive feature of the IBM-2, and the bar over $SU_\nu(3)$ denotes a particular choice in the sign of the second $SU(3)$ generator.

A very useful tool to describe the spectral statistics of a large set of quantum systems is the nearest neighbor level-spacing distribution, $P(s)$ [7,9,10,11,12]. This distribution evaluates the probability with which two adjacent eigenvalues are at the distance s. If the classical analog of a quantum system is integrable, then the $P(s)$ distribution is that of a Poisson ensemble with a high probability to have close degeneracies. Moreover, neither repulsions nor correlations between levels are observed. Instead, if the quantum systems are similar to chaotic classical systems (nonintegrable), the $P(s)$ statistics is approximated by the Wigner distribution (GOE), the levels aiming to repel each other ($P(s) \to 0$ as $s \to 0$).

It is a usual practice to quantify the chaotic behavior in terms of the Brody parameter, ω, interpolating these two limiting distributions by means of the follow-

[a]The F-spin is, by definition, the isotopic spin for the bosons, namely a quantum number which allows to distinguish between proton bosons ($F = -\frac{1}{2}$) and neutron bosons ($F = +\frac{1}{2}$).

ing:

$$P(s,\omega) = \alpha\,(\omega+1)\,s^{\omega}\,\exp\left(-\alpha\,s^{\omega+1}\right), \tag{2}$$

where

$$\alpha = \left[\Gamma\left(\frac{\omega+2}{\omega+1}\right)\right]^{\omega+1}. \tag{3}$$

It is easy to see that, if $\omega = 0$, the Poisson distribution is recovered, while when $\omega = 1$ one gets the Wigner distribution.

Theoretical IBM-2 nuclear spectra, obtained by means of a modified version of the NPBOS code [13,14] which calculates the excitation energies and the eigenvectors of the Talmi Hamiltonian [eq. (1)], have been analyzed in terms of Brody distribution. By this way, fourteen shape-phase transitions have been considered in this work. Six of them have been approximated by keeping constant the χ_π parameter, $\chi_\pi = -1.25$, and varying the χ_ν parameter in the $(-1.30, +1.30)$ range. The other six transitions have been carried out by putting χ_π equal to -0.80 and ranging χ_ν from -1.30 to $+1.30$[b]. Finally, the last two cases are concerned with the $SU_{\pi+\nu}(3) \to U_{\pi+\nu}(5)$ transition and have been calculated by fixing χ_π first to -1.20 and then to -1.10 and varying again the χ_ν parameter in the $(-1.30, +1.30)$ range. Hence, the calculated shape-phase transitions are the following:

$$\begin{aligned} SU_{\pi+\nu}(3) \to U_{\pi+\nu}(5) \quad &; \quad U_{\pi+\nu}(5) \to O_{\pi+\nu}(6)\,, \\ O_{\pi+\nu}(6) \to SU_{\pi+\nu}(3)^* \quad &; \quad SU_{\pi+\nu}(3)^* \to SU_{\pi+\nu}(3)\,, \\ SU_{\pi+\nu}(3)^* \to U_{\pi+\nu}(5) \quad &; \quad O_{\pi+\nu}(6) \to SU_{\pi+\nu}(3)\,. \end{aligned}$$

In every transition, the values 4 (5) for the proton (neutron) boson numbers N_π (N_ν) have been assumed, since the quantum fluctuations are independent of N_π (N_ν). The only exception is the $O_{\pi+\nu}(6) \to SU_{\pi+\nu}(3)$ transition since for it realistic calculations for the platinum and osmium isotopes have been carried out (the highest *Pt* isotopes have a $\gamma-$unstable feature, while the lightest *Pt* and *Os* isotopes exhibit an axially symmetric deformed character). In particular, use has been made of the IBM-2 Hamiltonian parameters given in ref. [15] by putting $N_\pi = 2\,(3)$ for *Pt* (*Os*) isotopes, while N_ν ranges from 4 (^{194}Os, ^{196}Pt) to 9 (^{184}Os, ^{186}Pt).

In order to consider only the information about energy levels which can significantly contribute to the spectral analysis, it is necessary to fit the staircase function, $N(E)$, by means of a suitable expression for the integral of the level density, $\rho(E)$. In the literature, a number of phenomenological nuclear level density models exists, the most widely used being the *Gilbert-Cameron* (GC) expression [16], the ***back-shifted Fermi gas*** (BSFG) model [17], the microscopic formalism of ref[18,19]. All of them have been used in the present analysis. Moreover, a finite-temperature extension of the interacting boson model to the description of nuclear statistical properties [20] has been also adopted. Fits to the theoretical IBM-2 spectra obtained

[b]The fact of keeping constant χ_π, instead of χ_ν, is a purely convenient choice and has no physical relevance.

by means of all the previous approaches do not show remarkable differences as for final results concerning the deduced Brody parameters.

The unfolded level density does not show an exponential character, like the original levels, once the smoothed trend has been removed by means of the level density fitting procedure. The actual distances, s_i $(i = 1, \ldots, N)$, between adjacent energy levels are computed as follows:

$$s_i = \left(\widetilde{E}_{i+1} - \widetilde{E}_i\right) \rho_C, \tag{4}$$

where $\widetilde{E}_i$ is the unfolded energy and ρ_C the corresponding constant level density. The $P(s)$ distribution is the histogram of the number of distances s_i, included in the $(s, s + \Delta s)$ range where Δs is equal to one half of the average distance between the levels [21]. The Brody distribution is calculated by a suitable fitting procedure of the $P(s)$ histogram.

3 Numerical results

As shown in fig.1, concerning the $SU_{\pi+\nu}(3) \rightarrow U_{\pi+\nu}(5)$ shape-phase transition, the presence of a nearly regular region is clearly seen in agreement with the results previously obtained by Whelan and Alhassid (1993) [22] within the framework of IBM-1. The chaoticity parameter ω assumes values larger than or equal to $+0.50$ when the χ_π parameter ranges from -1.25 to -1.10. The uncertainty in the ω parameter is less than or equal to 16 per cent: If a new value ω' of the Brody parameter is calculated within this error, the two curves $P(s, \omega)$ and $P(s, \omega')$ are quite near each other.

For $\chi_\pi < -1.10$, the dynamical behavior of the nuclear spectra becomes completely regular, the maximum value of the ω parameter being $\omega = +0.06$. Moreover, the fact that $\omega \geq +0.50$ when χ_π ranges from -1.25 to -1.10 points out that the nearly regular region connecting the $SU_{\pi+\nu}(3)$ and $U_{\pi+\nu}(5)$ dynamical limits is not a narrow band [22], but is a broad region. An explanation of the presence of a nearly regular region even far from the $SU_{\pi+\nu}(3)$ and $U_{\pi+\nu}(5)$ dynamical limits can be connected to the existence of a partial dynamical $SU_{\pi+\nu}(3)$ symmetry [23,24].

A further confirmation of the important role played by the partial dynamical $SU_{\pi+\nu}(3)$ symmetry in keeping regular motion patterns in nuclear spectra is supplied by the completely regular behavior of the $SU_{\pi+\nu}(3)^* \rightarrow U_{\pi+\nu}(5)$ transition. Since $SU_{\pi+\nu}(3)$ and $SU_{\pi+\nu}(3)^*$ dynamical limits have the same generators up to a sign, the presence of a nearly regular region like that observed in the $SU_{\pi+\nu}(3) \rightarrow U_{\pi+\nu}(5)$ case was expected. Present results correspond to a completely regular behavior as indicated by the maximum value assumed by the Brody parameter, $\omega = +0.12$. This complete regularity observed in the $SU_{\pi+\nu}(3)^* \rightarrow U_{\pi+\nu}(5)$ shape-phase transition is a clear indication that the $SU_{\pi+\nu}(3)^*$ limit has an attraction basin deeper than that of the $SU_{\pi+\nu}(3)$ symmetry. This fact is strengthened by the analyses of the $O_{\pi+\nu}(6) \rightarrow SU_{\pi+\nu}(3)^*$ and $O_{\pi+\nu}(6) \rightarrow SU_{\pi+\nu}(3)$ transitions, the latter results being listed in the following table.

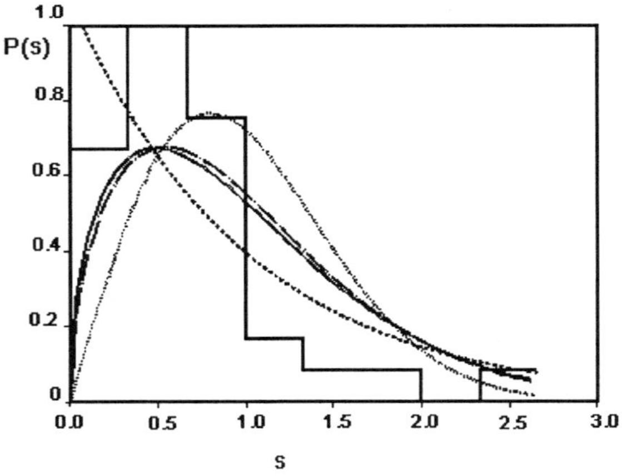

Figure 1. $P(s)$ distribution for the $SU_{\pi+\nu}(3) \to U_{\pi+\nu}(5)$ transition with $\chi_\pi = -1.25$ and $\chi_\nu = -1.05$. The dashed, dotted, and dot-dashed curves correspond with Poisson, GOE, and Brody [Eq.(2), with $\omega = 0.52$] distributions, respectively. The solid curve refers to the Brody distribution calculated within the uncertainty of 15 per cent in the ω parameter [Eq.(2), with $\omega' = 0.45$].

Table I shows the presence of a nearly regular region, since $\omega = +0.64$ is the maximum estimated value, while in the former case a completely regular behavior is found [maximum value is $\omega = -0.23$]. The appearing of $\omega = +0.64$ in correspondence with $\chi_\nu = 0.0$ in the $O_{\pi+\nu}(6) \to SU_{\pi+\nu}(3)$ case is acceptable since Bijker *et al.* [15] stopped their investigations at $\chi_\nu = -0.50$ [remember that in the $SU_{\pi+\nu}(3)$ limit $\chi_\rho = -1.30$ $(\rho = \pi, \nu)$]. Hence, $\omega = +0.64$ comes into sight at intermediate values of the χ_ν parameter as it is usually expected. The presence of the nearly regular region in the $O_{\pi+\nu}(6) \to SU_{\pi+\nu}(3)$ transition is a partial confirmation of the results previously obtained by Alhassid *et al.* [25], since it does not indicate a chaotic feature, as observed by these authors (see Fig. 1 in the first paper of ref[25]), but it only expresses a not completely regular motion pattern. As far as the $U_{\pi+\nu}(5) \to O_{\pi+\nu}(6)$ transition is concerned, it displays a completely regular motion pattern, due to the fact that the $U_{\pi+\nu}(5)$ and $O_{\pi+\nu}(6)$ limits have $O_{\pi+\nu}(5)$ as common subalgebra, thus two of their generators being the same. Finally, the $SU_{\pi+\nu}(3)^* \to SU_{\pi+\nu}(3)$ shape-phase transition shows an expected complete suppression of chaos, since the $SU_{\pi+\nu}(3)^*$ and $SU_{\pi+\nu}(3)$ dynamical symmetries have the same generators up to a sign.

These results can be summarized in the following diagram, where the shape transitions exhibiting chaotic properties are represented by dashed lines. Full regularity corresponds to solid lines connecting the limiting symmetries, where the nuclear system is completely integrable.

Table 1. $O_{\pi+\nu}(6) \rightarrow SU_{\pi+\nu}(3)^*$ transition. Hamiltonian parameters in eq.(1) are expressed in MeV; $c_0^\nu = c_2^\nu = c_4^\nu = 0$; $N_\pi = 4$ and $N_\nu = 5$. The Brody parameter, ω, is listed together with the relevant uncertainty in percentage.

Dynamical symmetry	ε	k	χ_ν	χ_π	α_1	α_2	α_3	ω
$O_{\pi+\nu}(6)$	0.430	−0.080	0.00	−1.25	0.900	0.090	−0.010	+0.05 (15)
⋮	0.470	−0.055	+0.30		0.900	0.090	−0.010	+0.05 (14)
⋮	0.480	−0.055	+0.50		0.900	0.090	−0.010	+0.09 (14)
⋮	0.480	−0.055	+0.90		0.900	0.090	−0.010	−0.17 (15)
⋮	0.340	−0.050	+1.20		0.200	0.040	−0.010	−0.29 (14)
$SU_{\pi+\nu}(3)^*$	0.350	−0.050	+1.30		0.200	0.040	−0.010	−0.31 (15)
$O_{\pi+\nu}(6)$	0.400	−0.050	+0.10	−0.80	0.900	0.150	−0.050	−0.17 (16)
⋮	0.450	−0.050	+0.30		0.900	0.150	−0.050	−0.15 (15)
⋮	0.450	−0.070	+0.50		0.900	0.050	−0.050	−0.20 (14)
⋮	0.445	−0.070	+0.70		0.900	0.040	−0.050	−0.23 (16)
⋮	0.455	−0.070	+0.90		0.900	0.040	−0.050	−0.14 (15)
$SU_{\pi+\nu}(3)^*$	0.395	−0.040	+1.30		0.300	0.060	−0.030	−0.08 (16)

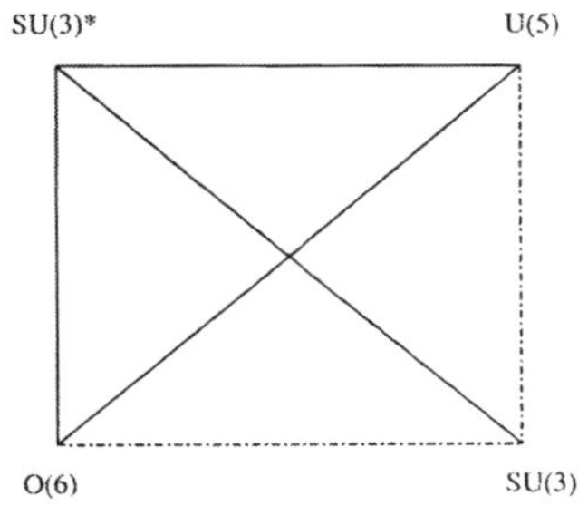

Figure 2. Schematic representation of shape transitions. Solid lines correspond to regular motion; dashed lines to presence of quantum chaos.

They show a persistence of regular motion patterns even far from dynamical symmetry limits of the IBM-2 model, fact whose explanation can be connected with the existence - investigated in the last few years [26,27] - of the so-called partial dynamical symmetries, since they may cause suppression (namely, reduction) of the chaotic behavior, as shown in ref[28]. In this connection extremely interesting is the $SU_{\pi+\nu}(3) \rightarrow U_{\pi+\nu}(5)$ transition which displays a broad *nearly regular region* whose presence can be explained in terms of partial dynamical $SU_{\pi+\nu}(3)$ symmetry, like previously made by Alhassid, Leviatan and coworkers [26],[24] in the frame of the original version of the interacting boson model (IBM-1). The fundamental role played by $SU_{\pi+\nu}(3)$ partial dynamical symmetry in maintaining regular motion patterns even far from the usual dynamical limits of the IBM model [24] is further confirmed by the unexpected completely regular behavior observed in the $SU_{\pi+\nu}(3)^* \rightarrow U_{\pi+\nu}(5)$ transition. A deeper investigation of this new feature of the $SU_{\pi+\nu}(3)^*$ limit needs a classical analysis, to be presented in a forthcoming paper, and possibly may be linked to the existence of critical point symmetries.

Acknowledgments

It is a pleasure to dedicate this work to Franco Iachello on the occasion of his 60th birthday. Twenty years ago he introduced me to the beautiful mathematical structure of the interacting boson model (a remarkable example of the unreasonable effectiveness of mathematics in the natural sciences, in Eugene Wigner's words). Since then, I benefited of valuable suggestions as well as many discussions for which I am deeply indebted and grateful to Franco.

References

1. A.Giannatiempo, A.Nannini, P.Sona and D.Cutoiu, Phys. Rev. **C52**, 2969 (1995), Phys. Rev. **C53**, 2770 (1995).
2. J.L.M.Duarte, T.Borello-Lewin, G.Maino and L.Zuffi, Phys. Rev. **C57**, 1539 (1998).
3. A. Gianantiempo, A.Nannini, A.Perego and P.Sona, in *Challenges of Nuclear Structure*, A.Covello ed. (World Scientific, Singapore, 2002), p.475.
4. E.Canetta and G.Maino, Phys. Lett. **B483**, 55 (2000).
5. F.Iachello, Phys. Rev. Lett. **85**, 3580 (2000).
6. F. Iachello and A. Arima, *The Interacting Boson Model* (Cambridge University Press, Cambridge, 1987).
7. E. Wigner, SIAM Rev. **9**, 1 (1967).
8. P. Van Isacker, Rep. Prog. Phys. **62**, 1661 (1999).
9. F. J. Dyson, J. Math. Phys. **3**, 140 (1962).
10. F. J. Dyson and M. L. Mehta, J. Math. Phys. **4**, 701 (1963).
11. O. Bohigas, M. J. Giannoni and C. Schmit, Phys. Rev. Lett. **52**, 1 (1984).
12. O. Bohigas and H. A. Weidenmüller, Ann. Rev. Nucl. Part. Sci. **38**, 421 (1988).
13. T. Otsuka and N. Yoshida, *User's manual of program NPBOS*, Japan Atomic Energy Research Institute Report JAERI-M/85-094, 1985.
14. G. Maino, Int. J. Mod. Phys. E **6**, 287 (1997).
15. R. Bijker, A. E. L. Dieperink, O. Scholten and R. Spanhoff, Nucl. Phys. **A344**, 207 (1980).
16. A. Gilbert and A. G. W. Cameron, Can. J. Phys. **43**, 1446 (1965).
17. W. Dilg, W. Schantl, H. Vonach and M. Uhl, Nucl. Phys. **A217**, 269 (1973).

18. A. V. Ignatyuk, Yad. Fiz. **21**, 20 (1975) [Sov. J. Nucl. Phys. **21**, 10 (1975)].
19. A. V. Ignatyuk, K. K. Istekov and G. N. Smirenkin, Yad. Fiz. **30**, 1205 (1979) [Sov. J. Nucl. Phys. **29**, 450 (1979)].
20. G. Maino, A. Mengoni and A. Ventura, Phys. Rev. C **42**, 988 (1990).
21. V. Paar and D. Vorkapić, Phys. Lett. B **205**, 7 (1988).
22. N. Whelan and Y. Alhassid, Nucl. Phys. **A556**, 42 (1993).
23. A. Leviatan, in *Symmetries in Science VII*, edited by B. Gruber and T. Otsuka (Plenum Press, New York, 1994), p. 383.
24. A. Leviatan, in *Perspectives for the Interacting Boson Model*, edited by R. F. Casten *et al.* (Word Scientific Publ., Singapore, 1994), p.129.
25. Y. Alhassid, A. Novoselsky and N. Whelan, Phys. Rev. Lett. **65**, 2971 (1990); Y. Alhassid and N. Whelan, Phys. Rev. C **43**, 2637 (1991); Y. Alhassid and A. Novoselsky, Phys. Rev. C **45**, 1677 (1992).
26. Y. Alhassid and A. Leviatan, J. Phys. A **25**, L1265 (1992).
27. A. Leviatan and N. Whelan, Phys. Rev. Lett. **77**, 5202 (1996).
28. N. Whelan, Y. Alhassid and A. Leviatan, Phys. Rev. Lett. **71**, 2208 (1993).

SOME REMARKS ON THE SYMMETRY OF THE SUPERCONDUCTING WAVEFUNCTION IN THE CUPRATES

KARL ALEXANDER MÜLLER
Physics Institute, University of Zürich
Winterthurerstr. 190, CH–8057 Zürich, Switzerland

A large part of the community considers the macroscopic superconducting wavefunction in the cuprates to be of near pure d–symmetry. The pertinent evidence has been obtained by experiments in which mainly surface phenomena have been used such as tunneling or the well known tricrystal or tetracrystal experiments[1]. However recently, data probing the property in the bulk gave mounting evidence that inside the cuprate cuperconductor a substantial $s-$component is present, and therefore a changing symmetry from pure d at the surface to more s inside, at least, was proposed[2]. The suggestion was made to reconcile the observations stemming from the surface and bulk. But such a behaviour would be at variance with the accepted classical symmetry properties in condensed matter[1,3]. In this respect, Iachello, applying the Interacting Boson–Model, successful in nuclear theory, to the C_{4v} symmetry of the cuprates, showed that indeed a crossover from a $d-$phase at the surface, over a $d+s$, to a pure $s-$phase could be present[4]. An attempt to estimate this crossover from known NMR experiments will be presented. It makes also plausible why the phase stiffness of the $d-$component is preserved over the a whole sample, i.e. in a Superconducting Quantum Interferometer Device (SQUID). Of interest is also the compatibility of the Interacting Boson Model with supersymmetry: in the hole doped cuprate superconductors there is since a number of years substantial evidence that there are two types of quasiparticles present, one of more bosonic and one of more fermionic character[5]. Due to the dynamic state in which one type transforms into the other the presence of a supersymmetry could be real.

References

1. C.C.Tsuei and J.R.Kirtley, Rev. Mod. Phys. **72** (2000) 969.
2. K.A.Müller, Phil. Mag. Lett. **82** (2002) 270.
3. J.F.Annett, N.D.Goldenfeld and A.J.Legett, in *Physical Properties of High–Temperature Superconductors V*, edited by D.M.Ginsberg, World Scientific, Singapore, 1996 p. 571.
4. F.Iachello, Phil. Mag. Lett. **82** (2002) 289.
5. D.Mihailovic and K.A.Müller, in *High–T, Superconductivity. Ten Years after the discovery. NATO ASI Ser. E: Appl. Sci.*, Vol **343**, Kulver 1997, p. 243.

ALGEBRAIC DESCRIPTION OF N-ALKANE MOLECULES

STEFANO OSS

Physics Department, University of Trento
and INFM, Unità di Trento, 38050 Povo (Trento) – Italy

We apply in a systematic fashion the vibron model to obtain the complete description of vibrational spectra of molecular chains of finite size, such as *n*-alkanes (paraffin molecules). The one-dimensional model is extended to include infrared spectra of both CH stretching and bending modes. We describe the possible effect of anharmonic (Fermi) resonances in the spectra of fundamental and overtone energy regions. In the present framework, we show that in such molecular systems the algebraic treatment leads to a reliable, consistent set of parameters providing a fair description of the infrared spectrum *without* including Fermi resonances, even if they can have an appreciable role at higher resolution. It also seems that the parameters are applicable to extend the computation to longer and longer chains eventually describing a polymer chain (polyethylene).

1 Introduction

In the last 15 years algebraic (vibron) models have been applied to an increasing number of molecular systems of actual interest[1]. A particularly intriguing field of application is that involving large molecules, eventually constituting a chained sequence of simple chemical units, such as *n*-alkane molecules (also known as *n*-paraffins, CH_3-$(CH_2)_{n-2}$-CH_3). In the limit of very large *n*, such molecules are better known as polymers (such as polyethylene). From an experimental point of view, there has been a wide interest during the last 40 years in obtaining and assigning infrared and Raman spectra of solid and liquid samples of such molecules[2]. The assignment of observed levels usually refers to the force-field representation of vibrational states which, for molecular chains, are characterized by specific normal mode dispersion. The single site vibrational modes get coupled and splitted following non-linear dispersion laws which are well-known for simple chemical units (such as the CH_2 methylene group in *n*-alkane molecules). Depending on the total number of groups in the molecule, one can face very challenging sizes of Hamiltonian matrixes when trying to address and solve the vibrational problem. Moreover, anharmonic terms become important as soon as one goes to overtone and combination levels. Finally, infrared and Raman transition intensities require very a demanding and careful treatment of dipole and quadrupole operators to achieve consistent and reliable results. All these matters, and more, can be directly and properly addressed within the framework of the vibron model in one-dimension[3], i.e. based on the single bond spectrum generating algebra $U(2)$. As often shown in several papers[4], this model has the dominant advantage of being definitely a simple one from the mathematical standpoint. Also, it can be easily extended to include a really large number of eventually coupled anharmonic oscillators of both stretching and bending nature, since the corresponding potential functions (Morse and Poeschl-Teller) have the same bound-state spectrum. In this contribution to the Conference, the latest advances in this field of molecular, algebraic spectroscopy are briefly described. The inspiring guidance and wide experience of Franco Iachello in this field (as well as in many others) is gratefully and deeply acknowledged.

2 One-dimensional vibron model and n-alkane molecules

In this section a brief summary of the most relevant aspects of the one-dimensional algebraic formulation of the vibrational Hamiltonian for the n-alkane molecular chain is provided. As it is well-known[5], each local (bond-based) coordinate of vibrational relevance is replaced by a $U(2)$ algebra (i.e. by its generating operators). The overall spectrum generating algebra for the physical system constituted by a finite number n of (interacting) oscillators is then given by

$$U_1(2) \otimes U_2(2) \otimes \cdots \otimes U_n(2) \tag{1}$$

and the specific anharmonic nature of the oscillators is introduced by considering the dynamic symmetry restriction to subalgebras $O_k(2)$, $| U_k(2) \supset O_k(2) >$, and its invariant (Casimir) operators. Upon restriction under the product (1), one is lead to consider algebraic lattices containing products $U_{hk}(2)=U_h(2)\otimes U_k(2)$ and $O_{hk}(2) = O_h(2)\otimes O_k(2)$ and the corresponding invariant operators. Casimir operators associated with coupled $U(2)$ algebras are usually referred to as Majorana interactions and they are used to realize inter-mode anharmonic (non-diagonal) couplings, i.e. to provide a *normal* picture of the molecular system. We are thus lead to the Hamiltonian operator for n oscillators in the one-dimensional model given by

$$\hat{H} = \sum_k a_k \hat{C}_k + \sum_{h<k} \lambda_{hk} \hat{M}_{hk} + \sum_{h<k} b_{hk} \hat{C}_{hk}\ , \tag{2}$$

in which C_k and C_{hk} are Casimir operators of groups $O_k(2)$ and $O_{hk}(2)$, respectively, and M_{hk} are invariant operators of groups $U_{hk}(2)$ (Majorana interactions). Upon introduction of the *local* vibrational basis $|v_1 v_2 \ldots v_n>$, strictly associated with irreducible representations of the algebraic lattice starting from (1) and closing to the product of orthogonal groups $O_k(2)$, one can compute the eigenvalues of the Hamiltonian (2) by diagonalization of a matrix representation in which $(v_1+v_2+\ldots+v_n)$ is a good quantum number. The actual computation is straightforward since analytical expressions of expectation values of Casimir and Majorana operators are available in simple forms. One has, for example, that the single-bond Casimir invariant leads to the expectation value

$$\langle v | \hat{C} | v' \rangle = v(N-v)\delta_{vv'}\ , \tag{3}$$

in which the parameter N (vibron number) is the label of the irreducible representation of the leading $U(2)$ algebra. It can be put in close correspondence with the anharmonicity of the vibrational sequence (3). Operators C_{hk} are also diagonal in the local basis and they can be used to describe cross-anharmonicities involving pairs of oscillators. The Majorana operator is *not* diagonal in the local representation since its non-zero matrix elements are given by

$$\langle v_h+1, v_k-1 | \hat{M}_{hk} | v_h, v_k \rangle = [v_k(v_h+1)(N-v_h)(N-v_h+1)]^{1/2}/N. \tag{4}$$

So, one can see that only those pairs of oscillators whose vibrational quantum numbers differs by one are effectively coupled through the action of Majorana terms, as expected in the local-to-normal representation of interacting degrees of freedom in a molecular system.

In the specific case of the n-alkane molecule, one has to construct the Hamiltonian operator for CH stretching modes including the full set of CH pairs of methylene groups (there are $2(n\text{-}2)$ of them) and of CH_3 (methyl) end-effects (there are always two of such groups). The Hamiltonian operator can thus be written as

$$\hat{H} = \hat{H}_{CH_2} + \hat{H}_{CH_3} + \hat{H}_{CH_2/CH_3}\ , \tag{5}$$

in which the CH_2 part is given by

$$\hat{H}_{CH_2} = a\sum_{k=1}^{n-2}\hat{C}_i + \lambda\sum_{k=1}^{n-2}\hat{M}_{k_ak_b} + + f_d\sum_{k=1}^{n-3}\hat{M}_{k_ak+1_a} + f_c\sum_{k=1}^{n-3}\hat{M}_{k_ak+1_b} + + s_d\sum_{k=1}^{n-4}\hat{M}_{k_ak+2_a} + s_c\sum_{k=1}^{n-4}\hat{M}_{k_ak+2_b} + \ldots \qquad (6)$$

where one has intrasite (Majorana, λ parameter) interactions, first-neighbour, direct (f_d) and cross (f_c) terms and second-neighbour, direct (s_d) and cross (s_c) terms, whose physical meaning is schematically made explicit in Fig.1. If one starts by the Hamiltonian operator (6) in which only diagonal (Casimir) terms are present, the resulting spectrum will be given by $2(n-2)$ degenerate levels. Under the action of the Majorana coupling, this manifold will split into two sub-manifolds of degenerate levels of distinct symmetry under specific operations of the molecular point group. Their energy difference will be directly related to the strength of the Majorana interaction. The action of subsequent first- and second-neighbour interactions (of either direct- or cross-nature) will be that of introducing a further splitting of the symmetric and antisymmetric manifolds of levels according to relatively simple dispersion laws. An approximate form of such laws is given by

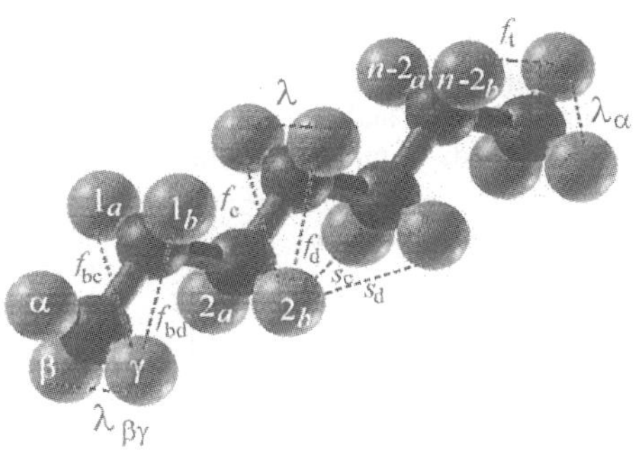

Fig.1 Schematic representation of labelling and couplings in the *n*-alkane molecule.

$$E_k^{(\pm)} \cong \varepsilon^{(\pm)} + 2(f_d \pm f_c)\cos\frac{k\pi}{n-1} + 2(s_d \pm s_c)\cos\frac{2k\pi}{n-1}, \quad k = 1,2,\cdots,n-2. \qquad (7)$$

In this expression we observe that the quantitative nature of the dispersion law is strictly related to the values of the parameters f and s: this allows one to fit their values by means of a direct comparison with observed energy levels. A throughout discussion of such problems can be found in Ref.6.

The CH_3 part in the Hamiltonian operator (5) is instead given by

$$\hat{H}_{CH_3} = a_{CH_3}\sum_{k=\alpha,\beta,\gamma}\hat{C}_k + \lambda_{\beta\gamma}\hat{M}_{\beta\gamma} + \lambda_\alpha(\hat{M}_{\beta\alpha} + \hat{M}_{\gamma\alpha}), \qquad (8)$$

whose terms are once again easily recognized in Fig.1. The CH_2/CH_3 interaction terms (stopped to first-neighbour couplings) are given by

$$\hat{H}_{CH_2\text{-}CH_3} = f_t(\hat{M}_{\alpha a} + \hat{M}_{\alpha b}) + f_{bd}(\hat{M}_{\beta a} + \hat{M}_{\gamma b}) + f_{bc}(\hat{M}_{\beta b} + \hat{M}_{\gamma a}). \qquad (9)$$

Expressions (8) and (9) come in pairs each referring to left and right CH_3 groups in a given *n*-alkane molecule. In Ref.6 the full set of parameters used to optimize the computation of CH stretches in solid and liquid samples of *n*-alkanes is provided.

3 Applications: fundamental CH stretches of gaseous *n*-alkanes

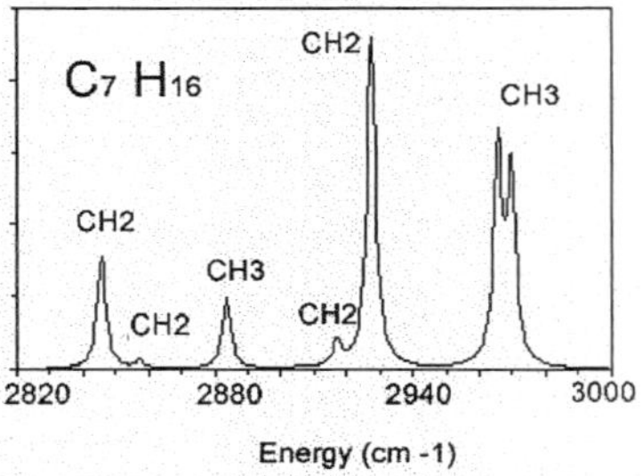

Fig.2 Computed infrared CH stretching spectrum in the *n*-alkane molecule (*n*=7, fitted to a solid state sample).

Since the pioneering works by several research groups[7], short molecular chains based on CH_2 (methylene) units have played a determinant role in the definition of important benchmarks for new theories and classes of experimental procedures. The most relevant aspect is found in the construction of a reliable set of force-field parameters (of either algebraic nature or not) to obtain energies and transitions intensities of such kinds of molecules with arbitrary length. As a welcome side result, this would allow one to predict the dynamics of infinite length chains, such as polymer chains (polyethylene is the polymer corresponding to *n*-alkane or *n*-paraffin finite chains). With the one-dimensional algebraic approach, we have been able to construct such a set of force-field equivalent parameters. We have obtained a first version of parameters by adapting the algebraic Hamiltonian operator to solid-state samples of liquid paraffins, as reported in Ref.6 and schematically shown in Fig.2 in which an example of infrared computed spectrum is given. The computations have been done for both CH stretching and bending vibrational modes, inclusive of terminal methyl CH_3 groups (quite a difficult task in more conventional approaches because of the increasingly high dimensionality of the problem at issue). In this contribution to the Conference we briefly report on the extension of our model to gas-phase samples of *n*-

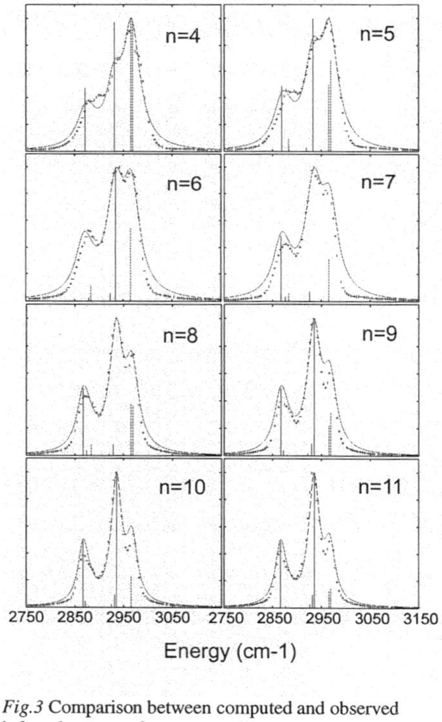

Fig.3 Comparison between computed and observed infrared spectra of gaseous *n*-alkanes ($n = 4,\ldots,11$) in the fundamental CH stretching region. Continuous sticks, CH_2 modes; dashed sticks, CH_3 modes. Lorentzian profile with 20 cm^{-1} FWHM.

alkanes, for which data exist at lower resolution but covering a larger number of molecular sizes[8]. More specifically, we show in Fig.3 infrared spectra of n-alkanes (n = 4, ..., 11) for fundamental CH stretches, computed according to the general prescriptions given in the previous section and by applying standard procedures to obtain infrared transition intensities[4] (based on a very simple algebraic rendition of the electric dipole operator once again in a local – bond based – picture). The agreement is a very good one: we observe that CH_3 modes (sticks represented by dashed vertical lines, whilst CH_2 modes are represented by sticks with continuous lines) have a less and less important role in the spectra for increasing n but they also give non-negligible contributions in the whole set of the studied molecules. As an important result, it is quite evident (from the quality of computed spectra, which are Lorentzian shapes superimposed to the computed stick spectra) that anharmonic coupling (à la Fermi) are not of dramatic importance at the present level of accuracy and resolution. Our previous analysis of 1:2 interactions in solid and liquid n-paraffin samples for fundamental CH stretching/CH bending manifolds of levels showed that a certain amount of coupling could be required to reproduce some minor features in the observed spectra[6]. Such features are however partially masked by stronger vibrational bands. We think that this is not anymore the case when considering gaseous samples. A further confirmation of this result will be given in the following section devoted to the study of the first overtone region of CH stretching modes.

4 Applications: overtone and combination bands of CH stretches in gaseous *n*-alkanes

In our previous work on n-paraffin molecules, we showed how to extend the Hamiltonian operator to include the v=2 region of CH vibrations[6]. In this contribution to the Conference (and preliminarily suggested in Ref.9) we show that our v=1 Hamiltonian operator can be used without any relevant change to compute v=2 CH stretching modes (arbitrary n). Such an extension is particularly interesting since it would allow one to establish the concreteness and reliability of the algebraic force field. In this section we also focus the attention to some preliminary issues concerning a systematic study of 1:2 anharmonic resonances in the v=2 region. It is in fact well-known that *Raman* transition intensities happen to be heavily influenced by the action of such coupling terms in the v=1 region[10], while, as already mentioned in the previous section, their effects is much smaller in the case of infrared transitions[11]. We will not discuss Raman processes in this paper. Yet, we claim that present results are clear enough to exclude any dominant anharmonic resonant effect in the infrared v=2 region. We are now able to compare our predictions with some more recent experimental data[12] once again in the liquid and solid state rather than with gaseous samples.

The v=2 computation is readily performed in terms of the same Hamiltonian operator (5) with the important difference than now one has two separate families of vibrational species, i.e. (20)- and (11)-types. These are based on (local) vibrations in which the two quanta are either localized in a single bond or shared by two of them, respectively. Strictly speaking, such kinds of modes differ in energy because of the anharmonicity factor which, we remember, is automatically accounted in the algebraic version of the vibrational Hamiltonian. More specifically, we expect to obtain that CH_2 modes, in absence of intermode couplings, become splitted in six groups of levels, where (20) antisymmetric and (20)/(11) symmetric combination of local modes get doubled via the anharmonicity factor. By extension of the algebraic Hamiltonian operator to the v=2 energy region (i.e. without any change of the v=1 parameters) we obtain infrared spectra of n-alkane molecules which are in quite a poor agreement with the available observed data (in liquid or solid state samples). The lack of agreement mainly concerns the intensity of those vibrational transitions involving states of (11) character (for both CH_2 and CH_3 groups). Following well-settled procedures of algebraic analysis of the electric dipole operator, such modes should have a negligible infrared activity (the algebraic transition operator has non-zero matrix elements only for localized excitations, such as in modes $|00 \ldots 0v0 \ldots 0\rangle$. This implies that in the first overtone region only modes with character (20) should be infrared active). We suggest[9] that such a (simplified) procedure should be revised in order to make "collective" modes with character (11) infrared active. This is readily accomplished by including an *ad-hoc* term in the dipole operator devoted to describe (11)-to-ground state mode transitions. More specifically, (11) excitations induce a dipole variation along the axis perpendicular to the "heavy" molecular axis (parallel to the CC skeletal axis) and belonging to the plane which contains the CC chain itself. One is thus left with an algebraic dipole which depends on two arbitrary parameters, κ and ξ, related to (11) excitations of CH_2 and CH_3 sites. The final form of the electric dipole can thus be written as

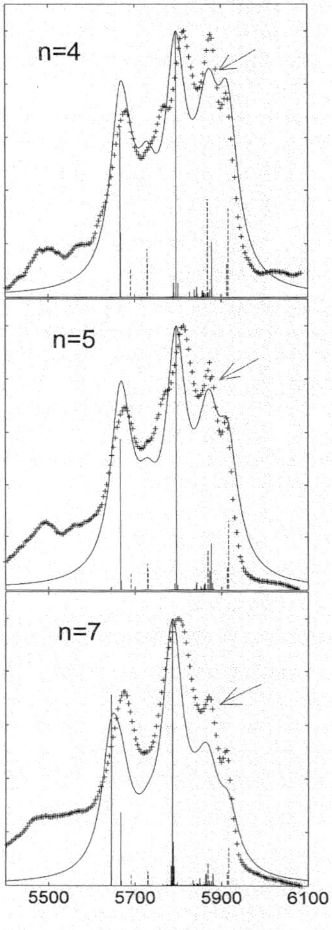

Fig. 4 Comparison between computed and observed (liquid state) infrared spectra of n-alkanes (n=4,5,7) in the first overton region of CH stretches. Continuous sticks, CH_2 modes; dashed sticks, CH_3 modes. Lorentzian profile with 30 cm^{-1} FWHM. The arrows point to the 1:2 resonant groups of combination bands.

$$\hat{T}_x = \sum_i \hat{t}_{x_i}^{CH_2} + \hat{t}_x^{CH_3}, \quad \hat{T}_y = \sum_i \hat{t}_{y_i}^{CH_2} + \hat{t}_y^{CH_3} + \kappa \hat{t}_{(11)}^{CH_2} + \xi \hat{t}_{(11)}^{CH_3},$$

in which the sums extend to the Cartesian components of CH_2 dipoles and $\hat{t}_{x,y}^{CH_3}$ refer to end-effect dipole operators. By adding such terms to the calculation of the v=2 region of CH stretches, we obtain a much better agreement with the available experimental data[12] (consisting of liquid and solid state infrared spectra of *n*-butane, pentane and heptane). Yet, we also find that the computed spectra can be further improved (some minor features are not well reproduced at this stage of the computation) by adding to the Hamiltonian operator a complete tier of resonating modes given by CH_S (v = 2) ~ CH_S (v = 1) + CH_B (v = 2), i.e. driven by a 1:2 anharmonic mechanism in which one quantum of CH stretching excitation is exchanged in favour of two quanta of CH (out-of-plane) bending excitation. Doing so, one remains (within anharmonic corrections) in the v=2 stretching region (~6000 cm^{-1}) but the number of involved levels grows quite rapidly (for *n*-heptane, the total number of combination bands is of the order of 1600). We have considered three different kinds of possible resonances depending on the specific geometry of the molecular bonds[6]. We think that just one of such mechanisms is effective in defining the actual excitation scheme in *n*-alkane molecules. More specifically, we have studied to some extent resonant (Fermi) couplings in terms of which the exchange of vibrational quanta happens between CH bonds in a single methylene site: it is only when such an exchange involves *adjacent* bonds that the interaction gives non-negligible contributions to the overall spectrum. By adding a fixed amount of anharmonic 1:2 coupling of ~14 cm^{-1}, the comparison between experimental and computed spectra gets even a more convincing one. We show in Fig.4 such a comparison for the studied *n*-alkanes (n = 4,5,7).

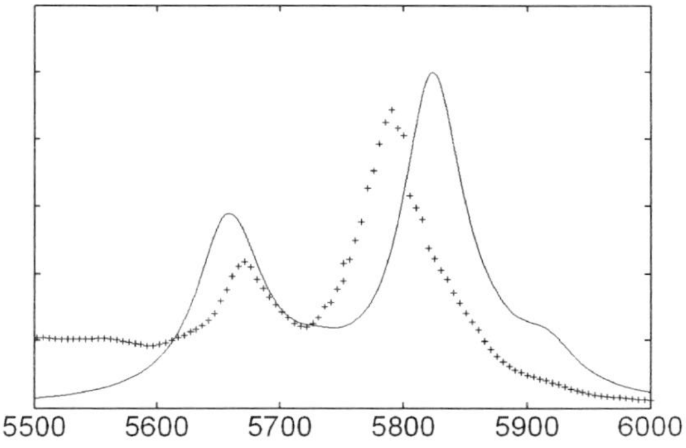

Fig. 5 Comparison between computed (continuous) infrared spectrum of $C_{15}H_{32}$ and observed (crosses) spectrum of crystalline polyethylene in the first overtone CH stretching region.

It is also possible to compute the spectrum of a much longer molecular chain in order to simulate the possible behaviour of a polyethylene polymeric chain. In Fig.5 we show the result of such a computation for n = 15. It is possible to see that CH_3 methyl groups give a negligible contribution to the calculated infrared spectrum which is fairly close to the measured one (in the solid, crystalline state[12]).

5 Conclusion

In this contribution to the Conference we have briefly discussed some novel applications of algebraic models to the infrared spectroscopy of simple molecular chains such as *n*-paraffins or *n*-alkanes. We show how such methods can be extended in a straightforward way to provide, in terms a common set of adapted parameters, a fairly detailed description of fundamental, overtone, combination and resonant vibrational modes in the infrared regime. More specifically, we show that (i) both v=1 (gas phase) and v=2 (liquid phase) CH stretching regions of n-alkanes are well described in terms of a one-dimensional Hamiltonian operator which does not include any complex, anharmonic Fermi resonance; in order to do so we have introduced a specific "collective" electric dipole activity involving combination modes of (11) character; (ii) the v=2 region is even more convincingly reproduced by adding a (relatively small) anharmonic 1:2 interaction involving combination/overtone modes of CH stretch/bend degrees of freedom; (iii) with increasing n, our model seems to reproduce well the infrared spectrum of the (solid state, crystalline) polyethylene chain.

Our work can (and will) be further improved and extended by considering several aspects, among which the most important are: (i) the calculation of CC skeletal modes and their coupling to CH modes, especially bending vibrations; (ii) the calculation of Raman transition intensities in terms of a new, bond-localized electric quadrupole operator which is at present being developed; (iii) the inclusion of eventually more complex resonant anharmonic terms in order to perform a systematic study of intramolecular redistribution of vibrational energy (IVR) to finally address a time-dependent analysis of the molecular dynamics.

References

1. F. Iachello and S.Oss, European J. Phys D, **19**, 307 (2002)
2. G. Zerbi (Ed.), Modern Polymer Spectroscopy, Wiley, VCH, Weinheim, 1999; A. H. Fawcett (Ed.), Polymer Spectroscopy, J. Wiley & Sons, Chichester, 1997; D. I. Bower and W. F. Maddams, The Vibrational Spectroscopy of Polymers, Cambridge University Press, Cambridge, 1989.
3. F. Iachello and P. Truini, Ann. Phys. (NY), **276**, 120 (1999)
4. F. Iachello and R.D. Levine, Algebraic Theory of Molecules, Oxford University Press, Oxford, 1994; S. Oss, Adv. Chem. Phys., **XCIII**, 455 (1996); A. Franck and P. Van Isacker, Algebraic Methods in Molecular and Nuclear Structure Physics, Wiley, New York, 1994.
5. F. Iachello and S. Oss, Phys.Rev.Lett. **66**, 2976 (1991)
6. T. Marinković and S. Oss, Phys. Chem. Comm. **5**, 66 (2002)
7. R. G. Snyder, J. Mol. Spectrosc. **4**, 411 (1960); R. G. Snyder, J. Mol. Spectrosc. **7**, 116 (1961); J. H. Schachtschneider and R. G. Snyder, Spectrochim. Acta, **19**, 117 (1963)
8. NIST Standard Reference Data Program, Collection (C) 2001 copyright by the U.S. Secretary of Commerce on behalf of the U.S.A.; CAS Registry no.106-97-8, 109-66-0, 142-82-5
9. T. Marinković and S. Oss, Chem. Phys. Lett., (2003), submitted
10. R. G. Snyder, J. R. Scherer, J. Chem. Phys. **71**, 3221 (1979); R. G. Snyder, H. L. Strauss and C. A. Elliger, J. Phys. Chem. **86**, 5145 (1982); L. Ricard, S. Abbate and G. Zerbi, J. Phys. Chem. **89**, 4793 (1985)
11. R. G. Snyder, S. L. Hsu and S. Krimm, Spectrochim. Acta, **34A**, 395 (1978); S. Abbate, G. Zerbi and S. L. Wunder, J. Phys. Chem. **86**, 3140 (1982)
12. L. Ricard-Lespade, G. Longhi and S. Abbate, Chem. Phys. **142**, 245 (1990)

ALGEBRAIC APPROACH TO VIBRATIONALLY HIGHLY EXCITED POLYATOMIC MOLECULES

KAORU YAMANOUCHI AND TOKUEI SAKO

Department of Chemistry, School of Science, The University of Tokyo, 7-3-1 Hongo, Bunkyo-ku, Tokyo 113-0033, Japan
E-mail: kaoru@chem.s.u-tokyo.ac.jp

Advantages of adopting algebraic approaches to vibration of polyatomic molecules are demonstrated. It is shown that the algebraic force-field expansion developed recently in our group can reproduce the experimental vibrational term values of H_2O and CO_2 in the wide energy range with a much smaller number of basis functions than the conventional force-field expansion.

1 Introduction

In molecular physics and chemistry, we gain rich information about molecular dynamics such as rotation, vibration, and chemical reaction through spectroscopic measurements of discrete and/or quasi-discrete energy levels of molecules. It was only very recently when spectroscopists succeeded to observe vibrationally highly excited molecules, that is, molecules having large vibrational energy.

The investigation of vibrationally highly excited molecules has been regarded to be important in many respects. For example, when chemical reaction is induced by collision or by absorption of a photon the molecular system is excited to its transition state in most cases, where nuclei within a molecule move with large relative kinetic energies, and as described by a statistical theory of chemical reaction, the energy randomization among the vibrational degrees of freedom is achieved before the system moves into the product side valley [1].

It should be challenging to observe the transition state of molecules spectroscopically. It is expected that we could examine how the energy randomization proceeds from one vibrational mode to another through the assignment of vibrational quantum numbers to the quasi-bound levels of the transition state. On the other hand, vibrationally highly excited molecules have also been regarded as an ideal dynamical systems for investigating a quantum mechanical correspondence of the transition of dynamics from regular to chaos in classical mechanics, and the pursuit of the signature of chaos in quantum systems has been an attractive issue in the field of nonlinear dynamics [2].

Originally molecular vibration was a research target in infrared and Raman spectroscopy. As described in a classical textbook written by Wilson et al. [3] vibrational frequencies characteristic of respective species of molecules were interpreted in terms of normal modes, and the vibrational spectroscopy was regarded as an efficient tool for identification of molecular species from spectra. In most cases, however, the vibrational excitation was achieved by only one unit of a vibrational quantum. Consequently, only the low-lying vibrational levels were the targets of the investigation, and the normal mode treatment based on harmonic oscillators afforded a sufficient basis for the spectral analysis.

After the introduction of visible and UV laser light into vibrational spectroscopy, the energy range of a spectrum was greatly widened. This was achieved by laser-based spectroscopic techniques such as dispersed fluorescence (DF) and stimulated emission pumping (SEP) [4-8]. For example, the DF spectrum of SO_2 in the electronic ground $\tilde{X}$ state shown in Fig. 1 exhibits dense vibrational levels in the entire observed range from low energy (5,000 cm^{-1}) to high energy (18,000 cm^{-1}) [6].

Figure 1. The DF spectra of SO_2 obtained via the (0, 2, 0) vibrational level in the $\tilde{C}$ state.

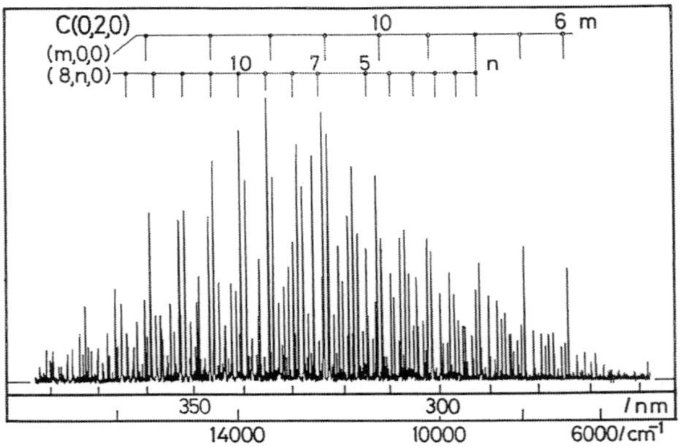

The major issue at that time was to develop a procedure to extract the information of vibrating molecules from such complex spectra covering a wide vibrational energy range. As shown in Fig. 1, a set of quantum numbers (v_1, v_2, v_3) can be assigned to respective peaks in the spectra, and the energy levels can be represented readily by a Dunham-type expansion as

$$E(v_1, v_2, v_3) = \sum_{i=1}^{3} \omega_i (v_i + \tfrac{1}{2}) + \sum_{i \geq j}^{3} x_{ij} (v_i + \tfrac{1}{2})(v_j + \tfrac{1}{2})$$
$$+ \sum_{i \geq j \geq k}^{3} y_{ijk} (v_i + \tfrac{1}{2})(v_j + \tfrac{1}{2})(v_k + \tfrac{1}{2}) + \cdots, \quad (1)$$

where v_1, v_2 and v_3 denote the vibrational quantum numbers of the ν_1

(symmetric stretch), v_2 (bend), and v_3 (antisymmetric stretch) modes, respectively. However, from this type of expansion, even though the energy levels are fitted well by the expansion, we could not derive vibrational wavefunctions. This is because we circumvent solving the Schrödinger equation to derive eigenvalues (vibrational energies) and eigenfunctions (vibrational wavefunctions), and simply assign the quantum numbers based on the spectroscopic intuition. In the case of the electronic ground state of SO_2, the vibrational assignment was possible because the anharmonicities of the respective three normal modes are very small, but we need to realize that the analysis stops here. Indeed, even though we could imagine the nodal structure of vibrational wavefunctions by the vibrational quantum numbers, neither quantitative information about the multi-dimensional potential energy surface nor about the shape of vibrational wavefunctions could be obtained.

When we hope to solve the Schrödinger equation, a standard procedure is to expand a potential energy part of the Hamiltonian as

$$\hat{H} = \sum_{i=1}^{3} \tfrac{1}{2}\omega_i (p_i^2 + q_i^2) + \sum_{i\geq j\geq k}^{3} k_{ijk} q_i q_j q_k + \sum_{i\geq j\geq k\geq l}^{3} k_{ijkl} q_i q_j q_k q_l + \cdots. \tag{2}$$

As shown in this expansion, the number of the expansion coefficients, k_{iji} and k_{ijil}, becomes much larger than x_{ij} and y_{ijk} in eq. (1). Furthermore, it is required to construct a large size hamiltonian matrix so that the convergence of eigenenergies is achieved. Since a product of harmonic oscillators representing respective normal modes is a basis function, an extremely large number of basis functions are required in most cases for achieving the convergence.

2 Algebraic approach to molecular vibrations

There is another approach to express the hamiltonian for molecular vibration. This is called an algebraic approach developed by Iachello, Levine, Oss and their coworkers [9]. One of the advantageous points of the algebraic approach is that vibrational level energies of molecules could be fitted using much smaller number of parameters than when using the conventional spectroscopic expansion in eq.(2). Even though an idea of dynamical symmetry seems abstract, it has been clarified that the dynamical chain,

$$U_1(4)\otimes U_2(4) \supset \begin{pmatrix} O_1(4)\otimes O_2(4) \\ U_{12}(4) \end{pmatrix} \supset O_{12}(4), \tag{3}$$

and the associated algebraic hamiltonian,

$$\hat{H} = a_1^{(1)}\hat{C}_1 + a_2^{(1)}\hat{C}_2 + a_{12}^{(1)}\hat{C}_{12} + {}_{12}^{(1)}a\hat{M}_{12} + \cdots, \quad (4)$$

describe properly the essential feature of the vibration of triatomic molecules, where $\hat{C}_i$ (i = 1, 2, 12) and $\hat{M}_{12}$ represent the Casimir operators of the $O_i(4)$ and $U_{12}(4)$ algebra, respectively, and the latter operator $\hat{M}_{12}$ is often called the Majorana operator [9].

In 1996, we applied the hamiltonian developed by Iachello and Oss to fit a large number (~ 300) of the experimental lever energy data of SO_2 in the electronic ground state, and showed that the fit was done satisfactorily and that the wave functions can be derived as the eigenfunctions [10]. On the basis of the wave functions of vibrational levels lying in the 20,000 cm^{-1} range, it was demonstrated that even a typical normal-mode molecule like SO_2 exhibits local-mode type vibration in the high energy range, and that the gradual transition from normal to local was exemplified clearly as the gradual deformation of vibrational wave functions. Moreover, the onset of the local-to-normal transition was identified from a bifurcation of the semiclassical trajectories constructed from the algebraic hamiltonian [11].

It should be noted that the least-squares fit to the large number of level energies in the wide energy range is made possible by adopting the algebraic approach. That is, when we adopt the conventional force field expansion to construct the vibrational hamiltonian, the size of the basis set needs to be extremely large to achieve the convergence, and the implementation of this hamiltonian into the least-squares routine is hopelessly time-consuming.

In general, as the vibrational energy increases, adjacent level energy spacings become narrower due to the anharmonicity in molecular vibration. Since the algebraic hamiltonian is described by the Casimir operators composed of the creation and annihilation operators of an *anharmonic* oscillator, it is suited intrinsically to describe such an anharmonicity. Therefore, a rapid convergence of the hamiltonian expansion is expected regardless of the energy range of the vibrational levels to which the hamiltonian is fitted.

3 Algebraic force-field expansion

After showing applicability of the algebraic hamiltonian to SO_2 in the electronic ground state, we applied it to a heavily anharmonic system. As far as an algebraic hamiltonian is described by using the Casimir and Majorana operators, we could describe only one type of resonance, i.e. the n : n resonance. Even though this resonance describes properly the local-normal transition, other types of resonances such as n : m ($n \neq m$)

could not be incorporated into the framework of the original algebraic theory. This could cause a serious problem when a molecule has a non-negligible cross ahnarmonicity that generates an extensive network of anharmonic resonances.

We encountered this type of resonances when we investigated the vibrational levels of the $\tilde{C}$ state of SO_2, which exhibits a strong Fermi-type coupling between ν_1 and ν_3 as well as between ν_1 and ν_2 [12]. In order to describe the level energies of a coupled oscillator having such a heavily anharmonic potential energy surface, we introduce an approach called an algebraic force-field expansion [12-14] whose hamiltonian is written as

$$\hat{H}_{alg} = \sum_i a_i \hat{C}_i + \sum_{i\le j\le k} b_{ijk} \hat{S}_i \hat{S}_j \hat{S}_k + \sum_{i\le j\le k\le l} b_{ijkl} \hat{S}_i \hat{S}_j \hat{S}_k \hat{S}_l + \cdots, \tag{5}$$

where $\hat{S}_i$ is defined as

$$\hat{S}_i = \frac{1}{2}(\hat{f}_i^- + \hat{f}_i^+), \tag{6}$$

using the *normalized* creation and annihilation operators, $\hat{f}_i^+$ and $\hat{f}_i^-$, of an anharmonic oscillator. These creation and annihilation operators of an anharmonic oscillator in eq.(6) are represented using the U(2) generators, $\hat{J}_x$, $\hat{J}_y$ and $\hat{J}_z$ as [11,15]

$$\hat{f}^- = \frac{1}{\sqrt{N}}(\hat{J}_z - i\hat{J}_y), \tag{7a}$$

$$\hat{f}^+ = \frac{1}{\sqrt{N}}(\hat{J}_z + i\hat{J}_y). \tag{7b}$$

It is worthwhile to compare eq.(5) with the conventional force-field expansion of a vibrational hamiltonian,

$$\hat{H} = \tfrac{1}{2}\sum_i \omega_i(\hat{P}_i^2 + \hat{Q}_i^2) + \sum_{i\ge j\ge k} k_{ijk}\hat{Q}_i\hat{Q}_j\hat{Q}_k + \sum_{i\ge j\ge k\ge l} k_{ijkl}\hat{Q}_i\hat{Q}_j\hat{Q}_k\hat{Q}_l + \cdots. \tag{8}$$

The algebraic force-field expansion of eq.(5) becomes eq.(8) at the harmonic limit. This means that the algebraic force-field expansion affords a general framework of a vibrational Hamiltonian of molecules.

The advantage of this algebraic force field expansion resides not only in its generality but also in its feasibility in describing experimental vibrational term values. This can be clearly demonstrated in a form of the residual plot in Fig.2. When we fit 20 vibrational term values of H_2O

using the conventional force-field expansion, the sum of the squared residuals obtained using seven parameters becomes an order of magnitude larger when the total number of the basis functions decreases from 250 to 70. However, when we adopt the algebraic force-field expansion with the same number of fitting parameters, the sum of the squared residuals stays almost constant even at the number of the basis functions of as small as 50, indicating that the algebraic Hamiltonian in which anharmonic oscillators are intrinsically incorporated is suited to describe polyatomic vibration in the wide vibrational energy range [13]. It should also be stressed that there is a close correspondence between the resultant parameters of the algebraic hamiltonian and those determined for the conventional force-field hamiltonian.

Figure 2. The sum of squared residuals in the least-squares fits to the 20 experimental vibrational term values of H_2O.

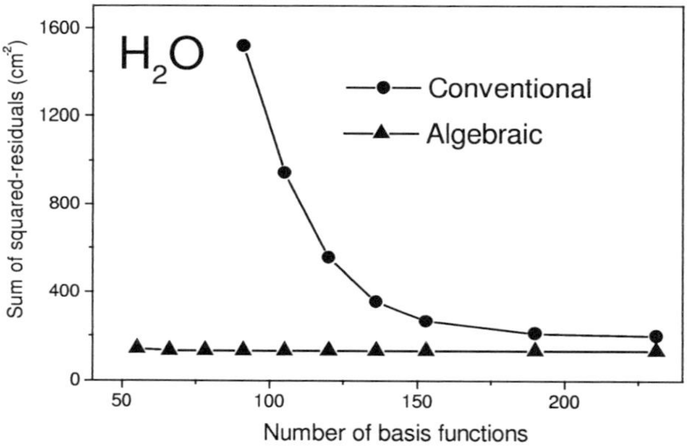

In the aforementioned example of bent triatomic molecules, we adopted an algebraic force-field hamiltonian composed of the generators of the U(2)⊗U(2)⊗U(2) space, i.e., the U(2) algebra is assigned to respective local mode vibrations. When we treat vibration of linear triatomic molecules like CO_2, the U(3) algebra is appropriate for describing the degenerated bending vibration. We constructed the algebraic hamiltonian of a linear triatomic molecule by using the U(3) generators for the bending vibration and the U(2) generators for respective bond stretches. The advantage of this algebraic force field expansion can be seen clearly in the residual plot in Fig. 3. Even at the basis size of 700, the convergence is achieved when the algebraic force-field expansion is adopted [14].

Now that the remarkable applicability of the algebraic force field expansion has been demonstrated for both bent and linear triatomic molecules, the next stage would be to extend this algebraic force field expansion for larger molecules. We realize that the extension is straightforward. For example, if you describe the hamiltonian for linear tetra-atomic molecule like acetylene (C_2H_2), the algebraic force-field expansion hamiltonian is readily constructed by the generators spanning the U(2)⊗U(3)⊗U(2)⊗U(3)⊗U(2) space. Considering the extremely rapid convergence demonstrated for H_2O and CO_2, the present algebraic force-field expansion hamiltonian could be regarded as an ultimate devise to characterize quantum mechanically individual vibrational levels of highly excited polyatomic molecules instead of introducing a statistical treatment.

Figure 3. The sum of squared residuals in the least-squares fits to the 60 experimental vibrational term values of CO_2.

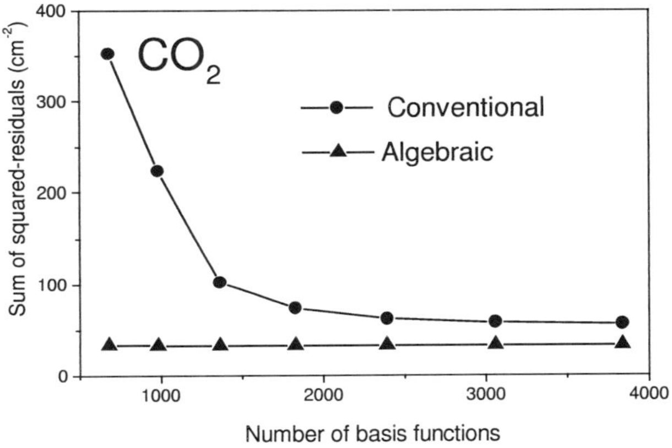

Acknowledgement

The authors thank Prof. Francesco Iachello for his cooperation, which enabled us to develop the algebraic force-field expansion hamiltonian for vibration of polyatomic molecules.

References

1. Baer T. and Hase W.L., *Unimolecular Reaction Dynamics: Theory and*

Experiments (Oxford University Press, Oxford, 1996).
2. Gutzwiller M.C., *Chaos in Classical and Quantum Mechanics* (Springer, New York, 1990).
3. Wilson, Jr. E.B., Decius J.C. and Cross P.C., *Molecular Vibrations: The Theory of Infrared and Raman Vibrational Spectra* (Dover, New York, 1980).
4. Abramson E., Field R.W., Imre D., Innes K.K. and Kinsey J., *J. Chem. Phys.* **83** (1985) pp.453-465.
5. Yamanouchi K., Yamada H. and Tsuchiya S., *J. Chem. Phys.* **88** (1988) pp.4664-4670.
6. Yamanouchi K., Takeuchi S. and Tsuchiya S., *J. Chem. Phys.* **92**, (1990) pp.4044-4054.
7. Yamanouchi K., Ikeda N., Tsuchiya S., Jonas D.M., Lundberg J.K., Adamson G.W. and Field R.W., *J. Chem. Phys.* **95** (1991) pp.6330-6342.
8. Jonas D.M., Solina S.A.B., Rajaram B., Silbey R.J., Field R.W., Yamanouchi K. and Tsuchiya S., *J. Chem. Phys.* **99** (1993) pp.7350-7370.
9. Iachello F. and Levine R.D., *Algebraic Theory of Molecules* (Oxford University Press, Oxford, 1995).
10. Sako T. and Yamanouchi K., *Chem. Phys. Lett.* **264** (1997) pp.403-410.
11. Sako T., Yamanouchi ·K. and Iachello F., *J. Chem. Phys.* **114** (2001) pp.9441-9452.
12. Sako T., Yamanouchi K. and Iachello F. *Chem. Phys. Lett.* **299** (1999) pp.35-41.
13. Sako T., Yamanouchi K. and Iachello F., *J. Chem. Phys.* **113** (2000) pp.7292-7305.
14. Sako T., Aoki D., Yamanouchi K. and Iachello F., *J. Chem. Phys.* **113** (2000) pp.6063-6069.
15. Sako T., Yamanouchi K. and Iachello F., *J. Chem. Phys.* **117** (2002) pp.1641-1648.

MOLECULAR QUASILINEARITY UNDER THE PRISM OF DYNAMICAL SYMMETRY BREAKING: A DETAILED STUDY OF THE METHINOPHOSPHIDE (HCP) $\tilde{A}-\tilde{X}$ SYSTEM

F. PÉREZ-BERNAL

Departamento de Física Aplicada, Universidad de Huelva, Huelva 21071 SPAIN
E-mail: francisco.perez@dfaie.uhu.es

F. IACHELLO

Dpts. of Physics and Chemistry, Yale University, New Haven, CT 06520-8107, USA

P.H. VACCARO

Department of Chemistry, Yale University, New Haven, CT 06520-8107, USA

H. ISHIKAWA, H. TOYOSAKI, AND N. MIKAMI

Department of Chemistry, Tohoku University, Aoba-ku, Senday 980-8578, JAPAN

An algebraic description of molecular non rigidity and the shape phase transition associated to the vibrational bending degree of freedom is presented. The algebraic approach, based on the dynamical symmetry breaking of an underlying $U(3)$ Lie algebra, allows for an unified description of radically different situations, which range from the rigid linear to the rigid bent limit. The use of the intrinsic state formalism sheds light on the connection to other approaches to this problem. The study of the methinophosphide (HCP) $\tilde{A}\,^1\mathrm{A}'' - \tilde{X}\,^1\Sigma^+$ emission spectrum, where the $\tilde{A}\,^1\mathrm{A}''$ electronic manifold shows quasi-rigid features and the $\tilde{X}\,^1\Sigma^+$ state is rigidly-linear, is presented as an example. The underlying dynamical algebra is $U(2) \otimes U(3) \otimes U(2)$, which embodies both stretching and bending degrees of freedom and permits the simultaneous quantitative description of term energies and Franck-Condon vibronic intensities.

1 Introduction

The quantification of vibronic band intensities, under the guidance of the Franck–Condon principle,[1] remains a cumbersome task for polyatomic molecules. The main cause of difficulty is the multidimensional nature of the involved potentials, which cannot be desentangled due to the mixing between the normal coordinates defined for each electronic manifold.[2] An algebraic analysis has proven successful in overcoming this difficulty and has been applied to the study of emission[3] and absorption[4] spectra in the $\tilde{C}\ ^1A' - \tilde{X}\ ^1A'$ ($\pi^* - \pi$) system, providing a very good agreement to an intensive set experimental data.

The problem of computing multidimensional Franck–Condon factors is specially complex when the molecule undergoes significant geometry changes upon electronic excitation. This is the case we are dealing with in the present work, as methinophosphide is linear in its ground electronic state ($\tilde{X}\,^1\Sigma^+$) and bent in the considered excited state ($\tilde{A}\ ^1A''$). A first approach to this problem has been accomplished using algebraic techniques,[5] a formalism that is extended in the present work.

Besides the snags associated to the geometry change, the analysis of the HCP vibronic spectra is further hindered by the quasi-rigid character of the excited electronic manifold. Recently we have shown that the algebraic method[6] allows for an elegant and simple way of modelling the spectra of non-rigid molecules.[7] The algebraic approach to this problem permits a simple, though robust and elegant, description of all the situations ranging from the rigidly-linear to the rigidly-bent cases, including the phase-shape transition that occurs between these limiting cases.

The present work is structured in two main sections, the first of which concentrates on the elucidation of the main points involved in the algebraic description on non-rigidity while the second presents the model and results for the HCP vibronic analysis. Finally we present in a last section the summary and conclusions.

2 Dynamical symmetry breaking in bent–linear triatomic molecular structures

Lie algebras have accomplished an important role in the description of complex physical systems. In particular, our interpretation is based upon the vibron scheme[6,9] that was inspired in a previous and very successful application of Lie algebraic methods to the description of nuclear structure.[10] Under this algebraic approach molecules are treated as coupled assemblies of bosonic excitations, the so called vibrons, which generate the appropriate dynamical algebra of the problem. The $U(4)$ Lie algebra was chosen as the dynamical algebra for the description of vibrorrotational excitations in a diatomic molecule.[9] The one-dimensional limit of the problem, leaving aside the rotational degrees of freedom, makes use of the $U(2)$ algebra.[6] The bending degree of freedom in triatomic linear molecules is doubly degenerate, thus the appropriate dynamical algebra is $U(3)$.[11] In the following two subsections we describe the mathematical framework of the $U(3)$ approach and elucidate the important subject of phase-shape transitions.

2.1 Dynamical algebra and dynamical symmetries

The algebraic formulation of two-dimensional problems in the framework of the vibron model was introduced in 1996.[11] The $U(3)$ dynamical algebra can be constructed by introducing boson creation ($\sigma^\dagger$, $\tau_+^\dagger$, $\tau_-^\dagger$) and annihilation (σ, τ_+, τ_-) operators where σ and $\tau_\pm$ denote scalar and circular bosons, respectively.[11] A generic Hamiltonian containing up to two-body terms can be written as:

$$\hat{H} = E_0 + \epsilon\hat{n} + \alpha\hat{n}(\hat{n}+1) + \beta\hat{l}^2 + A\hat{P} \quad , \tag{1}$$

where the operators $\hat{n}$, $\hat{l}$, and $\hat{P}$ are constructed from the boson creation and annihilation operators: $\hat{n} = \tau_+^\dagger\tau_+ + \tau_-^\dagger\tau_-$, $\hat{l} = \tau_+^\dagger\tau_+ - \tau_-^\dagger\tau_-$, and $\hat{P} = N(N+1) - (\tau_+^\dagger\sigma + \sigma^\dagger\tau_-)(\tau_-^\dagger\sigma + \sigma^\dagger\tau_+) - (\tau_-^\dagger\sigma + \sigma^\dagger\tau_+)(\tau_+^\dagger\sigma + \sigma^\dagger\tau_-) - \hat{l}^2$. The quantities E_0, ϵ, α, β, and A are adjustable modeling parameters.[5,11] The $U(3)$ algebra embodies two distinct dynamical symmetries, each of which corresponds to a situation in which $\hat{H}$ can be expressed in terms of the Casimir operators of a chain of subalgebras originating from $U(3)$. For the specific example under consideration, $(I) : U(3) \supset U(2) \supset SO(2)$ and $(II) : U(3) \supset SO(3) \supset SO(2)$ are the two limiting cases. The basis states for chain (I) are characterized by $|N, n, l\rangle$ where N, n, and l denote the total number of bound states, the number of vibrational quanta, and the magnitude of the vibrational angular momentum, respectively.[11] For a given N, the allowed values of the quantum numbers n and l are $n = N, N-1, \ldots, 1, 0$ and $l = \pm n, \pm(n-2), \ldots, \pm 1$ or 0 (n odd or even). The eigenvalues of a generic Hamiltonian with $U(2)$ symmetry are found to be

$$E^{(I)}(N,n,l) = E_0 + \epsilon n + \alpha n(n+1) + \beta l^2 \quad . \tag{2}$$

This Hamiltonian describes bending vibrations for linear molecules in either the harmonic ($\alpha = 0$) or the anharmonic ($\alpha \neq 0$) limit with l-splitting determined by β.

The basis states for chain (II) are characterized by $|N, v, l\rangle$ where N, v, and l signify the total number of bound states, the number of vibrational quanta, and the body-fixed projection of rotational angular momentum, respectively. For a given N, the vibrational index can assume the values $v = 0, 1, \ldots, \mathrm{mod}(N/2)$ while, for a specified v, the angular momentum quantum number is restricted to $l = 0, \pm 1, \pm 2, \ldots, \pm(N-2v)$. In this case l denotes the projection of angular momentum on the body-fixed figure axis, more commonly referred to as K. The eigenvalues of a generic Hamiltonian with $SO(3)$ symmetry are found to be

$$E^{(II)}(N,v,l) = E_0' - 4A\left[(N+1/2)v - v^2\right] + \beta l^2 \quad . \tag{3}$$

This Hamiltonian describes bending vibrations in non linear molecules. The mapping of the Eq. (2) spectrum to rigidly-linear species and Eq. (3) to rigidly-bent species is further clarified in the next subsection using the intrinsic state formalism.

Since our present interest is in non–rigid molecules,[8] we return to the most general two–body Hamiltonian of Eq. (1) which, excluding the zero–point energy (E_0), is characterized by the four parameters ϵ, α, β, and A. In order to further simplify ensuing analyses, the terms containing α and β are ignored and the resulting Hamiltonian is recast into a scaled form

$$\hat{H} = \epsilon\left[(1-\xi)\hat{n} + \frac{\xi}{N-1}\hat{P}\right] \quad . \tag{4}$$

The full Eq. (1) will be employed in the HCP analysis, but since the Eq. (4) model Hamiltonian contains terms belonging to both algebraic chains, it breaks the dynamical symmetries of $\hat{H}$ with the degree of symmetry breaking being characterized by the value of the control parameter ξ and provides in a simple and direct way all the necessary physical ingredients to draw a clear picture of the problem at hand. The $U(2)$ symmetry (representing rigidly–linear molecules) is recovered when $\xi = 0$ while the $SO(3)$ symmetry (representing rigidly–bent molecules) is obtained for $\xi = 1$. The situation corresponding to an arbitrary value of ξ ($0 \leq \xi \leq 1$) is depicted in Fig. 1 which represents the linear–bent correlation diagram deduced when the energy scale is set by fixing $\epsilon = 1$. The eigenvalues reported in Fig.1 were obtained by diagonalizing $\hat{H}$ in the $U(2)$ rigid–bender basis $|N, n, l\rangle$. Thus the parameter ϵ is a simple scale for the problem, while ξ id called the *control parameter* as its variation leads us from dynamical symmetry (I) to dynamical symmetry (II), with the corresponding phase transition in between.

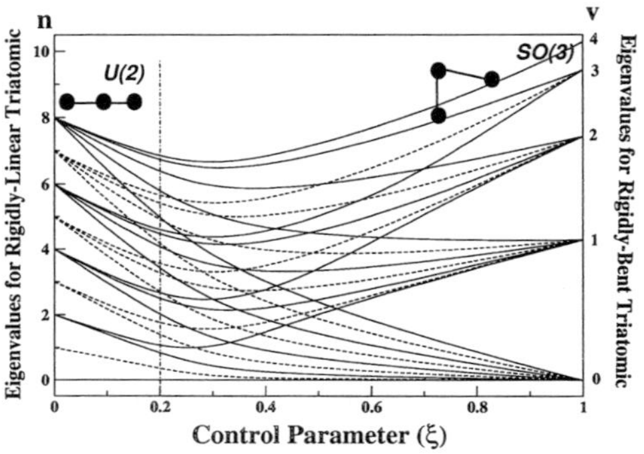

Figure 1. Linear–bent correlation diagram for vibrational energy level patterns as evaluated from the model Hamiltonian of Eq. (4) with $N = 8$ and $\epsilon = 1$. Even and odd values of the vibrational angular momentum quantum number l are depicted by continuous and dashed curves, respectively, while the critical value of the control parameter ξ (ξ_c, see text) has been marked with a dot–dashed vertical line.

2.2 Phase–shape transition

The situation described by the Hamiltonian of Eq. (4) can be clarified further by considering the classical limit of its second–quantized form. This can be accomplished succinctly by exploiting the method of intrinsic or coherent states.[12] Introducing the coherent state,

$$|N,r\rangle = \frac{1}{\sqrt{N!}}\left(b_c^\dagger\right)^N|0\rangle \ , \quad b_c^\dagger = \frac{1}{\sqrt{1+r^2}}\left[\sigma^\dagger + \frac{r}{\sqrt{2}}\left(\tau_+^\dagger + \tau_-^\dagger\right)\right] \ , \qquad (5)$$

as defined in terms of the boson condensate operator ($b_c^\dagger$) one can evaluate the corresponding expectation value of the Hamiltonian (4) in this state to obtain

$$\mathcal{E}(r) = \frac{1}{\epsilon N}\langle N,r|\hat{H}|N,r\rangle = (1-\xi)\frac{r^2}{1+r^2} + \xi\frac{(1-r^2)^2}{(1+r^2)^2} \ , \qquad (6)$$

where r represents a dimensionless distance that specifies the degree of distortion of the molecular framework from linearity (where $r = 0$). Aside from representing the ground state energy as a function of coordinate r, the energy functional $\mathcal{E}(r)$, when multiplied by ϵN, gives the potential energy function, $V(r)$, deduced from the algebraic method. Recently we have proposed a procedure to estimate the adequate scaling in order to compare the intrinsic state results to phase space calculations.[5] Minimizing $\mathcal{E}(r)$ with respect to r gives the equilibrium configuration and associated energy denoted by $\mathcal{E}_{min}$. As the control parameter ξ varies from zero to one, the equilibrium geometry is found to shift from $r = 0$ (rigidly–linear) to $r = 1$ (rigidly–bent). By examining the behavior of $\mathcal{E}_{min}$ and its derivatives with respect to ξ, the nature of the phase transition taking place between the two dynamical symmetries under consideration can be elucidated. The critical value of the control parameter is found to be $\xi_c = 1/5$. The second derivative of the energy functional (6) presents a discontinuity at this point, hence there is a second-order phase transition at this

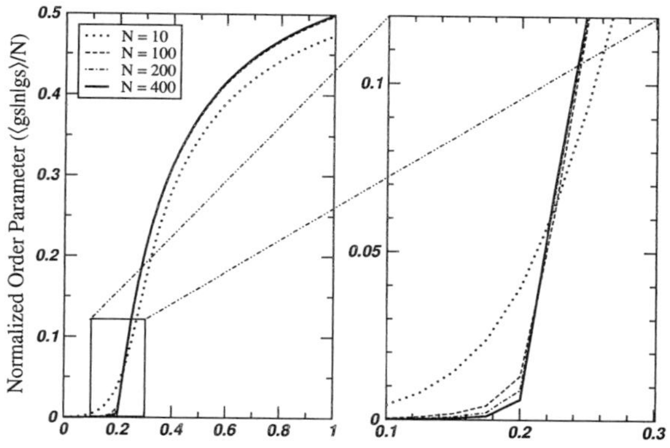

Figure 2. The order parameter defined as $\langle gs|\hat{n}|gs\rangle$ as a function of the control parameter ξ for different values of N. The right panel is a zoom of the region around the critical point, showing the increasingly abrupt behaviour as N takes larger values.

point.[10] Depicting the energy functionals[7] it can be noticed that for $\xi \leq \xi_c$ the $r = 0$ point is a minimum of the energy functional, while for $\xi > \xi_c$ it swaps to a maximum, with the appearence of a hump. For $\xi = \xi_c$ the potential has a very wide minimum for $r = 0$, causing the appearance of positive anharmonicities and anomalous l-splittings.[7] It is of great relevance the wide variety of physical situations that can be modeled through this simple algebraic formalism.

Equivalent situations to the one previously described have recently received a great deal of attention in the Nuclear Physics community, where both first[13] and second-order[14] phase transitions signatures have been found in experiment.[15] It should be remarked that the situation discussed here is similar to that encountered in a classical phase transition though these phase transitions are between two different shapes and not between different states of aggregation for matter (e.g., liquid–gas). However, the underlying formalism is common and compelling analogies can be drawn.[16] The control parameter introduced above (ξ) represents the strength of the transition–inducing interactions, rather than a thermodynamical variable (e.g., the temperature). Also, when N is finite, the order parameter (here we take as the order parameter the expectation value of the number of circular bosons in the ground state obtained after diagonalization of the Hamiltonian in Eq. (4), $\langle gs|\hat{n}|gs\rangle$, varies in an abrupt but continuous manner at the critical point $\xi = \xi_c$ as it is shown in Fig. 2. Only as $N \to \infty$ does the change in molecular shape becomes discontinuous and analogous to a true phase transition.

The theoretical approach outlined above can be used to interpret laboratory spectra, thus enabling the value of ξ, as well as the corresponding location on the linear–bent correlation diagram, to be determined for a given molecular system.[7]

3 Application to the vibronically resolved methinophosphide spectrum

A first algebraic approach to the analysis of HCP vibronic spectra has been recently carried out with promising results.[5] The present work will enlarge this previous analysis in two main ways. The first is the consideration of the full vibrational spectrum of the excited $\tilde{A}\ ^1A''$ electronic state, instead of isolating the bending overtones. The second novelty is the inclusion of the explicit dependence of the nuclear coordinate in the calculation of the wave function overlap integral for the bending degree of freedom (See below).

The scheme followed is the same than in the previous algebraic calculations of Franck-Condon intensities,[3,4,5] which basically can be resumed in two steps and that we present in the following two subsections. The first step consists on the election of a suitable dynamical algebra and the construction of an algebraic Hamiltonian for each of the involved electronic manifolds. The term energies and associated wavefunctions are obtained through diagonalization of the algebraic Hamiltonian, optimizing the accordance to the available experimental data through a least-square regression procedure. The second step involves the calculation of the relevant overlap integrals between the previously computed wave function which provides the relative transition intensities. In this case we manually adjust the potential concavities and geometrical offsets in order to maximize the agreement to the experimental spectra.

3.1 Theoretical analysis: Vibrational frequencies

The appropriate dynamical algebra for the problem under consideration is $U(2) \otimes U(3) \otimes U(2)$, where the $U(2)$ algebras model the stretching vibrations while the $U(3)$ algebra models the degenerate bending vibration in the case of linear species or the bending degree of freedom and rotation around the figure axis for bent molecules.[5] The calculations will be carried out in the linear–local basis, associated to the chain of algebras

$$\begin{array}{ccccccc} U_1(2)\otimes & U_2(3)\otimes & U_3(2) \supset & SO_1(2)\otimes & (U_2(2) \supset & SO_2(2))\otimes & SO_3(2) \\ \downarrow & \downarrow & \downarrow & \downarrow & \searrow & \swarrow & \downarrow \\ \mid [N_1] & [N_2] & [N_3]; & v_1 & & n_2{}^l & v_3 \rangle \end{array} , \qquad (7)$$

where the subscript "1" refers to the CH stretching, "2" to the bending and "3" to the CP stretching degree of freedom. For the sake of brevity we will represent the basis states as $\{|v_1 n_2{}^l v_3\rangle\}$. The quantum numbers N_i, which corresponds to the labels of the totally symmetric representations for each algebra, are related to the number of bound states; besides, to the depth of the corresponding potential.[6,11] The quantum numbers $v_i = 0, 1, \ldots, \text{mod}(N_i/2)$ is the number of excitation quanta in the local stretching degree of freedom, while $n_2 = 0, 1, \ldots, N_2$ and $l = n, n-2, \ldots, 1(\text{or } 0)$ denote the number of vibrational quanta in the bending degree of freedom and the associated vibrational angular momentum respectively.

The algebraic Hamiltonian employed, once the relevant parameters were sifted out, has the following form

$$\hat{H} = A_1\hat{C}_1 + A_3\hat{C}_3 + A_{13}\hat{C}_1\hat{C}_3 + \epsilon\,\hat{n} + \alpha\,\hat{n}(\hat{n}+1) + \beta\,\hat{l}^2 + A\,\hat{P}$$

$$+ A_{12}\hat{C}_1\hat{n} + A_{23}\hat{C}_3\hat{n} + f_{23}\hat{F}_{23} + f_{231}\hat{F}_{23,1} + f_{232}\hat{F}_{23,2} \quad , \tag{8}$$

which is block diagonal in l and v_1. The $\hat{C}_i$ operators are $SO_i(2)$ Casimir operators diagonal in the Eq. (7) basis with matrix elements $\langle v_1 n_2{}^l v_3|\hat{C}_i|v_1 n_2{}^l v_3\rangle = v_i - v_i^2/N_i$ for $i = 1, 3$. The $\hat{n}$ and $\hat{l}^2$ operators are Casimir operators of $U_2(2)$ and $SO_2(2)$, therefore they are diagonal in the selected basis with matrix elements $\langle v_1 n_2{}^l v_3|\hat{n}|v_1 n_2{}^l v_3\rangle = n_2$ and $\langle v_1 n_2{}^l v_3|\hat{l}^2|v_1 n_2{}^l v_3\rangle = l^2$. The $\hat{P}$ operator is related to the $U_2(3)$ pairing operator, and thus is not diagonal in the bending labels of Eq. (7) basis, having matrix elements

$$\begin{aligned}\langle v_1 n_2'{}^l v_3|\ \hat{P}\ |v_1 n_2{}^l v_3\rangle &= \langle v_1 n_2'{}^l v_3|\hat{N}_2(\hat{N}_2+1) - \hat{W}^2|v_1 n_2{}^l v_3\rangle \\ &= (N_2(\ N_2\ +1) - (N_2 - n_2)(n_2+2) - (N_2 - n_2 + 1)n_2 - l^2)\,\delta^{n_2}_{n_2'} \\ &\quad + \sqrt{(N_2 - n_2)(N_2 - n_2 - 1)(n_2 - l + 2)(n_2 + l + 2)}\,\delta^{n_2+2}_{n_2'} \\ &\quad + \sqrt{(N_2 - n_2 + 2)(N_2 - n_2 + 1)(n_2 + l)(n_2 - l)}\,\delta^{n_2-2}_{n_2'} \ .\end{aligned} \tag{9}$$

The $\hat{F}_{23}$, $\hat{F}_{23,1} = \hat{F}_{23}\hat{C}_1$, and $\hat{F}_{23,2} = \hat{F}_{23}\hat{n} + \hat{n}\hat{F}_{23}$ are off–diagonal $2:1$ Fermi interactions between the bending and CP stretching. The $\hat{F}_{23}$ operator has matrix elements

$$\begin{aligned}&\langle v_1 n_2'{}^l v_3'|\ \hat{F}_{23}\ |v_1 n_2{}^l v_3\rangle = \\ &\frac{1}{2}\sqrt{(1-\frac{v_3}{N_3})(v_3+1)(1-\frac{n_2-1}{N_2})(1-\frac{n_2-2}{N_2})(n_2+l)(n_2-l)}\,\delta^{n_2-2,v_3+1}_{n_2',v_3'} \\ &+\frac{1}{2}\sqrt{(1-\frac{v_3-1}{N_3})v_3(1-\frac{n_2}{N_2})(1-\frac{n_2+1}{N_2})(n_2+l+2)(n_2-l+2)}\,\delta^{n_2+2,v_3-1}_{n_2',v_3'} \quad ,\end{aligned} \tag{10}$$

while the other two connect the same states and can be trivially computed from this one. The Fermi operator are vital to reproduce the appropriate polyad scheme in the ground electronic state. The diagonalization of Eq. (8) provides explicit values for the vibrational expansion coefficients in both electronic states involved

$$|\Gamma;(\nu_1,\nu_2^l,\nu_3)\rangle = \sum_i a_i^{\Gamma;(\nu_1,\nu_2^l,\nu_3)}|\Gamma; v_1, n_2^l, v_3\rangle_i \ , \quad \Gamma = \tilde{X}, \tilde{A} \quad . \tag{11}$$

3.2 Theoretical analysis: Franck-Condon factors and intensity calculation

The intensity of the vibronic $|\Gamma';(\nu_1',\nu_2^{ll'},\nu_3')\rangle \rightarrow |\Gamma;(\nu_1,\nu_2^l,\nu_3)\rangle$ is defined as a function of the corresponding Franck-Condon factors $I[\Gamma';(\nu_1',\nu_2^{ll'},\nu_3') \rightarrow \Gamma;(\nu_1,\nu_2^l,\nu_3)] \propto \nu^4 \left|T_{\Gamma';(\nu_1',\nu_2^{ll'},\nu_3')\rightarrow\Gamma;(\nu_1,\nu_2^l,\nu_3)}\right|^2$, where $\nu = T_{00} + G_{\Gamma',\nu'} - G_{\Gamma,\nu}$ is a phase space factor and the Franck-Condon factor are defined as the squared overlap integrals of the vibrational wave functions belonging to different electronic manifolds, $\tilde{X}$ and $\tilde{A}$. In the present case $T_{\tilde{A}_{(\nu_1',\nu_2',\nu_3',K)}\rightarrow\tilde{X}_{(\nu_1,\nu_2^l,\nu_3)}} = \langle\tilde{A},(\nu_1',\nu_2',\nu_3',K)|\tilde{X},(\nu_1,\nu_2^l,\nu_3)\rangle$.

Making use of the wave functions obtained after diagonalization of the Hamiltonian in Eq. (8), and taking into account the local character of the defined basis, the Franck-Condon factor can be rewritten as

$$T_{\tilde{A};(\nu_1',\nu_2',\nu_3',K)\to\tilde{X};(\nu_1,\nu_2^l,\nu_3)} = \sum_{(v_1',n_2',v_3')_j}\sum_{(v_1,n_2,v_3)_i} a_j^{\tilde{A};(\nu_1',\nu_2'^K,\nu_3')} a_i^{\tilde{X};(\nu_1,\nu_2^l,\nu_3)} \times \langle\tilde{A};v_{1j}'|\tilde{X};v_{1i}\rangle\langle\tilde{A};n_{2j}',K|r|\tilde{X};n_{2i},l\rangle\langle\tilde{A};v_{3j}'|\tilde{X};v_{3i}\rangle \quad . \tag{12}$$

The overlap has been splitted into one-dimensional overlaps for the stretches, for which we use the same analytic formula as in previous cases,[3] and a two-dimensional bending overlap where we include the explicit dependence of the coordinate as a first approximation to the behavior of the dipole function in such linear-to-bent transitions. The two dimensional overlap formula is analytical and equal to

$$\begin{aligned}\langle v,l|r^p|\bar{v}',l'\rangle &= \int_0^\infty \psi_{v,l}^*(\alpha,r)r^p\psi_{v',l'}(\alpha',r)rdr = \mathcal{N}_{v,l}\mathcal{N}_{v',l'}\alpha^{l+1}\alpha'^{l'+1} \\ &\times \frac{\left(\frac{v+l}{2}\right)!\left(\frac{v'+l'}{2}\right)!}{2}\sum_{i=0}^{\frac{v-l}{2}}(-1)^i\frac{\alpha^{2i}}{\left(\frac{v-l}{2}-i\right)!(i+l)!i!} \\ &\times \sum_{j=0}^{\frac{v'-l'}{2}}(-1)^j\frac{\alpha'^{2j}}{\left(\frac{v'-l'}{2}-j\right)!(j+l')!j!}\left(\frac{\alpha^2+\alpha'^2}{2}\right)^{-\left(\frac{l+l'+p}{2}+i+j+1\right)} f(i,j) \quad ,\end{aligned} \tag{13}$$

where

$$f(i,j) = \begin{cases}(i+j+\frac{l+l'+p}{2})! & l+l'+p = \text{even} \\ \frac{\sqrt{\pi}(2i+2j+l+l'+p)!!}{2^{i+j+(l+l'+p+1)/2}} & l+l'+p = \text{odd}\end{cases} \quad . \tag{14}$$

The concavities α and α' for the optically connected bending potentials are in general different and can be estimated as $\alpha = \sqrt{\mu\omega/\hbar}$.

3.3 Results and discussion

The least square optimization of the modelling parameters in Eq. (8) for both electronic states lead to the parameters in table 1. We found an agreement in accordance to the experimental accuracy. In the ground electronic state the resulting rms is 1.23 for a fit to 111 experimental levels —where the experimental levels with energies higher than 12000 cm^{-1} were excluded in order to avoid the effects linked to molecular isomerization, for experimental data references see Ref. [5] and references therein— appropriatedly weighted in order to take into account the experimental errors. The table 1 parameters for the excited electronic state provide a $rms = 6.98\,cm^{-1}$ for the 63 available experimental levels. The results for intensities are depicted in Fig. 3. The $K = 1$ excited states can undergo perpendicular transitions to $l = 0$ and $l = 2$ levels of the ground electronic state, and these pathways cannot be distiniguished at the present resolution. Therefore their intensities have been summed in the experimental and calculated data sets.

The intensitiy parameters, with values given in Fig. 3 caption were manually adjusted in order to optimize the agreement with experiment. The abscence of activity in the CH stretch mode precludes from their inclusion in the theoretical analysis. In all the cases plotted in Fig. 3 a semi–quantitative agreement has been

Table 1. Hamiltonian parameters for the algebraic description of the HCP vibrational structure (units: cm^{-1}); $N_1 = 60$ for both electronic states, while $N_2 = 50$, $N_3 = 240$ for the $\tilde{X}$ and $N_2 = 29$, $N_3 = 160$ for the $\tilde{A}$ state.

Parameter	A_1	A_3	A_{13}	ϵ	α	β
$\tilde{X}$	3270.6(4)	1283.7(3)	−5.1(9)	673.2(1)	−2.02(2)	4.66(8)
$\tilde{A}$	2986.8(5)	997.6(5)	15.6(5)	620.6(15)	−13.05(7)	13.77(9)
Parameter	A	A_{12}	A_{23}	f_{23}	f_{231}	f_{232}
$\tilde{X}$	–	−16.4(4)	6.80(8)	−1.3(2)	5.1(3)	−0.63(1)
$\tilde{A}$	7.78(1)	–	5.40(6)	−6.3(3)	–	–

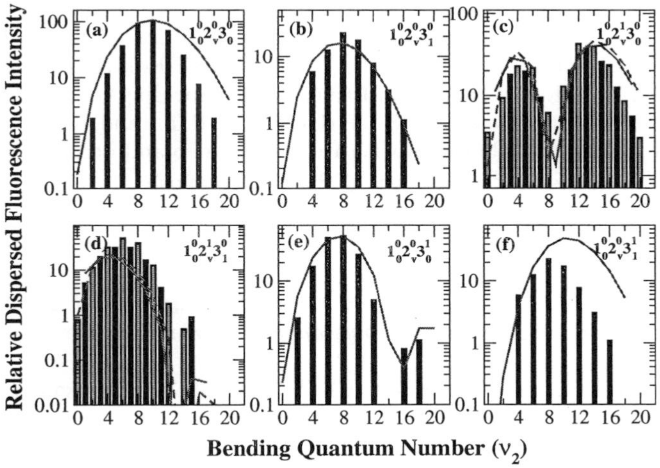

Figure 3. Algebraic analysis of the HCP $\tilde{A}\,^1A'' - \tilde{X}\,^1\Sigma^+$ emission spectrum. In the six panels the experimental data are represented by bars and the calculated transition strengths by curves. It is indicated in each panel which is the depicted progression. Panels **(c)** (2^1_v progression) and **(d)** ($2^1_v 3^0_1$) include the results for selective excitation of the $|\tilde{A};(010), K=1\rangle$ and $|\tilde{A};(010), K=0\rangle$ with full and dashed curves and full and empty bars respectively. The optimized intensity parameters values are $\alpha_2^{\tilde{X}} = 3.21\,\text{Å}^{-1}$, $\alpha_2^{\tilde{A}} = 3.02\,\text{Å}^{-1}$, $\alpha_3^{\tilde{X}} = 18\,\text{Å}^{-1}$, $\alpha_3^{\tilde{A}} = 14\,\text{Å}^{-1}$, and $\Delta R_3 = 0.106$ Å.

realized between theory and experiment, reproducing correctly the nodal patterns, which is a notable achievement considering the difficulties that plague the system.

4 Summary and conclusions

A novel algebraic formalism based on the $U(3)$ Lie algebra that permits the inclusion of non-rigidity in the study of molecular structure has been presented, stressing the possibility of modelling the phase–shape transition that takes place between the two possible limiting cases.

Experimental data for the HCP $\tilde{A} - \tilde{X}$ system has been analyzed using a $U(2) \otimes U(3) \otimes U(2)$ dynamical algebra that exploits the capabilities of the previously mentioned algebraic model. The vibrational term energies for the electronic states involved have been calculated, and a good agreement has been found with experiment. In particular the ground electronic state has shown rigidly-linear behaviour while the excited $\tilde{A}$ state has a more complicated structure with evidence of a quasi-rigid bent character. Semiquantitative agreement has been realized between experimentally measured and theoretically modelled dispersed fluorescence spectra induced through excitation of the $(000)^{K=1}$, $(010)^{K=0,1}$, and $(001)^{K=1}$ levels in the $\tilde{A}$ manifold.

Acknowledgments

This work was conducted under the auspices of grants provided by the US D.O.E. (DE-FG01-91ER40608) and the US N.S.F. (CHE-9975428). The authors wish to thank Prof. K.K. Lehmann of Princeton for providing copies from the Ph.D dissertation of M.A. Mason. F.P.B., P.H.V., H.I., H.T., and N.M. wish to ackowledge the ongoing collaboration with Prof. F.Iachello and dedicate to him their work on behalf of his 60th aniversary. ¡Feliz Cumpleaños!

References

1. G. Herzberg, *Molecular Spectra and Molecular Structure: III. Electronic Spectra and Electronic Structure of Polyatomic Molecules* (Van Nostrand Reinhold, NY, 1966).
2. F. Duschinsky, *Acta Physicochim.* URSS **7**, 551 (1937); I. Özkan, *J. Mol. Spectrosc.* **139**, 147 (1990).
3. T. Müller *et al.*, *Chem. Phys. Lett.* **292**, 243 (1998); T. Müller *et al.*, *J. Chem. Phys.* **111**, 5038 (1999); *J. Chem. Phys.* **112**, 6507 (2000).
4. T. Müller, *et al.*, *Chem. Phys. Lett.* **329**, 271 (2000).
5. H. Ishikawa, *et al.*, *Chem. Phys. Lett.* **329**, 271 (2000).
6. F. Iachello and R. D. Levine, *Algebraic Theory of Molecules*, (Oxford University Press, New York, 1995), further references therein.
7. F. Iachello, F. Pérez-Bernal, and P.H. Vaccaro, submitted to *Chem. Phys. Lett.*.
8. B. P. Winnewisser, in *Molecular Spectroscopy: Modern Research*, edited by K. Narahari Rao, Vol. III, p. 321 (Academic Press, New York, 1985).
9. F. Iachello, *Chem. Phys. Lett.* **78**, 581 (1981).
10. F. Iachello and A. Arima, *The Interacting Boson Model*, Cambridge University Press, Cambridge, (1987).
11. F. Iachello y S. Oss, *J. Chem. Phys.* **104**, 6956 (1996).
12. A. E. L. Dieperink, O. Scholten, and F. Iachello, *Phys. Rev. Lett.* **44**, 1747 (1980); J. Ginocchio and M. W. Kirson, *Phys. Rev. Lett.* **44**, 1744 (1980); A. Bohr and B. R. Mottelson, *Phys. Scr.* **22**, 468 (1980).
13. F. Iachello, *Phys. Rev. Lett.* **85**, 3580 (2000).
14. F. Iachello, *Phys. Rev. Lett.* **87**, 052502 (2001).
15. R.F. Casten and N.V. Zamfir, *Phys. Rev. Lett.* **85**, 3582 (2000); *Phys. Rev. Lett.* **87**, 052503 (2001).
16. J. Jolie *et al.*, *Phys. Rev. Lett.* **89**, 182502 (2002); J.E. García-Ramos *et al.*, submitted to *Phys. Rev. Lett.*.

ALGEBRAIC METHODS FOR THE QUANTITATIVE INTERPRETATION OF VIBRONICALLY-RESOLVED MOLECULAR SPECTRA: THE STRUCTURE AND DYNAMICS OF DISULFUR MONOXIDE (S_2O)

P. H. VACCARO AND T. MÜLLER
Department of Chemistry, Yale University, New Haven, CT 06520 USA
E-mail: patrick.vaccaro@yale.edu

F. IACHELLO
Departments of Physics & Chemistry, Yale University, New Haven, CT 06520 USA

F. PÉREZ-BERNAL
Departamento de Física Aplicada, Universidad de Huelva, Huelva 21071 SPAIN

A coupled $U(2)$ algebraic theory has been employed to perform detailed analyses on absorption and emission spectra recorded for the $\tilde{C}\,{}^1\mathrm{A}' - \tilde{X}\,{}^1\mathrm{A}'$ $(\pi^* \leftarrow \pi)$ electronic system of jet-cooled disulfur monoxide (S_2O) molecules. Vibronically-resolved features possessing up to 20 quanta of excitation in the ν_2 S−S stretching mode of the $\tilde{X}$ state ($E_{\mathrm{vib}} < 14000\,\mathrm{cm}^{-1}$) and up to 8 quanta of excitation in the analogous ν_2' vibration of the $\tilde{C}$ state ($E_{\mathrm{vib}} < 3500\,\mathrm{cm}^{-1}$) have been examined. Aside from providing an economical description for the inherently anharmonic and strongly coupled patterns of energy levels that distinguish highly-excited polyatomic species, the algebraic approach enables facile evaluation of multidimensional Franck-Condon factors required for the interpretation of spectral intensities. This ability to extract wavefunction information directly from spectroscopic data sets has revealed pronounced differences in the vibrational dynamics supported by the $\tilde{C}\,{}^1\mathrm{A}'$ and $\tilde{X}\,{}^1\mathrm{A}'$ manifolds, with the latter found to be substantially more "local" in character than the former.

1 Introduction

Franck-Condon factors [1-3] play pivotal roles for the modeling of molecular transition probabilities whenever a "sudden approximation" is evoked in the framework of the adiabatic (Born-Oppenheimer) approximation and dependence of the electronic transition moment on nuclear degrees of freedom can be neglected (the Condon approximation). Aside from unraveling the structural/dynamical information contained in molecular spectra, [4-6] these quantities provide a valuable tool for describing non-radiative relaxation phenomena. [7] Despite such diverse applications, the calculation of Franck-Condon factors for polyatomic species remains a formidable task owing to the multidimensional nature of vibrational coordinates and difficulties incurred by their transformation among electronic states (the Duschinsky effect [8]). A novel Lie algebraic approach has been proposed for the facile evaluation of polyatomic Franck-Condon factors and related transition moment matrix elements. [9] This coupled $U(2)$ formalism allows the effects of mechanical anharmonicity to be incorporated from the onset and avoids complications associated with the mixing of vibrational coordinates by introducing a basis of local oscillators (or "vibrons"). The resulting vibron scheme has been used to interpret jet-cooled emission and absorption spectra obtained for transient disulfur monoxide (S_2O) molecules through selective excitation of isolated vibronic bands comprising the near-ultraviolet $\tilde{C}\,{}^1\mathrm{A}' - \tilde{X}\,{}^1\mathrm{A}'$ $(\pi^* \leftarrow \pi)$ electronic transition. [10-12]

2 Lie Algebraic Analyses

2.1 The Coupled U(2) Scheme

The $U(2)$ algebraic scheme enables the vibrational structure of a polyatomic system to be treated as an assembly of η coupled Morse (and/or Pöschl-Teller) oscillators, [13,

14] where η equals the number of vibrational degrees of freedom. This representation has the form of a tensor product vector space: $U_1(2) \otimes U_2(2) \otimes U_3(2) \otimes \ldots \otimes U_\eta(2)$ with each subscript ($i = 1,2,3,\ldots\eta$) denoting a distinct *local* degree of freedom. [15-17] The vibron Hamiltonian for a particular vibrational manifold is constructed from invariant Casimir operators belonging to the possible subalgebras of the encompassing dynamical algebra. Subsequent diagonalization is facilitated by use of a "local" (uncoupled) representation that can be identified as the eigenbasis for an assembly of non-interacting *anharmonic* oscillators and follows from the complete chain of branching algebras. [13, 14] For the nonlinear triatomic S_2O species this leads to basis states of the form:

$$\begin{array}{ccccccccccccc} U_a(2) & \otimes & U_b(2) & \otimes & U_c(2) & \supset & SO_a(2) & \otimes & SO_b(2) & \otimes & SO_c(2) & \supset & SO(2) \\ \downarrow & & \downarrow & & \downarrow & & \downarrow & & \downarrow & & \downarrow & & \downarrow \\ \|[N_a] & & [N_b] & & [N_c] & & \upsilon_a & & \upsilon_b & & \upsilon_c & & V\rangle \\ \Updownarrow & & \Updownarrow & & \Updownarrow & & \Updownarrow & & \Updownarrow & & \Updownarrow & & \Updownarrow \\ N_{SO} & & N_{SS} & & N_{SSO} & & \upsilon_{SO} & & \upsilon_{SS} & & \upsilon_{SSO} & & \sum_i \upsilon_i \end{array} \tag{1}$$

where the subscripts a, b, and c signify the *local* S – O stretching, S – S stretching, and S – S – O bending degrees of freedom, respectively. Each $U_i(2)$ irreducible representation is characterized by a boson number, N_i, which is related to the depth of the potential well and specifies the number of bound levels as $1 + N_i/2$ or $1 + (N_i - 1)/2$ for N_i even or odd. The associated $SO_i(2)$ labels, υ_i, serve as a vibrational index for the corresponding anharmonic local oscillator and are defined such that $\upsilon_i = 0,1,\ldots N_i/2$ or $(N_i - 1)/2$. Finally, the muliplet number arising from the terminal $SO(2)$ subalgebra, V, denotes the total number of excited vibrational quanta.

The interactions taking place between "local oscillators" are a crucial aspect of intramolecular dynamics and can be partitioned into two broad categories depending upon whether they are diagonal or non-diagonal in the uncoupled (tensor product) basis. These essential characteristics can be incorporated into a model vibron Hamiltonian given by:

$$\hat{H} = E_0 + \sum_{i=1}^{\eta} A_i \hat{C}_i + \sum_{i \le j}^{\eta} A_{ij} \hat{C}_i \hat{C}_j + \sum_{i \le j}^{\eta} \lambda_{ij} \hat{M}_{ij} + \cdots \tag{2}$$

where terms involving isolated Casimir operators, $\hat{C}_i$, define an uncoupled set of η anharmonic oscillators while the product Casimir operators, $\hat{C}_i\hat{C}_j$, and Majorana operators, $\hat{M}_{ij}$, describe the diagonal and non-diagonal interactions, respectively, which occur in the local basis. [15, 17] The effective zero-point energy is denoted as E_0, with the symbols A_i, A_{ij}, and λ_{ij} signifying adjustable parameters that reflect the detailed vibrational behavior of a particular system. Within the local tensor product representation, the matrix elements of operators $\hat{C}_i$ and $\hat{C}_i\hat{C}_j$ follow from: [13, 14]

$$\left\langle N_i, \upsilon_i \middle| \hat{C}_i \middle| N_i, \upsilon_i' \right\rangle = \left(\upsilon_i - \frac{\upsilon_i^2}{N_i} \right) \delta_{\upsilon_i, \upsilon_i'} \tag{3}$$

while the matrix elements for the Majorana operators, $\hat{M}_{ij}$, can be formulated as: [18]

$$\langle N_i,\upsilon_i;N_j,\upsilon_j|\hat{M}_{ij}|N_i,\upsilon_i';N_j,\upsilon_j'\rangle = \left[\left(1-\frac{\upsilon_j}{N_j}\right)(\upsilon_j+1)\left(1-\frac{\upsilon_i-1}{N_i}\right)\upsilon_i\right]^{1/2}\delta_{\upsilon_j,\upsilon_j'+1}\,\delta_{\upsilon_i,\upsilon_i'-1}$$
$$+\left[\left(1-\frac{\upsilon_i}{N_i}\right)(\upsilon_i+1)\left(1-\frac{\upsilon_j-1}{N_j}\right)\upsilon_j\right]^{1/2}\delta_{\upsilon_i,\upsilon_i'+1}\,\delta_{\upsilon_j,\upsilon_j'-1} \tag{4}$$

where δ_{ij} denotes the canonical Kronecker Delta function. The matrix elements in Eqs. (3) and (4) have been given in a form that reduces to conventional normal mode coupling expressions in the harmonic limit (*i.e.*, when $N_i \to \infty$). [17] Owing to their non-diagonal representation in the uncoupled algebraic basis, the Majorana operators mediate an interaction of local oscillators that eventually leads to the manifestation of collective vibrational behavior. [13, 14] When incorporated into the vibron Hamiltonian as part of a higher-order term, the Majorana interaction introduces a quantum-number-dependent coupling of local oscillators that allows for the modeling of more complex vibrational dynamics such as normal-to-local or local-to-normal transitions. Diagonalization of $\hat{H}$ yields desired vibrational eigenvalues (*i.e.*, fundamental frequencies and overtones) and their corresponding eigenvectors, with the latter expressed as an expansion over (uncoupled) tensor product basis states.

2.2 *Transition Intensities and Franck-Condon Factors*

The algebraic evaluation of multidimensional Franck-Condon factors and associated vibronic transition intensities requires detailed knowledge of the vibrational eigenstates supported by two distinct potential surfaces. Such information can be obtained readily through least-squares adjustment of modeling parameters embodied in the appropriate vibron Hamiltonian for each electronic manifold so as to reproduce experimentally-observed patterns of vibrational energy levels. [13, 14] In particular, this procedure yields expressions for the corresponding vibrational eigenvectors in terms of the local anharmonic basis. Independent diagonalization of vibron Hamiltonians for the $\tilde{X}\,^1A'$ and $\tilde{C}\,^1A'$ states of S_2O provide explicit values for the vibrational expansion coefficients:

$$|\tilde{X};\upsilon_1,\upsilon_2,\upsilon_3\rangle = \sum_{\upsilon_a,\upsilon_b,\upsilon_c} d^{\upsilon_1,\upsilon_2,\upsilon_3}_{\upsilon_a,\upsilon_b,\upsilon_c}|\tilde{X};\upsilon_a,\upsilon_b,\upsilon_c\rangle \qquad |\tilde{C};\upsilon_1',\upsilon_2',\upsilon_3'\rangle = \sum_{\upsilon_a',\upsilon_b',\upsilon_c'} c^{\upsilon_1',\upsilon_2',\upsilon_3'}_{\upsilon_a',\upsilon_b',\upsilon_c'}|\tilde{C};\upsilon_a',\upsilon_b',\upsilon_c'\rangle \tag{5}$$

where the numerical subscripts appearing on vibrational quantum numbers signify the conventional mode ordering for a bent triatomic system [10] and the N_i labels of the local basis have been dropped for notational convenience. Assuming that the transition-inducing electric dipole moment operator does not exhibit a significant dependence upon nuclear coordinates (the Condon approximation), the probability or intensity, $I_{\upsilon_1'\upsilon_2'\upsilon_3'-\upsilon_1\upsilon_2\upsilon_3}$, of a vibronic absorption or emission process connecting electronically-excited state $|\tilde{C};\upsilon_1',\upsilon_2',\upsilon_3'\rangle$ with ground state $|\tilde{X};\upsilon_1,\upsilon_2,\upsilon_3\rangle$ is proportional to the corresponding Franck-Condon factor, $q_{\upsilon_1'\upsilon_2'\upsilon_3'-\upsilon_1\upsilon_2\upsilon_3}$:

$$I_{\upsilon_1'\upsilon_2'\upsilon_3'-\upsilon_1\upsilon_2\upsilon_3} \propto \nu^n\, q_{\upsilon_1'\upsilon_2'\upsilon_3'-\upsilon_1\upsilon_2\upsilon_3} \tag{6}$$

where the exponent n assumes the values of 1 or 4 to account for the frequency scaling of photon absorption or emission intensity, respectively. The Franck-Condon factor of Eq. (6) is defined by the square modulus of a vibrational overlap integral: [1-3]

$$q_{\upsilon_1'\upsilon_2'\upsilon_3'-\upsilon_1\upsilon_2\upsilon_3} = \left|\langle\tilde{X};\upsilon_1,\upsilon_2,\upsilon_3|\tilde{C};\upsilon_1',\upsilon_2',\upsilon_3'\rangle\right|^2. \tag{7}$$

These overlap integrals are inherently multidimensional; however, the local anharmonic basis of the vibron model introduces both conceptual and practical simplifications. In particular, the eigenvector expansions of Eq. (5) yields:

$$\left\langle \tilde{X};\upsilon_1,\upsilon_2,\upsilon_3 \middle| \tilde{C};\upsilon_1',\upsilon_2',\upsilon_3' \right\rangle = \sum_{\substack{\upsilon_a',\upsilon_b',\upsilon_c' \\ \upsilon_a,\upsilon_b,\upsilon_c}} \left(d^{\upsilon_1,\upsilon_2,\upsilon_3}_{\upsilon_a,\upsilon_b,\upsilon_c}\right)^* c^{\upsilon_1',\upsilon_2',\upsilon_3'}_{\upsilon_a',\upsilon_b',\upsilon_c'} \left\langle \tilde{X};\upsilon_a,\upsilon_b,\upsilon_c \middle| \tilde{C};\upsilon_a',\upsilon_b',\upsilon_c' \right\rangle \tag{8}$$

where the tensor product nature of the uncoupled representation leads to a further partitioning into computationally-expedient products of one-dimensional integrals:

$$\left\langle \tilde{X};\upsilon_a,\upsilon_b,\upsilon_c \middle| \tilde{C};\upsilon_a',\upsilon_b',\upsilon_c' \right\rangle = \left\langle \tilde{X};\upsilon_a \middle| \tilde{C};\upsilon_a' \right\rangle \left\langle \tilde{X};\upsilon_b \middle| \tilde{C};\upsilon_b' \right\rangle \left\langle \tilde{X};\upsilon_c \middle| \tilde{C};\upsilon_c' \right\rangle. \tag{9}$$

Furthermore, the intrinsic coordinates utilized by the vibron scheme permit changes in equilibrium geometry between the excited ($\tilde{C}\,^1\mathrm{A}'$) and ground ($\tilde{X}\,^1\mathrm{A}'$) electronic states to be specified completely in terms of simple translational displacements, [9-11] thereby eliminating the complications (the Duschinsky effect [8]) incurred during analogous normal mode treatments (which also neglect the influence of mechanical anharmonicity).

2.3 Calculation of One-Dimensional Overlap Integrals

The local basis integrals of Eq. (9) entail the eigenstates of one-dimensional Morse (or Pöschl-Teller) potentials and can be evaluated by a variety of methods. Rather than resorting to numerical procedures, our Franck-Condon calculations make use of an analytical formula [9, 19] derived through perturbative correction of the analogous overlap expression for two harmonic oscillators characterized by different frequencies, ω (ω'), and concavities, α (α'), as well as relative spatial displacement, Δ:

$$\begin{aligned} \langle \upsilon | \upsilon' \rangle &= \int_{-\infty}^{+\infty} \psi_\upsilon^*(\alpha;x)\psi_{\upsilon'}(\alpha';(x-\Delta))\,dx \\ &= \exp\left(\frac{-(\alpha\alpha'\Delta)^2}{2(\alpha^2+\alpha'^2)}\right)\left(\frac{\alpha\alpha'\upsilon!\upsilon'!}{2^{\upsilon+\upsilon'}}\frac{2}{\alpha^2+\alpha'^2}\right)^{1/2}\left(\frac{1}{\alpha^2+\alpha'^2}\right)^{\frac{\upsilon+\upsilon'}{2}} \sum_{l=0}^{\min\{\upsilon,\upsilon'\}} \frac{1}{l!\left(-\frac{(\alpha\alpha'\Delta)^2}{\alpha^2+\alpha'^2}\right)^l} \\ &\times \sum_{\substack{\upsilon\ge j\ge l \\ j=\upsilon \bmod 2}} \sum_{\substack{\upsilon'\ge j'\ge l \\ j'=\upsilon' \bmod 2}} \frac{(\alpha^2-\alpha'^2)^{\frac{\upsilon-j}{2}}(\alpha'^2-\alpha^2)^{\frac{\upsilon'-j'}{2}}}{(\alpha^2+\alpha'^2)^{\frac{j+j'}{2}}} \frac{\alpha^j\alpha'^{j'}(\alpha'^2)^j(-\alpha^2)^{j'}(2\Delta)^{j+j'}}{\left(\frac{\upsilon-j}{2}\right)!\left(\frac{\upsilon'-j'}{2}\right)!(j-l)!(j'-l)!} \end{aligned} \tag{10}$$

with $\alpha = \sqrt{\mu\omega/\hbar}$ defining the concavity for a harmonic oscillator of reduced mass μ and frequency ω. This expression can be adapted to approximate the overlap integrals for anharmonic (*e.g.*, Morse-type) oscillators, with first-order perturbation theory yielding corrections for the concavity, α (α'), and displacement, Δ, of the form:

$$\alpha = \alpha_0(1-\xi\upsilon) \qquad \alpha' = \alpha_0'(1-\xi'\upsilon') \qquad \Delta = \Delta_0 - \eta\upsilon + \eta'\upsilon' \tag{11}$$

where the values of $\alpha_0, \xi, \text{and } \eta$ ($\alpha_0', \xi', \text{and } \eta'$) can be estimated from intrinsic parameters of the corresponding Morse or Pöschl-Teller potential.

Despite the simplicity engendered by Eq. (10), the explicit determination of overlap parameters still represents a formidable task. Detailed consideration of emission traces obtained through excitation of the 0_0^0 origin band enabled the influence of Δ_0 and α_0 to be examined in isolation since the vibrationless $\tilde{C}\,^1\mathrm{A}'$ level is neither mixed by Majorana coupling nor affected by the values of ξ and η. Consequently, the displacement and concavity for each local oscillator (Δ_0^i and α_0^i where $i = \mathrm{SO, SS, or\ SSO}$) could be quantified by analyzing the intensity envelopes of vibronic progressions involving the corresponding vibrational degree of freedom, with perturbative estimates of ξ and η

allowing initial calculation of dispersed fluorescence amplitudes for remaining excitation lines. An iterative, manual adjustment of the parameters α_0, ξ, η, and Δ_0 was then performed in order to obtain a global fit for all observed emission intensities, [10] with further refinements introduced by subsequent modeling of absorption data. [11]

3 Results and Discussion

3.1 Vibrational Energies for the $\tilde{X}\ {}^1A'$ Ground State

The long vibronic progressions observed in S_2O $\tilde{C}-\tilde{X}$ dispersed fluorescence spectra provide an extensive ground state data set consisting of 240 distinct vibrational levels which span $\sim 14000\,\text{cm}^{-1}$ of internal energy within the $\tilde{X}\ {}^1\text{A}'$ potential surface. Initial attempts to reproduce such laboratory findings through use of a minimal $U(2)$ vibron Hamiltonian involving only the three linear, diagonal Casimir operators, $\hat{C}_a$, $\hat{C}_b$, and $\hat{C}_c$, gave an unsatisfactory root-mean-square (*rms*) deviation between experiment and theory of $\sim 35\,\text{cm}^{-1}$. This discrepancy was reduced to $\sim 9\,\text{cm}^{-1}$ by including the complete set of six quadratic forms, $\hat{C}_i\hat{C}_j$; however, incorporation of non-diagonal (Majorana) interactions failed to realize any further improvement. Comparison of residual patterns with the energy shifts produced by various algebraic operators suggested the introduction of selective cubic and quartic combinations, leading to a vibron Hamiltonian of the form:

$$\begin{aligned}\hat{H}_{\tilde{X}} &= A_a\hat{C}_a + A_b\hat{C}_b + A_c\hat{C}_c \\ &+A_{aa}\hat{C}_a\hat{C}_a + A_{ab}\hat{C}_a\hat{C}_b + A_{ac}\hat{C}_a\hat{C}_c + A_{bb}\hat{C}_b\hat{C}_b + A_{bc}\hat{C}_b\hat{C}_c + A_{cc}\hat{C}_c\hat{C}_c \\ &+A_{abb}\hat{C}_a\hat{C}_b\hat{C}_b + A_{bbc}\hat{C}_b\hat{C}_b\hat{C}_c + A_{abbb}\hat{C}_a\hat{C}_b\hat{C}_b\hat{C}_b\end{aligned} \quad (12)$$

Least-squares adjustment of the modeling parameters in $\hat{H}_{\tilde{X}}$ produced the optimized quantities presented in Table II of Ref. [10], with the comparison of measured and calculated vibrational term values showing an *rms* deviation of $5.7\,\text{cm}^{-1}$.

The absence of non-diagonal (Majorana) interactions in the ground state algebraic Hamiltonian implies that the corresponding S_2O vibrational dynamics are described well by local stretching and bending modes which, although anharmonic in nature, remain essentially uncoupled for the set of 240 levels examined by our dispersed fluorescence studies. This assertion of orthogonal and local vibrational degrees of freedom over a span of internal energy in excess of 1.6 eV [*viz.*, $\sim 46\%$ of the $D_0^0(\text{S}-\text{SO})$ dissociation limit [20]] is remarkable, especially given the theoretical prediction of other bound species within the $\tilde{X}\ {}^1\text{A}'$ potential surface. [21]

3.2 Vibrational Energies for the $\tilde{C}\ {}^1A'$ Excited State

The algebraic interpretation of energy level patterns in the S_2O $\tilde{C}\ {}^1\text{A}'$ potential surface is hampered by the onset of rapid predissociation which takes place for internal energies in excess of $\sim 1200\,\text{cm}^{-1}$. [20] A total of 39 experimental term values are available (providing access to $\tilde{C}\ {}^1\text{A}'$ eigenstates which possess up to 2, 11, and 4 quanta of excitation, respectively, in the v_1', v_2', and v_3' modes); however, careful inspection of these data reveals a local perturbation in vibrational term values to occur for $\tilde{C}\ {}^1\text{A}'$ eigenstates having 6–7 quanta of excitation in the v_2' S – S stretching degree of freedom. This observation corroborates previous assertions [20] that the $\tilde{C}\ {}^1\text{A}'$ potential surface intersects a dissociative electronic manifold roughly $2400\,cm^{-1}$ above the $\tilde{C}-\tilde{X}$ band origin. Since the present work focuses on Franck-Condon transitions for bound states, analyses were restricted to 22 low-lying vibronic features ($V \leq 4$) which are all situated well below the predicted barrier crest for predissociation. The parameters included in the vibron Hamiltonian were limited to those that could be determined satisfactorily from this subset of spectral data:

$$\hat{H}_{\tilde{C}} = A_a \hat{C}_a + A_b \hat{C}_b + A_c \hat{C}_c + \lambda_{aa} \hat{M}_{ab} + \lambda_{bc} \hat{M}_{bc} . \qquad (13)$$

The optimized vibron parameters obtained from this analysis are compiled in Table II of Ref. [10] with the comparison of measured and calculated term values showing the resulting *rms* deviation to be $3.2\,cm^{-1}$. Non-diagonal Majorana coupling was found to be exceptionally strong between the S − O and S − S stretching degrees of freedom although less pronounced interactions could also be discerned between the local S − S stretching and S − S − O bending oscillators. As demonstrated below, this mixing of local vibrational character has a profound influence on $\tilde{C}\,^1A'$ eigenfunctions and accompanying $\tilde{C}-\tilde{X}$ Franck-Condon intensities.

3.3 Franck-Condon Analyses of S_2O Emission Spectra

The algebraic Franck-Condon treatment provides a viable approach for disentangling the wealth of geometrical and dynamical information contained in vibronically-resolved emission spectra such as those depicted in Fig. 1. While the expansion coefficients used to describe molecular eigenstates reflect the distinct vibrational behavior supported by individual potential surfaces, overlap integrals evaluated in the uncoupled $U(2)$ representation reveal the changes in molecular structure incurred through electronic excitation. In particular, the precise relationship between (local) anharmonic oscillators comprising the ground and excited states can be ascertained from the analysis of Franck-Condon intensities, thereby enabling the extraction of equilibrium geometries and related structural constants. Given the known minimum energy configuration of the ground state, [22] an equilibrium geometry for the $\tilde{C}\,^1A'$ surface can be deduced from the algebraic displacement parameters, Δ_0^i, of Eq. (11). As suggested by the pronounced and extensive nature of the ν_2 progressions observed in all emission spectra, the S − S distance is found to elongate substantially from the $\tilde{X}\,^1A'$ value of 1.8845 Å to 2.168 Å. In contrast, only a slight increase in S − O bond length from 1.459 Å to 1.483 Å is induced by the $\pi^* \leftarrow \pi$ electron promotion. Conversion of the Δ_0^{SSO} displacement into an angular change

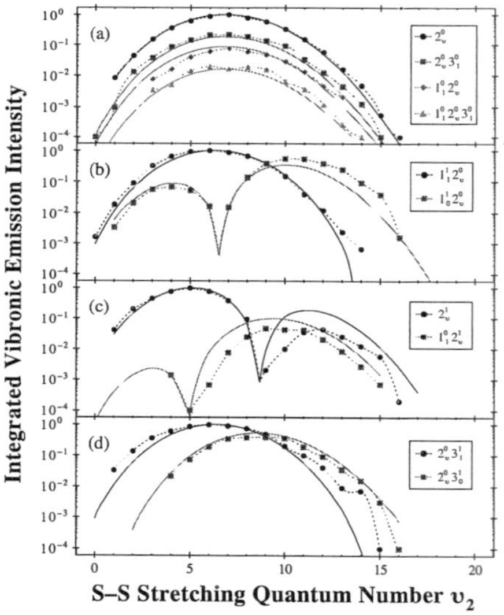

Figure 1. The integrated spectral irradiance of single vibronic bands in the S_2O emission spectrum are presented, with solid curves denoting the results of algebraic Franck-Condon calculations and symbols connected by dashed lines representing experimental data sets acquired through excitation of the (a) 0_0^0, (b) 1_0^1, (c) 2_0^1, and (d) 3_0^1 transitions.

indicates a decrease in equilibrium bond angle from $\angle SSO = 118.08°$ to either 106° or 110° depending upon the (indeterminate) sign of Δ_0^{SSO}. [10] While emission intensities alone are insufficient for a complete determination of the $\tilde{C}\,^1A'$ equilibrium geometry, modeling of analogous absorption data resolves this ambiguity and yields $\angle SSO = 110°$. [11] Inertial constants derived from vibron parameters are in good accord with published $\tilde{C}\,^1A'$ rotational constants upon which prior structural analyses have been based. [23]

The influence of intramolecular dynamics on S_2O intensity patterns is both dramatic and profound. This can be appreciated by considering the hypothetical case of purely "local" (uncoupled) vibrational behavior in *both* electronic manifolds. Under such circumstances, the changes in equilibrium geometry accompanying the $\pi^* \leftarrow \pi$ transition would still lead to extensive S−S stretching progressions where the number of minima observed in vibronically-resolved emission envelopes would reflect the nodal count along the S−S coordinate for the corresponding $\tilde{C}\,^1A'$ vibrational wavefunction. [5] Consequently, data derived through excitation of the 2_0^1 band would again be dominated by 2_υ^1 features which display a single minimum in the intensity envelope as depicted in Fig. 1c. However, the distribution of Franck-Condon intensity for all related progressions originating from the same $\tilde{C}\,^1A'$ $\upsilon_2' = 1$ level (*e.g.*, the $1_1^0 2_\upsilon^1$ progression) would be predicted to exhibit identical shapes as modified by a constant scaling factor depending on the values of displacement parameters Δ_0^{SO} and Δ_0^{SSO}. Clearly, the experimental results obtained for S_2O do not comply with this simplistic interpretation.

Excluding data sets acquired for the origin band (Fig. 1a), the S_2O dispersed fluorescence spectra do not display the identical (progression-independent) intensities envelopes that would be predicted for purely "local" vibrational dynamics within *both* electronic states. Of even greater significance is the complete alteration in nodal patterns obtained for the $1_0^1 2_\upsilon^0$ and $1_1^1 2_\upsilon^0$ bands of Fig. 1b. The vibron model has identified these effects as the signature of manifestly different behavior for the pertinent vibrational manifolds. While algebraic analyses reveal the $\tilde{X}\,^1A'$ ground state to be essentially local in nature, strong non-diagonal (Majorana) coupling within the excited $\tilde{C}\,^1A'$ surface leads to pervasive mixing of vibrational character as expressed in the local oscillator basis. Consequently, the eigenstate nominally assigned to the $\tilde{C}\,^1A'$ "$1^1 2^0$" level is found to involve an admixture of ~21% "$1^0 2^1$" character as reflected by the expansion coefficients $c_{100}^{100} = -0.8884$, $c_{010}^{100} = 0.4583$, and $c_{001}^{100} = 0.0263$. In contrast, the local dynamics suggested for the $\tilde{X}\,^1A'$ manifold imply that the corresponding $d_{\upsilon_a,\upsilon_b,\upsilon_c}^{\upsilon_1,\upsilon_2,\upsilon_3}$ expansion coefficients will vanish unless $\upsilon_a = \upsilon_1$, $\upsilon_b = \upsilon_2$, and $\upsilon_c = \upsilon_3$. Owing to the small magnitude of the Δ_0^{SO} translational displacement parameter, the overlap integrals evaluated in Eq. (10) will be negligible unless the number of quanta in the local S−O oscillator, υ_a, is conserved. Franck-Condon factors for the $1_1^1 2_\upsilon^0$ transitions in Fig. 1.3b thus are dominated by contributions from the local "$1^1 2^0$" component and do not display the node ascribed to the local S−S stretching degree of freedom. Conversely, emission intensities for the accompanying $1_0^1 2_\upsilon^0$ progression are governed primarily by the "$1^0 2^1$" basis state and therefore display a single pronounced node.

3.4 Franck-Condon Analyses of S_2O Absorption Spectra

In order to assess critically the predictive capabilities afforded by the vibron scheme, simulations were performed on S_2O absorption spectra (*cf.*, Fig. 2) recorded under rotationally-cold molecular beam conditions through use of sensitive cavity ring-down techniques. [11] The scope of measured absorption intensities is much more limited than corresponding emission data; however, substantial elongation of S – S bond length during the $\pi^* \leftarrow \pi$ transition has important ramifications for the wavefunction overlap integrals that determine Franck-Condon intensities. In particular, the amplitude of ν_2 S – S stretching progressions observed in absorption is dominated by contributions from the steep inner turning point of the local $\tilde{C}\,^1A'$ oscillator whereas $\tilde{C}-\tilde{X}$ emission profiles explores the much smaller slope (and larger amount of coordinate space) of the corresponding outer turning point in the $\tilde{X}\,^1A'$ ground potential surface.

The three panels in Fig. 2 summarize absorption intensities obtained by integrating data sets acquired for isolated members of the (a) $2_0^{\upsilon'}$, (b) $1_0^1 2_0^{\upsilon'}$ and $2_0^{\upsilon'} 3_0^1$, and (c) $2_1^{\upsilon'}$ progressions. Both experimental measurements (symbols) and algebraic calculations (curves) are depicted, with logarithmic ordinate scales serving to emphasize the shape of each spectral envelope. In assessing the quality of theoretical predictions, it should be remembered that the forms of the vibron Hamiltonians [*cf.*, Eqs. (12) and (13)], as well as the modeling parameters employed for their specification and for subsequent Franck-Condon analyses, were taken directly from the results of previous S_2O emission studies and used (essentially) without further adjustment. [11] Given that the only free variables were overall scaling factors (one each for all cold and hot band progressions), the agreement between experiment and theory is deemed to be quite satisfactory, thereby corroborating overall validity of our vibron analyses.

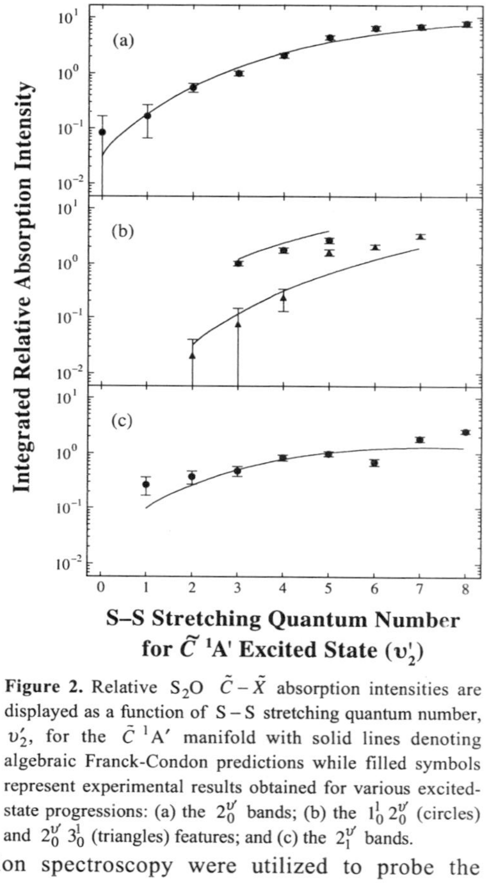

Figure 2. Relative S_2O $\tilde{C}-\tilde{X}$ absorption intensities are displayed as a function of S – S stretching quantum number, υ_2', for the $\tilde{C}\,^1A'$ manifold with solid lines denoting algebraic Franck-Condon predictions while filled symbols represent experimental results obtained for various excited-state progressions: (a) the $2_0^{\upsilon'}$ bands; (b) the $1_0^1 2_0^{\upsilon'}$ (circles) and $2_0^{\upsilon'} 3_0^1$ (triangles) features; and (c) the $2_1^{\upsilon'}$ bands.

4 Summary and Conclusions

Dispersed fluorescence and absorption spectroscopy were utilized to probe the $\tilde{C}\,^1A' \leftarrow \tilde{X}\,^1A'$ absorption system of jet-cooled S_2O molecules with the resulting vibronic transition frequencies and intensities interpreted quantitatively through use of a novel vibron scheme. In particular, our analyses exploit the unique capabilities afforded by an algebraic description of molecular structure and dynamics, thereby establishing a comprehensive framework for the extraction of detailed vibrational information from spectroscopic data sets. The success of this approach builds upon the inherent simplicity, economy of parameterization, and coupled *anharmonic* nature of the vibron Hamiltonian.

In particular, such features enable the facile evaluation of multidimensional Franck-Condon factors and related transition moment matrix elements.

The picture of S_2O vibrational dynamics that emerges from our algebraic analyses suggests radically different behavior for the ground and excited electronic states of the $\tilde{C}-\tilde{X}$ system. The term energies of $\tilde{X}\,^1A'$ vibrational levels over the observed $0-1.6\,\mathrm{eV}$ range of internal excitation could be reproduced satisfactorily without incorporating non-diagonal mixing into the local anharmonic (vibron) basis, thereby implying that the ground potential surface supports exceptionally local and uncoupled vibrational motion. In contrast, the $\tilde{C}\,^1A'$ manifold is distinguished by pervasive off-diagonal coupling between the local $S-S$ stretching and $S-O$ stretching degrees of freedom, as well as less pronounced interactions between the $S-S$ stretching and bending coordinates. A quantifiable signature for such distinct intramolecular behavior has been identified in vibronically-resolved $\tilde{C}-\tilde{X}$ emission envelopes which exhibit "non-intuitive" nodal patterns that shift in a characteristic manner depending upon the selected excitation band. This analysis is further confirmed by the successful modeling of experimentally-observed $\tilde{C}-\tilde{X}$ absorption intensities using parameter values deduced from emission studies. By refining determined structural parameters, the absorption studies lead to a complete prediction of $\tilde{C}\,^1A'$ equilibrium geometry which is in good accord with that inferred from prior measurements.

The methodology introduced in this work provides unique capabilities that readily can be extended and refined. Aside from obvious advantages afforded by the explicit incorporation of mechanical anharmonicity into the zero-order vibrational Hamiltonian, algebraic Franck-Condon analyses for polyatomic species can achieve exceptional computational efficiency owing to the separability that the underlying vibron basis introduces into an otherwise multidimensional problem. Moreover, the intrinsic coordinates employed by this approach permit changes in equilibrium geometry between initial and final states to be specified completely in terms of simple translational displacements. Consequently, the evaluation of overlap integrals is not encumbered by the coordinate transformation difficulties that often plague analogous normal mode treatments. The same advantage exists for calculations that move beyond the Condon approximation, [1-3] so as to include dependence of the transition moment operator on internal degrees of freedom. Indeed, the quantitative investigation of such non-Condon effects in the S_2O system has led to substantial improvements between predicted and measured $\tilde{C}\rightarrow\tilde{X}$ emission intensities (for high $\tilde{X}\,^1A'$ energies) and has revealed the detailed dependence of the transition dipole moment function on bond coordinates. [12]

5 Acknowledgments

This work was performed under the auspices of grants provided by the US National Science Foundation and the US Department of Energy. P.H.V., T.M., and F. P.-B. dedicate this work to Prof. Iachello on the occasion of his 60th birthday (*Tanti Auguri!*).

References

1. Franck, J., Elementary Processes of Photochemical Reactions, *Trans. Faraday Soc.* **21** (1925) pp. 536-542.
2. Condon, E.U., Nuclear Motions Associated with Electronic Transitions in Diatomic Molecules, *Phys. Rev.* **32** (1928) pp. 858-872.
3. Condon, E.U., The Franck–Condon Principle and Related Topics, *Am. J. Phys.* **15** (1947) pp. 365-374.

4. Herzberg, G., *Molecular Spectra and Molecular Structure: I. Spectra of Diatomic Molecules* (Van Nostrand Reinhold, New York, 1950).
5. Coon, J.B., DeWames, R.E., and Loyd, C.M., The Franck–Condon Principle and the Structures of Excited Electronic States of Molecules, *J. Mol. Spectrosc.* **8** (1962) pp. 285-299.
6. Moule, D.C., Vibrational Structure in Electronic Spectra: The Poly-Dimenisonal Franck–Condon Method, in *Vibrational Spectra and Structure: A Series of Advances*, J.R. Durig, Editor (Elsevier Scientific Publishing, Amsterdam, 1977) pp. 228-271.
7. Freed, K.F., Radiationless Transitions in Molecules, *Acc. Chem. Res.* **11** (1978) pp. 74-80.
8. Duschinsky, F., Meaning of the Electronic Spectrum of Polyatomic Molecules. I. The Franck–Condon Principle, *Acta Physicochim. U. R. S. S.* **7** (1937) pp. 551-566.
9. Müller, T., Dupré, P., Vaccaro, P.H., Pérez-Bernal, F., Ibrahim, M., and Iachello, F., Algebraic Approach for the Calculation of Polyatomic Franck–Condon Factors: Application to the Vibronically-Resolved Emission Spectrum of S_2O, *Chem. Phys. Lett.* **292** (1998) pp. 243-253.
10. Müller, T., Vaccaro, P.H., Pérez-Bernal, F., and Iachello, F., The Vibronically-Resolved Emission Spectrum of Disulfur Monoxide (S_2O): An Algebraic Calculation and Quantitative Interpretation of Franck-Condon Transition Intensities, *J. Chem. Phys.* **111** (1999) pp. 5038-5055.
11. Müller, T., Vaccaro, P.H., Pérez-Bernal, F., and Iachello, F., Algebraic Approach for the Calculation of Polyatomic Franck–Condon Factors: Application to the Vibronically-Resolved Absorption Spectrum of Disulfur Monoxide (S_2O), *Chem. Phys. Lett.* **329** (2000) pp. 271-282.
12. Iachello, F., Pérez-Bernal, F., Müller, T., and Vaccaro, P.H., A Quantitative Study of Non-Condon Effects in the S_2O $C \rightarrow X$ Emission Spectrum, *J. Chem. Phys.* **112** (2000) pp. 6507-6510.
13. Iachello, F. and Levine, R.D., *Algebraic Theory of Molecules* (Oxford University Press, New York, 1995).
14. Oss, S., Algebraic Models in Molecular Spectroscopy, *Adv. Chem. Phys.* **93** (1996) pp. 455.
15. Iachello, F. and Oss, S., Model of n Coupled Anharmonic Oscillators and Applications to Octahedral Molecules, *Phys. Rev. Lett.* **66** (1991) pp. 2976-2979.
16. Iachello, F. and Oss, S., Vibrational Modes of Polyatomic Molecules in the Vibron Model, *J. Mol. Spectrosc.* **153** (1992) pp. 225-239.
17. Frank, A., Lemus, R., Bijker, F., Pérez-Bernal, F., and Arias, J.M., A General Algebraic Model for Molecular Vibrational Spectroscopy, *Ann. Phys.* **252** (1996) pp. 211-238.
18. Pérez–Bernal, F., Arias, J.M., Frank, A., Lemus, R., and Bijker, R., Symmetry–Adapted Algebraic Description of Stretching and Bending Vibrations of Ozone, *J. Mol. Spectrosc.* **184** (1997) pp. 1-11.
19. Iachello, F. and Ibrahim, M., Analytic and Algebraic Evaluation of Franck–Condon Overlap Integrals, *J. Phys. Chem.* **102** (1998) pp. 9427-9432.
20. Zhang, Q., Dupré, P., Grzybowski, B., and Vaccaro, P.H., Laser–Induced Fluorescence Studies of Jet–Cooled S_2O: Axis–Switching and Predissociation Effects, *J. Chem. Phys.* **103** (1995) pp. 67.
21. Fueno, T. and Buenker, R.J., Electronic Structures of the S_2O and S_3 Isomers: an ab initio CI Study, *Theor. Chim. Acta* **73** (1988) pp. 123-134.
22. Lindenmayer, J., Rudolph, H.D., and Jones, H., The Equilibrium Structure of Disulfur Monoxide: Diode Laser Spectroscopy of ν_1 and ν_3 of $S_2{}^{18}O$ and ν_3 of $S_2{}^{16}O$, *J. Mol. Spectosc.* **119** (1986) pp. 56-67.
23. Hallin, K.-E.J., Merer, A.J., and Milton, D.J., Rotational Analysis of Bands of the 3400 Å System of Disulfur Monoxide (S_2O), *Can. J. Phys.* **55** (1977) pp. 1858-1867.

FRANCO, THE EARLY DAYS

R.H. SIEMSSEN
Kernfysisch Versneller Instituut, 9747 AA Groningen, The Netherlands

As this meeting is to honour Franco on the occasion of his 60 birthday I thought that it might be fitting to report on some early reminiscences of Franco of the pre-IBA days. Franco first came to Groningen in 1972 for a seminar on the invitation of Alex Lande. Alex and Franco had known each other from the Niels Bohr Institute in Copenhagen, where they had collaborated. In 1972 both Alex and I had been freshly appointed at Groningen, Alex on the Faculty of the Theory Department, and I myself as the new director of the KVI. A position for a Senior Scientist in theory had been newly created at the KVI with the aim to establish a strong in-house theory group. Needless to say that everyone who met Franco was deeply impressed by him. We thus were extremely happy to be able to entice Franco to join the KVI as a Senior Scientist in 1974, after he had spent a few weeks in Groningen in 1973 as a visitor. So characteristic of Franco he immediately took a strong interest in the experimental program as evidenced by the following publications on the weak-coupling description of three-nucleon pickup in the (p,α) reaction [1] and the spreading width of deep-hole states [2]. Both topics appear to have maintained their actuality, looking at the many papers that have been published since on these and related topics. But this brief citation of the "other Franco" would not do justice to him without mentioning the diverse palette of Franco's work also listed in the KVI 1974 Annual Report, reflecting Franco's extremely broad and diversified scientific interests. [3-10].

For the Amsterdam Conference on Nuclear Physics in the Fall of 1974 Franco had been asked to give an invited talk on collective excitations, and this could be viewed as the birth of the IBA as we know it. In his thesis as a graduate student of Herman Feshbach Franco had applied the boson approximation to ^{16}O. In the preparation of his talk for the Amsterdam Conference Franco had the idea to apply a related scheme to the transitional nuclei in the A=150 mass region. The group theoretical approach to the IBA had not yet been "invented" yet, and Franco thus had a young undergraduate student, Olaf Scholten, to numerically solve the problem. I still remember the great excitement that his results produced. In the Summer of the same year Akito Arima, whom I had known from my days at Argonne, decided to spend two months in Groningen. As Akito later confessed to me, he thought that the KVI would be a good hiding place, since only a few people would know where Groningen is. During his stay at the KVI Akito gave a series of lectures on nuclear structure in which he also discussed the SU(3) model of Elliott. At this moment the spark occurred, and the rest is history. In quick succession the SU(3), O(6) and SU(5) descriptions of the IBA were derived. These were intellectually very exciting times. Many visitors passed through the KVI, and Franco's keen interest in experiments was a great stimulus. He clearly fulfilled the task of "theorists being the salt in the soup" of an experimental institute. The KVI owed much to Franco, and I would thus like to take this opportunity to thank you, Franco, for this. The accompanying photograph shows Franco with among others Herman Feshbach in the cafeteria of the KVI during the early days."

Franco's clear lectures and lecture notes were also much appreciated. In writing out his lecture notes or the draft of his papers I was reminded of Mozart in the Amadeus movie, who would write his compositions out of his head without the need for any corrections. The same was true of Franco. Talking of music it might also be of interest here to mention a scholarly work Franco had co-authored on medieval music.

Franco stayed at the KVI until 1982, when he definitely moved to Yale. In 1976 we applied for a personal extraordinary professorship for Franco. This was already at a time, when the number of professorships at Dutch Universities began to be frozen, before they actually were reduced. In particular there were very stringent conditions on the extremely rare personal extraordinary professorships, which had to be approved by the minister himself. It might be of interest to mention that the other physicist for whom at the same time an application for such a position was handed in was Gerard t'Hooft from Utrecht, so Franco certainly was in good company. In order to succeed we had to get letters of recommendations from a large number of leading nuclear theorists. I just would like to quote from one of these the statement "You have with Franco a tiger at his tail".

If I frequently was asked by outsiders about the spin-offs of the KVI, I liked to mention aside of the PET Center at the Academic Hospital in Groningen, whose beginnings were at the KVI, the IBA. It also is an example that fundamental research cannot be planned. The KVI was originally conceived as a purely experimental cyclotron laboratory, and the plans to establish a strong in-house theory group were not everywhere received with enthusiasm. Out of the accidental meeting of Franco and Akito Arima the IBA in its present form has originated that is having such an impact on nuclear structure physics and as we have seen at this conference also on chemistry.

Fig. 1: Photograph taken at KVI in September 1975. Facing the camera from left to right: The author, Herman Feshbach, Franco Iachello and Siebren van der Werff. With the back to the camera: Alex Lande, André Zuker, Hiro Ejiri, Adriaan van der Woude.

Let me finally end this presentation with a personal note. During Franco's stay in Groningen my wife and I got close also in the private sphere with Franco and Irene, who lived only a few blocks away from our house. It thus has been a twofold pleasure for me to have had this opportunity here to talk on Franco, the early days.

References

1. "Coherence Effects in the (p,α) Reaction", J.W. Smits, F. Iachello, R.H. Siemssen and A. van der Woude, Phys. Lett. **53B** (1974) 337

2. “Deeply Bound Hole States as a Giant Resonance Phenomena”, S.Y. van der Werf, B.R. Kooistra, W.H.A. Hesselink, F. Iachello, L.W. Put and R.H. Siemssen, Phys. Rev. Lett. **33** (1974) 712
3. “The Interacting Boson Model”, H. Feshbach and F. Iachello, Ann. Phys. **84** (1974)211
4. “Clifford Algebra and Relativistic Wave Mechanics”, F. Iachello, Rend. Sem. Fac. Sci.. Cagliari **44** (1974)115
5. “Effects of Proton Pair Vibrations in the Low-Lying Spectrum of Closed Shell Nuclei: Structure of ^{48}Ca”, Nucl. Phys. **A228** (1974)356
6. “Finite Range Effects in Pionic Atoms”, Phys. Lett. **50B** (1974)313
7. “A Quark Model of High-Density Matter”, F. Iachello, W.D. Langer and A. Lande, Nucl. Phys. **A219** (1974) 612
8. “Effective Interaction in Liquid Helium”, F. Iachello and M. Rasetti, Nuovo Cim. Lett. **11** (1974) 477
9. “Effects of Intermediate Channels in Isospin Forbidden Reactions”, F. Iachello and P.P. Singh, Phys. Lett. **48B** (1974) 81

Happy 60th to Franco

The time has come to tell you this yarn
Of Franco, whose impact on science is big as a barn.

He is a scholar, scientist, and friend,
Whose ideas and achievements never end.

Franco is indeed a true Renaissance man
Like many another great Italian.

From racing cars to early music too
There is no challenge that Franco can't do.

Franco started by driving his racing car.
But he soon found that wouldn't get him very far.

He might have been driving at Indy and the Grand Prix
But he would never have known Andrea Vitturi.

In school he studied to be an engineer
But it was so boring that he shed a tear.

So he took up a physics career
And that is why we are gathered here.

His career has been long and great,
Even though he cannot stay up late.

Instead he gets up early to do his math
And so puts physics on a whole new path.

Early on he decided to give a try
To understand the structure of some nuclei.

The nucleus is so damn complicated
No easy solution was anticipated.

But Franco so loves the idea of symmetries
He even thinks of them when he skis.

So he worked with Akito both day and night
And finally one day they got it just right.

And a key part of their clever fix
Was throwing s and d bosons into the mix.

Indeed there was no easy physics
Until Franco then thought of U(6).

He calculates isoscalar factors with no fear
In fact he thinks of them as friends so dear.

Thus came the IBA, or is it IBM?
But how can one distinguish them?

But when Franco sought citation numbers as proof
Choosing the acronym IBM was no goof.

So Franco chose group chains like U(5)
And suddenly the model came alive.

And when SU(3) was something they could derive
their model then really began to thrive.

But it was the new symmetry O(6)
That was one of his most clever tricks.

When O(6) in platinum and barium was found
This finally showed that their model was sound.

When it also worked for erbium one sixty eight
Most everyone thought that that was just great.

Of course there was some dispute from the north
And arguments were made back and forth.

Nevertheless the IBA took flight like a ball
And after some time the model stood tall.

Franco's next physics coup
Came really out of the blue.

Tossing bosons and fermions into a SUSY
Made most nuclei simple, you see.

Of course, Franco's had help through the years
¿From his students, colleagues and peers.

With the likes of Roelof, Onno, Olaf, and Baha,
Other professors would just sigh: "ooh-la-la".

They all worked hard and stayed up late,
Checked his math and did physics great.

One has strayed and taken a different fork,
And spent his time calculating a quark.

He has come back and joined us tonight,
Rediscovered nuclei and seen the light.

But don't ever think that Franco is done
There is no end to his physics fun.

He turned to phase transitions next
Which were thought to be too complex.

He thought about ice and water
And realized they were first order.

The way most would attack this situation
Is to solve a differential equation.

But Franco said, "Oh what the hell
Let's just assume a square well."

Thus with critical point symmetries
He showed us the forest and not the trees.

Of course, Franco also applied his tools
To baryons and molecules.

And so I come to the end of this tale
Having spoken of Franco's lifelong trail.

Franco has given us many new paradigms
And brought us all such bright new times.

And so I end with - Franco, Franco, he's our man
There's no one else that can do all he can.

Rick Casten, March 27, 2003 (With help from Lee Riedinger)

List of participants

ALHASSID, Yoram
Yale University
Sloane Physics Laboratory
217, Prospect Street
New Haven, CT 06520, U.S.A.
yoram.alhassid@yale.edu

APRAHAMIAN, Ani
University of Notre Dame
Department of Physics
Notre Dâme, TN 46556 , U.S.A.
aprahamian.1@nd.edu

ARIAS, José, M.
University of Seville
Departamento FAMN
Facultad de Fisica
Aptdo 1065
41080 Sevilla, Spain
pepe@nucle.us.es

ÄYSTÖ, Juha
University of Jyväskylä
Department of Physics
P.O.Box 35
FIN-40351 Jyväskylä , Finland
juha.aysto@phys.jyu.fi

BALABANSKI, Dimiter
IKS, University of Leuven and
University of Sofia, Bulgaria
Celestijnenlaan 200 D
B-3001 Leuven, Belgium
dimiter.balabanski@fys.kuleuven.ac.be

BALANTEKIN, A. Baha
University of Wisconsin-Madison
Department of Physics
1150 University Ave.
Madison, WI 53706, U.S.A.
baha@nucth.physics.wisc.edu

BIZZETI, Piergiorgio
University of Firenze and INFN
Department of Physics
via G. Sansone, 1
I-50019 Sesto Fiorentino (FI), Italy
bizzeti@fi.infn.it

BLASI, Nives
INFN - Section of Milano
via Celoria, 16
I-20133 Milano, Italy
nives.blasi@mi.infn.it

BONSIGNORI, Giovanni C.
University of Bologna and INFN
Department of Physics
via Irnerio, 46
I-40127 Bologna, Italy
bonsignori@bo.infn.it

BRENTANO, von Peter
Univeristy of Koeln
Institut für Kernphysik
Zülpicker str. 77
D-50937 Köln, Germany
brentano@ikp.uni-koeln.de

CAPPUZZELLO, Francesco
INFN - Laboratorio Nazionale del Sud
via S.Sofia, 44
I-95125 Catania, Italy
cappuzzello@lns.infn.it

CAPRIO, Mark
Yale University
P.O.Box 208124
New Haven, CT 06520, U.S.A.
mark-caprio@yale.edu

CASTEN, Richard F.
Yale University
Physics Department-WNSL
P.O. Box 2 08124
New Haven, CT 06520-8124, U.S.A.
rick@riviera.physics.yale.edu

CATARA, Francesco
University of Catania and INFN
Department of Physics
Via S. Sofia, 64
I-95123 Catania, Italy
catara@ct.infn.it

CIZEWSKI, Jolie
Argonne National Laboratory
Bldg 203
9700 South Cass Avenue
Argonne, IL 60439-4843, U.S.A.
cizewski@physics.rutgens.edu

CUNSOLO, Angelo
INFN - Laboratorio Nazionale del Sud
and University of Catania
via S.Sofia, 44
I-95125 Catania, Italy
cunsolo@lns.infn.it

de ANGELIS, Giacomo
INFN – Laboratori Nazionali di
Legnaro
v.le dell'Università, 2
I-35020 Legnaro (PD), Italy
deangelis@lnl.infn.it

DEWALD, Alfred
Univeristy of K ln
Institut für Kernphysik
Zülpicher Str. 77
D-50931 Köln, Germany
dewald@ikp.uni-koeln.de

DIEPERINK, Alex
K.V.I.
Zernikelaan 25
NL-9747 AA Groningen, The Netherlands
dieperink@kvi.nl

DRAAYER, Jerry P.
Louisiana State University
Department of Physics and Astronomy
Baton Rouge, LA 70803-400, U.S.A.
draayer@sura.org

DUKELSKY, Jorge
Instituto de Estructura de la Materia
CSIC
Serrano 123
28006 Madrid, Spain
dukelsky@iem.cfmac.csic.es

EMLING, Hans
GSI
Planckstrasse 1
D-64291 Darmstadt, Germany
H.Emling@gsi.de

FAHLANDER, Claes
Lund University
Department of Physics
P.O. Box 118
SE-22100 Lund, Sweden
claes.fahlander@nuclear.lu.se

FORTUNA, Graziano
INFN - Laboratori Nazionali di Legnaro
v.le dell'Università, 2
I-35020 Legnaro (PD), Italy
fortuna@lnl.infn.it

FORTUNATO, Lorenzo
University of Padova and INFN
Department of Physics
via Marzolo, 8, I-35131 Padova, Italy
fortunat@pd.infn.it

FRANSEN, Christoph
University of Köln
Institut für Kernphysik
Zülpicker Str. 77
D-50937 Köln, Germany
fransen@ikp.uni-koeln.de

GARCIA-RAMOS, José Enrique
University of Huelva
Facultad de CC.EE.
21071 Huelva, Spain
jegramos@nucle.us.es

GELBERG, Adrian
University of Köln
Institut für Kernphysik
Zülpicker Str. 77
D-50937 Köln, Germany
gelberg@ikp.uni-koeln.de

GIANNATIEMPO, Angela
University of Firenze and INFN
Department of Physics
via G. Sansone, 1
I-50019 Sesto Fiorentino (FI), Italy
giannatiempo@fi.infn.it

GIANNINI, Mauro
Univeristy of Genova and INFN
Department of Physics
via Dodecaneso, 33
I-16146 Genova, Italy
giannini@ge.infn.it

GINOCCHIO, Joseph
Los Alamos National Laboratory
MS B283
Los Alamos, NM 87545, U.S.A.
gino@lanl.gov

GOUTTE, Dominique
GANIL
B.P. 55027
F-14076 Caen Cedex 5, France
goutte@ganil.fr

GRAW, Gerhard
LMU, München
Sektion Physik
Am Coulombwall 1
D-857 48 Garching, Germany
gerard.graw@physik.uni-muenchen.de

HARAKEH, Muhsin N.
K.V.I.
Zernikelaan 25
NL-9747 AA Groningen, The Netherland
harakeh@kvi.nl

HEYDE, Kris
University of Gent
Vakgroep Subatomian Fysica
Proeftuinstraat, 86
B-9000 Gent, Belgium
kris.heyde@rug.ac.be

IACHELLO, Francesco
Yale University
Sloane Physics Laboratory
217 Prospect Str.
New Haven, CT 06520-8120, U.S.A.
francesco.iachello@yale.edu

INSOLIA, Antonio
University of Catania and INFN
Department of Physics
Via S.Sofia, 64
I-95123 Catania, Italy
insolia@ct.infn.it

JOLIE, Jan
University of Köln
Institut für Kernphysik
Zülpicher Str. 77
50937 Köln, Germany
jlie@ikp.uni-koeln.de

JOLOS, Rostislav
J.I.N.R.
Bogoliubov Lab. of Theoretical Phys.
141 980 Dubna, Moscow Region, Russia
jolos@thsun1.jinr.ru

KIRSON, Michael W.
The Weizmann Institute
Department of Physics
P.O. Box 26
76100 Rehovot, Israel
michael.kirson@weizmann.ac.il

KNEISSL, Ulrich
University of Stuttgart
Institut für Strahlenphysik
Allmandring 3
D-70569 Stuttgart, Germany
kneissl@ifs.physik.uni-stuttgart.de

KOLLER, Noemi
Rutgers University
Dept. of Physics and Astronomy
136 Frelinghuysen Road
Piscataway, NJ 08854-8019 ,U.S.A.
nkoller@physics.rutgers.edu

KRÜCKEN, Reiner
University of München
TU-Physik Department E12
James-Franck Str.
D-857 48 Garching Germany
reiner.kruecken@ph.tum.de

LANDE, Alexander
University of Groningen
Institute for Theoretical Physics
Nijenborgh 4
NL-9747 AA Groningen, The Netherland
lande@kvi.nl

LENZI, Silvia Monica
University of Padova and INFN
Department of Physics
Via F.Marzolo, 8
I-35131 Padova, Italy
lenzi@pd.infn.it

LEONARDI, Renzo
University of Trento and ECT*
Department of Physics
via Sommarive, 14
I-38050 POVO (Trento), Italy
renzo@ect.it

LÉVAI, Géza
Hungarian Academy of Science
Institute of Nuclear Research (Atomki)
Mta Atomki, P.O.Box 51
H-4001 Debrecen, Hungary
levai@atomki.hu

LEVIATAN, Amiran
The Hebrew University
Racah Institute of Physics
Jerusalem, 91904, Israel
ami@vms.huji.ac.il

LIEB, Peter K.
II. Physikalisches Institut
Bunsenstr. 7-9
D-37073 Göttingen , Germany
lieb@physik2.uni-goettingen.de

LO BIANCO, Giovanni
University of Camerino and INFN
Department of Physics
via Madonna delle Carceri
I-62032 Camerino (MC) , Italy
giovanni.lobianco@unicam.it

LUNARDI, Santo
University of Padova and INFN
Department of Physics
Via F.Marzolo, 8
I-35131 Padova, Italy
lunardi@pd.infn.it

MAINO, Giuseppe
ENEA and University of Bologna
via Fiammelli 2
I-40129 Bologna, Italy
maino@bologna.infn.it

McCUTCHAN, Elizabeth
Yale University
Wright Nuclear Structure Lab.
P.O. Box 208124
New Haven CT 06520, U.S.A.
elizabeth.ricard-mccutchan@yale.edu

MOYA DE GUERRA, Elvira
IEM - CSIC
Consejo Superior Investigaciones Cientificas
Serrano 123
28006 Madrid, Spain
imtem22@pinar2.csic.es

MÜLLER, K. Alex
University of Zurich
CH-8057 Zurich, Switzerland
abdalla@physik.unizh.ch

OSS, Stefano
University of Trento and INFN
Department of Physics
via Sommarive, 14
I-38050 Povo (TN), Italy
stefano.oss@unitn.it

PÉREZ-BERNAL, Francisco
University of Huelva
Facultad de CC.EE.
21071 Huelva, Spain
curro@nucle.us.es

PETRACHE, Costel M.
University of Camerino and INFN
Department of Physics
via Madonna delle Carceri
I-62032 Camerino (MC), Italy
costel.petrache@unicam.it

PIETRALLA, Norbert
University of Köln
Institut für Kernphysik
Zülpicker Str. 77
D-50937 Köln, Germany
pietrall@ikp.uni-koeln.de

PITTEL, Stuart
University of Delaware
Bartol Research Institute
104 Center Mall, Rm. 217
Newark, DE 19716,U.S.A.
pittel@bartol.udel.edu

ROWE, David J.
University of Toronto
Department of Physics
60, St.George Str.
Toronto, ONT M5S 1A7, Canada
rowe@physics.utoronto.ca

SAKAI, Hideyuki
University of Tokyo
Department of Physics
7-3-1, Hongo, Bunkyo-ku
Tokyo 113-0033, Japan
sakai@phys.s.u-tokyo.ac.jp

SANTOPINTO, Elena
University of Genova and INFN
Department of Physics
via Dodecaneso, 33
I-16146 Genova, Italy
santopinto@ge.infn.it

SCHOLTEN, Olaf
K.V.I.
Zernikelaan 25
NL-9747 AA Groningen, The Netherlands
scholten@kvi.nl

SIEMSSEN, Rolf H.
K.V.I.
Zernikelaan 25
NL-9747 AA Groningen, The Netherlands
siemssen@kvi.nl

SONA, Anna Maria
University of Firenze and INFN
Department of Physics
via G.Sansone, 1
I-50019 Sesto Fiorentino (FI), Italy
annamaria@fi.infn.it

SONA, Pietro
University of Firenze and INFN
Department of Physics
Largo E.Fermi, 2
I-50125 Firenze, Italy
sona@fi.infn.it

TALMI, Igal
The Weizmann Institute
Department of Physics
P.O. Box 26
76100 Rehovot, Israel
igal.talmi@weizmann.ac.il

TOMASI-GUSTAFSSON, Egle
CEA/Saclay
Dapnia SPhN
F-91191 Gif-sur-Yvette Cedex, France
etomasi@cea.fr

VACCARO, Patrick
Yale University
Department of Chemistry
225 Prospect Str.
P.O. Box 208107
New Haven, CT 06520-8107, U.S.A.
patrick.vaccaro@yale.edu

VAN ISACKER, Piet
GANIL
B.P. 55027
F-14076 Caen Cedex 5, France
isacker@ganil.fr

VITTURI, Andrea
University of Padova and INFN
Department of Physics
Via F. Marzolo, 8
I-35131 Padova, Italy
vitturi@pd.infn.it

VRETENAR, Dario
University of Zagreb
Department of Physics, PMF
Bijenicka 32
10000 Zagreb, Croatia
vretenar@phy.hr

WARNER, David
CLRC Laboratory
Daresbury
Warrington, Cheshire WA4 4AD, U.S.A.
d.d.warner@dl.ac.uk

WEIDENMÜLLER, Hans
Max-Planck Institut für Kernphysik
P.O. Box 103980
D-69029 Heidelberg, Germany
haw@daniel.mpi-hd.mpg.de

WINFIELD, John
INFN - Laboratorio Nazionale del Sud
via S.Sofia, 44
I-95125 Catania, Italy
winfield@lns.infn.it

YOSHIDA, Nobuaki
University of Kansai
Faculty of Informatics
2-1-1 Ryozenji-cho
Takatsuki, 569-1095, Japan
yoshida@res.kutc.kansai-u.ac.jp

ZAMFIR, Victor
Yale University
Wright Nuclear Structure Laboratory
272 Whitney Ave
Box 208/2
New Haven, CT 06520-8124, U.S.A.
victor.zamfir@yale.edu

ZICHICHI, Antonino
Ettore Majorana Centre
via Guarnotta, 26
I-91016 Erice, Italy

ZILGES, Andreas
Technische Univ. Darmstadt
Institut für Kernphysik
Schlossgartenstr. 9
D-64289 Darmstadt , Germany
zilges@ikp.tu-darmstadt.de

ZUFFI, Lina
University of Milano and INFN
Department of Physics
via Celoria, 16
I-20133 Milano , Italy
zuffi@mi.infn.it

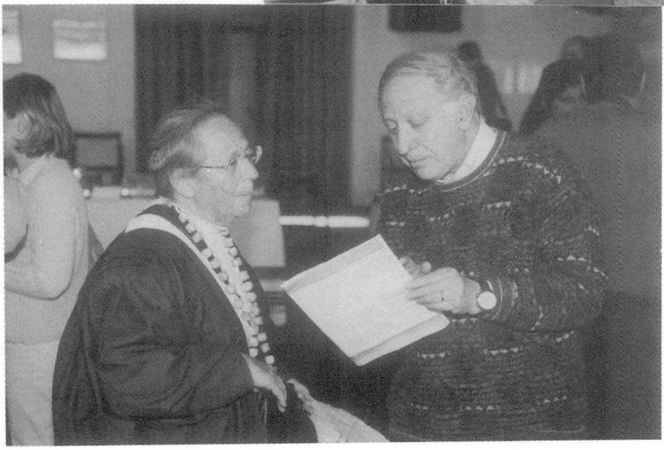